reaction diffusion systems

LECTURE NOTES IN PURE AND APPLIED MATHEMATICS

1. *N. Jacobson,* Exceptional Lie Algebras
2. *L.-Å. Lindahl and F. Poulsen,* Thin Sets in Harmonic Analysis
3. *I. Satake,* Classification Theory of Semi-Simple Algebraic Groups
4. *F. Hirzebruch et al.,* Differentiable Manifolds and Quadratic Forms
5. *I. Chavel,* Riemannian Symmetric Spaces of Rank One
6. *R. B. Burckel,* Characterization of C(X) Among Its Subalgebras
7. *B. R. McDonald et al.,* Ring Theory
8. *Y.-T. Siu,* Techniques of Extension on Analytic Objects
9. *S. R. Caradus et al.,* Calkin Algebras and Algebras of Operators on Banach Spaces
10. *E. O. Roxin et al.,* Differential Games and Control Theory
11. *M. Orzech and C. Small,* The Brauer Group of Commutative Rings
12. *S. Thomier,* Topology and Its Applications
13. *J. M. Lopez and K. A. Ross,* Sidon Sets
14. *W. W. Comfort and S. Negrepontis,* Continuous Pseudometrics
15. *K. McKennon and J. M. Robertson,* Locally Convex Spaces
16. *M. Carmeli and S. Malin,* Representations of the Rotation and Lorentz Groups
17. *G. B. Seligman,* Rational Methods in Lie Algebras
18. *D. G. de Figueiredo,* Functional Analysis
19. *L. Cesari et al.,* Nonlinear Functional Analysis and Differential Equations
20. *J. J. Schäffer,* Geometry of Spheres in Normed Spaces
21. *K. Yano and M. Kon,* Anti-Invariant Submanifolds
22. *W. V. Vasconcelos,* The Rings of Dimension Two
23. *R. E. Chandler,* Hausdorff Compactifications
24. *S. P. Franklin and B. V. S. Thomas,* Topology
25. *S. K. Jain,* Ring Theory
26. *B. R. McDonald and R. A. Morris,* Ring Theory II
27. *R. B. Mura and A. Rhemtulla,* Orderable Groups
28. *J. R. Graef,* Stability of Dynamical Systems
29. *H.-C. Wang,* Homogeneous Branch Algebras
30. *E. O. Roxin et al.,* Differential Games and Control Theory II
31. *R. D. Porter,* Introduction to Fibre Bundles
32. *M. Altman,* Contractors and Contractor Directions Theory and Applications
33. *J. S. Golan,* Decomposition and Dimension in Module Categories
34. *G. Fairweather,* Finite Element Galerkin Methods for Differential Equations
35. *J. D. Sally,* Numbers of Generators of Ideals in Local Rings
36. *S. S. Miller,* Complex Analysis
37. *R. Gordon,* Representation Theory of Algebras
38. *M. Goto and F. D. Grosshans,* Semisimple Lie Algebras
39. *A. I. Arruda et al.,* Mathematical Logic
40. *F. Van Oystaeyen,* Ring Theory
41. *F. Van Oystaeyen and A. Verschoren,* Reflectors and Localization
42. *M. Satyanarayana,* Positively Ordered Semigroups
43. *D. L Russell,* Mathematics of Finite-Dimensional Control Systems
44. *P.-T. Liu and E. Roxin,* Differential Games and Control Theory III
45. *A. Geramita and J. Seberry,* Orthogonal Designs
46. *J. Cigler, V. Losert, and P. Michor,* Banach Modules and Functors on Categories of Banach Spaces
47. *P.-T. Liu and J. G. Sutinen,* Control Theory in Mathematical Economics
48. *C. Byrnes,* Partial Differential Equations and Geometry
49. *G. Klambauer,* Problems and Propositions in Analysis
50. *J. Knopfmacher,* Analytic Arithmetic of Algebraic Function Fields
51. *F. Van Oystaeyen,* Ring Theory
52. *B. Kadem,* Binary Time Series
53. *J. Barros-Neto and R. A. Artino,* Hypoelliptic Boundary-Value Problems
54. *R. L. Sternberg et al.,* Nonlinear Partial Differential Equations in Engineering and Applied Science
55. *B. R. McDonald,* Ring Theory and Algebra III
56. *J. S. Golan,* Structure Sheaves Over a Noncommutative Ring
57. *T. V. Narayana et al.,* Combinatorics, Representation Theory and Statistical Methods in Groups

Additional Volumes in Preparation

reaction diffusion systems

edited by

Gabriella Caristi
Enzo Mitidieri
University of Trieste
Trieste, Italy

MARCEL DEKKER, INC.   NEW YORK · BASEL · HONG KONG

Library of Congress Cataloging-in-Publication Data

Reaction diffusion systems / edited by Gabriella Caristi, Enzo Mitidieri.
    p. cm. — (Lecture notes in pure and applied mathematics ; v. 194)
    Papers from a meeting held Oct. 2–7, 1995 at the University of Trieste, Italy.
    Includes bibliographical references.
    ISBN 0-8247-0125-9 (pbk. : acid-free paper)
    1. Reaction–diffusion equations—Congresses. I. Caristi, Gabriella. II. Mitidieri,
Enzo. III. Series.
AQ377.R423   1997
515'.353—dc21

                                          97-35931
                                            CIP

The publisher offers discounts on this book when ordered in bulk quantities. For more information, write to Special Sales/Professional Marketing at the address below.

This book is printed on acid-free paper.

MARCEL DEKKER, INC.
270 Madison Avenue, New York, New York 10016
http://www.dekker.com

Current printing (last digit):
10 9 8 7 6 5 4 3 2 1

PRINTED IN THE UNITED STATES OF AMERICA

Preface

The meeting Reaction Diffusion Systems was held at the Dipartimento di Scienze Matematiche dell'Università di Trieste, Trieste, Italia. The topics treated in this conference included several recent advances in the theory of elliptic, parabolic, and hyperbolic problems and related applications, in particular: existence and nonexistence of solutions for semilinear and quasilinear elliptic equations and systems, qualitative properties of solutions of elliptic and parabolic equations, maximum and comparison principles for linear and quasilinear operators, elliptic bifurcation problems, higher order semilinear elliptic equations, selfsimilarity and applications to nonlinear parabolic equations, existence, stability and blow-up of solutions to dissipative evolution equations, applications to nonlinear optical, chemical, and population biology models.

On behalf of the Organizing Committee (G. Caristi, E. Mitidieri, and K. P. Rybakowski) we express our sincere thanks to all speakers at the conference and to the contributors to and referees of this volume.

The organization of the conference was made possible by the financial support of Università degli Studi di Trieste, EC through the HCM Project Reaction Diffusion Equations, Consiglio Nazionale delle Ricerche, and Regione autonoma Friuli Venezia Giulia. In addition, we gratefully acknowledge the financial support of SAAB-MAZDA Girometta–Trieste. Special thanks are due to Ms. Maria Allegra of Marcel Dekker, Inc., for her kind cooperation and assistance during the preparation of this volume.

Gabriella Caristi
Enzo Mitidieri

Contents

Contributors

GIOVANNI ALESSANDRINI Dipartimento di Scienze Matematiche, Università di Trieste,Piazzale Europa 1, 34100 Trieste, Italy

MARIE-FRANÇOISE BIDAUT-VERON Département de Mathématique, Université de Tours, Parc Grandmont, 37200 Tours, France

PIOTR BILER Mathematical Institute, University Wrocław, pl. Grunwaldzki 2/4, 50-384 Wrocław, Poland

ISABEAU BIRINDELLI Istituto G.Castelnuovo, Università di Roma La Sapienza, 00100 Roma, Italy

LUCIO BOCCARDO Istituto G.Castelnuovo, Università di Roma La Sapienza, 00100 Roma, Italy

CARMEN CORTÁZAR Faculdad de Matemàticas, Universidad Catòlica, Casilla 306 Correo 22, Santiago, Chile

MABEL CUESTA Département de Mathématique, Université Libre de Bruxelles, Campus Plaine C.P.214, B-1050 Bruxelles, Belgium

DONATELLA DANIELLI Dipartimento di Metodi e Modelli Matematici per le Scienze Applicate, Università di Padova, via Belzoni 7, 35131 Padova, Italy

PANAGIOTA DASKALOPOULOS Department of Mathematics, University of California, Irvine CA 92717, USA

DJAIRO G. DE FIGUEIREDO IMECC-UNICAMP, Caixa Postal 6065, 13081-970 Campinas S.P., Brazil

MANUEL A. DEL PINO Departamento de Matematicás, Universidad de Chile,Casilla 653, Santiago, Chile

FRANÇOIS DE THÉLIN Université Toulouse III, route de Narbonne 118, 31062 Toulouse cedex, France

MANUEL ELGUETA Departamento de Matematicás, Faculdad de Matemáticas, Universidad Católica de Chile, Casilla 306 Correo 22, Santiago, Chile

PATRICIO FELMER Departamento de Ingenieria Matemática, F.C.F.M. Universidad de Chile, Casilla 170 Correo 3 Santiago, Chile

WILLIAM E. FITZGIBBON Department of Mathematics, University of Houston, Houston, Texas 77204-3476, USA

JACQUELINE FLECKINGER-PELLÉ Université des Sciences Sociales, 21 Allee de Brienne, F-31042 Toulouse Cedex, France

NICOLA GAROFALO Dipartimento di Metodi e Modelli Matematici per le Scienze Applicate, Università di Padova, via Belzoni 7, 35131 Padova, Italy

JEAN-PIERRE GOSSEZ Département de Mathematique, Université Libre de Bruxelles, C.P.214 Bvd de la Plaine 1050 Bruxelles, Belgium

HANS-CHRISTOPH GRUNAU Mathematisches Institut, Universität Bayreuth, D-95440 Bayreuth, Germany

JESUS HERNÁNDEZ Departamento de Matemáticas, Universidad Autónoma de Madrid, 28049 Madrid, Spain

JOSEPHUS HULSHOF, Mathematical Institute of the Leiden University, P.O. Box 9512, 2500 RA, Leiden, The Netherlands

ALESSANDRA LUNARDI Dipartimento di Matematica, Università di Parma, via D'Azeglio 85/A, 43100 Parma, Italy

R. MAGNANINI Dipartimento di matematica "U. Dini", Università di Firenze, Firenze, Italy

FRANCISCO J. MANCEBO Departamento de Fundamentos Matemáticos, Universidad Politécnica de Madrid, E.T.S.I. Aeronauticos, Plaza Cardenal Cisneros, 3, 28040 Madrid, Spain

A. MARCOS Institut de Mathematiques et des Sciences Physiques, B.P. 613, Porto Novo, Bènin

J.J. MORGAN Department of Mathematics, Texas A & M University, College Station, Texas 77843

TADEUSZ NADZIEJA Mathematical Institute, University Wrocław, pl. Grunwaldzki 2/4, 50-384 Wrocław, Poland

M.E. PARROTT Department of Mathematics, University of South Florida, Tampa, Florida 33620

S. I. POHOZAEV, Steklov Mathematical Institute, Moscow, Russia

HUMBERTO PRADO Universidad de Santiago de Chile, Casilla 307, Correo 2,Santiago, Chile

PATRIZIA PUCCI Dipartimento di Matematica, Università di Perugia, via Vanvitelli 1, 06123 Perugia, Italy

ANDRZEJ RACZYŃSKI Mathematical Institute, University Wrocław, pl. Grunwaldzki 2/4, 50-384 Wrocław, Poland

M. RAMOS Faculdade de Ciencias, Universidade de Lisboa, Av. Prof. Gama Pinto 2, 1699 Lisboa, Portugal

IAN SCHINDLER Ceremath, Université de Toulouse I, France

JAMES SERRIN Department of Mathematics, University of Minnesota, Minneapolis, MN 55455 Minnesota, USA

ILIA A. SHISHMAREV Department of Computational Matehematics and Cybernetics (BMK), Moscow State University, Moscow 119899, Russia

PAVEL E. SOBOLEVSKII Institute of Mathematics, Hebrew University of Jerusalem, Givat Ram Campus, 01904 Jerusalem, Israel

NIKOS STAVRAKAKIS Department of Mathematics, National Technical University of Athens, Zographou Campous 157-80, Athens, Greece

CHARLES STUART Departement de Matematique, Ecole Polytechnique Federale de Lausanne, CH 1015 Lausanne, Suisse

GUIDO SWEERS Department of Pure Mathematics, Delft University of Technology, P.O. Box 5031, 2600 GA Delft, The Netherlands

PETER TAKAC Fachbereich Mathematik, Universität Rostock, Universitätplatz 1, D 18055 Rostock, Germany

PEDRO UBILLA Universidad de Santiago, Casilla 307 Correo 2, Santiago, Chile

JOSÉ M. VEGA Departamento de Fundamentos Matemáticos, Universidad Politécnica de Madrid, E.T.S.I. Aeronauticos, Plaza Cardenal Cisneros, 3, 28040 Madrid, Spain

VINCENZO VESPRI Dipartimento di Matematica, Università dell'Aquila, via Vetoio loc. Coppito, 67100 L'Aquila, Italy

G.F. WEBB Department of Mathematics, Vanderbilt University, Nashville, Tennessee 37240

CECILIA YARUR Universidad de Santiago, Casilla 307 Correo 2, Santiago, Chile

reaction diffusion systems

Symmetry and Nonsymmetry in Some Overdetermined Boundary Value Problems

G. ALESSANDRINI   Università di Trieste, Dipartimento di Scienze Matematiche, Trieste, Italy.

R. MAGNANINI   Università di Firenze, Dipartimento di Matematica "U. Dini", Firenze, Italy.

1   INTRODUCTION

Consider the Stekloff eigenvalue problem:

$$\Delta u = 0 \quad \text{in } \Omega, \tag{1.1}$$

$$\frac{\partial u}{\partial \nu} = pu \quad \text{on } \partial\Omega. \tag{1.2}$$

Here, $\Omega \subset \mathbf{R}^n$ is a bounded domain with sufficiently smooth boundary $\partial\Omega$, and ν denotes the exterior unit normal vector to $\partial\Omega$. It is well-known that this problem has infinitely many eigenvalues $0 = p_1 < p_2 \leq p_3 \leq \ldots$ (see [16]).

Payne and Philippin [12] proved, for $n = 2$, that if there is an eigenfunction u of (1.1), (1.2) which also satisfies the overdetermined condition:

$$|Du| = 1 \quad \text{on } \partial\Omega, \tag{1.3}$$

Work partially supported by MURST 40 % and 60 %.

and corresponds to the second eigenvalue p_2, then u is linear and Ω must be a disk. They also provided an example of a non-circular domain Ω, for which an eigenfunction u of (1.1), (1.2) also satisfying (1.3) exists, but which corresponds to some higher eigenvalue $p = p_k, k > 2$. In their example $\partial\Omega$ is C^1-smooth but not C^2.

Thus, it is natural to pose the question:

*Suppose $\partial\Omega \in C^2$ and that, for some $p > 0$, there exists
a solution u to (1.1)–(1.3): does this imply that Ω is a ball?*

In this paper, we shall illustrate our work on this issue, [1], [2], providing a rather complete answer for the case $n = 2$ and some results for the higher dimensional case.

In [1], we examined the case of two dimensions and constructed a variety of non–symmetric domains for which a solution of (1.1)–(1.3) exists. We also determined an additional condition on the solution u to (1.1)–(1.3) which is satisfied if and only if Ω is a disk. See Theorems 2.1 and 2.2 in the next Section.

The problem shows quite different features in the case $n \geq 3$. In order to understand this, it is worth looking at solutions of (1.1)–(1.3) in the unit ball B_n of $\mathbf{R}^n$.

In this case, $\nu(x) = x$ on ∂B_n; by (1.2), since $x \cdot Du(x) - pu(x)$ is harmonic in B_n, we have that $x \cdot Du(x) = pu(x)$, $x \in B_n$, that is u must be a homogeneous harmonic polynomial of degree p. Therefore, (1.1)–(1.3) can be transformed into the problem:

$$(1.4) \qquad\qquad\qquad \tilde{\Delta}u = g(u),$$

$$(1.5) \qquad\qquad\qquad |\nabla u|^2 = f(u),$$

on $S^{n-1} = \partial B_n$, where $g(u) = -p(p + n - 2)u$, $f(u) = 1 - p^2u^2$. Here, $\tilde{\Delta}$ and ∇ denote the Laplace–Beltrami operator and tangential gradient on S^{n-1}, respectively.

Solutions of a system of type (1.4), (1.5), with f and g smooth, are well–known in the literature as *isoparametric functions*. Their level surfaces at regular values, the *isoparametric surfaces*, enjoy the nice geometric property of having all their principal curvatures constant.

Up to this date, a complete classification of these surfaces on the sphere is not available. Here, we want to stress the fact that they seem to be very rare. When $n = 3$, for example, it can be shown that the solutions of (1.4), (1.5) are just the restrictions to S^2 of linear functions on $\mathbf{R}^3$. More results and examples in this direction are contained in the works of E. Cartan [4], [5], who first considered the isoparametric surfaces on the sphere, Nomizu

[10], [11], Munzner [9], Ferus–Karcher–Munzner [6], and Wang Q. M. [17], [18], who examined them on a complete Riemannian manifold. We refer the reader to [18] for a survey on the subject.

Our main result, [2], is an analogue to Payne and Philippin's theorem.

THEOREM 1.1. *Let $\Omega \subset \mathbf{R}^n$ be a contractible bounded domain with boundary $\partial\Omega \in C^2$. Suppose that there exists a solution u of (1.1)–(1.3) which also satisfies:*

$$(1.6) \qquad \int_{\partial\Omega} (u - x \cdot Du)\, u \, d\sigma = 0.$$

Then, Ω is a ball.

The proof of this result is based on the analysis of an overdetermined problem for the Hamilton–Jacobi equation (1.5) on a Riemannian manifold M. Essentially, we have that if u is a solution of equation (1.5) on a Riemannian manifold M, then the sets $\{x \in M : u(x) = c\}$, at critical values c, are smooth submanifolds of M. This quite surprising result is proved in [17], in a slightly different setting. In [2], we produce an alternative proof, based on some elementary arguments and with a more analytical flavour. See Theorem 3.1 for a precise statement.

This paper is organized as follows. In Section 2, we outline our two-dimensional results and we illustrate them with some examples. In Section 3, we present a proof of Theorem 1.1 along with some auxiliary results which we believe may have some independent interest.

2 The two-dimensional case

Throughout this section we shall adopt the usual identification of $\mathbf{R}^2$ with $\mathbf{C}$.

Let us introduce the conjugate harmonic function v to u in Ω, chosen in such a way that

$$(2.1) \qquad \int_{\partial\Omega} v \, ds = 0,$$

here ds denotes the arclength element. We let

$$(2.2) \qquad F = u + iv$$

be the *complex potential associated to u*. We shall first prove the following necessary and sufficient condition for the symmetry.

THEOREM 2.1. *Let $u \in C^2(\overline{\Omega})$ be a solution of (1.1)–(1.3), and let F be the complex potential associated to u, as defined by (2.1) and (2.2).*
  Then, Ω is a disk if and only if F vanishes at only one point in Ω.

We stress the fact that the conclusion of Theorem 2.1 does not involve the vanishing rate of F at its zero. We shall give a proof of Theorem 2.1 through a sequence of statements which may be of some interest of their own. In particular, Theorem 2.5 shows an interesting connection with another symmetry problem, involving Green's function, which has already been treated by Payne and Schaefer [13] and Lewis and Vogel [8].

The combination of Theorem 2.1 and of Theorem 2.2 below shows that the disk is not the only domain with C^2 boundary for which a solution of (1.1)–(1.3) exists.

THEOREM 2.2. *Given the integers $K > 1$, and $m_1, \ldots, m_K \geq 1$, there exist a simply connected domain Ω with analytic boundary, and a function F holomorphic in Ω such that $u = Re(F)$ satisfies (1.1)–(1.3) and F has exactly K distinct zeros $z_1, \ldots, z_K \in \Omega$ with respective multiplicities $m_1, \ldots, m_K$.*

We shall denote by $|\partial\Omega|$ the perimeter of Ω, and by $[0, |\partial\Omega|] \ni s \mapsto z(s)$ the arclength parametrization of $\partial\Omega$ taken with the counterclockwise orientation, so that $\dot{z}(s)$ and $-i\dot{z}(s)$ are respectively the tangent and normal unit vector to $\partial\Omega$. As usual, a prime will denote the derivative with respect to the complex variable z, while we chose to indicate by the subscripts s, n respectively the tangential and normal partial derivatives at points of $\partial\Omega$.

THEOREM 2.3. *Let $u \in C^2(\overline{\Omega})$ satisfy (1.1)–(1.3). Then, there exists a positive integer N such that we have:*

$$(2.3) \qquad\qquad p = \frac{2\pi N}{|\partial\Omega|},$$

$$(2.4a) \qquad\qquad F(z(s)) = F(z(0))\, e^{ips}, \quad 0 \leq s \leq |\partial\Omega|,$$

$$(2.4b) \qquad\qquad |F(z(0))| = \frac{1}{p}.$$

Moreover, F and F' have respectively N and $N-1$ zeros in Ω, when counted according to their multiplicities.

COROLLARY 2.4. *There exists $u \in C^2(\overline{\Omega})$ satisfying (1.1)–(1.3) if and only if there exists a holomorphic function F in Ω such that*

$$(2.5a) \qquad\qquad |F| = \frac{1}{p} \quad on \quad \partial\Omega,$$

$$(2.5b) \qquad\qquad |F'| = 1 \quad on \quad \partial\Omega.$$

Moreover, F is the complex potential defined in (2.2).

The proof of the above two statements is essentially based on the observation that (1.2)–(1.3) induce $u(z(s))$ to satisfy an ordinary differential equation. The reader is referred to [1] for details.

THEOREM 2.5. *Let F be a holomorphic function in Ω satisfying (2.5) and suppose F vanishes at only one point $z_0 \in \Omega$.*

Then Ω is a disk $B_R(z_0)$ and, for some positive integer N, we have:

$$F(z) = \alpha(z - z_0)^N,$$

(2.6)
$$p = N/R,$$

$$|\alpha| = 1/NR^{N-1}.$$

Proof. Let z_0 be the only zero of F in Ω, and let N be its multiplicity. We may factor $F(z) = (z - z_0)\Phi(z)$ where Φ is holomorphic and never vanish in Ω. Then the function $w(z) = \log p|F(z)|$ satisfies $\Delta w = 2\pi N(\cdot - z_0)$ in Ω, $w = 0$ on $\partial\Omega$, and also $|Dw| = |F'|/|F| = p$ on $\partial\Omega$.

In other words, $w(z) = -2\pi N G(z, z_0)$ where $G(z, z_0)$ is the Green's function for Ω with pole at z_0. Therefore, $|\nabla G(\cdot, z_0)| = p/2\pi N$ on $\partial\Omega$. and, by Theorems III.1, III.2 in [13], we have that Ω is a disk centered at z_0, and (2.6) follows easily. □

REMARK. We observe that another proof of the spherical symmetry for the above mentioned overdetermined problem for the Green's function can be found in Lewis and Vogel [8]. As is observed in [13], still another proof could be obtained by the method of moving parallel planes of Serrin [15].

Proof of THEOREM 2.1. Let Ω be a disk of radius R centered at z_0. As is well–known (see [3]), the Stekloff eigenfunctions of Ω are given by the real or the imaginary part of the holomorphic functions $\alpha(z - z_0)^N$, where N is an integer and α is a complex number. Therefore, the complex potential associated to a solution of (1.1)–(1.3) in Ω takes the form $F = \alpha(z - z_0)^N$ with $|\alpha| = 1/(NR^{N-1})$.

Viceversa, if u satisfies (1.1)–(1.3) in Ω and the function F in (2.2) vanishes only at one point $z_0 \in \Omega$, then Corollary 2.3 and Theorem 2.4 imply that Ω is a disk centered at z_0.

REMARK We point out an interesting connection between the overdetermined problem (1.1)–(1.3) and the field of quadrature identities.

From Corollary 2.4 and by the arguments of Theorem 2.5, we readily see that, if u satisfies (1.1)–(1.3), and F has K distinct zeros $z_1, \ldots, z_K$ with respective multiplicities $m_1, \ldots, m_k$, then the function $w = \log p|F|$ satisfies $\Delta w = 2\pi \sum_{k=1}^{K} m_k(\cdot - z_k)$ in Ω, $w = 0$ on $\partial\Omega$, and also $|\nabla w| = p$ on $\partial\Omega$. Let f be any holomorphic function in a neighborhood of $\overline{\Omega}$. By applying Green's identity for f and w, we have:

$$\int_{\partial\Omega} f \, ds = \sum_{k=1}^{K} \frac{2\pi m_k}{p} f(z_k).$$

A standard density argument allows to extend the validity of the above identity to any holomorphic function f in Ω, whose non–tangential limit at the boundary exists in the space $L^1(\partial\Omega)$.

An identity of this kind is known as a "quadrature identity for the arc–length" (see [7]).

Proof of THEOREM 2.2. (Sketch, see [1] for details). We fix K, $K > 1$, distinct points $\zeta_1, \ldots, \zeta_K$, in the unit disk $B_1(0)$, and K positive integers $m_1, \ldots, m_k$. Our aim is to find a univalent function $\chi : B_1(0) \rightarrow \Omega$ and a holomorphic function Φ on B_1 such that the function $F = \Phi \circ \chi^{-1}$, defined on Ω, satisfies conditions (2.5) and vanishes at $z_1, \ldots, z_K$, $z_k = \chi(\zeta_k)$, $k = 1, \ldots, K$, with respective multiplicities $m_1, \ldots, m_K$.

Let $\Psi(\zeta) = F'(\chi(\zeta))$, $\zeta \in B_1(0)$. We have:

$$\Phi'(\zeta) = F'(\chi(\zeta))\chi'(\zeta) = \Psi(\zeta)\chi'(\zeta),$$

hence

$$\chi'(\zeta) = \frac{\Phi'(\zeta)}{\Psi(\zeta)}.$$

On $\partial B_1(0)$, we impose;

$$|\Phi| = \frac{1}{p} \quad \text{and} \quad |\Psi| = 1,$$

so that (2.5) will be satisfied. We also require that Φ vanishes at the points $\zeta_1, \ldots, \zeta_k$ with multiplicities $m_1, \ldots, m_K$. It follows that Ψ should vanish at $\zeta_1, \ldots, \zeta_k$ with multiplicities $m_1 - 1, \ldots, m_K - 1$ and at some other points $\eta_1, \ldots, \eta_L$ with multiplicities $n_1, \ldots, n_L$ such that $\sum_{\ell=1}^{L} n_\ell = K - 1$.

Thus, up to constant rotations, Φ and Ψ must have the following Blaschke product representation

(2.7) $$\Phi(\zeta) = \frac{1}{p} \prod_{k=1}^{K} \left(\frac{\zeta - \zeta_k}{1 - \zeta\overline{\zeta}_k}\right)^{m_k},$$

(2.8) $$\Psi(\zeta) = \prod_{k=1}^{K} \left(\frac{\zeta - \zeta_k}{1 - \zeta\overline{\zeta}_k}\right)^{m_k - 1} \prod_{\ell=1}^{L} \left(\frac{\zeta - \eta_\ell}{1 - \zeta\overline{\eta}_\ell}\right)^{n_\ell}.$$

Therefore

$$\Phi'(\zeta) = \frac{1}{p} \prod_{k=1}^{K} \left(\frac{\zeta - \zeta_k}{1 - \zeta \overline{\zeta}_k}\right)^{m_k} \sum_{k=1}^{K} m_k \frac{1 - |\zeta_k|^2}{(\zeta - \zeta_k)(1 - \zeta \overline{\zeta}_k)}.$$

Consequently

$$\chi'(\zeta) = \frac{1}{p} \prod_{k=1}^{K} \frac{\zeta - \zeta_k}{1 - \zeta \overline{\zeta}_k} \prod_{\ell=1}^{L} \left(\frac{1 - \zeta \overline{\eta}_\ell}{\zeta - \eta_\ell}\right)^{n_\ell} \sum_{k=1}^{K} m_k \frac{1 - |\zeta_k|^2}{(\zeta - \zeta_k)(1 - \zeta \overline{\zeta}_k)}.$$

Thus, it remains to show that for suitable choices of $n_1, \ldots, n_L, \eta_1, \ldots, \eta_L, \chi'$ is regular and nonvanishing on all $\overline{B}_1(0)$. Finally, if $\zeta_1, \ldots, \zeta_K$ are sufficiently close to zero, χ' turns out to have a univalent primitive χ which provides us with the desired conformal mapping.

EXAMPLES. The construction outlined above enables the analytical description of some examples of non-circular domains for which a solution (1.1)–(1.3) exists. The following four pictures represent some of these examples.

In Fig. 1, we chose $\zeta_1 = -\zeta_2 = 4/5$, and $m_1 = m_2 = 1$;
in Fig. 2, $\zeta_k = (3/4) \exp(2ik\pi/3), k = 0, 1, 2$, and $m_1, m_2, m_3 = 1$;
in Fig. 3, $\zeta_1 - -\zeta_2 = 1/6$, and $m_1 = 2, m_2 = 1$;
in Fig. 4, $\zeta_1 = -\zeta_2 = 2/9$, and $m_1 = 2, m_2 = 1$.
The pictures were obtained by the use of Mathematica.

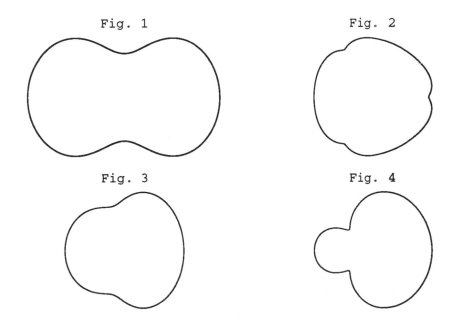

Fig. 1 Fig. 2

Fig. 3 Fig. 4

3   The higher dimensional case

We start with some preliminary notations. We consider a C^2 manifold M, without boundary, of dimension m, endowed with a Riemannian metric, represented by $\{g_{ij}(x)\}_{i,j=1,\dots,m}$ in the local coordinates $x = (x^1,\dots,x^m)$. If $v = (v^1,\dots,v^m)$ and $w = (w^1,\dots,w^m)$ are tangent vector fields on M, we define:

$$< v, w >= g_{ij}(x)v^i w^j, \quad |v| =< v, v >^{\frac{1}{2}}.$$

Here, we adopt the usual assumption on the sum over repeated indices.

Given a C^1 function on M, we introduce the gradient of u on M as

$$\nabla u = (\nabla_1 u, \dots, \nabla_m u), \qquad \nabla_i u = g^{ij}(x) u_{x^j}, \;\; i = 1,\dots,m;$$

here $\{g^{ij}(x)\}_{i,j=1,\dots,m}$ is the inverse of the matrix $\{g_{ij}(x)\}_{i,j=1,\dots,m}$.

We denote by $d: M \times M \to \mathbf{R}$ the geodetic distance on M; moreover, for any $x \in M$ and any closed subset $C \subset M$, it is well defined the number:

$$d(x, C) = \min\{d(x, y) : y \in C\}.$$

Let D be a bounded domain in M. We shall look at solutions of the following boundary value problem:

(3.1)
$$|\nabla \phi|^2 = f(\phi) \quad \text{in } D,$$
$$0 < \phi \le \Phi \quad \text{in } D,$$

(3.2)
$$\phi = 0 \quad \text{on } \partial D.$$

Here $f \in C^1((0, \Phi])$ is a function satisfying

(3.3)
$$f > 0 \quad \text{on } (0, \Phi),$$
$$f(\Phi) = 0, \quad f'(\Phi) < 0.$$

Notice that (3.1) and (3.3) easily imply that

(3.4)
$$\max_{\overline{D}} \phi = \Phi.$$

THEOREM 3.1. *Let $D \subset M$ be a bounded domain with boundary ∂D of class C^2. Let f be a $C^1((0, \Phi])$ function satisfying (3.3).*

If $\phi \in C(\overline{D}) \cap C^2(D)$ is a solution of (3.1)–(3.2), then for some integer h, $0 \le h \le m - 1$, the extremal level set

(3.5)
$$D_\Phi = \{x \in D : \phi(x) = \Phi\}$$

*is an h–dimensional C^1 connected compact submanifold without boundary of
M.*

   Moreover, D satisfies:

(3.6)
$$D = \{x \in M : d(x, D_\Phi) < L\},$$

where

(3.7)
$$L = \int_0^\Phi \frac{ds}{\sqrt{f(s)}} \, .$$

*Finally, if D is contractible, then D_Φ consists of a single point, and D is a
geodetic ball centered at D_Φ.*

   Proof.. The reader is referred to [17] and [2].   □

   We shall need the following result.

   PROPOSITION 3.2 *Let $u \in C^2(\Omega) \cap C^1(\overline{\Omega})$ satisfy (1.1) and (1.3). If*

(3.8)
$$\int_{\partial\Omega} [u - x \cdot Du] \, \frac{\partial u}{\partial \nu} \, d\sigma = 0,$$

then u is linear.

   Proof. By Rellich's identity (see [15]),

$$\int_{\partial\Omega} \left\{ 2(x \cdot Du)\frac{\partial u}{\partial \nu} - |Du|^2(x \cdot \nu) \right\} d\sigma = \int_{\Omega} \left\{ 2(x \cdot Du)\Delta u + (2 - N)|Du|^2 \right\} dx.$$

By (1.1) and (1.3), we obtain via the divergence theorem:

$$\int_{\partial\Omega} \left\{ 2(x \cdot Du)\frac{\partial u}{\partial \nu} - (x \cdot \nu) \right\} d\sigma = 2 \int_{\partial\Omega} u \, \frac{\partial u}{\partial \nu} \, d\sigma - N \int_{\Omega} |Du|^2 \, dx.$$

Thus, (3.8) yields:

(3.9)
$$N \int_{\Omega} |Du|^2 \, dx = \int_{\partial\Omega} x \cdot \nu \, d\sigma = N|\Omega|,$$

where $|\Omega|$ is the Lebesgue measure of Ω.

   Since $|Du|^2$ is subharmonic in Ω, by (1.3), we have $|Du| \leq 1$ in Ω, so that
(3.9) implies $|Du| \equiv 1$ in Ω. Therefore, $2\sum_{i,j=1}^n u_{ij}^2 = \Delta|Du|^2 \equiv 0$ in Ω, and
hence u is linear in Ω.

Theorem 3.1 will be a consequence of the following more general result. In the sequel, we will denote by $x = (x', x^n)$ a point of $\mathbf{R}^n$, where $x' \in \mathbf{R}^{n-1}$ has coordinates $(x^1, \ldots, x^{n-1})$; $\nu = (\nu_1, \ldots, \nu_n)$ will indicate the exterior normal unit vector to $\partial\Omega$.

THEOREM 3.3 *Let $\Omega \subset \mathbf{R}^n$ be a bounded domain with boundary $\partial\Omega$ in C^2. Suppose that u is a linear solution of (1.1)–(1.3).*

Then, up to a rigid change of coordinates, for some $h = 0, 1, \ldots, n - 2$, there exists a C^1 h-dimensional submanifold $D_\Phi \subset \{x \in \mathbf{R}^n : x^n = 0\}$, such that

$$(3.10) \qquad \Omega = \{x \in \mathbf{R}^n : \mathrm{dist}(x, D_\Phi) < \frac{1}{p}\}.$$

Furthermore, if Ω is contractible, then Ω is a ball.

Proof. We may assume that $u(x) = x^n$ up to a rigid change of coordinates. By (1.2), we have:

$$(3.11) \qquad \nu_n = px^n \quad \text{on } \partial\Omega.$$

If we consider $\Gamma = \{x \in \partial\Omega : x^n > 0\}$, we have that $\nu_n > 0$ on Γ, hence Γ is the graph of a function $\phi = \phi(x')$, where x' ranges over $D = \{x \in \Omega : x^n = 0\}$. The vector ν is then given by $\{1 + |\nabla\phi|^2\}^{-1/2}(-\nabla\phi, 1)$ on D, where ∇ denotes the gradient in the variable $x' \in D$.

Therefore, (3.11) yields $\{1 + |\nabla\phi|^2\}^{-1/2} = px^n = p\phi$, that is

$$(3.12) \qquad \begin{aligned} |\nabla\phi|^2 &= \frac{1}{p^2\phi^2} - 1 \quad \text{in } D, \\ \phi &= 0 \qquad \text{on } \partial D. \end{aligned}$$

Since $\partial\Omega \in C^2$, we also have that $\phi \in C(\overline{D}) \cap C^2(D)$; hence, by setting $m = n-1$, Theorem 3.1 applies to the (flat) domain D. Note that

$$(3.13) \qquad \phi(x') = \sqrt{\frac{1}{p^2} - \mathrm{dist}(x', D_\Phi)^2}$$

and that $D_\Phi = \{x' \in D : \phi(x') = \frac{1}{p}\}$. Therefore,

$$\{x \in \Omega : x^n > 0\} = \{x \in \mathbf{R}^n : x^n > 0, \mathrm{dist}(x, D_\Phi) < \frac{1}{p}\}$$

and, by the same argument,

$$\{x \in \Omega : x^n < 0\} = \{x \in \mathbf{R}^n : x^n < 0, \mathrm{dist}(x, D_\Phi) < \frac{1}{p}\}.$$

Consequently, we obtain (3.10).

Finally, one easily sees that D is a deformation retract of Ω, and hence D is contractible, if Ω is so. Theorem 3.3 implies that D is an $(n-1)$–dimensional ball, that is, by (3.13), Ω is an n–dimensional ball

Proof of THEOREM 1.1. By (1.1)–(1.3), (1.6), and Proposition 3.2, we have that u is linear, and hence Theorem 3.3 applies.

References

[1] G. Alessandrini and R. Magnanini, *Symmetry and non–symmetry for the over-determined Stekloff eigenvalue problem*, Journ. Appl. Math. Phys.(ZAMP), 45 (1994), pp. 44–52.

[2] ———, *Symmetry and non–symmetry for the overdetermined Stekloff eigenvalue problem II*, in Nonlinear Problems in Applied Mathematics, Ed. T.S. Angell et al., SIAM, Philadelphia 1996.

[3] C. Bandle *Isoperimetric Inequalities and Applications*, Pitman, London, 1980

[4] E. Cartan, *Familles de surfaces isoparamétriques dans les espaces à courbure constante*, Ann. Mat., 17 (1938), pp. 177–191.

[5] ———, *Sur des familles remarquables d'hypersurfaces isoparamétriques dans les espaces sphériques*, Math. Z., 45 (1939), pp. 335–367.

[6] D. Ferus, H. Karcher, and H. F. Munzner, *Cliffordalgcbrcn und neue isopara-metrische Hyperflachen*, Math. Z., 177 (1981), pp. 479–502.

[7] B. Gustafsson, *Application of half–order differentials on Riemann surfaces to quadrature identities for arc–length*, J. Analyse Math., 49 (1987), pp. 54–89.

[8] J.L. Lewis and A. Vogel, *On some almost everywhere symmetry theorems*, in Progr. Nonlinear Differential Equations Appl., 7, Birkhäuser, Boston 1992, pp. 347–374.

[9] H. F. Munzner, *Isoparametrische Hyperflachen in Sphären*, Math. Ann., 251 (1980), pp. 57–71.

[10] K. Nomizu, *Some results in E. Cartan's theory of isoparametric families of hypersurfaces* , Bull. AMS 79 (1973), pp. 1184–1188.

[11] ———, *Elie Cartan's work on isoparametric families of hypersurfaces* , Proc. Symp. Pure Math. 27 (1975), pp. 191–200.

[12] L. E. Payne and G. A. Philippin, *Some overdetermined boundary value prob-lems for harmonic functions*, Journ. Appl. Math. Phys.(ZAMP), 42 (1991), pp. 864–873.

[13] L.E. Payne and P.W. Schaefer, *Duality theorems in some overdetermined boundary value problems*, Math. Meth. in Appl. Sci. 11 (1989), pp. 805–819.

[14] F. Rellich, *Darstellung der eigenwerte $\Delta u + \lambda u = 0$ durch ein Randintegral*, Math. Z., 46 (1940), pp. 635–646.

[15] J.B. Serrin, *A symmetry problem in potential theory*, Arch. Rat. Mech. Anal. 43 (1971), pp. 304–318.

[16] M. W. Stekloff, *Sur les problèmes fondamentaux en physique mathématique*, Ann. Sci. Ecole Norm. Sup., 19 (1902), pp. 455–490.

[17] Q. M. Wang, *Isoparametric functions on Riemannian manifolds. I*, Math. Ann., 277 (1987), pp. 639–646.

[18] ——, *Isoparametric maps of Riemanniann manifolds and their applications*, Advances in science of China. Mathematics, 2 (1986),pp. 79–103.

Asymptotical Behavior of Solutions of Some Reaction-Diffusion Systems

MARIE-FRANÇOISE BIDAUT-VERON   Département de Mathématiques, Université de Tours, Faculté des Sciences, Parc de Grandmont, 37200 Tours, France

1 INTRODUCTION

Here we give the most recent results about the local behaviour near a punctual singularity or near infinity for the nonnegative solutions of some nonlinear elliptic reaction-diffusion systems in a domain Ω of $\mathbf{R}^N$, $N \geq 3$:

$$\Delta u + u^s v^p = 0 \tag{1}$$

$$\Delta v + u^q v^t = 0 \tag{2}$$

where p, q, s, $t \in \mathbf{R}$, with s, $t \geq 0$ and p, $q > 0$. In case of a singularity we can assume it is located at 0 and $\Omega = B$, where $B = \{x \in \mathbf{R}^N : |x| < 1\}$. For the study at infinity we can assume $\Omega = \mathbf{R}^N \setminus \overline{B}$.

Our purpose is to extend to system (1)-(2) the results relative to the scalar case of equation

$$\Delta U + U^Q = 0, \ Q > 0. \tag{3}$$

Two critical values of Q play an important role in (3), in the superlinear case $Q > 1$: $Q_1 = \frac{N}{N-2}$ and $Q_2 = \frac{N+2}{N-2}$. This equation admits a radially symmetric solution $U^*(x) = \lambda |x|^{-\frac{2}{Q-1}}$, where $\lambda = \lambda(N, Q)$, whenever $Q > Q_1$. When $Q < Q_1$ any solution satisfies an estimate $U(x) = O(|x|^{2-N})$ near 0, and it behaves like an harmonic function. When $Q_1 < Q < Q_2$, it satisfies the estimate $U(x) = O(|x|^{-\frac{2}{Q-1}})$ and if $Q \neq Q_2$ it behaves like U^* or like an harmonic function. We refer to [18], [15] for the first undercritical case $Q < Q_1$, [1] when $Q = Q_1$, [14] in the second undercritical case $Q_1 < Q < Q_2$ and in fact when $1 < Q < Q_2$, [10], [8], [9] when $Q = Q_2$, [5], [23] for results beyond Q_2 , and [17] in the sublinear case $Q < 1$. A global survey of this equation is given in [21].

In particular we are concerned by two kinds of systems, which have been studied by several authors, and in the sequel we only mention the nonradial results. The first one is the case of the *gradient-type system*

$$\Delta u + u^s v^{t+1} = 0, \tag{4}$$

$$\Delta v + u^{s+1}v^t = 0, \tag{5}$$

where $p = t+1$, $q = s+1$, which was studied in [4]. We find two critical conditions:

$$s + t + 1 = Q_1 \tag{6}$$
$$s + t + 1 = Q_2; \tag{7}$$

that is not surprising, since (4)-(5) admits solutions under the form (U, U), with $\Delta U + U^{s+t+1} = 0$. The second one is the case of the *Hamiltonian system*

$$\Delta u + v^p = 0 \tag{8}$$
$$\Delta v + u^q = 0 \tag{9}$$

where $s = t = 0$. In particular the case $p = 1$ corresponds to the biharmonic problem of the superharmonic solutions of equation $\Delta^2 u = u^q$, $q > 0$. Here also two critical conditions appear when $pq > 1$:

$$\max\{2\frac{p+1}{pq-1}, 2\frac{q+1}{pq-1}\} = N - 2, \tag{10}$$

$$\frac{1}{p+1} + \frac{1}{q+1} = \frac{N-2}{N}. \tag{11}$$

Notice that those two systems are variational ones : system (4)-(5) is the Euler system of the functional $L(u,v) = \int_\Omega ((t+1)|\nabla u|^2 + (s+1)|\nabla v|^2 - 2u^{s+1}v^{t+1})$ and system (8)-(9) the Euler system of $L(u,v) = \int_\Omega (\nabla u \cdot \nabla v - \frac{1}{q+1}u^{q+1} - \frac{1}{p+1}v^{p+1})$. Conditions (10)-(11) correspond to a lack of compacity for the functionals This allowed to give existence results for the Dirichlet regular problem up to these second barriers by using Palais-Smale condition, see [12] and [11], [22], [7]. But excepted those cases, system (1)-(2) is non variational.

Our study is guided by the existence of radially symmetric solutions of system (1)-(2): assuming in the sequel that $\delta = pq - (1-t)(1-s) \neq 0$ for simplification, we find

$$u^*(x) = A|x|^{-\gamma}, \quad v^*(x) = B|x|^{-\gamma}, \tag{12}$$

where

$$\gamma = 2\frac{p+1-t}{\delta}, \quad \xi = 2\frac{q+1-s}{\delta}, \tag{13}$$

(and A, B depend on γ, ξ, N), under the condition

$$0 < \min\{\gamma, \xi\} \leq \max\{\gamma, \xi\} < N - 2. \tag{14}$$

We shall call *superlinear* the case ($\delta > 0$ or s, $t > 1$) and *sublinear* the case ($\delta < 0$ and s, $t < 1$). We define the *first undercritical case* by the conditions

$$\max\{\gamma, \xi\} > N - 2 \text{ if } s, t \leq 1 \text{ and}$$
$$q + t < Q_1 \text{ if } t > 1, \ p + s < Q_1 \text{ if } s > 1. \tag{15}$$

2 NONEXISTENCE RESULTS FOR THE EXTERIOR PROBLEM

We prove in [3] some non existence results including the following :

THEOREM 1 [3] System (1)-(2) admits no nonnegative solution except 0 whenever one of the following conditions is fulfilled :

(i) s, $t > 1$ and $\min\{p+s, q+t\} \leq \frac{N}{N-2}$;

(ii) $\delta > 0$ and s, $t \leq 1$ and $(\max\{\gamma, \xi\} > N-2$ or $\gamma = \xi = N-2)$;

(iii) $t \leq 1 < s$ (hence $\delta > 0$) and $(p+s < \frac{N+a}{N-2}$ or $\gamma > N-2$ or $\gamma = N-2 = \frac{N}{p+s})$;

(iv) $s \leq 1 < t$ (hence $\delta > 0$) and $(q+t < \frac{N+b}{N-2}$ or $\xi > N-2$ or $\xi = N-2 = \frac{N}{q+t})$;

(v) $\delta < 0$ and s, $t < 1$.

This applies in particular to the gradient-type system (4)-(5) whenever $s+t+1 \leq Q_1$; it applies also to the Hamiltonian system (8)-(9) whenever $pq < 1$, or $pq > 1$ and

$$\max\{2\frac{p+1}{pq-1}, 2\frac{q+1}{pq-1}\} > N-2 \text{ or } p = q = \frac{N}{N-2}. \tag{16}$$

This improves the preceeding result of Souto [20], which was not optimal. Our proof of theorem 1 lies on the same idea : we consider the function $f = u^m v^{1-m}$ with $m \in (0,1)$. We prove that it satisfies an inequality $-\Delta f \geq Cf^\eta$ for some $\eta \neq 1$ and $C > 0$, under some conditions on parameter m. But such a supersolution cannot exist when $\eta \leq Q_1$, from [2],[13], see also [17] [3].

3 FUNDAMENTAL PROPERTY OF THE SOLUTIONS OF (1)-(2)

Consider the case of an isolated singularity. From Brezis-Lions theorem [6], since the functions are superharmonic, there exist some α, $\beta \geq 0$ such that

$$-\Delta u = u^s v^p + \alpha \delta_0 \tag{17}$$
$$-\Delta v = u^q v^t + \beta \delta_0 \tag{18}$$

in $\mathcal{D}'(B)$, where δ_0 is the Dirac mass at the origin. In the particular case of system (4)-(5) we made in [4] a simple but crucial observation : the difference $y = v - u$ satisfies an equation with the other sign :

$$-\Delta y + u^s v^t y = (\beta - \alpha)\delta_0, \tag{19}$$

hence from the maximum principle y is bounded from below near 0 if $\beta \geq \alpha$, and in any case there exists a constant M such that

$$u(x) \leq c_N \alpha |x|^{2-N} + v(x) + M \text{ near 0}. \tag{20}$$

In fact, such a comparison property holds for the general system (1)-(2), which is remarkable :

THEOREM 2 [3] Let $(u,v) \in (C^2(B \setminus \{0\}))^2$ be any nonnegative subsolution of (17)-(18). Assume for example that $p+t \leq q+s$. Then there exist some constants K, M such that

$$u(x) \leq 2c_N \alpha |x|^{2-N} + Kv^d(x) + M \text{ near 0}. \tag{21}$$

where $d = \frac{p+1-t}{q+1-s}$ if $p+1-t > 0$, and $d > 0$ is arbitrary if not.

The proof is more complicated; we show that the maximum principle can be applied to the function $Y = u^e v^{d(1-e)} - u$, where $e \in [0,1)$ is carefully chosen ($e = 0$ if $q \geq s$). See also [3] for similar properties in any domain Ω, which can be applied also in the regular Dirichlet problem.

4 BEHAVIOUR OF SYSTEM (1)-(2): 1st UNDERCRITICAL AND SUBLINEAR CASES

Using this comparison property and we can give the precise estimates and the exact behaviour of the solutions of system (1)-(2) in the first undercritical case, and in the sublinear case . For simplicity we consider only the first undercritical case for systems (4)-(5) and (8)-(9).The complete results can be found in [3].

THEOREM 3 [4], [3] Let $(u,v) \in (C^2(B \setminus \{0\}))^2$ be any nonnegative subsolution of (4)-(5) with $s + t + 1 < Q_1$. Then up to the change from u into v and s into t,

(i) either $\lim_{x \to 0} |x|^{N-2} u(x) = a > 0$ and $\lim_{x \to 0} |x|^{N-2} v(x) = b > 0$;

(ii) or $(u,v) \in (C^2(B))^2$, i.e. (u,v) can be extended as a C^2 solution of (4)-(5) in whole B;

(iii) or $N = 3$, $t \geq 1$, $s < 1$ and

$$\lim_{x \to 0} |x| u(x) = a > 0 \text{ and } \lim_{x \to 0} v(x) = c > 0;$$

(iv) or $0 \leq t < 1$ and ($N \geq 4$ or ($N = 4$, $s > 0$) or ($N = 3$, $s > 1$)) and

$$\lim_{x \to 0} |x|^{N-2} u(x) = a > 0 \text{ and } \lim_{x \to 0} |x|^k v(x) = \lambda_a > 0,$$

where $k = \frac{N - 4 + (N-2)s}{1-t} > 0$ and $\lambda_a = (a^{-(s+1)} k(N - 2 - k))^{\frac{1}{t-1}}$;

(iv) or $0 \leq t < 1$ and (($N = 4$, $s = 0$) or ($N = 3$, $s = 1$)) and

$$\lim_{x \to 0} |x|^{N-2} u(x) = a > 0 \text{ and } \lim_{x \to 0} |\log|x||^{-\frac{1}{1-t}} v(x) = \left(\frac{N - 2}{(1-s)a^{s+1}} \right)^{\frac{1}{t-1}}.$$

THEOREM 4 [3] Let $(u,v) \in (C^2(B \setminus \{0\}))^2$ be any nonnegative subsolution of (8)-(9) with $\max\{\gamma, \xi\} > N - 2$.

(i) either $\lim_{x \to 0} |x|^{N-2} u(x) = a > 0$ and $\lim_{x \to 0} |x|^{N-2} v(x) = b > 0$;

(ii) or $(u,v) \in (C^2(B))^2$;

(iii) or up to the change from u into v and p into q, $\lim_{x \to 0} |x| u(x) = a > 0$, and

$$\lim_{x \to 0} |x|^{(N-2)q-2} v(x) = a^q / ((N-2)q - 2)(N - (N-2)q) > 0$$

if $q > \dfrac{2}{N-2}$,

$$\lim_{x \to 0} v(x) = c > 0 \text{ if } q < \frac{2}{N-2},$$

$$\lim_{x \to 0} |\log|x||^{-1} v(x) = \frac{a^q}{N-2} \text{ if } q = \frac{2}{N-2}.$$

In any case we see that the asymptotical behaviour is asymptotically radial; and observe the eventuality of dissymmetrical solutions, where and present different types of behaviour. See [16] for the proof of existence of such solutions. Notice that theorem 5 gives in particular the behaviour of the biharmonic problem ($p = 1$) when $q < N/(N-4)$. The result was claimed in [19], but in fact the proof only worked when $q < N/(N-2)$ and not in the general case. Our method extends to more general systems involving powers of $|x|$, see [3].

The main idea of our proof is to plugg inequality (21) in the equation satisfied by v, and use the new inequality satisfied by v in order to prove that it satisfies Harnack inequality. For a simpler proof of theorem 3, see [2], theorem 4.1 and [16]; see also Section 5 for an alternative proof.

5 THE CASE OF THE GRADIENT TYPE SYSTEM

In case of system (4)-(5) we can go further in our study, and give the behaviour up to the second barrier (7), namely when $s + t + 1 < Q_2$. We use the very delicate technique introduced in the scalar case in [14] and developed in [5]. It comes from Bernstein methods. The idea is to write the equations satisfied by the functions $|\nabla(u^\sigma)|^2$, $|\nabla(v^\tau)|^2$ for suitable powers σ, τ, in order to give a priori estimates of the gradients $|\nabla u|^2$, $|\nabla v|^2$. By integrating by parts in (4)-(5) system (3) this gives estimates on some powers u^ω, v^θ, with ω, θ large enough. In turn in the simpler case it gives suitable estimates in some $\mathcal{L}^m$ space with $m > N/2$ of the coefficient $H = u^s v^t$ of the equations $\Delta u + Hu = 0$, $\Delta v + Hv = 0$, satisfied by u and v. In fact a great difficulty in the system comes from crossed terms $\nabla u \cdot \nabla v$, and our comparison property is essential to the proof. Finally we get the local behaviour when $s + t + 1 < Q_2$, in particular :

THEOREM 5 [4] Let $(u, v) \in (C^2(B \setminus \{0\}))^2$ be any nonnegative subsolution of (4)-(5) with $Q_1 < s + t + 1 < Q_2$. Then, up to the change from u into v and s into t,

(i) either

$$\lim_{x \to 0} |x|^{\frac{2}{s+t}} u(x) = \lim_{x \to 0} |x|^{\frac{2}{s+t}} v(x) =$$

$$= \left(\frac{2}{s+t} (N - 2 - \frac{2}{s+t}) \right)^{\frac{1}{s+t}};$$

(ii) or $(u, v) \in (C^2(B))^2$;

(iii) or $N = 3$, $t > 1 > s$ and

$$\lim_{x \to 0} |x| u(x) = a > 0 \text{ and } v(x) = O(|x|^{-\frac{1-s}{t-1}}) \text{ near to } 0;$$

moreover if $t + 2s \neq 3$, then

$$\lim_{x \to 0} v(x) |x|^{\frac{1-s}{t-1}} = \left((1 - s)(s + t)/(t - 1)^2 a \right)^{\frac{1}{t-1}} \text{ or } \lim_{x \to 0} v(x) = c > 0.$$

Notice the eventuality of solutions with a growth of linear type, greater that the nonlinear one , in oppostion to the scalar case. See [16] for the proof of their existence.

6 OPEN PROBLEMS

The first question is to obtain estimates for system (8)-(9) up to the second barrier, i.e. when $1/(p+1)+1/(q+1) > (N-2)/N$. Very likely the technique of Section 5 will work in the case $p = q$; but so far we cannot cover the general case. Another possible approach, even more delicate, would be the use of moving hyperplanes technique as in [10], [8], [9]. It would give the estimates for system (4)-(5) when $s + t + 1 \leq Q_2$. But it seems that its extension to the case of system (8)-(9) will not give optimal results : using the methods introduced in [11] we could cover at the best the case where $p \leq Q_2$ and $q \leq Q_2$, and not the whole case $1/(p+1)+1/(q+1) \geq (N-2)/N$.

The second question is relative to the nonvariational system (1)-(2): without any Pohozaev relation and no Euler functional, can we define a second barrier ? A possible candidate is the condition $\gamma + \xi = N - 2$, but the problem is still open.

References

[1] P. AVILES, Local behavior of solutions of some elliptic equation, Comm. Math. Phys., 108, 1987, p. 177-192.

[2] M.-F. BIDAUT-VERON, Local and global behaviour of solutions of quasilinear equations of Emden-Fowler type, Arc. for Rat. Mech. and Anal., 107, 1989, p. 293-324.

[3] M.-F. BIDAUT-VERON, Local behaviour of the solutions of a class of nonlinear elliptic systems, preprint.

[4] M.-F. BIDAUT-VERON and T. RAOUX, Asymptotics of solutions of some nonlinear elliptic systems, Comm. in Part. Diff. Equ., to appear.

[5] M.-F. BIDAUT-VERON and L. VERON, Nonlinear elliptic equations on compact Riemannian manifolds and asymptotics of Emden equations, Invent. Math., 106, 1991, p. 489-539.

[6] H. BREZIS and P.-L. LIONS, A note on isolated singularities for linear elliptic equations, Math. Anal. and Appl., 7A, 1981, p. 263-266.

[7] P. CLEMENT, R. MANASEVITCH and E. MITIDIERI, Positive solutions for a quasilinear system via blow-up, Comm. Part. Diff. Eq., 17, 1992, p. 923-940.

[8] W.X. CHEN and C.M. LI, A priori estimates for solutions to nonlinear elliptic equations, Arc. Rat. Mech. Anal., 122, 1993, p. 145-157.

[9] C.C CHEN and C.S. LIN, Local behaviour of positive solutions of semilinear elliptic equations with Sobolev exponents, preprint.

[10] L.-A. CAFFARELLI, B. GIDAS and J. SPRUCK, Asymptotic symmetry and local behavior of semilinear equations with critical Sobolev growth, Comm. Pure Applied Math., 42, 1989, p. 271-297.

[11] DE FIGUEIREDO and FELMER, A Liouville-type theorem for elliptic systems, Ann. Scu. Norm. Sup. Pisa,21,1994, p. 387-397.

[12] F. DE THELIN et J. VELIN, Existence et non-existence de solutions non triviales pour des systèmes elliptiques non linéaires, C.R. Acad. Sci. Paris, 313, série I,1991, p. 589-592.

[13] B. GIDAS, Symmetry properties and isolated singularities of positive solutions of nonlinear elliptic equations, in Nonlinear Part. Diff. Equ. in Engineering and Appl. Sc.,Sternberg,Kalinovski and Papadakis, Dekker Inc., 1980.

[14] B. GIDAS and J. SPRUCK, Global and local behaviour of positive solutions of nonlinear elliptic equations, Comm. Pure Applied Math., 34, 1981, p. 525-598.

[15] P.-L. LIONS, Isolated singularities in semilinear problems, Jl Diff. Eq., 38, 1980, p. 441-450.

[16] T. RAOUX, Comportement asymptotique d'équations et de systèmes d'équations elliptiques semi-linéaires, Thesis, Université de Tours, 1995.

[17] T. RAOUX, Local and global behaviour for an elliptic equation with a sublinearity,Adv. in Math. Sci. and Appl., to appear.

[18] J. SERRIN, Isolated singularities of solutions of quasilinear equations, Acta Math.,113,1965, p.219-240.

[19] R. SORANZO, Isolated singularities of positive solutions of a superlinear biharmonic equation,Potential Anal., 1996.

[20] M. SOUTO, A priori estimates and existence of positive solutions of nonlinear cooperative elliptic systems, Diff. and Int. Eq., to appear.

[21] L.VERON, Singularities of solutions of second order quasilinear equations, Research Notes in Math., Addison Wesley Longman, 1996.

[22] R. Van der VORST, Variational problems with a strongly indefinite structure, Thesis, Leiden University, 1994.

[23] H. ZOU, Symmetry of positive solutions of $\Delta u + u^p = 0$ in $\mathbf{R}^N$, J. Diff.Eq., 120, 1995, p. 46-88.

Nonlinear Singular Parabolic Equations

PIOTR BILER Mathematical Institute, University of Wrocław, Wrocław, Poland

TADEUSZ NADZIEJA Mathematical Institute, University of Wrocław, Wrocław, Poland

ANDRZEJ RACZYŃSKI Mathematical Institute, University of Wrocław, Wrocław, Poland

1 INTRODUCTION

Parabolic–elliptic systems of partial differential equations

$$u_t = \nabla \cdot (\nabla u + u \nabla \varphi),\tag{1}$$

$$\Delta(\varphi - V) = \pm u\tag{2}$$

appear in statistical mechanics in several contexts.

First, with the + sign in (2), they are used as models of time evolution of a cloud of particles moving under the influence of their mutual gravitational interaction and a given external potential V. In this interpretation $u = u(x, t)$, $u : \Omega \times [0, T) \to I\!R^+$, $\Omega \subset I\!R^n$, is the spatial density of the particles and $\varphi = \varphi(x, t)$, $\varphi : \Omega \times [0, T) \to I\!R$, is the gravitational potential. For a more detailed physical background of this model we refer the reader to (Wolansky, 1992; Biler, Nadzieja, 1996b).

Second, (1)–(2) with the − sign in (2) is a simplified version of the system of equations describing the evolution of densities of charged particles in a solute. Here u is the density of ions moving in an electrolyte and φ is the electric potential, see e.g. (Biler, Nadzieja 1996a; 1996b; Rubinstein 1990). Such systems have been introduced much earlier (a hundred years ago) to describe the electrolysis phenomena, see e.g. (Rubinstein, 1990; Biler, Hebisch, Nadzieja, 1994; Biler, Nadzieja 1996b) and references therein.

When the system (1)–(2) is considered in a bounded smooth domain Ω in $\mathbb{R}^n$, a natural boundary condition is the (nonlinear) no-flux condition

$$\frac{\partial u}{\partial \nu} + u\frac{\partial \varphi}{\partial \nu} = 0, \tag{3}$$

where ν denotes the exterior unit vector normal to $\partial\Omega$. (3) guarantees the conservation of the total mass (charge, resp.) $M_0 = \int_\Omega u(x,t)\,dx$ (q_0, resp.) in time.

In the first case we assume for the potential φ a physically relevant "free" condition (replacing (2))

$$\varphi = E_n * u + V, \tag{4.1}$$

where E_n is the fundamental solution of the Laplacian on $\mathbb{R}^n$ ($E_2(z) = (2\pi)^{-1}\log|z|$, $E_n(z) = -((n-2)\sigma_n)^{-1}|z|^{2-n}$ for $n \geq 3$, σ_n — the area of the unit sphere in $\mathbb{R}^n$).

In the second case, the wall $\partial\Omega$ of the container Ω can be isolated, i.e.

$$\varphi = -E_n * u + V, \tag{4.2}$$

or grounded, i.e.

$$\varphi = 0 \quad \text{on} \ \partial\Omega. \tag{4.3}$$

To determine completely the evolution, the system (1)–(4) is supplemented with the initial condition

$$u(x,0) = u_0(x) \geq 0. \tag{5}$$

Recently much attention has been paid to mathematical questions related to solvability of the initial-boundary value problems for related parabolic-elliptic systems, regularity and asymptotic behavior of solutions, see e.g. (Biler, 1992; 1995a; 1995b; Biler, Nadzieja, 1993; 1995a; 1996a). The problem (1)–(5) with $V \equiv 0$ has been studied in (Biler, Nadzieja, 1994; Biler, Hilhorst, Nadzieja, 1994; Biler, 1995a; 1995c), and in a greater generality in (Biler, Nadzieja, 1993) (with sufficiently regular V's). Our aim here is to study (1)–(2) for two particular potentials V which are highly singular: either $V(x) = M^*E_n(x)$, $M^* > 0$, or $V(x) = q^*E_n(x)$, $q^* \neq 0$. We suppose that $0 \in \Omega$ so the singularity does occur.

Physical interpretations of these singular systems come from a remark in Lemma 4.5 in (Wolansky, 1992) and problem 4 in Sec. 2.4 in (Rubinstein, 1990). In the first case, the steady state radial version of (1)–(4) has been mentioned as a model of gravitational equilibrium of a "nebula" of self-attracting particles surrounding a "star" of mass $M^* > 0$ fixed at the origin. Below we refer to this problem as to the "star problem".

In the second case, the problem (1)–(5) generalizes the equations proposed to describe the situation when a charge q^* is fixed at the origin $0 \in \Omega$ and surrounded by ions of opposite sign moving in a solute. The original formulation of the Rubinstein question concerns the radially symmetric situation in the unit ball $\Omega = B(0,1) \subset \mathbb{R}^3$ with the condition $q^* = q_0 \equiv \int_{B(0,1)} u_0$ expressing the global electroneutrality assumption. The two-dimensional version of (1)–(5) appears when a charged thin wire is plunged in an electrolyte and the cylindrical symmetry is assumed. The system (1)–(5) with $-$ sign in (2) and $V = q^*E_n$, $q^* > 0$, will be called in the sequel the "Rubinstein problem". Of course, it is also of interest to consider the case $V = q^* E_n$, $q^* < 0$, when the charges of the electrode and ions in the solute are of the same sign (Nadzieja, Raczyński, 1996).

Similar parabolic-elliptic systems are also used to describe a biological phenomenon of chemotaxis (Jäger, Luckhaus, 1992; Nagai, 1995; Biler, 1996b). In this case the conditions (3)–(4) are replaced by the homogeneous Neumann conditions

$$\frac{\partial u}{\partial \nu} = \frac{\partial \varphi}{\partial \nu} = 0, \tag{4.4}$$

and $V \equiv 0$ (however, in early texts on chemotactic movement, other potentials V are also admitted, including source type $V \sim E_n$). The condition (4.4) is also relevant for (1)–(2) in the $-$ (Coulomb), electroneutral, radially symmetric case, i.e. when Ω is a ball and $q^* \equiv \int_{\partial \Omega} \nabla V \cdot \nu = q_0$.

Since the potential V in our situation is singular: ΔV is a multiple of the Dirac mass δ_0, the analysis of (1)–(5) becomes particularly delicate compared to the case of a bounded smooth V. Besides the nonexistence of steady states and global-in-time solutions in the $+$ (Newtonian) case with large initial mass, known for $V \equiv 0$, see (Biler, Nadzieja, 1994), a new phenomenon is present. Namely, when V is "too singular", the nonexistence of solutions at all may occur. This kind of difficulties is not only connected with a troublesome definition of a solution to (1)–(2) (ΔV enters into (1)!), but also with the appearance of an instantaneous blow-up. As we will see further, "too singular" means in the two-dimensional case that $M^* \geq 4\pi$, or $q^* > 4\pi$, and $M^* > 0$, $q^* > 0$ in the higher dimensional case ($n \geq 3$).

In the sequel, we will concentrate on the radially symmetric case, see (6)–(8), (10)–(11) below, which is representative enough for these difficulties. For illustrating the general case we recall from the recent preprints (Biler, Nadzieja, 1995b; 1996a) results (Theorems 3, 4) on the convergence of solutions to approximating nonsingular parabolic-elliptic problems with regularized potentials converging to the singular ones $M^* E_n$, $q^* E_n$.

Our results provide a generalization of an observation from (Wolansky, 1992) and give an answer to the question raised by (Rubinstein, 1990) concerning the situation when no steady state exists for (1)–(5).

Evidently, the analysis of (1)–(5) in the radial case is simpler than without symmetry assumptions. Nevertheless, this involves (besides standard extensions of the theory of nonlinear nonsingular parabolic equations) the Feller test for the nonexistence of solutions to singular linear diffusion equations, cf. (Nagasawa, 1993), and tools from the theory of dynamical systems.

The proofs of blow-up, the most intriguing phenomenon for these equations, are indirect (Biler, 1995b; Nagai, 1995; Jäger, Luckhaus, 1992; Biler, Nadzieja, 1995a). Some information on the way solutions blow up has been recently obtained in (Herrero et al., 1996) for a related problem of chemotactic collapse. Their results can be adopted to study the gravitational collapse problem (1)–(5).

Our presentation is based on two companion preprints (Biler, Nadzieja, 1995b; 1996a), with some new results added (Th. 2 (i), (iii) from Nadzieja, Raczyński, 1996, and Proposition 2).

2 RADIALLY SYMMETRIC SOLUTIONS

We restrict our attention here to the case of radially symmetric potentials and densities $u(x,t) = u(|x|,t)$ in the ball $B(0,R) \subset I\!\!R^n$. In this situation the nonlocal parabolic-elliptic problem (1)–(5) can be reformulated as a nonlinear parabolic equation with singular coefficients

$$Q_t = Q_{rr} - (n-1)r^{-1}Q_r + \sigma_n^{-1}r^{1-n}F(Q)Q_r. \tag{6}$$

Here the new unknown function $Q(r,t) = \int_{B(0,r)} u(x,t)\,dx$ is the integrated density, $\frac{\partial}{\partial r}Q(r,t) = \sigma_n r^{n-1}u(r,t)$, compare (Biler, Hilhorst, Nadzieja, 1994; (6)–(7)). The function F is either $F(Q) = M^* + Q$ or $F(Q) = q^* - Q$, for the star problem or the Rubinstein problem respectively (Biler, Nadzieja 1995b, 1996a). The equation (6) is supplemented with the boundary conditions

$$Q(0,t) = 0, \quad Q(R,t) = M_0 \text{ (or } q_0, \text{ resp.)}, \tag{7}$$

and the initial condition

$$Q(r,0) = Q_0(r), \quad 0 \le r \le R, \tag{8}$$

which is a positive nondecreasing function. The obvious compatibility condition is

$$Q_0(0) = 0, \quad Q_0(R) = M_0 \ (q_0, \text{ resp.}). \tag{9}$$

Such a formulation allows us to consider some singular densities (e.g. measures) $u(r,t) = \sigma_n^{-1}r^{1-n}\frac{\partial}{\partial r}Q(r,t)$, so the problem (6)–(9) is not a priori equivalent to (1)–(5) with radial density u.

The scaling properties of (6) permit us to assume, without loss of generality, that $R = 1$. Indeed, together with $Q(r,t)$, the function $R^{2-n}Q(Rr, R^2t)$ is a solution of (6) with F as above. The problem (6)–(8) can be transformed, using a new independent variable $y = r^n$, into

$$Q_t = n^2 y^{2-2/n}Q_{yy} + n\sigma_n^{-1}F(Q)Q_y, \tag{10}$$

$$Q(0,t) = 0, \quad Q(1,t) = M_0 \ (q_0, \text{ resp.}), \quad Q(y,0) = Q_0(y), \tag{11}$$

with suitably rescaled constants M_0, M^* or q_0, q^*.

For $n = 1$ and $F(Q) = \text{const} \pm Q$ the equation (10) becomes the well known Burgers equation, which can be transformed into the heat equation using the Hopf–Cole substitution. In this way one can prove that for $n = 1$ (10)–(11) has a unique solution which tends to the unique stationary solution as times goes to infinity (Biler, Hilhorst, Nadzieja, 1994; Krzywicki, Nadzieja, 1992; Nadzieja, 1995; Nadzieja, Raczyński, 1996).

Theorems 1, 2 below recall a few results, mainly from (Biler, Nadzieja, 1995a; 1995b; 1996a), on the existence and nonexistence of solutions of the problem (10)–(11) in two dimensions.

Let us begin with the star problem where $F(Q) = M^* + Q$.

THEOREM 1. (i) For $n = 2$ and $M_0 + 2M^* < 8\pi$ there exists a unique positive nondecreasing stationary solution Q to (10)–(11).

(ii) If either $n = 2$ and $M_0 + 2M^* \geq 8\pi$, or $n \geq 3$ and $M_0 > 0$, $M^* > 0$, then there is no steady state of (10)–(11).

(iii) For $n = 2$ and $M_0 + 2M^* < 8\pi$, the problem (10)–(11) admits a global-in-time solution for each Hölder continuous initial condition Q_0 such that $\frac{dQ_0}{dy}(0) < \infty$.

(iv) If either $n = 2$ and $M^* \geq 4\pi$, or $n \geq 3$ and $M^* > 0$, then there is no (even local-in-time) positive nondecreasing solution $Q \not\equiv 0$ of (10) satisfying the boundary condition $Q(0, t) = 0$.

(v) For $n = 2$, $M^* < 4\pi$ and each Hölder continuous initial condition Q_0, there exists a local-in-time solution to (10)–(11).

(vi) Under the assumption $n = 2$, $M_0 + 2M^* \geq 8\pi$ but $M^* < 4\pi$, the local-in-time solutions cease to exist after a finite time.

Next we consider the electric case with $F(Q) = q^* - Q$.

THEOREM 2. (i) For $n = 2$ and all $q^* \in (-\infty, 4\pi]$, $q_0 \geq 0$, there exists a unique stationary solution Q of the system (10)–(11), in particular, there is a unique steady state of the Rubinstein problem in the unit disk $B(0, 1) \subset \mathbb{R}^2$.

(ii) If either $q^* > 4\pi$ and $q_0 > 0$, or $n \geq 3$ and $q^* > 0$, $q_0 > 0$, there is no stationary solution Q of the problem (10)–(11) in the n-dimensional ball.

(iii) If $n = 2$ and $q^* \in (-\infty, 4\pi]$, then the problem (10)–(11) has a global-in-time solution for any Hölder continuous initial condition Q_0.

(iv) If either $n = 2$ and $q^* > 4\pi$, or $n \geq 3$ and $q^* > 0$, then there is no positive nondecreasing solution $Q \not\equiv 0$ of (10) satisfying the boundary condition $Q(0, t) = 0$.

The parts (i) of Theorems 1, 2 follow from the exact integrability of the stationary problems (10)–(11). For the star problem the stationary solution of (10)–(11) is

$$Q(y) = (8\pi - 2M^*)y^\gamma(y^\gamma + c)^{-1} \tag{12}$$

with $\gamma = 1 - M^*/(4\pi) \in (0, 1]$, $c = 8\pi\gamma/M_0 - 1$, so that $Q(1) = M_0$ (and necessarily $M_0 + 2M^* < 8\pi$). For the electric case when $q^* < 4\pi$ the stationary solution has the form

$$Q(y) = (8\pi - 2q^*)cy^\gamma/(1 - cy^\gamma)$$

with $\gamma = 1 - q^*/(4\pi)$, $c = q_0/(q_0 + 8\pi - 2q^*)$, and for $q^* = 4\pi$

$$Q(y) = \left(q_0^{-1} - (8\pi)^{-1}\log y\right)^{-1}. \tag{13}$$

The general strategy of proofs of the existence statements is based on an approximation of the problem (10)–(11) by nonsingular ones. The essential difficulty is the singular diffusion coefficient $4y$ in (10) for $n = 2$. To overcome this we consider parabolic regularizations of the original problem (10)–(11) like in (Biler, Hilhorst, Nadzieja, 1994; Th. 2)

$$Q_t = 4(y + \varepsilon)Q_{yy} + \pi^{-1}F(Q)Q_y, \quad \varepsilon > 0,$$
$$Q(0, t) = 0, \quad Q(1, t) = M_0 \quad (q_0, \text{ resp.}), \qquad (\mathcal{P}_\varepsilon)$$
$$Q(y, 0) = Q_0(y).$$

The regularized problem $(\mathcal{P}_\varepsilon)$, with the initial (nondecreasing) condition $Q_0 \in C^\alpha([0, 1])$ for some $\alpha > 0$, has a unique (nondecreasing) solution $Q = Q_\varepsilon \in C^{2+\alpha, 1+\alpha/2}([0, 1] \times [0, T])$ for each $T > 0$. This follows from the standard theory of uniformly parabolic problems, see (Ladyženskaja et al., 1988; Ch. VI, Th. 5.2). Moreover, the estimate of the Hölder norm

$$\|Q_\varepsilon\|_{C^{\alpha, \alpha/2}([\delta, 1] \times [0, T])} \leq C(\delta) \tag{14}$$

for each $\delta > 0$ with $C(\delta)$ *independent* of ε is a consequence of the result cited above.

The comparison principle holds for solutions of the problem $(\mathcal{P}_\varepsilon)$, hence suitable supersolutions $\bar{Q}_\varepsilon$ can supply us with bounds on Q_ε showing the compactness of the family of approximating solutions $\{Q_\varepsilon : \varepsilon > 0\}$ in $C([0, 1] \times [0, T])$. Of course, the form of supersolutions $\bar{Q}_\varepsilon$ depends on F in a very sensitive manner.

For instance, if $F(Q) = M^* + Q$ then we may choose

$$\bar{Q}_\varepsilon(y) = A(y + \varepsilon)^\gamma \left((y + \varepsilon)^\gamma + b\right)^{-1}$$

with $\gamma = 1 - M^*/(4\pi) \in (0, 1]$, $0 \leq A \leq 8\pi - 2M^*$, $b > 0$. This form of the supersolution is suggested by the stationary solution (12) to (10)–(11).

If $F(Q) = q^* - Q$ then we consider

$$\bar{Q}_\varepsilon(y) = A(b - \log(y + \varepsilon))^{-1}$$

with $A \geq 8\pi$, $b > 0$, which is reminiscent of the stationary solution (13) to (10)–(11).

These supersolutions, combined with the estimate (14), imply the existence of a common modulus of continuity for $\{Q_\varepsilon : \varepsilon > 0\}$. Thus, there is a sequence $\varepsilon_k \to 0$ and a (nondecreasing) function $Q \in C([0, 1] \times [0, T])$ such that $Q_{\varepsilon_k} \to Q$ as $\varepsilon_k \to 0$, uniformly in (y, t). Again from the bound $Q_\varepsilon \leq \bar{Q}_\varepsilon$ we infer that Q satisfies the boundary condition (11).

Next let $(y, t) \in (0, 1) \times (0, T)$. From standard results in (Ladyženskaja et al., 1988; Ch. III, Th. 10.1) it follows that $\{Q_\varepsilon : \varepsilon > 0\}$ is bounded in the space $C^{m+\alpha, (m+\alpha)/2}([y/2, 1] \times [t/2, T])$ for all $m \in \mathbb{N}$, $\alpha \in (0, 1)$. Then, since Q_ε solves $(\mathcal{P}_\varepsilon)$, Q satisfies (10) in $[y/2, 1] \times [t/2, T]$. The point (y, t) was chosen arbitrarily in $(0, 1) \times (0, T)$, so Q is a solution of (10) in $(0, 1) \times (0, T)$ for each $T > 0$.

The nonexistence of solutions results can be shown in a completely elementary, although somewhat tedious manner. However, we prefer a quick argument involving the Feller test for the existence of evolution solutions to one-dimensional singular diffusion equations, see e.g. (Nagasawa, 1993; Ch. II, Th. 2.6). Namely, observe that a hypothetical solution $Q \not\equiv 0$ of (10) with the boundary condition $Q(0,t) = 0$ would be a solution of a *linear* singular parabolic equation

$$n^2 y^{2-2/n} q_{yy} + n\sigma_n^{-1} F(Q) q_y - q_t = 0$$

whose diffusion coefficient $n^2 y^{2-2/n}$ (i.e. $4y$ for $n = 2$) and the drift coefficient $n\sigma_n^{-1} F(Q)$ correspond to the case of an entrance point in the Feller test, i.e. Q cannot preserve the boundary condition $Q(0,t) = 0$.

To have a better understanding of the phenomenon of nonexistence of solutions for highly singular diffusion coefficients and thus explain the meaning of an instantaneous blow-up, we recall a result (Theorem 2 in Biler, Nadzieja, 1996b) refining Th. 1 (vi) on finite time blow-up for (10)–(11) with the function F generalizing that for the star problem. The hypotheses in Proposition 1 below are weaker than those for the parts (ii) and (iv) of Theorem 1.

PROPOSITION 1. Let $G(s) = \int_0^s F(\tau)\, d\tau$. Assume that $G(s)/s \geq 2\sigma_n(n-1) + a$ for some $a > 0$ and all $s > \tilde{M}$. If $M_0 > \tilde{M}$, then there are no global solutions of (10)–(11) such that $\lim_{y \to 0} y^{2-2/n} Q_y = 0$ for each $t > 0$.

Proof: Integrating the equation (10) over $[\delta, 1]$, $\delta > 0$, we get

$$\left(\int_\delta^1 Q(y,t)\, dy \right)_t = -2n(n-1) \int_\delta^1 y^{1-2/n} Q_y(y,t)\, dy + n^2 y^{2-2/n} Q_y(y,t)|_\delta^1$$

$$+ \frac{n}{\sigma_n} \int_\delta^1 (G(Q(y,t)))_y\, dy.$$

After another integration by part, letting $\delta \to 0$ we obtain

$$\left(\int_0^1 Q(y,t)\, dy \right)_t \geq 0 - 2n(n-1) y^{1-2/n} Q|_0^1 + n\sigma_n^{-1} G(M_0)$$

$$= -2n(n-1)M_0 + n\sigma_n^{-1} G(M_0) \geq a n \sigma_n^{-1} M_0$$

for all $t > 0$. But this is absurd in view of $\int_0^1 Q(y,t)\, dy \leq M_0$.

Remark that for the gravitational problem with radial symmetry the blow-up has been proved in (Biler, Hilhorst, Nadzieja, 1994; Th. 3(ii)) (in balls) and in (Nadzieja, Th. 4) ($M^* = 0$, in the whole space $\mathbb{R}^2$). Note that in the n-dimensional ($n \geq 3$) case arbitrarily small $M_0 > 0$ and $M^* \geq 0$ can lead to the nonexistence of global solutions, cf. a related computation in (Biler, 1995b). All these proofs are based on calculations of integral functionals of Q. Another method is to use suitable subsolutions cf. (Biler, Nadzieja, 1995a).

PROPOSITION 2. For $F(Q) \geq Q$, $n \geq 2$ the function
$Ky + Cy(y^{2/n} + (b - ct)^2)^{-1}$, and for $n \geq 3$ the functions
$\max(Ky + Cy(y + |b - ct|^3)^{-1}, 2\sigma_n y^{1-2/n}), 2\sigma_n y^{1-2/n} + Cy(T - t)^{-1}$
for suitably chosen constants K, C, b, c, T, are the subsolutions of (10)–(11).

We skip the (easy) proof of this statement, noting that the first subsolution may suggest for $n = 2$ the blow-up of solution Q with a concentration of mass $C \geq 8\pi$ at the origin. For $n \geq 3$ the derivative $\frac{\partial}{\partial y} Q$, hence the density u blows up at the origin, without concentration of mass. The second one suggests that mass concentrating at the origin is equal to $C > 0$. The third one blows up with an infinite mass. Observe that $2\sigma_n y^{1-2/n}$ is an exact solution of (10) on $(0, \infty)$. Of course, it is probable that solutions with the initial data larger than that of the subsolution explode *before* time $T = b/c$. Nevertheless, these families of subsolutions show in a suggestive manner a great variety of possible explosion mechanisms for higher dimensional models ($n \geq 3$) (10)–(11).

Actually, the mechanism of the blow-up for (10)–(11) is quite intriguing. Solutions that cease to exist after a concentration of mass at the origin (in our case this is 8π, when $M_0 > 8\pi$, $M^* = 0$) are shown to exist in (Herrero, Velázquez, 1996). Their construction is based on matched asymptotic expansions (the formal part), and on a delicate topological argument showing the existence of solutions with prescribed asymptotics in different regions (the rigorous part of the proof). In particular, this shows that near the blow-up time the diffusion is negligible compared to the nonlinear transport effects. In the next paper (Herrero, Medina, Velázquez, 1996) the three-dimensional case is also studied. The solutions blowing up without the concentration of mass are constructed as well as those exploding after a concentration at the origin of an arbitrarily prescribed mass. Such a phenomenon can be heuristically understood using different classes of subsolutions to (10) mentioned above. We will return to the nonexistence of global-in-time solutions in Section 4.

3 REGULARIZED SYSTEMS AND THEIR GLOBAL-IN-TIME SOLUTIONS

In order to confirm the role of the critical values of mass or charge for (6)–(8) also for the two-dimensional problem (1)–(5), we formulate below two results on the (global-in-time) approximation of singular systems by regular ones. They give a possible way to obtain a suitable solution to (1)–(5) when weak solutions cannot be defined in an usual way (because of singular terms like $\Delta V \sim \delta_0$ in the equation (1)). A new ingredient compared to the preceding analysis of radial solution is a physically motivated functional which will play the role of a Lyapunov function. A priori estimates obtained with the use of this function will provide compactness of the family of approximating solutions, and thus will replace direct comparison arguments applied previously – unavailable for the boundary conditions (3)–(4) more complicated than (7). The case of the Coulomb repulsion between particles is substantially simpler than that of the Newtonian attraction.

THEOREM 3. Suppose that $\Omega \subset I\!R^2$ is a bounded domain, $0 \in \Omega$, $\partial\Omega \in C^{1+\epsilon}$ for some $\epsilon > 0$, $q^* \in [0, 4\pi)$, and $0 \leq u_0 \in L^p(\Omega)$ for some $p > 1$. Then there exists a function $u = u(t)$ defined for all $t \geq 0$, and $u(x, t) \geq 0$ a.e. in x, t, which is the weak–$L^1(\Omega)$ limit of a subsequence of solutions $u_k = u_k(t)$ to (1)–(5) with regular potentials V_k suitably approximating V. The function $u(t)$ has an accumulation point in $L^1(\Omega)$ as $t \to +\infty$.

Proof: We begin with recalling the definition of the *weak* solution of the problem (1)–(5) when $0 \leq u_0 \in L^2(\Omega)$ and the potential V is assumed to be smooth and bounded, see e.g. (Biler, Hebisch, Nadzieja, 1994; Th. 1). We will apply it to the truncated potential $V = V_k = q^* E_2 \chi_k(q^* E_2)$, where $\chi_k(x) = \chi(x/k)$, $k \in \mathbb{N}$, and χ is a smooth function with compact support, $\chi(x) \equiv 1$ for $x \in [-1,1]$, $0 \leq \chi \leq 1$. The definition is based on a standard one (Ladyženskaja et al., 1988) and reads

$u \in L^\infty((0,T); L^2(\Omega)) \cap L^2((0,T); H^1(\Omega))$ is a weak solution to (1)–(5) on $(0,T)$ if the identity

$$\int_\Omega u(x,t)\eta(x,t)\,dx - \int_0^t \int_\Omega u\eta_t + \int_0^t \int_\Omega (\nabla u + u\nabla\varphi)\cdot\nabla\eta = \int_\Omega u_0(x)\eta(x,0)\,dx$$

holds for each $\eta \in H^1(\Omega \times (0,T))$, a.e. $t \in (0,T)$, and $\varphi = \varphi(t) \in H^1(\Omega)$ is a weak solution of (2), (4), i.e.

$$-\int_\Omega \nabla\varphi \cdot \nabla\xi + \int_\Omega (u - \Delta V)\xi = 0$$

for each $\xi \in H^1(\Omega)$ and a.e. $t \in (0,T)$; for (4.2) we simply put $\varphi = -E_2 * u + V$, for (4.3) we assume $\varphi(t) \in H_0^1(\Omega)$.

The proof of Theorem 3 is divided into several steps.

LEMMA 1. If V is bounded and smooth, then there exists $T > 0$ and a unique weak solution u of (1)–(5) on $(0,T)$ in the sense of the definition above. Moreover, $\int_\Omega u(x,t)\,dx = \int_\Omega u_0(x)\,dx$ and $u(x,t) \geq 0$ a.e.

The idea of the construction, based on regularizing properties of parabolic operators, follows from (Biler, 1992; Biler, Hebisch, Nadzieja, 1994).

Next we recall from the latter reference the existence of solutions with less regular initial data, and fine regularity properties of them due to parabolic character of the problem (1)–(5).

LEMMA 2. Local in time solutions to (1)–(5) with bounded V can be defined whenever $u_0 \in L^p(\Omega)$, $p > 1$. They enjoy some regularity properties including $u \in L_{loc}^\infty((0,T); L^\infty(\Omega))$.

We note that, because of the nonlinear no-flux condition on u in (3), we cannot expect the parabolic maximum principle for u be useful; no a priori control on u is available on the parabolic boundary of $\Omega \times (0,T)$. This makes the question of the continuation of solutions in time quite delicate, even for smooth V. The global-in-time existence for the full Debye system of electrolysis equations in two dimensions has been proved in (Biler, Hebisch, Nadzieja, 1994; Th. 3), see also (Biler, Nadzieja, 1994; Th. 1(iv)) for a similar construction in the gravitational case, using a supplementary a priori estimate suggested by the physical origin of the problem.

LEMMA 3. The function

$$W(t) = W(u(t)) = \int_\Omega u(x,t)\log u(x,t)\,dx$$
$$+\frac{1}{2}\int_\Omega |\nabla(-\varphi(x,t) + V(x) + \Gamma(x))|^2\,dx + \int_\Omega u(x,t)(V(x) + \Gamma(x))\,dx \tag{15}$$

is a Lyapunov function for the problem (1)–(5) with a bounded potential V. Here Γ is a bounded harmonic function correcting the boundary values of the potential V according to (4) (in particular, for (4.2) $\Gamma \equiv 0$).

For (3)–(4.3) or (4.4) W – the free energy – is a Lyapunov function. Indeed, multiplying formally (1) by $(\log u + \varphi)$, integrating by parts and integrating on $[0, t]$, we arrive at the inequality

$$W(t) + \int_0^t \int_\Omega u |\nabla(\log u + \varphi)|^2 \leq W(0), \tag{16}$$

i.e. $W(t)$ is a nonincreasing function on the trajectories of (1)–(5): $W(t) \leq W(0) = W(u_0)$. This computation is made rigorous by approximating u by $u + \delta$ ($\delta > 0$), using solvability results in Lemmas 1, 2, and passing to the limit $\delta \to 0$, cf. (Biler, 1992; Lemma 3). The case of (3)–(4.2) is slightly different. The function W is an approximative Lyapunov function, i.e. $W(t) \leq W(u_0) + C(\Omega, u_0)$.

LEMMA 4. For $n = 2$ and bounded smooth V the solutions of (1)–(5) can be continued globally in time, and they satisfy the estimate $\sup_{t \geq 0} |u(t)|_2 < \infty$.

By Lemma 1 this can be obtained by proving an a priori bound on the $L^2(\Omega)$ norm of u. This is done combining a local-in-time estimate from Lemma 1, a non-linear imbedding inequality in (Biler, Hebisch, Nadzieja, 1994; (22))

$$|u|_3^3 \leq \varepsilon \|u\|_1^2 \, |u \log |u||_1 + C_\varepsilon |u|_1$$

valid for every $\varepsilon > 0$, some C_ε and each $u \in H^1(\Omega)$, $\Omega \subset I\!R^2$, and the uniform in time bound on $\int_\Omega u \log u$ and $\int_\Omega |\nabla \varphi|^2$ from the Lyapunov function W (16).

LEMMA 5. Suppose that $0 \leq u_0 \in L^p(\Omega)$ for some $p > 1$, $V_k = q^* E_2 \chi_k(q^* E_2)$ is a bounded smooth truncation of the potential $V = q^* E_2$ with $q^* \in (0, 4\pi)$, and u_k are the solutions of (1)–(5) corresponding to $V = V_k$. Then $W_k(t) = W_k(u_k(t)) \leq \sup_k W_k(u_0) < \infty$ for each $k \in I\!N$ and all $t \geq 0$, and

$$\sup_{k,t} \int_\Omega u_k(x, t) \log u_k(x, t) \, dx < \infty. \tag{17}$$

For the proof it is sufficient to establish a lower bound for the last term in the Lyapunov function $W = W_k$ in (15). Using the Young inequality, or simply the inequality $st \leq s \log s + \exp(t - 1)$ valid for all $s \geq 0$, $t \in I\!R$, we estimate

$$\left| \int_\Omega u_k(x, t) V_k(x) \, dx \right| \leq \int_\Omega u_k(x, t) q^* (2\pi)^{-1} \, |\log |x|| \, dx$$

$$\leq (1 - \delta) \int_\Omega |u_k| \log |u_k| + \int_\Omega |x|^{-q^*/(2\pi(1-\delta))} \, dx.$$

Here $\delta > 0$ is chosen small enough so that $q^* < 4\pi(1 - \delta)$. The integrals $\int_\Omega u_k \Gamma_k$ pose no problem since Γ_k's are uniformly bounded. Finally, we have

$$\delta \int_\Omega u_k \log u_k - C(\Omega, q_0) \leq W_k(t) \leq \sup_k W_k(u_0)$$

$$\leq 2 \int_\Omega u_0 \log u_0 + \frac{1}{2} \int_\Omega |\nabla(-\varphi_0 + V + \Gamma)|^2 + C(\Omega, q_0) < \infty,$$

as $L^p(\Omega) \subset L \log L(\Omega)$.

LEMMA 6. Under the assumptions of Lemma 5 there exists a function $u = u(t)$ which is for every $t \geq 0$ the weak–$L^1(\Omega)$ limit of a subsequence of $u_k = u_k(t)$ solving (1)–(5) with the regularized potentials V_k. The potentials $-\varphi_k(t)+V_k$ corresponding to the approximating solutions $u_k(t)$ converge in $L^q(\Omega)$ for each $1 \leq q < \infty$ and $t \geq 0$. The trajectory $u(t)$ has an accumulation point as $t \to +\infty$.

The uniform in $k \in I\!N$ bound (17) on $|u_k(t) \log u_k(t)|_1$ proved in Lemma 5 enables us to pass to the weak–$L^1(\Omega)$ limit for each $t \geq 0$ thanks to the compactness criterion of de la Vallée–Poussin and the Dunford–Pettis theorem. Concerning the behavior of potentials $-\varphi_k+V_k$, the uniform boundedness of the Lyapunov functions W_k gives

$$\sup_{k,t} |\nabla(-\varphi_k(t) + V_k + \Gamma_k)|_2 < \infty,$$

so by the Rellich lemma a subsequence of $-\varphi_k(t) + V_k$ converges in $L^2(\Omega)$ for each $t \geq 0$. Moreover, for any solution of (2)–(3) if $u_k \in L \log L(\Omega)$, then $(-\varphi_k + V_k) \in L^\infty(\Omega)$. Finally, the convergence of potentials in $L^2(\Omega)$ together with their uniform boundedness implies for each $1 \leq q < \infty$ the $L^q(\Omega)$ convergence.

The bound (17) is also uniform in $t \geq 0$, hence by a diagonal argument $u(t)$ is weakly convergent in $L^1(\Omega)$ along a sequence of t's, $t \to +\infty$, to a limit, which is expected to be a steady state of (1)–(4).

For the gravitational problem (1)–(3), (4.1), (5) we have the following counterpart of Theorem 3

THEOREM 4. Suppose that $\Omega \subset I\!R^2$ is a bounded domain, $0 \in \Omega$, $\partial\Omega \in C^{1+\epsilon}$ for some $\epsilon > 0$, and $0 \leq u_0 \in L^p(\Omega)$ for some $p > 1$. If $M_0 + M^* < 4\pi$, then there exists a function $u = u(t)$ defined for all $t \geq 0$, and $u(x,t) \geq 0$ a.e. in x, t. This function (obtained as a limit in the weak–$L^1(\Omega)$ sense of solutions to suitably regularized problems) has an accumulation point in $L^1(\Omega)$ as $t \to +\infty$.

The proof of Theorem 4 is similar to that of Theorem 3. The form of the Lyapunov function is different; in general, this is no longer bounded from below. However, under the assumptions of Theorem 4 such a lower bound holds as a consequence of rather delicate estimations involving e.g. the Moser–Trudinger inequality. We refer the reader for the details to (Biler, Nadzieja, 1994 and 1995b, Lemma 5) recalling only that the function (replacing (15) in the Newtonian case)

$$W(t) = W(u(t)) = \int_\Omega u(x,t) \log u(x,t) \, dx$$

$$+\frac{1}{2} \int_\Omega u(x,t)(\varphi(x,t) - V(x)) \, dx + \int_\Omega u(x,t)V(x) \, dx$$

is a Lyapunov function for the problem (1)–(5) with a bounded potential V.

REMARK. We may resume the contents of Lemma 5 speaking of a supplementary bound on $u(t) \geq 0$ in the (local) Hardy space $\mathcal{H}^1(I\!R^2)$. Recently, there appeared numerous papers with similar constructions for two-dimensional elliptic problems.

REMARK. The global-in-time solvability condition $M_0 + M^* < 4\pi$ in Theorem 4 would be improved to $M_0 + 2M^* < 8\pi$ (the same as for the radial problem) if (4.1) were replaced by the (nonphysical) Dirichlet condition (4.3).

4 FINITE TIME BLOW-UP

We present in this section some results on the nonexistence of global-in-time solutions to (1)–(5) in the two-dimensional case. First, the blow-up in the gravitational case ($M^* = 0$) without symmetry assumptions on u has been proved in (Biler, Nadzieja, 1994). More refined results in the general n-dimensional case in bounded domains have been presented in (Biler, 1995b). For the Cauchy problem with suitably localized $u_0 \in L^1\left(I\!R^n, (1 + |x|^2)\, dx\right)$ of high concentration and $M^* = 0$ this phenomenon has been shown in (Biler, 1996a) for solutions constructed in the scale of Morrey spaces instead of the usual $L^p(\Omega)$ framework for parabolic equations (cf. Theorems 3, 4 above).

Let us start with the star problem. The result below shows that for large M_0, M^* any reasonable weak solution u (irrespective of actual difficulties of its construction, cf. Theorems 3, 4) in a star-shaped domain $\Omega \subset I\!R^2$ of the problem (1)–(5) ceases to exist after a finite time. In particular, there are no stationary solutions with large M_0.

PROPOSITION 3. If Ω is a star-shaped (with respect to the origin) domain in $I\!R^2$ and $M_0 + 2M^* > 8\pi$, then solutions to (1)–(5) cannot be defined globally in time.

The proof is based on virial calculations (i.e. concerning the evolution of moments of the density u) similar to those in (Biler, Nadzieja, 1994), (Biler, 1995b). Define an auxiliary function

$$w(t) = \int_\Omega u(x,t)|x|^2\, dx \geq 0.$$

We have (writing for simplicity of notation differential inequalities instead of the integral ones — consequences of the definition of weak solutions)

$$\frac{dw}{dt} = -2\int_\Omega \nabla u \cdot x - 2\int_\Omega u\nabla\varphi \cdot x$$

$$= -2\int_{\partial\Omega} u\, x \cdot \nu + 4\int_\Omega u + 2\iint_{\Omega\times\Omega} u(x,t)\,(\nabla_x E_2(x-y)) \cdot x\, u(y,t)\, dy\, dx$$

$$-M^*\pi^{-1}\int_\Omega u(x,t)|x|^{-2}x \cdot x\, dx.$$

After dropping the first term on the right-hand side ($x \cdot \nu \geq 0$ holds for star-shaped domains Ω), and after the symmetrization of the third integral leading to

$$(2\pi)^{-1}\iint_{\Omega\times\Omega} u(x,t)\, u(y,t)|x-y|^{-2}((x-y)\cdot x + (y-x)\cdot y)\, dy\, dx$$

$$= (2\pi)^{-1}\iint_{\Omega\times\Omega} u(x,t)\, u(y,t)\, dx\, dy = (2\pi)^{-1}M_0^2,$$

we arrive at

$$\frac{dw}{dt} \leq 4M_0 - \frac{M_0^2}{2\pi} - \frac{M_0 M^*}{\pi} = M_0 \left(4 - \frac{M_0}{2\pi} - \frac{M^*}{\pi} \right).$$

Hence if $M_0 + 2M^* > 8\pi$, then $w(t)$ becomes negative in a finite time, a contradiction.

Similarly, for q^* large compared to q_0 any reasonable weak solution u of the Rubinstein problem with the condition (4.2) cannot exist for all $t \geq 0$.

PROPOSITION 4. If Ω is a star-shaped domain of $I\!\!R^2$, $0 \in \Omega$ and $2q^* - q_0 > 8\pi$, then solutions to (1)–(3), (4.2) cannot be defined globally in time.

The proof is again based on virial calculations and completely analogous to that of Proposition 3 above.

5 SELF-SIMILAR SOLUTIONS

It can be inferred from the formula (12) that in the gravitational case there exist steady states of (10) on $(0, \infty)$ with a finite total mass $M_0 = 8\pi - 2M^*$, i.e. radially symmetric solutions of the star problem in the whole plane $I\!\!R^2$. In the next proposition we show the existence of other particular solutions of (10) on $(0, \infty)$, namely the self-similar radial densities satisfying (1)–(2) on $I\!\!R^2$ and the Boltzmann scaling relation $u(x, t) = t^{-1} U(xt^{-1/2})$. They correspond to the solutions $Q(r, t) = 2\pi\zeta(r^2/t)$ of (6), cf. (Biler, 1995a; (21), (31)–(34)). The importance of self-similar solutions is connected with their role in describing the long time behavior of arbitrary global solutions, as was discussed in (Biler, 1995a; Sec. 3).

The nondecreasing positive function ζ defined above solves the problem

$$\zeta'' + \frac{1}{4}\zeta' + \frac{1}{2y}\left(\zeta + \frac{M^*}{2\pi}\right)\zeta' = 0, \quad y = r^2/t, \quad ' = \frac{d}{dy}, \tag{18}$$
$$\zeta(0) = 0.$$

The third section of (Biler, 1995a) deals with the existence of (not necessarily radial) self-similar solutions for the nonsingular system (1), (4.1) with $M^* = 0$ in arbitrary space dimensions. In particular, for $n = 2$ the functional analytic methods in Prop. 2(i) of the above reference, first introduced by Y. Meyer and his collaborators in 1994, give the existence of self-similar solutions with $u_0 = M_0\delta$ and small M_0, that is with small $\zeta(\infty) > 0$. This result was refined in (Biler, 1995c; Prop. 3(i)), using the dynamical systems approach and comparison techniques for O.D.E., to the whole expected range of M_0's corresponding to all $\zeta(\infty) \in [0, 4)$.

PROPOSITION 5. For each $Z \in [0, 4 - M^*/\pi)$ there exists a nondecreasing solution of the problem (18) such that $\lim_{y \to \infty} \zeta(y) = Z$, i.e. for $n = 2$ and M_0, M^* satisfying the assumption $M_0 + 2M^* < 8\pi$ there is a self-similar radial solution of (1), (4.1).

Proof: For an analysis of the asymptotics of solutions as $y \to \infty$ we adopt the phase plane approach from (Biler, 1995c; Prop. 3). Using the change of variables $s = \frac{1}{2} \log y$, $v(s) = 2y \frac{d\zeta}{dy}(y)$, $w(s) = \zeta(y)$, we transform (18) into the problem

$$v' = \left(2 - w - \frac{M^*}{2\pi}\right) v - \frac{e^{2s}}{2} v, \quad w' = v, \quad ' = \frac{d}{ds}, \tag{19}$$
$$v(-\infty) = 0, \quad w(-\infty) = 0.$$

(19) is a nonautonomous system in the plane $\mathbb{R}^2 \ni (v, w)$ whose solutions can be compared with those of the autonomous integrable systems

$$\underline{v}' = (2 - \underline{w} - M^*/(2\pi))\underline{v} - \varepsilon\underline{v}, \quad \underline{w}' = \underline{v}, \tag{20}$$

where $\varepsilon > 0$, $\underline{v} = \underline{v}_\varepsilon$, $\underline{w} = \underline{w}_\varepsilon$, with the same condition at $s = -\infty$.

Evidently, $\lim_{s \to \infty} w(s) < 4 - M^*/\pi$ because the function $(w + M^*/(2\pi) - 2)^2 + 2v \equiv L$ is strictly decreasing along the phase trajectories of (19). Indeed, $\frac{dL}{ds}(s) = -e^{2s} v(s) < 0$ since $v(s) > 0$ for every $s \in (-\infty, \infty)$, and $L(-\infty) > L(\infty)$.

A comparison of the vector fields in (19) and (20) gives the relation $\underline{w}(s) \leq w(s)$ for all $s \leq s_\varepsilon$ with $e^{2s_\varepsilon} = 2\varepsilon$. Since $\underline{w}(s) = 2\alpha A e^{\alpha s} (1 + A e^{\alpha s})^{-1}$ with $\alpha = 2 - M^*/(2\pi) - \varepsilon$ and an arbitrary $A > 0$ is a solution of the auxiliary system (20), so $\underline{w}(s_\varepsilon) = 2\alpha A(2\varepsilon)^{\alpha/2} \left(1 + A(2\varepsilon)^{\alpha/2}\right)^{-1}$. Therefore, taking sufficiently small $\varepsilon > 0$ we obtain $\sup Z = \sup w(s) \geq \limsup_{\varepsilon \to 0, s \leq s_\varepsilon, A > 0} \underline{w}_\varepsilon(s) = 4 - M^*/\pi$ which concludes the proof.

Similarly, for the Rubinstein problem in the whole plane, we look for solutions to (6) of the form $Q(r, t) = 2\pi\zeta(r^2/t)$. This self-similar Q leads to an ordinary differential equation problem

$$\zeta'' + \frac{1}{4}\zeta' - \frac{1}{2y}\left(\zeta - \frac{q^*}{2\pi}\right)\zeta' = 0, \quad y = r^2/t, \quad ' = \frac{d}{dy}, \tag{21}$$

for nondecreasing ζ such that $\zeta(0) = 0$.

We remark that functional analytic methods developed in (Biler, 1995a; Prop. 2(i)) for the gravitational case can be applied to the problem of the existence of self-similar solutions when $n = 2$, $q^* = 0$ and $u_0 = q_0 \delta$ with small $q_0 > 0$; the $-$ sign in (1) does not matter compared to the $+$ sign. Using O.D.E. methods ("shooting"), one can prove

PROPOSITION 6. For $n = 2$ and $q^* \in [0, 4\pi)$ there exist self-similar solutions of the Rubinstein problem with a finite total charge $q_0 = \zeta(\infty)/(2\pi)$, i.e. solutions of (21) with $\zeta(\infty) \in (0, \infty)$.

REFERENCES

1. P. Biler, *Existence and asymptotics of solutions for a parabolic-elliptic system with nonlinear no-flux boundary conditions*, Nonlinear Analysis T. M. A. **19** (1992), 1121–1136.

2. P. Biler, *The Cauchy problem and self-similar solutions for a nonlinear parabolic equation,* Studia Math. **114** (1995a), 181–205.

3. P. Biler, *Existence and nonexistence of solutions for a model of gravitational interaction of particles, III* , Colloq. Math. **68** (1995b), 229–239.

4. P. Biler, *Growth and accretion of mass in an astrophysical model,* Applicationes Math. **23** (1995c), 179–189.

5. P. Biler, *Local and global solutions of a nonlinear nonlocal parabolic problem,* Proceedings of the Banach Center minisemester "Nonlinear Analysis and Applications, Warsaw 1994", N. Kenmochi, M. Niezgódka eds., Gakuto Int. Series Math. Sciences and Its Appl. **7** (1996a), 49–66.

6. P. Biler, *Local and global solvability of some parabolic systems modelling chemotaxis,* Adv. Math. Sci. Appl., (1996b), 1–29, to appear.

7. P. Biler, W. Hebisch, T. Nadzieja, *The Debye system: existence and long time behavior of solutions,* Nonlinear Analysis T. M. A. **23** (1994), 1189–1209.

8. P. Biler, D. Hilhorst, T. Nadzieja, *Existence and nonexistence of solutions for a model of gravitational interaction of particles, II,* Colloq. Math. **67** (1994), 297–308.

9. P. Biler, T. Nadzieja, *A class of nonlocal parabolic problems occurring in statistical mechanics,* Colloq. Math. **66** (1993), 131–145.

10. P. Biler, T. Nadzieja, *Existence and nonexistence of solutions for a model of gravitational interaction of particles, I,* Colloq. Math. **66** (1994), 319–334.

11. P. Biler, T. Nadzieja, *Growth and accretion of mass in an astrophysical model, II,* Applicationes Math. , **23** (1995a), 351–361.

12. P. Biler, T. Nadzieja, *A nonlocal singular parabolic problem modelling gravitational interaction of particles,* Adv. Diff. Eq., (1995b), 1–19, to appear.

13. P. Biler, T. Nadzieja, *A singular problem in electrolytes theory,* Math. Methods Appl. Sci., (1996a), to appear.

14. P. Biler, T. Nadzieja, *Nonlocal parabolic problems in statistical mechanics,* Proceedings of WCNA '96 – Athens, (1996b), to appear.

15. M. A. Herrero, E. Medina, J. J. L. Velázquez, *Analysis of a model of chemotactic aggregation,* preprint Universidad Complutense Madrid (1996).

16. M. A. Herrero, J. J. L. Velázquez, *Singularity patterns in a chemotaxis model,* Math. Annalen, **306** (1996), 583–623.

17. W. Jäger, S. Luckhaus, *On explosions of solutions to a system of partial differential equations modelling chemotaxis* , Trans. Amer. Math. Soc. **329** (1992), 812–824.

18. A. Krzywicki, T. Nadzieja, *A nonstationary problem in the theory of electrolytes* Quart. Appl. Math. **50** (1992), 105–107.

19. O. A. Ladyženskaja, V. A. Solonnikov, N. N. Ural'ceva, *Linear and Quasilinear Equations of Parabolic Type*, Amer. Math. Soc., Providence, R.I., 1988.

20. T. Nadzieja, *A model of radially symmetric cloud of self-attracting particles*, Applicationes Math. **23** (1995), 169–178.

21. T. Nadzieja, A. Raczyński, *A singular radially symmetric problem in electrolytes theory*, (1996), in preparation.

22. T. Nagai, *Blow-up of radially symmetric solutions to a chemotaxis system*, Adv. Math. Sci. and Appl. **5** (1995), 581–601.

23. M. Nagasawa, *Schrödinger Equations and Diffusion Theory*, Birkhäuser, Basel, Boston, Berlin, 1993.

24. I. Rubinstein, *Electro-Diffusion of Ions*, SIAM, Studies in Appl. Math. vol. 11, Philadelphia, 1990.

25. G. Wolansky, *On steady distributions of self-attracting clusters under friction and fluctuations* Arch. Rational Mech. Anal. , **119** (1992), 355–391.

Nonlinear Liouville Theorems

ISABEAU BIRINDELLI   Dipartimento di Matematica, Università di Roma
"La Sapienza", Italia

1 A SURVEY DISGUISED AS AN INTRODUCTION

Existence of positive solutions for semilinear equations of the type

$$\begin{cases} u > 0 & \text{in } \Omega, \\ -Lu = f(x, u), & \text{in } \Omega, \\ u = \phi & \text{on } \partial\Omega, \end{cases} \tag{1}$$

where

$$L := a_{ij}(x)\partial_{ij} + b_i(x)\partial_i + c(x), \tag{2}$$

satisfying, for some positive constants c_o, C_o, and b

$$c_o|\xi|^2 \leq a_{ij}(x)\xi_i\xi_j \leq C_o|\xi|^2, \tag{3}$$

$$\begin{aligned} a_{ij} \in C(\overline{\Omega}), \quad b_i, \quad c \in L^\infty, \\ (\textstyle\sum b_i^2)^{\frac{1}{2}}, \ |c| \leq b, \end{aligned} \tag{4}$$

where $f(x, u) \cong a(x)u^p$ with $p > 1$ and Ω is a bounded domain has been thoroughly studied either for $a(x) > 0$ or for $a(x) \leq 0$ or with $a(x)$ changing sign.

   If (1) doesn't have a variational structure then a natural approach to prove existence of a solution is to give L^∞ a priori bounds. Hence, using degree theory, under the right hypothesis on f, one proves existence of a classical solution.

To prove a priori bounds in the case $a(x) > 0$, Gidas and Spruck, in [15], used the so called "blow-up" method, which we are going to illustrate in the simple case:

$$\begin{cases} u > 0 & \text{in } \Omega, \\ -\Delta u = a(x)u^p, & \text{in } \Omega, \\ u = \phi & \text{on } \partial\Omega, \end{cases} \tag{5}$$

where $a(x) > 0$ is a continuous function in $\overline{\Omega}$.

Suppose by contradiction that there exists a sequence of solutions $\{u_k\}$ such that

$$M_k := \sup_{x \in \Omega} u_k(x) := u_k(P_k) \to +\infty,$$

as $k \to +\infty$.

$\{P_k\}$ is in $\overline{\Omega}$, we can therefore assume, eventually passing to a subsequence, that

$$P_k \to P_o \in \overline{\Omega}.$$

It is quite easy to see that for the right value of α the sequence of functions $v_k(y) = M_k^{-1}u_k(yM_k^\alpha + P_k)$, with $v_k(0) = 1$ converges to v which is, depending on whether P_o is on the boundary of Ω or not, either the solution of

$$\begin{cases} v \geq 0 & \text{in } \mathbb{R}^n, \\ -\Delta v = a(P_o)v^p & \text{in } \mathbb{R}^n, \end{cases} \tag{6}$$

or the solution of

$$\begin{cases} v \geq 0 & \text{in } \Pi, \\ -\Delta v = a(P_o)v^p & \text{in } \Pi, \end{cases} \tag{7}$$

where Π is a half space.

We have reached a contradiction because $v(0) = 1$ but $v \equiv 0$ since the following theorem holds (see [14]):

THEOREM 1 *If $a(P_o) > 0$ and*

$$1 \leq p < \frac{n+2}{n-2},$$

then the only solution of (6) and of (7) is $v \equiv 0$.

The case of indefinite non linearity i.e. when $a(x)$ changes sign, has been treated by Berestycki, Capuzzo Dolcetta and Nirenberg in [3].

Under some hypothesis on $a(x)$, the existence proof in [3] follows the scheme above illustrated, but, beside Theorem 1, Liouville theorems for semilinear problems in cones are required.

Precisely, consider the following problem

$$\begin{cases} v \geq 0 & \text{in } \Sigma, \\ -\Delta v \geq h(x)v^p & \text{in } \Sigma, \end{cases} \tag{8}$$

where Σ is a cone and $h(x)$ is a positive function. Without loss of generality we suppose that the vertex of Σ is the origin.

Let λ and ϕ be respectively the principal eigenvalue and the corresponding eigenfunction of the Laplacian with Dirichlet boundary conditions restricted to $S = \Sigma \cap S_1$ where S_1 is the sphere of radius 1 and let α be the positive solution of $\alpha(\alpha + n - 2) = \lambda$. Furthermore let h be a function such that for r sufficiently large, the function

$$k(r) = \left[\int_S h^{\frac{1}{1-p}}(r,\omega)\phi(\omega)d\omega \right]^{1-p},$$

is well defined.

Then the result of [3], stated in a slightly general form, is the following

THEOREM 2 *If $v \in C^2(\Sigma)$ is a positive solution of (8), bounded near the origin and*

$$\lim_{R \to +\infty} CR^{-2\frac{p}{p-1}} \int_{r_o}^{R} k(r)^{\frac{1}{1-p}} r^{\alpha+n-1} dr = 0, \tag{9}$$

then the only solution of (8) is $v \equiv 0$.

REMARK: 1: In [3], the authors consider the case where for large $|x|$, h satifies $h(x) = |x|^\gamma$. In this hypothesis, condition (9) which coincides with the condition in [3], becomes

$$1 < p \leq \frac{n + \alpha + \gamma}{n + \alpha - 2}. \tag{10}$$

REMARK 2: This bound is optimal in the sense that for any $p > \frac{n+\alpha+\gamma}{n+\alpha-2}$, one can construct an explicit solution of (8) with $h(x) = |x|^\gamma$ see [6]. It still is an open problem if, for

$$\frac{n + \alpha + \gamma}{n + \alpha - 2} < p < \frac{n + \alpha + \gamma + 2}{n + \alpha - 2},$$

the only positive solution of

$$-\Delta u = r^\gamma u^p \quad \text{in } \Sigma,$$

is the trivial one.

REMARK 3: Let me emphasise that in Theorem 1 and 2 there are no conditions on the behaviour of u at infinity or conditions on the boundedness of the solution. Indeed, for example, Dancer in [9] proves that if $1 < p < \frac{n+1}{n-3}$ then the only positive bounded solutions of

$$\begin{cases} -\Delta u = u^p, & \text{in } x_n > 0, \\ u = 0 & \text{in } x_n = 0, \end{cases} \tag{11}$$

are $u \equiv 0$. Let us observe that the range of p for which the non-existence result improves the result of Gidas and Spruck where u is not supposed a priori bounded.

Let us also mention that for solutions u of semilinear equations in unbounded domains , decaying to zero as $|x| \to +\infty$, M. Esteban & P.L. Lions in [11] proved non-existence results for domains Ω satisfying the geometric condition:

there exists $X \in \mathbb{R}^n$, $|X| = 1$ such that $n(x).X \geq 0$, $n(x).X \not\equiv 0$ on $\partial\Omega$.

Non-linear Liouville theorems such as Theorem 1 and 2 have been extended in many directions. For example non existence results have been given for systems of equations. Precisely the following theorem holds.

THEOREM 3 *Suppose $u \geq 0$ and $v \geq 0$ are solutions of*

$$\begin{cases} -\Delta u \geq v^p & in \ \Sigma, \\ -\Delta v \geq u^q & in \ \Sigma. \end{cases} \tag{12}$$

If

$$\frac{(n + \alpha - 2)}{2}(pq - 1) \leq \max\{p+1, q+1\}, \quad p > 1, \ q > 1, \tag{13}$$

then $u \equiv 0$ and $v \equiv 0$.

The case $\Sigma = \mathbb{R}^n$ (i.e. $\alpha = 0$) was solved by Mitidieri [18] for $p > 1$ and $q > 1$ and by Serrin and Zou for the case $pq > 1$ but either p or q are smaller than 1. The case of the cone i.e. Theorem 3 was treated by Birindelli and Mitidieri in [6] in a more general setting. See also De Figuereido, Felmer [10] when Σ is a half space.

In [6] the non-existence result are used to prove a priori bounds for systems of equations with indefinite non linearity.

On the other hand Birindelli, Capuzzo Dolcetta and Cutrì in [4] proved some similar non existence results for the degenerate elliptic operator Δ_H called Heisenberg laplacian (see section 2 for the definition), in $\mathbb{R}^{2n+1}$, in half spaces and some other unbounded sets.

Let us call H^n the space $\mathbb{R}^{2n+1}$ endowed with the Heisenberg group and $|.|_H$ the intrinsic norm in H^n (for more details see section 2). Then, in [4], the following theorem is proved

THEOREM 4 *Suppose* $u \geq 0$ *is a solution of*

$$-\Delta_H u \geq h(\xi) u^p \quad in \ \mathbb{R}^{2n+1}, \tag{14}$$

where, for $|\xi|_H$ *large* h *satisfies* $h(\xi) \geq |\xi|^\gamma$, *then if*

$$1 < p \leq \frac{2n + 2 + \gamma}{2n} \tag{15}$$

then $u \equiv 0$ *in* H^n.

In [5] in particular this result and other non-existence results are used to prove a priori bounds and existence of a solution for equation with indefinite non linearity and principal part equal to the Δ_H.

REMARK 4: It is a well known fact that the homogenous dimension of H^n is $Q = 2n + 2$ (see (19) in section 2). Hence (15) is just (10) where $\alpha = 0$ and n is replaced by the homogenous dimension Q.

REMARK 5: Results of non existence for semilinear equation involving the Heisenberg laplacian have also been proved by Garofalo and Lanconelli in [13]. Their results are the equivalent, in H^n, of the result of Esteban and Lions [11] mentioned in Remark 3.

We are going to state now the main result of this paper. Precisely we are going to extend the results in [4] to systems of inequality.

THEOREM 5 *Suppose* $u \geq 0$ *and* $v \geq 0$ *are solutions of*

$$\begin{cases} -\Delta_H u \geq v^p & in \ \Sigma, \\ -\Delta_H v \geq u^q & in \ \Sigma, \end{cases} \tag{16}$$

$$\frac{(Q-2)}{2}(pq - 1) \leq \max\{p + 1, q + 1\}, \quad p > 1, \ q > 1 \tag{17}$$

then $u \equiv 0$ *and* $v \equiv 0$.

In the last section we will also prove results in half spaces. The case of cones is still an open problem.

2 PRELIMINARY

In this section we collect for the convenience of the reader some known facts about the Heisenberg group H^n and the operator Δ_H which will be useful later on. For their proof and more information we refer for example to [12, 13, 16, 17].

The Heisenberg group H^n is the Lie group whose underlying manifold is $\mathbb{R}^{2n+1}$ ($n \geq 1$), endowed with the group action,

$$\xi_0 \circ \xi = (x + x_0, y + y_0, t + t_0 + 2 \sum_{i=1}^{n} (x_i y_{0_i} - y_i x_{0_i})),$$

for $\xi = (x_1, \ldots, x_n, y_1, \ldots, y_n, t) := (x, y, t)$.

The corresponding Lie Algebra of left-invariant vector fields is generated by

$$X_i := \frac{\partial}{\partial x_i} + 2y_i \frac{\partial}{\partial t} \quad \text{for} \quad i = 1, \ldots, n,$$

$$Y_i := \frac{\partial}{\partial y_i} - 2x_i \frac{\partial}{\partial t} \quad \text{for} \quad i = 1, \ldots, n,$$

$$T = \frac{\partial}{\partial t}.$$

It is easy to check that X_i and Y_i satisfy $[X_i, Y_j] = -4T\delta_{i,j}$, $[X_i, X_j] = [Y_i, Y_j] = 0$ for any $i, j \in \{1, \ldots, n\}$. Therefore, the vector fields X_i, Y_i (i=1,..., n) and their first order commutators span the whole Lie Algebra. Hence, the Heisenberg laplacian, defined by

$$\Delta_H := \sum_{i=1}^{n} X_i^2 + Y_i^2$$

satisfies the Hormander condition (see [17]); this implies its hypoellipticity (i.e. if $\Delta_H u \in C^\infty$ then $u \in C^\infty$ (see [17])) and the validity of Bony's maximum principle (see [7]).

A distance from the origin can be defined on H^n by setting

$$|\xi|_H := \left(\sum_{i=1}^{n} (x_i^2 + y_i^2)^2 + t^2 \right)^2,$$

see e.g. [12].

Clearly in the corresponding metric, the open ball of radius R centred at ξ_o, is the set:

$$B_H(\xi_o, r) = \{\eta \in H^n : |\eta \circ \xi_o^{-1}|_H < r\},$$

where ξ_o^{-1} is the inverse with respect to the group action and $\xi_o^{-1} = -\xi_o$.

It is also important to observe that $\xi \to |\xi|_H$ is homogenous of degree one with respect to the natural group of dilation (see [12]):

$$\delta_\lambda(\xi) = (\lambda x, \lambda y, \lambda^2 t). \tag{18}$$

Finally let us recall that

$$|B_H(\xi_o, R)| = |B_H(0,1)|R^Q, \tag{19}$$

where $Q = 2n + 2$ and $|\cdot|$ denotes the Lebesgue measure. As mentioned in the introduction Q is called the homogenous dimension of H^n (see [16]).

For later purposes let us introduce the function

$$\psi(\omega) = \sum_{i=1}^{n} \frac{x_i^2 + y_i^2}{\rho^2}, \tag{20}$$

homogenous of degree 0 with respect to the intrinsic dilation, here $\omega = \frac{\xi}{\rho}$.

3 REMARKS AND STATEMENTS OF LIOUVILLE THEOREMS

Before stating the results we start by introducing a few notions and notations.

Let us denote by ϕ_R a cut-off function satisfying:

$$\begin{cases} \phi_R(\rho) := \phi(\frac{\rho}{R}) \text{ with } \phi \in C^\infty[0, +\infty), \\ 0 \le \phi \le 1, \ \phi \equiv 1 \text{ on } [0, \frac{1}{2}], \ \phi \equiv 0 \text{ on } [1, +\infty), \\ -\frac{C}{R} \le \frac{\partial \phi_R}{\partial \rho} \le 0, \text{ and } |\frac{\partial^2 \phi_R}{\partial \rho^2}| \le \frac{C}{R^2} \text{ for some constant } C > 0. \end{cases} \tag{21}$$

Let $D \subset H^n$ be an unbounded domain. Assume that η satisfies

$$\begin{cases} \eta > 0 & \text{in } D, \\ \Delta_H \eta \ge 0 & \text{in } D, \\ \eta = 0 & \text{on } \partial D, \end{cases}$$

Let u and v be two solutions of

$$\begin{cases} u \ge 0, & -\Delta_H u \ge g(\xi)v^p & \text{in } D, \ p > 1, \\ v \ge 0, & -\Delta_H v \ge h(\xi)u^q & \text{in } D, \ q > 1, \end{cases} \tag{22}$$

with $g > 0$ and $h > 0$ in D.

LEMMA 1 *Suppose u and v are solutions of (22), continuous up to ∂D. If we define*

$$I_R(u^q, h, s) := \int_D h u^q \, (\phi_R)^s \, \eta^\gamma d\xi,$$

where γ and s are positive constants, then the following estimate holds

$$I_R(u^q, h, s) \le \int_{\Omega_R} g v^p \, (\phi_R)^{(s-1)p} \, \eta^\gamma d\xi^{\frac{1}{p}}. \tag{23}$$

$$\cdot \left(\frac{C}{R^2} \left[\int_{\Omega_R} \eta^\gamma g^{-\frac{p'}{p}} d\xi \right]^{\frac{1}{p'}} + \frac{C}{R} \left[\int_{\Omega_R} \eta^{\gamma - p'} |\nabla_H \eta \cdot \nabla_H \rho|^{p'} g^{-\frac{p'}{p}} d\xi \right]^{\frac{1}{p'}} \right),$$

where $\Omega_R := (B_H(0, R) \setminus B_H(0, \frac{R}{2})) \cap D$, and p' is the conjugate exponent of p.

Clearly an equivalent result holds for $I_R(v^p, g, s)$.

The proof of the lemma is left to the reader as it is similar to the proof of Lemma 3.1 of [5].

Let us define the following radial functions

$$\bar{g}(\rho) = \left[\int_{S_1} g^{\frac{1}{1-p}}(\rho, \omega) d\omega \right]^{1-p} \quad \text{and} \quad \bar{h}(\rho) = \left[\int_{S_1} h^{\frac{1}{1-q}}(\rho, \omega) d\omega \right]^{1-q},$$

where $S_1 = \partial B_H(0, 1)$. Now we can state are the first theorem.

THEOREM 6 *Let (u, v) be a solution of (22) with $D = H^n$. Then, if one of the following limit:*

$$\lim_{R \to \infty} \frac{C}{R^{2 + \frac{2}{p}}} \left[\int_{r_o}^R \bar{g}(\rho)^{-\frac{1}{p-1}} \rho^{Q-1} d\rho \right]^{\frac{p-1}{p}} \cdot \left[\int_{r_o}^R \bar{h}(\rho)^{-\frac{1}{q-1}} \rho^{Q-1} d\rho \right]^{\frac{q-1}{qp}},$$

or

$$\lim_{R \to \infty} \frac{C}{R^{2 + \frac{2}{q}}} \left[\int_{r_o}^R \bar{g}(\rho)^{-\frac{1}{p-1}} \rho^{Q-1} d\rho \right]^{\frac{p-1}{pq}} \cdot \left[\int_{r_o}^R \bar{h}(\rho)^{-\frac{1}{q-1}} \rho^{Q-1} d\rho \right]^{\frac{q-1}{q}}$$

is bounded then $u \equiv 0$ and $v \equiv 0$.

REMARK 6: Theorem 5 is a particular case of Theorem 6. Indeed if $g \equiv 1$ and $h \equiv 1$, then, for example

$$\frac{C}{R^{2 + \frac{2}{p}}} \left[\int_{r_o}^R \bar{g}(\rho)^{-\frac{1}{p-1}} \rho^{Q-1} d\rho \right]^{\frac{p-1}{p}} \cdot \int_{r_o}^R \bar{h}(\rho)^{-\frac{1}{q-1}} \rho^{Q-1} d\rho \right]^{\frac{q-1}{qp}} =$$

$$= C R^{\frac{1}{pq}[(Q-2)(pq-1) - 2(q+1)]}.$$

Hence theorem 6 implies that if

$$\frac{Q-2}{2}(pq - 1) \leq \max\{p + 1, q + 1\}, \quad p > 1, \ q > 1$$

then $u \equiv 0$ and $v \equiv 0$.

Before giving the proof of Theorem 6 let us state another non existence result. Let us define $\gamma(\omega) := \frac{t}{\rho^2}$ which is independent of ρ because t is homogenous of degree 2. We introduce the radial functions

$$\hat{g}(\rho) := \left[\int_{S_1 \cap D} (g(\rho, \omega))^{\frac{1}{1-p}} \gamma d\omega\right]^{1-p} \quad \text{and} \quad \hat{h}(\rho) := \left[\int_{S_1 \cap D} (h(\rho, \omega))^{\frac{1}{1-q}} \gamma d\omega\right]^{1-q}.$$

We will prove the following

THEOREM 7 *Let* (u, v) *be a solution (22) with* $D = \{\xi \text{ such that } t > 0\}$. *Then if one of the following limit:*

$$\lim_{R \to \infty} \frac{C}{R^{2+\frac{2}{p}}} \left[\int_{r_o}^{R} \hat{g}(\rho)^{-\frac{1}{p-1}} \rho^{Q+1} d\rho\right]^{\frac{p-1}{p}} \cdot \left[\int_{r_o}^{R} \hat{h}(\rho)^{-\frac{1}{q-1}} \rho^{Q+1} d\rho\right]^{\frac{q-1}{qp}},$$

or

$$\lim_{R \to \infty} \frac{C}{R^{2+\frac{2}{q}}} \left[\int_{r_o}^{R} \hat{g}(\rho)^{-\frac{1}{p-1}} \rho^{Q-1} d\rho\right]^{\frac{p-1}{pq}} \cdot \left[\int_{r_o}^{R} \hat{h}(\rho)^{-\frac{1}{q-1}} \rho^{Q-1} d\rho\right]^{\frac{q-1}{q}}$$

is bounded then $u \equiv 0$ *and* $v \equiv 0$.

REMARK 7: If, for $|\xi|_H \geq r_o$, $g(\rho, \omega) = \rho^\alpha$ and $h(\rho, \omega) = \rho^\beta$ for some constant α and β, then $\hat{g}(\rho) = \rho^\alpha$ and $\hat{h}(\rho) = \rho^\beta$ and by a simple calculation we get:

$$\frac{C}{R^{2+\frac{2}{p}}} \left[\int_{r_o}^{R} \hat{g}(\rho)^{-\frac{1}{p-1}} \rho^{Q+1} d\rho\right]^{\frac{p-1}{p}} \cdot \left[\int_{r_o}^{R} \hat{h}(\rho)^{-\frac{1}{q-1}} \rho^{Q+1} d\rho\right]^{\frac{q-1}{qp}} =$$

$$= R^{\frac{1}{pq}[(Q)(pq-1)-(q(\alpha+2)+2+\beta)]}$$

Hence the condition of the theorem is satisfied if

$$(Q)(pq - 1) \leq \max\{p(\beta + 2) + \alpha + 2, q(\alpha + 2) + \beta + 2\}, \quad p > 1, \ q > 1.$$

REMARK 8. Theorem 7 can be extended to solutions of (22) with D any half space defined by

$$D := \{\xi \text{ such that } a \cdot x + b \cdot y + ct > d \text{ with } c \neq 0\}.$$

Indeed observe that Δ_H is invariant with respect to the group action $\circ$, and it is easy to see that there exists $\xi_o \in H^n$ such that

$$\xi_o \circ D = \{\xi \text{ such that } t > 0\}.$$

Proof of Theorem 6 and 7. The proof of both theorems relies on lemma 1 and on the following observation.

Suppose that we call

$$K_R(g,p) := \frac{C}{R^2}[\int_{\Omega_R} \eta^\gamma g^{-\frac{p'}{p}} d\xi]^{\frac{1}{p'}} + \frac{C}{R}[\int_{\Omega_R} \eta^{\gamma-p'}|\nabla_H\eta \cdot \nabla_H\rho|^{p'} g^{-\frac{p'}{p}} d\xi]^{\frac{1}{p'}}.$$

Then applying lemma 1 twice we get

$$\begin{aligned}
I_R(u^q, h, s) &\leq I_R(v^p, g, (s-1)p)^{\frac{1}{p}} K_R(g,p) \\
&\leq I_R(u^q, h, s)^{\frac{1}{pq}}(K_R(h,q))^{\frac{1}{p}} K_R(g,p). \quad (24)
\end{aligned}$$

Here we have chosen s such that $((s-1)p-1)q = s$ i.e. $s = \frac{p+q}{p-1}$.

Hence it is clear that from (24) if

$$\lim_{R\to\infty} (K_R(h,q))^{\frac{1}{p}} K_R(g,p) = 0, \quad (25)$$

then

$$\int_D hu^q\eta^\gamma d\xi = \lim_{R\to\infty}\int_D hu^q (\phi_R)^s \eta^\gamma d\xi = 0.$$

Hence $u \equiv 0$ and therefore $v \equiv 0$.

On the other hand if the limit (25) is bounded then we see from (24) that $I_R(u^q, h, s)$ is also bounded. In particular we obtain

$$\lim_{R\to\infty}\int_{\Omega_R} hu^q (\phi_R)^s \eta^\gamma d\xi^{\frac{1}{q}} = 0.$$

Applying lemma 1 again and passing to the limit for $R \to \infty$ we can conclude as above that

$$\int_D hu^q\eta^\gamma d\xi = 0.$$

Hence we only have to check that with the right choice of η the hypothesis of the Theorems 6 and 7 imply that the limit in (25) is bounded.

Let us first consider the case of Theorem 6 where $D = H^n$ then we can choose $\eta \equiv 1$. And then

$$K_R(g,p) \leq \frac{C}{R^2}[\int_{r_o}^R g(\rho,\omega)^{-\frac{1}{p-1}}\rho^{Q-1}d\rho d\omega]^{1-\frac{1}{p}}$$

$$= \frac{C}{R^2}[\int_{r_o}^R \bar{g}(\rho)^{-\frac{1}{p-1}}\rho^{Q-1}d\rho],$$

and similarly for $K(h,q)$.

On the other hand in the case of Theorem 7 where $D = \{\xi$ such that $t > 0\}$ we can choose $\eta = t$ and $\gamma = 1$. Let us observe that by a simple computation we have

$$\nabla_H \eta \cdot \nabla_H \rho = \frac{t\psi}{\rho},$$

where ψ is the function defined in (20).

Hence, using the fact that $t = \rho^2 \gamma(\omega)$ we obtain

$$K_R(g,p) \leq \frac{C}{R^2} [\int_{r_o}^{R} g(\rho,\omega)^{-\frac{1}{p-1}} t \rho^{Q-1} d\rho d\omega]^{1-\frac{1}{p}}$$

$$+ \frac{C}{R} [\int_{\Omega_R} t(\frac{\psi(\omega)}{\rho})^{p'} g^{-\frac{p'}{p}} d\xi]^{\frac{1}{p'}}$$

$$= \frac{C}{R^2} [\int_{r_o}^{R} \hat{g}(\rho)^{-\frac{1}{p-1}} \rho^{Q+1} d\rho].$$

This concludes the proof.

For the half spaces of H^n unconcerned by Theorem 7 we have a slightly weaker result.

Indeed let us consider the system of equaitons (22) with $D := \{\xi$ such that $x_1 > 0\}$ and with $g(\rho,\omega) \geq C_1 \rho^\alpha$ and $h(\rho,\omega) \geq C_2 \rho^\beta$ for $\rho \geq r_o$ for some constants α, β $C_1 > 0$ and $C_2 > 0$. Then the following theorem holds true:

THEOREM 8 *Suppose $u \geq 0$ and $v \geq 0$ are solutions of*

$$\begin{cases} -\Delta_h u \geq C_1 \rho^\alpha v^p & in \ D, \\ -\Delta_H v \geq C_2 \rho^\beta u^q & in \ D. \end{cases} \quad (26)$$

If

$$(Q-2)(pq-1) < \max\{p(\beta+1)+\alpha+1, q(\alpha+1)+\beta+1\}, \quad p > 1, \ q > 1, \quad (27)$$

then $u \equiv 0$ and $v \equiv 0$.

Proof of Theorem 8. From the symmetry in u and v we can for example suppose that

$$(Q - 2)(pq - 1) < q(\alpha + 1) + \beta + 1.$$

We procede as in the proofs of the Theorems 6 and 7.

Hence we choose in I_R, $\eta := x_1$.

We only have to show that, choosing the right γ,

$$(K_R(h,q))^{\frac{1}{p}} K_R(g,p)$$

is bounded for $R \to \infty$.

Observe that

$$K_R(g,p) := \frac{C}{R^2}[\int_{\Omega_R} x_1^\gamma \rho^{-\frac{\alpha p'}{p}} d\xi]^{\frac{1}{p'}} + \frac{C}{R}[\int_{\Omega_R} x_1^{\gamma-p'}|\nabla_H x_1 \cdot \nabla_H \rho|^{p'} \rho^{-\frac{\alpha p'}{p}} d\xi]^{\frac{1}{p'}}.$$

(28)

It is immediate to see that $\nabla_H x_1 \cdot \nabla_H \rho = \frac{x_1(x^2+y^2)+y_1 t}{\rho^3}$, which is bounded in Ω_R, therefore

$$\int_{\Omega_R} x_1^{\gamma-p'}|\nabla_H x_1 \cdot \nabla_H \rho|^{p'} \rho^{-\frac{\alpha p'}{p}} d\xi := \int_{\frac{R}{2}}^{R} \rho^{\gamma-p'-\frac{\alpha p'}{p}+Q-1} \int_S \varphi(\omega)^{\gamma-p'} d\omega \quad (29)$$

where $\varphi(\omega) := \frac{x_1}{\rho}$ and $S = D \cap S_1$. Clearly,

$$\int_S \varphi(\omega)^{\gamma-p'} d\omega$$

is integrable if $\gamma - p' > -1$, hence we are going to choose $\gamma = p' - 1 + \varepsilon$, with ε a positive constant sufficiently small that the following inequality still holds:

$$(Q - 2 + \varepsilon)(pq - 1) < q(\alpha + 1) + \beta + 1. \quad (30)$$

Using this choice of γ, and (29), (28) becomes

$$K_R(g,p) \leq CR^{\left(Q-1+\varepsilon-\alpha\frac{p'}{p}\right)\left(\frac{1}{p'}\right)-1}.$$

Therefore

$$(K_R(h,q))^{\frac{1}{p}} K_R(g,p) \leq C.R^\Pi,$$

where

$$\Pi = \frac{1}{pq}\left((Q - 2 + \varepsilon)(pq - 1) - q(\alpha + 1) - \beta - 1\right).$$

Clearly from (30), $\Pi \leq 0$ and $(K_R(h,q))^{\frac{1}{p}} K_R(g,p)$ is bounded. This concludes the proof of Theorem 8.

REFERENCES

1. H. Berestycki, I. Capuzzo Dolcetta, L. Nirenberg, Problèmes elliptiques indéfinis et théorèmes de Liouville non-linéaires, *C. R. Acad. Sci. Paris*, Série I, 317, 945-950, (1993).

2. H. Berestycki, I. Capuzzo Dolcetta, L. Nirenberg, Variational methods for indefinite superlinear homogeneous elliptic problems, *Nonlinear Differential Equations and Applications*, Vol 2, 553-572, (1995).

3. H. Berestycki, I. Capuzzo Dolcetta, L. Nirenberg, Superlinear indefinite elliptic problems and nonlinear Liouville theorems. *Topological Methods in Nonlinear Analysis*, Vol. 4.1, 59-78, (1995).

4. I. Birindelli, I. Capuzzo Dolcetta, A. Cutrì, Liouville theorems for semilinear equations on the Heisenberg group, To appear in *Annales de l'Institut Henri Poincaré-Analyse non linéaire.*

5. I. Birindelli, I. Capuzzo Dolcetta, A. Cutrì, Indefinite semi-linear equations on the Heisenberg group: a priori bounds and existence, preprint.

6. I. Birindelli, E. Mitidieri, Liouville Theorems for Elliptic Inequalities and Applications, preprint.

7. J.M. Bony, Principe du Maximum, Inégalité de Harnack et unicité du problème de Cauchy pour les opérateurs elliptiques dégénérés, *Ann. Inst. Fourier Grenobles* **19,1** 277-304 (1969).

8. P. Clement, R. Mansasevich, E. Mitidieri, Positive solutions for a quasilinear system via blow up, *Comm. Partial Diff. Eq.*, Vol 18, No12, 2071-2106, (1993).

9. E.N. Dancer, Some notes on the method of moving plancs, *Bull. Austral. Math. Soc.* **46** 425-434 (1992).

10. D.G. De Figueiredo, L. Felmer. A Liouville-type theorem for Elliptic systems, *Annali di Pisa*, Vol 21, 387-397, (1994).

11. M. Esteban, P.L. Lions, Existence and non-existence results for semilinear elliptic problems in unbounded domains, *Proc. R.S.E. (A)*, 93A, 1-14 (1982).

12. G.B. Folland, E.M.Stein, Estimates for the ∂_h complex and analysis on the Heisenberg Group, *Comm. Pure Appl.Math.* **27** 492-522 (1974).

13. N. Garofalo, E. Lanconelli, Existence and non existence results for semilinear equations on the Heisenberg group, *Indiana Univ. Math. Journ.***41** 71-97 (1992).

14. B. Gidas, J. Spruck, Global and local behavior of positive solutions of nonlinear elliptic equations, *Comm. Pure Appl. Math* **35** 525-598 (1981).

15. B. Gidas, J. Spruck, A priori bounds for positive solutions of nonlinear elliptic equations, *Comm. Partial Diff. Eq.* 6(8), 883-901 (1981).

16. P.C. Greiner, Spherical harmonics in the Heisenberg group, *Canad. Math. Bull.* **23** (4) 383-396 (1980).

17. L. Hormander, Hypoelliptic second order differential equations, *Acta Math.*,Uppsala, **119**, 147-171 (1967).

18. E. Mitidieri, Non existence of positive solutions for semilinear elliptic systems in $\mathbb{R}^n$. *Differential and Integral Eq.* Vol. 9, 465-480 (1996).

19. P. H. Rabinowitz,*Minimax methods in critical point theory with applications to differential equations,* CBMS Reg. Conf. vol. 65,(1986).

20. J. Serrin, H. Zou, Non existence of positive solutions of Lane-Emden Systems, *Differential and Integral Eq.* (1996).

Existence and Regularity Results for Quasilinear Parabolic Equations

L. BOCCARDO   Dipartimento di Matematica, Università di Roma I, Piazzale A. Moro 2, 00185, Roma, Italy

I present here some results obtained in a joint work with Andrea Dall'Aglio (Università di Firenze), Thierry Gallouët (ENS Lyon) and Luigi Orsina (Università di Roma I).

Let Ω be a bounded domain in $\mathbf{R}^N$, $N \geq 2$. For $T > 0$, let us denote by Q the cylinder $\Omega \times (0, T)$, and by Γ the lateral surface $\partial\Omega \times (0, T)$. We will consider the following quasilinear parabolic Cauchy-Dirichlet problem:

$$\begin{cases} u' - \operatorname{div}\left((A(x, t, u)\nabla u)\right) = f & \text{in } Q, \\ u(x, 0) = 0 & \text{in } \Omega, \\ u(x, t) = 0 & \text{on } \Gamma. \end{cases} \tag{1}$$

Here $A(x, t, \sigma)$ is a bounded, elliptic, Carathéodory matrix. We will be concerned with some regularity results for the solutions of problem (1), depending on the summability of the datum f. If f belongs to $L^r(0, T; L^q(\Omega))$, and if r and q are large enough, that is, if

$$\frac{2}{r} + \frac{N}{q} < 2,$$

then in Aronson and Serrin (1967) it is proved that u belongs to $L^\infty(Q)$.

On the other hand, if A does not depend on u, and if r and q satisfy the opposite inequality, that is if

$$2 < \frac{2}{r} + \frac{N}{q} \leq \min\left\{2 + \frac{N}{r}, 2 + \frac{N}{2}\right\}, \qquad r \geq 1, \tag{2}$$

then in Ladyženskaja, Solonnikov and Ural'ceva (1968), Theorem 9.1, it is proved that any weak solution of (1) belongs to $L^s(Q)$, with s given by

$$s = \frac{(N+2)qr}{Nr + 2q - 2qr}, \tag{3}$$

and this easily implies that such solutions belong to $L^2(0, T; H_0^1(\Omega))$.

Observe that if $r \geq 2$, then (2) becomes

$$2 < \frac{2}{r} + \frac{N}{q} \leq 2 + \frac{N}{r}.$$

If r and q satisfy this latter inequality, then the function f belongs to the space $L^2(0, T; H^{-1}(\Omega))$, so that there exists at least a solution of (1) (which is indeed unique since A does not depend on u).

Conversely, if $1 \leq r < 2$, (2) can be rewritten as

$$2 < \frac{2}{r} + \frac{N}{q} \leq 2 + \frac{N}{2},$$

and in this case the function f is not in $L^2(0, T; H^{-1}(\Omega))$; anyway, the *a priori* estimate of Ladyženskaja, Solonnikov and Ural'ceva (1968), together with the linearity of the operator, allow to prove the existence of at least a solution of (1) also in this case.

The *a priori* estimates in $L^2(0, T; H_0^1(\Omega))$ can be obtained, always in the linear case, by means of different techniques, such as norm estimates using the heat kernel, duality properties and embeddings between Lebesgue spaces (see Brezis, 1996).

If A depends on u, then the *a priori* estimate proved in Ladyženskaja, Solonnikov and Ural'ceva (1968) still holds true, since the linearity of the operator is never used in the proof. Thus, if $r \geq 2$, their result is a regularity result in $L^s(Q)$ for solutions of (1), which are known to exist in $L^2(0, T; H_0^1(\Omega))$ (see Lions, 1969).

On the other hand, if $1 \leq r < 2$, then the *a priori* estimates in $L^s(Q)$ and in $L^2(0, T; H_0^1(\Omega))$ can be used in order to prove existence of solutions for (1) proceeding by approximation. Indeed, one can approximate the datum f with a sequence of smooth functions f_n, solve problem (1) with datum f_n, and use the linearity of the operator with respect to the gradient, together with compactness results (see Simon, 1987), in order to pass to the limit in the approximate equations (see also the proof of Theorem 1, below).

If the operator does not depend linearly on the gradient, as is for example the case of the following operator

$$-\operatorname{div}\left(A(x,t,u)|\nabla u|^{p-2}\nabla u\right), \qquad p > 1,$$

then the a *priori* estimates in $L^s(Q)$ and in $L^p(0,T;W_0^{1,p}(\Omega))$ can be obtained in a similar way (see Boccardo, Dall'Aglio, Gallouët et al., 1997), but are no longer enough in order to prove existence of solutions in the case of data not belonging to $L^{p'}(0,T;W^{-1,p'}(\Omega))$ (which is the correct "dual" space). Indeed, a result of strong convergence of approximating solutions is needed.

The a *priori* estimates, the strong convergence result, and so the existence of solutions, can be found in Boccardo, Dall'Aglio, Gallouët et al. (1997). Other related convergence results can be found in Boccardo and Murat (1992) and Boccardo, Dall'Aglio, Gallouët et al. (1996).

We state here the existence result in the particular case $p = 2$, with a quasilinear principal part.

Let us make our assumption on A more precise. Let $A = (A_{i,j})_{i,j=1,\ldots,N}$, with $A_{i,j} : Q \times \mathbf{R} \to \mathbf{R}$ Carathéodory functions (i.e., $A_{i,j}(x,t,\cdot)$ is continuous on $\mathbf{R}$ for almost every $(x,t) \in Q$, and $A_{i,j}(\cdot,\cdot,\sigma)$ is measurable on Q for every $\sigma \in \mathbf{R}$) such that the following holds:

$$A(x,t,\sigma) \geq \alpha\, I \qquad |A_{i,j}(x,t,\sigma)| \leq \beta\,, \tag{4}$$

for almost every $(x,t) \in Q$, for every $\sigma \in \mathbf{R}$, for every i and j in $1,\ldots,N$, where α and β are positive constants, and $I = (\delta_{i,j})_{i,j=1,\ldots,N}$ is the identity matrix.

Our results is the following.

THEOREM 1 Let f be in $L^r(0,T;L^q(\Omega))$ with r and q such that

$$2 < \frac{2}{r} + \frac{N}{q} \leq 2 + \frac{N}{2}, \qquad 1 \leq r < 2\,. \tag{5}$$

Then there exists at least a solution u of problem (1), with u belonging to $L^2(0,T;H_0^1(\Omega)) \cap L^s(Q)$, with

$$s = \frac{(N+2)qr}{Nr + 2q - 2qr}\,.$$

Before proving the result, let us recall the Gagliardo-Nirenberg inequality.

LEMMA 2 Let u be a function in $L^2(0,T;H_0^1(\Omega)) \cap L^\infty(0,T;L^m(\Omega))$, with $m \geq 1$. Then u belongs to $L^{\bar{s}}(Q)$, with $\bar{s} = 2\frac{N+m}{N}$.

Proof . See DiBenedetto (1993), Proposition 3.1. ∎

We state now the *a priori* estimate which can be achieved using essentially the same technique of Ladyženskaja, Solonnikov and Ural'ceva (1968).

LEMMA 3   Let r and q be real numbers satisfying (5). Let f be a function in $L^r(0,T;L^q(\Omega)) \cap L^2(0,T;H^{-1}(\Omega))$. Let u be a solution in $L^2(0,T;H_0^1(\Omega))$ of (1) with datum f, and let s be given by (3). Then the norms of u in $L^s(Q)$ and in $L^2(0,T;H_0^1(\Omega))$ are bounded by a constant which depends on α, N, q, r, Q and on the norm of f in $L^r(0,T;L^q(\Omega))$.

Proof . In the following, we will denote by c any constant depending only on α, N, q, r and $\|f\|_{L^r(0,T;L^q(\Omega))}$. The value of c may vary from line to line.

Let n be a positive integer, and suppose that $r > 1$.

For $\gamma \geq 1$, and $\tau \in (0,T)$, we can take $\varphi = |T_n(u)|^{2(\gamma-1)}T_n(u)\chi_{(0,\tau)}(t)$ as test function. This is possible since $T_n(u) \in L^\infty(Q) \cap L^2(0,T;H_0^1(\Omega))$, so that φ belongs to the same space. Integrating by parts, recalling that $u(x,0) = 0$, and setting $\Psi(s) = \int_0^s |T_n(t)|^{2(\gamma-1)} T_n(t)\, dt$, we obtain

$$\int_\Omega \Psi(u(\tau))\, dx + (2\gamma - 1) \int_0^\tau \int_\Omega A(x,t,u)|\nabla T_n(u)|^2 |T_n(u)|^{2(\gamma-1)}$$
$$\leq \int_0^\tau \int_\Omega |f|\,|T_n(u)|^{2\gamma-1}.$$

Observing that $\Psi(s) \geq c\,|T_n(s)|^{2\gamma}$ for some positive constant c, and recalling that $A(x,t,u) \geq \alpha\,I > 0$, we obtain

$$\int_\Omega |T_n(u)(\tau)|^{2\gamma}\, dx \leq \int_0^\tau \int_\Omega |f|\,|T_n(u)|^{2\gamma-1},$$

and

$$\int_0^\tau \int_\Omega |\nabla |T_n(u)|^\gamma|^2 \leq \int_0^\tau \int_\Omega |f|\,|T_n(u)|^{2\gamma-1},$$

In order to simplify the notations, we will set $v = T_n(u)$ from now on. Taking the supremum of both terms for τ in $(0,T)$ and using Hölder's inequality, we obtain

$$\|v\|_{L^{\infty}(0,T;L^{2\gamma}(\Omega))}^{2\gamma} + \int_{Q} |\nabla|v|^{\gamma}|^{2} \leq c \int_{Q} |f| \, |v|^{2\gamma-1}$$

$$\leq \int_{0}^{T} \|f(t)\|_{L^{q}(\Omega)} \|v(t)\|_{L^{(2\gamma-1)q'}(\Omega)}^{2\gamma-1} \, dt$$

$$\leq \|f\|_{L^{r}(0,T;L^{q}(\Omega))} \left[\int_{0}^{T} \|v(t)\|_{L^{(2\gamma-1)q'}(\Omega)}^{(2\gamma-1)r'} \, dt \right]^{\frac{1}{r'}} \quad (6)$$

$$\leq c \left[\int_{0}^{T} \|v(t)\|_{L^{(2\gamma-1)q'}(\Omega)}^{(2\gamma-1)r'} \, dt \right]^{\frac{1}{r'}}.$$

Observe that both r' and q' are real numbers since $r > 1$ and since (2) implies $q > 1$. Let us now assume that the following inequalities are satisfied:

$$2\gamma \leq (2\gamma - 1)q' \leq 2^{*}\gamma, \quad (7)$$

and let $\theta \in [0, 1]$ be such that

$$\frac{1}{(2\gamma - 1)q'} = \frac{1 - \theta}{2\gamma} + \frac{\theta}{2^{*}\gamma}. \quad (8)$$

Then one has, for almost every $t \in (0, T)$,

$$\|v(t)\|_{L^{(2\gamma-1)q'}(\Omega)} \leq \|v(t)\|_{L^{2\gamma}(\Omega)}^{1-\theta} \|v(t)\|_{L^{2^{*}\gamma}(\Omega)}^{\theta}$$

$$\leq \|v\|_{L^{\infty}(0,T;L^{2\gamma}(\Omega))}^{1-\theta} \|v(t)\|_{L^{2^{*}\gamma}(\Omega)}^{\theta}.$$

Therefore from (6) we obtain, using Young's inequality:

$$\|v\|_{L^{\infty}(0,T;L^{2\gamma}(\Omega))}^{2\gamma} + \|\nabla|v|^{\gamma}\|_{L^{2}(Q;\mathbf{R}^{N})}^{2}$$

$$\leq c \|v\|_{L^{\infty}(0,T;L^{2\gamma}(\Omega))}^{(2\gamma-1)(1-\theta)} \left[\int_{0}^{T} \|v(t)\|_{L^{2^{*}\gamma}(\Omega)}^{(2\gamma-1)\theta r'} \, dt \right]^{\frac{1}{r'}}$$

$$\leq \frac{1}{2} \|v\|_{L^{\infty}(0,T;L^{2\gamma}(\Omega))}^{2\gamma} + c \left[\int_{0}^{T} \|v(t)\|_{L^{2^{*}\gamma}(\Omega)}^{(2\gamma-1)\theta r'} \, dt \right]^{\frac{2\gamma}{r'[1+\theta(2\gamma-1)]}}$$

$$(9)$$

If we now choose γ satisfying

$$(2\gamma - 1)\theta r' = 2\gamma, \quad (10)$$

from (9) and Sobolev's imbedding theorem we obtain

$$\|v\|_{L^\infty(0,T;L^{2\gamma}(\Omega))}^{2\gamma} + \|\nabla|v|^\gamma\|_{L^2(Q;\mathbf{R}^N)}^2 \leq c\|v\|_{L^{2\gamma}(0,T;L^{2^*\gamma}(\Omega))}^{2\gamma\frac{2\gamma}{r'+2\gamma}}$$

$$\leq c\|\nabla|v|^\gamma\|_{L^2(Q;\mathbf{R}^N)}^{2\frac{2\gamma}{r'+2\gamma}}.$$

Since $\frac{2\gamma}{r'+2\gamma} < 1$, we have proved that

$$\|v\|_{L^\infty(0,T;L^{2\gamma}(\Omega))}^{2\gamma} + \|\nabla|v|^\gamma\|_{L^2(Q;\mathbf{R}^N)}^2 \leq c,$$

and therefore, by the Gagliardo-Nirenberg inequality, $\||v|^\gamma\|_{L^2\frac{N+2}{N}(Q)} \leq c$, so that, for every $n \in \mathbf{N}$,

$$\|T_n(u)\|_{L^{2\gamma\frac{N+2}{N}}(Q)} \leq c, \tag{11}$$

where c depends on N, α, γ, $\|f\|_{L^r(0,T;L^q(\Omega))}$. ¿From (8) and (10) we obtain:

$$\theta = \frac{N(2\gamma - q)}{2q(2\gamma - 1)}, \qquad 2\gamma = \frac{Nqr}{Nr + 2q - 2qr},$$

which implies

$$2\gamma\frac{N+2}{N} = \frac{(N+2)qr}{Nr + 2q - 2qr},$$

which is s, given by (3). Now we have to check that condition (7) is satisfied. It is easy to see that this is equivalent to

$$2 < \frac{N}{q} + \frac{2}{r} \leq 2 + \frac{N}{r}.$$

Then we have to check that $\gamma \geq 1$. This is equivalent to

$$2 < \frac{N}{q} + \frac{2}{r} \leq 2 + \frac{N}{2},$$

and so the conditions on r and q are

$$2 < \frac{N}{q} + \frac{2}{r} \leq \min\left\{2 + \frac{N}{r}, 2 + \frac{N}{2}\right\}.$$

Since $r < 2$, this condition becomes exactly (5). Passing to the limit as n tends to infinity in (11), we obtain the desired estimate for the norm of u in $L^s(Q)$ if $r > 1$.

If $r = 1$, and q satisfies (2) with $r = 1$ (that is, if $2 \leq q < +\infty$), we choose $\varphi = |T_n(u)|^{q-2} T_n(u)$ as test function in (1). By means of the same technique we used before, we arrive after straightforward passages to

$$|T_n(u)|^{\frac{q}{2}} \in L^\infty(0, T; L^2(\Omega)) \cap L^2(0, T; H_0^1(\Omega)),$$

with a bound on the norms of $T_n(u)$ which is uniform with respect to n. By the Gagliardo-Nirenberg inequality, this implies (again after a passage to the limit as n tends to infinity) that u is bounded in $L^s(Q)$, with s given by $q\frac{N+2}{N}$. This value is the one given by (3) in the case $r = 1$.

To prove the estimate in $L^2(0, T; H_0^1(\Omega))$, we take u as test function in (1). Throwing away positive terms we obtain

$$\alpha \|\nabla u\|^2_{L^2(Q; \mathbf{R}^N)} \leq \int_Q |f||u|,$$

so that it will be enough to show that the last integral can be estimated using the norm of f in $L^r(0, T; L^q(\Omega))$. Indeed, for every $\gamma \geq 1$ one has

$$\int_Q |fu| = \int_{\{(x,t): |u| \leq 1\}} |f||u| + \int_{\{(x,t): |u| > 1\}} |f||u|$$
$$\leq \|f\|_{L^1(Q)} + \int_Q |f| \, |u|^{2\gamma - 1}.$$

If we choose γ as in the first part of the proof, then the last integral is bounded. On the other hand, since $r \geq 1$, and $q \geq 1$, the first norm is bounded by a constant depending only on Q, q, r, $\|f\|_{L^r(0,T;L^q(\Omega))}$, and this concludes the proof. ∎

Proof of Theorem 1 . We proceed by approximation. Let $\{f_n\}$ be a sequence of $L^\infty(Q)$ functions that converges to f strongly in $L^r(0, T; L^q(\Omega))$. Then for every n there exists a solution u_n of the parabolic problem with datum f_n. From Lemma 3 it follows that the sequence $\{u_n\}$ is bounded in $L^2(0, T; H_0^1(\Omega)) \cap L^s(Q)$, with s given by (3). Therefore we can extract a subsequence which is weakly convergent to some function u in $L^2(0, T; H_0^1(\Omega))$. Moreover, since from the equation one obtains that $\{u_n'\}$ is bounded in $L^1(Q) + L^2(0, T; H^{-1}(\Omega))$, using compactness arguments (see Simon, 1987), it is easy to see that u_n converges strongly to u in $L^1(Q)$. This fact, and the continuity hypotheses on $A(x, t, \cdot)$, imply that

$$A(x, t, u_n) \to A(x, t, u) \quad *\text{-weakly in } L^\infty(Q) \text{ and almost everywhere in } Q.$$

Furthermore, standard results (see again Simon, 1987) yield the compactness of u_n in $C^0([0,T]; W^{-1,1}(\Omega))$, so that the initial datum is assumed. These facts, together with the weak convergence of u_n, allows a passage to the limit in the approximate equations, and so the existence of a solution for (1). ∎

REFERENCES

Aronson, D. G., and Serrin, J. (1967). Local behavior of solutions of quasilinear parabolic equations, *Arch. Rational Mech. Anal.*, **25**: 81.

Boccardo, L., Dall'Aglio, A., Gallouët, T., and Orsina, L. (1996). Nonlinear parabolic equations with measure data, *J. Funct. Anal.*, to appear.

Boccardo, L., Dall'Aglio, A., Gallouët, T., and Orsina, L. (1997). Existence and regularity results for nonlinear parabolic equations, preprint.

Boccardo, L., and Murat, F. (1992). Almost everywhere convergence of the gradients of solutions to elliptic and parabolic equations, *Nonlinear Anal.*, **19** n. 6: 581.

Brezis, H. (1996). Personal communication.

DiBenedetto, E. (1993), *Degenerate parabolic equations*, Springer-Verlag, New York.

Ladyženskaja, O., Solonnikov, V. A., and Ural'ceva, N. N. (1968). *Linear and quasilinear equations of parabolic type*, Translations of the American Mathematical Society, American Mathematical Society, Providence.

Lions, J. L. (1969). *Quelques méthodes de resolution des problèmes aux limites non linéaires*, Dunod, Gauthier-Villars, Paris.

Simon, J. (1987). Compact sets in the space $L^p(0,T;B)$, *Ann. Mat. Pura Appl.*, **146**: 65.

Elliptic Systems with Various Growth

LUCIO BOCCARDO, Dipartimento di Matematica,
Universitá di Roma I, Piazza A.Moro 2, 00185 ROMA

JACQUELINE FLECKINGER-PELLÉ, CEREMATH
Université Toulouse I, 31042 TOULOUSE

FRANÇOIS de THELIN, Laboratoire M.I.P.,
Université Paul Sabatier, 31062 TOULOUSE

1 INTRODUCTION

In a recent work [2], the authors have studied the existence of solutions $u \in (W_0^{1,p}(\Omega))^n$ for elliptic systems of the form:

$$-\Delta_p u_i = \sum_{j=1}^{n} a_{ij}|u_j|^{p-2}u_j + f_i, \ \ \forall i = 1, \cdots, n, \qquad (S)$$

where

$$a_{ij} \geq 0 \ \ \text{pour} \ i \neq j, \ \ \Delta_p u = div(|\nabla u|^{p-2}\nabla u), \ \ f_i \in L^{p'}(\Omega), \ p' = \frac{p}{p-1}.$$

Denote by $\lambda_1(p)$ the principal eigenvalue of the Dirichlet p-Laplacian defined on Ω; it is characterized by :

$$\lambda_1(p) = Inf_{\{u \in W_0^{1,p}(\Omega) \int_\Omega |u|^p = 1\}} \int_\Omega |\nabla u|^p,$$

(S) has a solution if and only if $\lambda_1(p)I - A$ is a non singular M-matrix. The aim of this work is to establish existence results for such a system when the differential operators in the left-handside members and when the nonlinearities in the right-handside members have different polynomial growth. For simplicity we consider only systems of 2 equations.

We want to find a solution $u \in W_0^{1,p}(\Omega), v \in W_0^{1,q}(\Omega)$ such that, in Ω:

$$\begin{cases} -\Delta_p u = a(x)|u|^{\alpha-2}u + b(x)|v|^{\beta-2}v + f \\ \\ -\Delta_q v = c(x)|u|^{\gamma-2}u + d(x)|v|^{\delta-2}v + g \end{cases} \qquad (P)$$

where Ω is a smooth bounded domain in $\mathbf{R}^N, p > 1$ and $q > 1$.

2 RESULTS

Set: $p^* = \frac{Np}{N-p}$ if $p < N$, and $p^* = +\infty$ if $p \geq N$.
We assume that:

$$a, b, c, d, \in L^\infty(\Omega); \qquad (H1)$$

$$\begin{cases} \alpha \geq 1, 1 \leq \beta < q^* + 1, \delta \geq 1, 1 \leq \gamma < p^* + 1 \\ \\ \text{and } (\beta-1)(\gamma-1) < (p-1)(q-1). \end{cases} \qquad (H2)$$

One of the three following assertions is satisfied: $\qquad (H3)$

$$\alpha < p \text{ and } \delta < q \qquad (*)$$

$$\alpha < p, \delta = q \text{ and } \forall x \in \Omega, d(x) < \lambda_1(q) \qquad (**)$$

$$\alpha = p, \delta = q \text{ and } \forall x \in \Omega, a(x) < \lambda_1(p) \text{ and } d(x) < \lambda_1(q) \qquad (***)$$

THEOREM 1 *We assume that hypotheses (H1), (H2) and (H3) are satisfied; then for all $f \in L^{p'}(\Omega)$ and for all $g \in L^{q'}(\Omega)$, problem (P) admits at least one solution $(u, v) \in W_0^{1,p}(\Omega) \times W_0^{1,q}(\Omega)$.*

The proof relies on the following lemma which is established in [3].

LEMMA 1 : *Let $(u_n)_n \in W_0^{1,p}(\Omega)$ and $(g_n)_n \in L^1(\Omega)$ such that :*

$$-\Delta_p u_n = g_n$$

$$u_n \rightharpoonup u \text{ weakly in } W_0^{1,p}(\Omega)$$

$$g_n \rightharpoonup g \text{ weakly in } L^1(\Omega)$$

Then $\nabla u_n \to \nabla u$ strongly in $(L^q(\Omega)^N, \forall q < p$ and u satisfies:

$$-\Delta_p u = g$$

Proof of Theorem 1 For any given $\varepsilon > 0$, we associate to (P) the following problem:

$$-\Delta_p u_\varepsilon = a\frac{|u_\varepsilon|^{\alpha-2}u_\varepsilon}{1+|\varepsilon^s u_\varepsilon|^{\alpha-1}} + b\frac{|v_\varepsilon|^{\beta-2}v_\varepsilon}{1+|\varepsilon^t v_\varepsilon|^{\beta-1}} + f \qquad (1_\varepsilon)$$

$$-\Delta_q v_\varepsilon = c\frac{|u_\varepsilon|^{\gamma-2}u_\varepsilon}{1+|\varepsilon^s u_\varepsilon|^{\gamma-1}} + d\frac{|v_\varepsilon|^{\delta-2}v_\varepsilon}{1+|\varepsilon^t v_\varepsilon|^{\delta-1}} + g \qquad (2_\varepsilon)$$

where s et t are chosen such that :

$$\frac{\beta-1}{p-1} < \frac{s}{t} < \frac{q-1}{\gamma-1}. \qquad (3)$$

This is possible by (H2).
We deduce from Schauder theorem that the equations $(1_\varepsilon), (2_\varepsilon)$ have at least one solution $(u_\varepsilon, v_\varepsilon) \in W_0^{1,p} \times W_0^{1,q}$.
Multiplying (1_ε) by $\varepsilon^{sp} u_\varepsilon, (2_\varepsilon)$ by $\varepsilon^{tq} v_\varepsilon$, and integrating, we get:

$$\int_\Omega |\nabla \varepsilon^s u_\varepsilon|^p =$$

$$\varepsilon^{s(p-\alpha)} \int_\Omega \frac{a|\varepsilon^s u_\varepsilon|^\alpha}{1+|\varepsilon^s u_\varepsilon|^{\alpha-1}} + \varepsilon^{s(p-1)-t(\beta-1)} \int_\Omega \frac{b|\varepsilon^t v_\varepsilon|^{\beta-2}\varepsilon^t v_\varepsilon \varepsilon^s u_\varepsilon}{1+|\varepsilon^t v_\varepsilon|^{\beta-1}} + \varepsilon^{sp} \int_\Omega f u_\varepsilon, \quad (4_\varepsilon)$$

$$\int_\Omega |\nabla \varepsilon^t v_\varepsilon|^q =$$

$$\varepsilon^{t(q-1)-s(\gamma-1)} \int_\Omega c\frac{|\varepsilon^s u_\varepsilon|^{\gamma-2}\varepsilon^s u_\varepsilon \varepsilon^t v_\varepsilon}{1+|\varepsilon^s u_\varepsilon|^{\gamma-1}} + \varepsilon^{t(q-\delta)} \int_\Omega \frac{d|\varepsilon^t v_\varepsilon|^\delta}{1+|\varepsilon^t v_\varepsilon|^{\delta-1}} + \varepsilon^{tq} \int_\Omega g v_\varepsilon. \quad (5_\varepsilon)$$

The proof is established with 4 steps.
(i) **Step 1** Assume $\alpha < p$.
It follows from (4_ε) and from the caracterization of $\lambda_1(p)$ that:

$$\lambda_1(p) \int_\Omega |\varepsilon^s u_\varepsilon|^p \le K\varepsilon^{r_1} \int_\Omega |\varepsilon^s u_\varepsilon| + \varepsilon^{sp} \int_\Omega f u_\varepsilon \le K'\varepsilon^r \{\int_\Omega |\varepsilon^s u_\varepsilon|^p\}^{\frac{1}{p}}$$

where, by (3): r_1 and $r > 0$. Hence $\varepsilon^s u_\varepsilon \to 0$ in $L^p(\Omega)$ strongly, and using again (4_ε), we have: $\varepsilon^s u_\varepsilon \to 0$ in $W_0^{1,p}(\Omega)$ strongly.
We do analoguous calculations for the second equation when $\delta < q$.
(ii) **Step 2** Assume $\alpha = p$ and $\delta = q$.
It follows from (4_ε), that:

$$\lambda_1(p) \int_\Omega |\varepsilon^s u_\varepsilon|^p \le K_1 \int_\Omega |\varepsilon^s u_\varepsilon| + K_2\varepsilon^r \int_\Omega |\varepsilon^s u_\varepsilon| + \varepsilon^{sp} \int_\Omega |f u_\varepsilon| \le K'\{\int_\Omega |\varepsilon^s u_\varepsilon|^p\}^{\frac{1}{p}}.$$

We have analoguous calculations for the second equation.
If (***) is satisfied, we deduce from (4_ε) that $(\varepsilon^s u_\varepsilon, \varepsilon^t v_\varepsilon)$ is bounded in $W_0^{1,p} \times W_0^{1,q}$.
By extraction of subsequences, we derive:

$$\varepsilon^s u_\varepsilon \to \bar{u} \text{ strongly in } L^p, \text{ weakly in } W_0^{1,p};$$

$$\varepsilon^t v_\varepsilon \to \bar{v} \text{ strongly in } L^q, \text{ weakly in } W_0^{1,q};$$

$$|\varepsilon^s u_\varepsilon|^{p-2}\varepsilon^s u_\varepsilon \to |\bar{u}|^{p-2}\bar{u} \text{ in } L^{p'};$$

$$|\varepsilon^t v_\varepsilon|^{q-2}\varepsilon^t v_\varepsilon \to |\bar{v}|^{q-2}\bar{v} \text{ in } L^{q'}.$$

Therefore

$$-\Delta_p(\varepsilon^s u_\varepsilon) = f_\varepsilon;$$
$$-\Delta_q(\varepsilon^t v_\varepsilon) = g_\varepsilon,$$

with

$$f_\varepsilon = \frac{a|\varepsilon^s u_\varepsilon|^{p-2}\varepsilon^s u_\varepsilon}{1 + |\varepsilon^s u_\varepsilon|^{p-1}} + \varepsilon^{s(p-1)-t(\beta-1)}\frac{b|\varepsilon^t v_\varepsilon|^{\beta-2}\varepsilon^t v_\varepsilon}{1 + |\varepsilon^t v_\varepsilon|^{\beta-1}} + \varepsilon^{s(p-1)}f,$$

and

$$f_\varepsilon \to \frac{a|\bar{u}|^{p-2}\bar{u}}{1 + |\bar{u}|^{p-1}} \text{ in } L^m, \forall m < \infty.$$

On the same way:

$$g_\varepsilon = \varepsilon^{t(q-1)-s(\gamma-1)}\frac{c|\varepsilon^s u_\varepsilon|^{\gamma-2}\varepsilon^s u_\varepsilon}{1 + |\varepsilon^s u_\varepsilon|^{\gamma-1}} + \frac{d|\varepsilon^t v_\varepsilon|^{q-2}\varepsilon^t v_\varepsilon}{1 + |\varepsilon^t v_\varepsilon|^{q-1}} + \varepsilon^{t(q-1)}g,$$

and

$$g_\varepsilon \to \frac{d|\bar{v}|^{q-2}\bar{v}}{1 + |\bar{v}|^{q-1}} \text{ in } L^1.$$

Passing through the limit, we deduce from the lemma that:

$$-\Delta_p\bar{u} = \frac{a|\bar{u}|^{p-2}\bar{u}}{1 + |\bar{u}|^{p-1}} \tag{6}$$

$$-\Delta_q\bar{v} = \frac{d|\bar{v}|^{q-2}\bar{v}}{1 + |\bar{v}|^{q-1}} \tag{7}$$

Multiplying (6) by $\bar{u}$ we obtain when $\bar{u} \neq 0$:

$$\lambda_1(p)\int |\bar{u}|^p \leq \int |\nabla\bar{u}|^p \leq \int_\Omega a|\bar{u}|^p < \lambda_1(p)\int_\Omega |\bar{u}|^p.$$

Hence we have a contradiction and necessarily $\bar{u} = 0$ and $\bar{v} = 0$.

If (**) is satisfied, we combine the results of (i) and (ii); for all cases we have:

$$\varepsilon^s u_\varepsilon \to 0 \text{ strongly in } L^p, \text{ weakly in } W_0^{1,p}.$$

$$\varepsilon^t v_\varepsilon \to 0 \text{ strongly in } L^q, \text{ weakly in } W_0^{1,q}.$$

(iii) **Step 3** We prove now by contradiction that $(u_\varepsilon, v_\varepsilon)$ is bounded in $W_0^{1,p} \times W_0^{1,q}$.

Assume that this is not true; consider

$$M_\varepsilon = \max(\| u_\varepsilon \|^{\frac{1}{s}}, \| v_\varepsilon \|^{\frac{1}{t}}) \to +\infty.$$

Set $z_\varepsilon = \frac{u_\varepsilon}{M_\varepsilon^s}$, $w_\varepsilon = \frac{v_\varepsilon}{M_\varepsilon^t}$, for s and t satisfying (3). Of course $\| z_\varepsilon \| = 1$ or $\| w_\varepsilon \| = 1$. Hence, there exists a subsequence such that:

$$z_\varepsilon \to z \text{ strongly in } L^p, \text{ weakly in } W_0^{1,p}$$

$$w_\varepsilon \to w_\varepsilon \text{ strongly in } L^q, \text{ weakly in } W_0^{1,q}.$$

$$|w_\varepsilon|^{\beta-2} w_\varepsilon \to |w|^{\beta-2} w \text{ in } L^1 \qquad (\beta - 1 < q^*)$$

$$|z_\varepsilon|^{\gamma-2} z_\varepsilon \to |z|^{\gamma-2} z \text{ in } L^1 \qquad (\gamma - 1 < p^*)$$

$(z_\varepsilon, w_\varepsilon)$ satisfies :

$$-\Delta_p z_\varepsilon = \tilde{f}_\varepsilon \tag{8}$$

$$-\Delta_q w_\varepsilon = \tilde{g}_\varepsilon \tag{9}$$

with

$$\tilde{f}_\varepsilon = \frac{a M_\varepsilon^{s(\alpha-p)}}{1 + |\varepsilon^s u_\varepsilon|^{\alpha-1}} |z_\varepsilon|^{\alpha-2} z_\varepsilon + \frac{b M_\varepsilon^{t(\beta-1)-s(p-1)}}{1 + |\varepsilon^t v_\varepsilon|^{\beta-1}} |w_\varepsilon|^{\beta-2} w_\varepsilon + \frac{f}{M_\varepsilon^s}.$$

Combining this with (3), we derive:

$$\tilde{f}_\varepsilon \to a|z|^{p-2} z \text{ in } L^1 \text{ if } \alpha = p \text{ et } \tilde{f}_\varepsilon \to 0 \text{ in } L^1 \text{ if } \alpha < p;$$

analoguously:

$$\tilde{g}_\varepsilon = \frac{c M_\varepsilon^{s(\gamma-1)-t(q-1)}}{1 + |\varepsilon^s u_\varepsilon|^{\gamma-1}} |z_\varepsilon|^{\gamma-2} z_\varepsilon + \frac{d M_\varepsilon^{t(\delta-q)}}{1 + |\varepsilon^t v_\varepsilon|^{\delta-1}} |w_\varepsilon|^{\delta-2} w_\varepsilon + \frac{g}{M_\varepsilon^t}$$

and, by (3):

$$\tilde{g}_\varepsilon \to d|w|^{q-2} w \text{ in } L^1 \text{ if } \delta = q \text{ and } \tilde{g}_\varepsilon \to 0 \text{ in } L^1 \text{ if } \delta < p.$$

Following (i) if $\alpha < p$ and (ii) if $\alpha = p$, we deduce from Lemma 1 that $z = 0$ and $w = 0$, so that we have a contradiction with $\| z_\varepsilon \| = 1$ or $\| w_\varepsilon \| = 1$. Therefore, u_ε and v_ε is bounded.

(iv) **Step 4** We can extract a subsequence so that, by (iii), we have :

$$u_\varepsilon \to u \text{ strongly in } L^p, \text{ weakly in } W_0^{1,p}$$

$$v_\varepsilon \to v \text{ strongly in } L^q, \text{ weakly in } W_0^{1,q}$$

$$|u_\varepsilon|^{\gamma-2}u_\varepsilon \to |u|^{\gamma-2}u \text{ in } L^1 \qquad (\gamma - 1 < p^*)$$

$$|v_\varepsilon|^{\beta-2}v_\varepsilon \to |v|^{\beta-2}v \text{ in } L^1 \qquad (\beta - 1 < q^*)$$

Passing through the limit in (1_ε), and (2_ε), with the help of Lemma 1, we obtain a solution of (P).

We show now that in some cases, it is possible to avoid the double constraint:

$$\beta - 1 < q* \text{ and } \gamma - 1 < p* .$$

This can be done by use of the following result ([4] (Théorème 3.2, p329)):

LEMMA 2 *Let* $u \in W_0^{1,p}(\Omega)$ *solution of*

$$-\Delta_p u = h(x, u)$$

where

$$|h(x, u)| \leq a|u|^{\sigma-2}u + b(x)$$

with

$$a > 0, \sigma \in [1, p^*[, \quad b \geq 0, b \in L^r(\Omega), \quad r > \frac{N}{p} \text{ if } N > p \text{ and } r > 1 \text{ if } N \leq p;$$

then $u \in L^\infty(\Omega)$ *and* $\| u \|_{L^\infty} \leq C$ *where* C *is a constant which depends only on* $a, \sigma, p, N, r, \| b \|_{L^r}$ *and* $\| u \|_{L^{p_0}}$ *with* $p_0 = p^*$ *if* $p^* < +\infty$, $p_0 = 2\max(pr', \sigma)$ *if* $p^* = +\infty$.

By use of Lemma 1, we derive directly from Theorem 1

COROLLARY *We assume that the hypotheses of Theorem 1 are satisfied. Then, any solution (u,v) of (P) is in* $(L^\infty(\Omega))^2$ *if one of the 3 following hypotheses is satisfied :*

$$p \geq N, \ q \geq N, \ f \in L^{p'}, \ g \in L^{q'} \tag{i}$$

$$p < N, \ q \geq N, \ f \in L^r \text{ with } r > \frac{N}{p}, \ g \in L^{q'} \tag{ii}$$

$$p \geq N, \ q < N, \ f \in L^{p'}, \ g \in L^\varrho \text{ with } \varrho > \frac{N}{q} . \tag{iii}$$

Proof. i) We have $b|v|^{\beta-2}v + f \in L^{p'}$ and we apply lemma 2 to the first equation ; we do the same for the 2nd equation.
ii) We have $b|v|^{\beta-2}v + f \in L^r$ with $r > \frac{N}{p}$ so that $u \in L^\infty$; then by the 2nd equation, $c|u|^{\gamma-2}u + g \in L^{q'}$ and we get the result.

iii) We proceed exactly as for *ii*).

REMARK 1 In case *i*) there is no limitation for β et γ; we just have to verify $(\beta - 1)(\gamma - 1) < (p - 1)(q - 1)$ for having existence of a solution.
In case *ii*), there is no limitation on β but we must have $\gamma - 1 < p^*$ and of course $(\beta - 1)(\gamma - 1) < (p - 1)(q - 1)$.

THEOREM 2 *Assume that (H1) and (H3) are satisfied and that:*

$$\alpha \geq 1, \beta \geq 1, \gamma \geq 1, \delta \geq 1, (\beta - 1)(\gamma - 1) < (p - 1)(q - 1); \qquad (H'2)$$

moreover we suppose that

(*) either $p < N, q < N$ and $\beta - 1 < q^* \frac{p}{N}$

(**) or $p < N$ and $q \geq N$.

Then, for all $f \in L^r$ with $r > \frac{N}{p}$ and for all $g \in L^{q'}$, there exists a solution to Problem (P): $(u, v) \in W_0^{1,p}(\Omega) \times W_0^{1,q}(\Omega)$; moreover $(u, v) \in (L^\infty(\Omega))^2$.

Proof. As for Theorem 1, we shaw that:

$$\varepsilon^s u_\varepsilon \to 0 \text{ strongly in } L^p, \text{ weakly in } W_0^{1,p}$$

$$\varepsilon^t v_\varepsilon \to 0 \text{ strongly in } L^q, \text{ weakly in } W_0^{1,q}$$

For step (iii), we suppose that $(u_\varepsilon, v_\varepsilon)$ is not bounded in $W_0^{1,p} \times W_0^{1,q}$ and we introduce again z_ε and w_ε.
In case (**) $|w_\varepsilon|^{\beta - 1}$ is bounded in L^ϱ.
In case (*), there exists $\varrho > \frac{N}{p}$ such that we have simultaneously $f \in L^\varrho$ and $|w_\varepsilon|^{\beta - 1}$ bounded in L^ϱ (choose $\varrho < \frac{q^*}{\beta - 1}$).
Finally (8) can be written as:

$$-\Delta_p z_\varepsilon = a'_\varepsilon(x)|z_\varepsilon|^{\alpha - 2} z_\varepsilon + b'_\varepsilon(x)$$

where a'_ε is bounded in L^∞ , z_ε is bounded in L^{p^*} and b'_ε is bounded in L^ϱ. By Lemma 2, z_ε is bounded in L^∞.
By dominated convergence theorem:

$$|z_\varepsilon|^{\gamma - 2} z_\varepsilon \to |z|^{\gamma - 2} z \text{ in } L^1$$

Since $\beta - 1 < q^*$, we also have:

$$|w_\varepsilon|^{\beta - 2} w_\varepsilon \to |w|^{\beta - 2} w \text{ in } L^1$$

Passing through the limit in (8) and (9), as for Theorem 1, we obtain a contradiction which shows that indeed $(u_\varepsilon, v_\varepsilon)$ is bounded.

For proving step (iv), we consider $(u_\varepsilon, v_\varepsilon)$ bounded in $W_0^{1,p} \times W_0^{1,q}$. Choosing a subsequence, we have:

$$u_\varepsilon \to u \text{ strongly in } L^p, \text{ weakly in } W_0^{1,p}$$

$$v_\varepsilon \to v \text{ strongly in } L^q, \text{ weakly in } W_0^{1,q}$$

In case (**), $|w_\varepsilon|^{\beta-1}$ is bounded in any L^ϱ.

In case (*), there exists $\varrho > \frac{N}{P}$ such that $f \in L^\varrho$ and $|w_\varepsilon|^{\beta-1}$ are both bounded in L^ϱ.

Finally (1_ε) can be written as :

$$-\Delta_p u_\varepsilon = a_\varepsilon(x)|u_\varepsilon|^{\alpha-2}u_\varepsilon + b_\varepsilon''(x)$$

where a''_ε is bounded in L^∞, u_ε is bounded in L^{p*} et b''_ε is bounded in L^ϱ. By Lemma 2, u_ε is bounded in L^∞.

By dominated convergence theorem:

$$\frac{|u_\varepsilon|^{\gamma-2}u_\varepsilon}{1 + |\varepsilon^s u_\varepsilon|^{\gamma-1}} \to |u|^{\gamma-2}u \text{ in } L^1.$$

Since we always have $\beta - 1 < q^*$, we also have:

$$\frac{|v_\varepsilon|^{\beta-2}v_\varepsilon}{1 + |\varepsilon^t v_\varepsilon|^{\beta-1}} \to |v|^{\beta-2}v \text{ in } L^1.$$

We can now make use of Lemma 1 and go to the limit in $(1_\varepsilon), (2_\varepsilon)$; we have obtained a solution to (P). The estimates in L^∞ for the solution u and v can be derived from Lemma 2 again.

REMARK 2

When (**) holds, there is no limitation for β and γ except that $(\beta - 1)(\gamma - 1) < (p-1)(q-1)$; that is the announced improvement of Theorem 1 and of its corollary.

Acknowledgment The first results of this work have been presented by L.B. to the Conference on Nonstandard Problems organized by A. Cellina and P. Marcellini in Firenze, June 1994. J.F. and F.T. were supported by the European network HCM93409 and L.B. was "Professeur Invité" at CEREMATH, University Toulouse 1.

REFERENCES

1. A.Anane "Etudes des valeurs propres et de la résonance pour l'opérateur p-Laplacien" Thèse, Université Libre de Bruxelles, 1988.

2. L.Boccardo, J.Fleckinger-Pellé and F.de Thélin "Existence of solutions for some non-linear cooperative systems", *Diff. and Int. Eq 7*: 689-698, (1994).

3. L.Boccardo and F.Murat "Almost everywhere convergence of the gradients of solutions to elliptic and parabolic equations". *Nonlinear Analysis T.M.A. 19*: 581-587, (1994).

4. O.A.Ladyzhenskaya and N.Uraltseva "Linear and quasilinear elliptic equations", *Acad. Press*, (1968).

Some Uniqueness Results for $\Delta u + f(u) = 0$ in $\mathbf{R}^N$, $N \geq 3$

CARMEN CORTÁZAR [1] and MANUEL ELGUETA [1] Facultad de Matemáticas, Universidad Católica, Casilla 306 Correo 22, Santiago, Chile.

PATRICIO FELMER [2] Departamento de Ing. Matemática, F.C.F.M., Universidad de Chile, Casilla 170 Correo 3, Santiago, Chile.

1 INTRODUCTION

In this note we present some uniqueness results for positive solutions of the ordinary differential equation

$$u''(r) + \frac{N-1}{r}\, u'(r) + f(u) = 0$$

$$u'(0) = 0, \qquad u(r) \to 0 \text{ as } r \to \infty.$$

$$(1.1)$$

This equation is satisfied by the radially symmetric solutions of the semilinear elliptic partial differential equation

$$\Delta u + f(u) = 0, \qquad x \in \mathbf{R}^N, \tag{1.2}$$

decaying to zero at infinity. In general, nonnegative solutions of (1.2), decaying to zero at infinity are called ground states. An important question is whether they are unique.

In the classical work of Gidas, Ni and Nirenberg [7] and [8] it is shown that all ground states of (1.2) are radially symmetric, when f satisfies some mild conditions. Thus, the uniqueness question for (1.2) is essentially that for (1.1).

The first step in the study of uniqueness for (1.1) was given by Coffman [1], in the case of the model nonlinearity

$$f(u) = -u + u^p \tag{1.3}$$

[1] Partially supported by FONDECYT under grant No 1940705.
[2] Partially supported by FONDECYT under grant No 1980698.

with $p = 3$ and $N = 3$. Coffman introduces the function

$$\varphi(r) = \frac{\partial u(r, \alpha)}{\partial \alpha}, \qquad (1.4)$$

where $u(r, \alpha)$ is the solution of the initial value problem, starting at $r = 0$ with the values $u(0) = \alpha$ and $u'(0) = 0$. Through a careful analysis of the zeroes of φ, Coffman was able to prove his uniqueness result (this approach was already used by Kolodner [9]).

A different approach was taken later by Peletier and Serrin for the case of a class of nonlinearities having a sublinear growth at infinity. In [14] and [15], Peletier and Serrin introduced a Monotone Separation Theorem for proving uniqueness assertions. In these articles the nonlinearity f is merely locally Lipschitz continuous in $[0, \infty)$.

Still in the context of sublinear nonlinearities, and considering very general quasi-linear elliptic operators, Franchi, Lanconelli and Serrin [6], proved in a recent article a remarkable general uniqueness theorem. Even in the case of equation (1.1), their result is surprising. If the nonlinearity f is locally Lipschitz continuous in (β, ∞), continuous in $[0, \infty)$, and the primitive F of f is negative in the interval $(0, \beta)$, then (1.1) has a unique solution (other conditions, were also needed, but need not concern us here). Franchi, Lanconelli and Serrin introduce a key identity for solutions of (1.1), which together with the Monotone Separation Theorem gives the uniqueness assertion.

Meanwhile, the study of uniqueness in the super linear case was considered by McLeod and Serrin in [13]. They improved Coffman's method to include more general cases of f as in (1.3), in particular for $1 < p \leq N/(N-2)$ and $N = 3, 4$. The next important idea in this context was introduced by Kwong in [10], allowing him to extend the uniqueness results to f of the type (1.3), for $N \geq 3$ and $1 < p \leq (N+2)/(N-2)$.

After [10] more general nonlinearities with superlinear growth where studied by Kwong and Zhang [11], Chen and Lin [2], McLeod [12] and Yanagida [15]. However, in contrast with the results for sublinear nonlinearity, in all these cases the function f is assumed to be differentiable on $[0, \infty)$.

A first case where the function f is not differentiable at zero was studied by the authors in [3]. More precisely, we obtained a uniqueness theorem for the function

$$f(u) = -u^q + u^p,$$

with $0 < q < 1 < p < (N+2)/(N-2)$. In this case the function f is not Lipschitz continuous at 0, and solutions of (1.1) have compact support.

In [4], the authors obtained uniqueness results for more general nonlinearities. More precisely, we consider the following conditions on f:

(f1) $f(0) = 0$, *and there exists $u_0 > 0$ such that $f(u) > 0$ for $u > u_0$, and $f(u) \leq 0$, $f(u) \not\equiv 0$ for $u \in [0, u_0)$.*

(f2) *f is continuous in $[0, \infty)$, and locally Lipschitz in $[u_0, \infty)$.*

If the function f satisfies (f2), then f is differentiable in $[u_0, \infty)$ a.e. We denote by D the subset of $[u_0, \infty)$ where f' is defined.

Throughout the paper we write

$$F(s) = \int_0^S f(t)dt$$

and denote by β the unique positive zero of $F(u)$ when $u > u_0$.

The next two hypotheses describe the behavior of $f(u)$ for $u > u_0$.

(f3) $f(u) \le f'(u)(u - u_0)$ *for all* $u \in D$.

(f4) *The function* $S(u) = uf'(u)/f(u)$ *is monotone decreasing in* D.

Note that (f1)-(f4) are satisfied for the canonical case (1.3).

The main theorem in [4] is the following.

THEOREM 1.1 *Assume the function f satisfies hypotheses (f1), (f2), (f3) and (f4). Then equation (1.1) possesses at most one non-trivial non-negative solution.*

Its proof consists of a combination of the arguments of Coffman and Kwong, and of the Monotone Separation Theorem. The former are used to handle the range when the solution u is above u_0, and the latter to deal with the case when u is below u_0. The link that connects both type of arguments is the following energy like functional similar to the one in [6].

Given a solution $u(r, \alpha)$ of the initial value problem associated to (1.1), with $u(0, \alpha) = \alpha$ and $u'(0, \alpha) = 0$, we define

$$J = r\{(u'(r, \alpha))^2 + 2F(u(r, \alpha))\}^{\frac{1}{2}}.$$

In this note we present a variation of Theorem 1.1. We will modify hypothesis (f4) by imposing monotonicity of S only above β and adding a stronger growth condition.

More precisely we consider

(f4') *The function* $S(u) = \frac{uf'(u)}{f(u)}$ *is monotone decreasing in* $D \cap [\beta, +\infty)$, $\inf\{S(u)/$ $u \in [u_0, \beta] \cap D\} \ge \sup\{S(u)/u \in [\beta, +\infty) \cap D\}$ *and* $S(\beta) \le \frac{N}{N-2}$.

We can state now the following

THEOREM 1.2 *Assume the function f satisfies hypotheses (f1), (f2), (f3) and (f4'). Then equation (1.1) possesses at most one non-trivial non-negative solution.*

The proof of this theorem follows the idea of that of Theorem 1.1, but it avoids several arguments, making more clear the role of the functional J in connecting the behavior of u above and below u_0.

We note that a function f such that restricted to $[u_0, +\infty)$ is piecewise linear does not fit the hypotheses of Theorem 1.1. unless it is a straight line. On the other hand some of these functions do fit the hypotheses of Theorem 1.2.

We devote Section §2 to present the proof of Theorem 1.2. Section §3 is devoted to some words on uniqueness for elliptic systems.

PROOF OF THEOREM 1.2

We will give the proof under the extra assumption that $f(s)$ is differentiable for $s \geq u_0$, the general case can be handled as in [4].

We denote by $u(r, \alpha)$ a C^2 solution of

$$u''(r) + \frac{N-1}{r} u'(r) + f(u) = 0 \tag{2.1}$$

$$u(0) = \alpha \qquad u'(0) = 0,$$

for $\alpha \in (0, \infty)$.

Let us define the functional

$$I(r, \alpha) = (u'(r, \alpha))^2 + 2F(u(r, \alpha)). \tag{2.2}$$

It is easy to see that $I(r, \alpha)$ is decreasing in r. A standard consequence of this is that if $\alpha \leq u_0$, then there exists $c > 0$ such that $u(r, \alpha) > c$ for all $r \in (0, \infty)$; hence $u(r, \alpha)$ is not a solution of (1.1). Consequently, in what follows we will always assume $\alpha > u_0$.

It is easily checked that, for any $\alpha \in (u_0, \infty)$, one has $u(r, \alpha) > 0$ and $u'(r, \alpha) < 0$ for small positive values of r, so we can define

$$R(\alpha) = \sup\{r \mid u(s, \alpha) > 0 \text{ and } u'(s, \alpha) < 0 \text{ for all } s \in (0, r)\}.$$

Thus, given $\alpha \in (u_0, \infty)$, there exists a unique solution $u(r, \alpha)$ defined on $[0, R(\alpha))$. This solution, $u(r, \alpha)$, and its derivative $u'(r, \alpha)$ depend continuously on α, and $u(r, \alpha)$ is invertible in $0 < r < R(\alpha)$. We denote the inverse of $u(r, \alpha)$ by $r(s, \alpha)$.

Let us define now the sets, (as in [6], [13], [14])

$$\begin{aligned}
\mathcal{N} &= \{\alpha \mid u(R(\alpha), \alpha) = 0 \text{ and } u'(R(\alpha), \alpha) < 0\}, \\
\mathcal{G} &= \{\alpha \mid u(R(\alpha), \alpha) = 0 \text{ and } u'(R(\alpha), \alpha) = 0\}, \\
\mathcal{P} &= \{\alpha \mid u(R(\alpha), \alpha) > 0\}.
\end{aligned}$$

We observe that the sets $\mathcal{N}$, $\mathcal{G}$ and $\mathcal{P}$ form a partition of (u_0, ∞) and also that $\mathcal{N}$ and $\mathcal{P}$ are open. See [4].

We set for $r \in [0, r_0(\alpha)]$,

$$\varphi(r, \alpha) = \frac{\partial u}{\partial \alpha}(r, \alpha) \text{ and } \varphi'(r, \alpha) = \frac{\partial u'}{\partial \alpha}(r, \alpha). \tag{2.3}$$

Note that φ satisfies the linear differential equation

$$(r^{N-1}\varphi'(r))' + r^{N-1}f'(u(r))\varphi(r) = 0 \qquad \text{in} \quad [0, r_0(\alpha)],$$

and   (2.4)

$$\varphi(0, \alpha) = 1 \text{ and } \varphi'(0, \alpha) = 0.$$

Next we give a variation of the monotone separation argument that appears in [6] and [14]. For real numbers r, u and p, we define

$$J(r, u, p) = r(p^2 + 2F(u))^{\frac{1}{2}}.$$

We recall that $u(r, \alpha)$ is invertible in $r \in [0, R(\alpha))$, with $r(s, \alpha)$ its inverse. Define

$$W(s, \alpha) = J(r(s, \alpha), s, u'(r(s, \alpha), \alpha)). \tag{2.5}$$

If $\mathcal{G} = \emptyset$ then Theorem 1.2 holds trivially, so we assume from now on $\mathcal{G} \neq \emptyset$. Given $\overline{\alpha} \in \mathcal{G}$ and $u_* \in (0, \overline{\alpha})$, the functions $r_*(\alpha) \equiv r(u^*, \alpha)$ and $W(u_*, \alpha)$ are well defined in a neighborhood $(\overline{\alpha} - \varepsilon, \overline{\alpha} + \varepsilon)$, for some $\varepsilon > 0$.

PROPOSITION 2.1 *Let $\overline{\alpha} \in \mathcal{G}$. Assume that there exist $u_* \in (0, \beta)$ and $\delta > 0$ such that $r_*(\alpha)$ is decreasing and $W(u_*, \alpha)$ is strictly increasing in α on $(\overline{\alpha} - \delta, \overline{\alpha} + \delta)$. Then*

$$(\overline{\alpha}, \overline{\alpha} + \delta) \subset \mathcal{N} \ \text{and} \ (\overline{\alpha} - \delta, \overline{\alpha}) \subset \mathcal{P}.$$

PROOF. We repeat the proof given in [4] for the sake of completeness. Let $\alpha \in (\overline{\alpha}, \overline{\alpha} + \delta)$. Then from the hypothesis, we have

$$r_*(\alpha) \leq r_*(\overline{\alpha}) \ , \quad W(u_*, \alpha) > W(u_*, \overline{\alpha}),$$

and consequently

$$u'(r_*(\alpha), \alpha) < u'(r_*(\overline{\alpha}), \overline{\alpha}). \tag{2.6}$$

To simplify the notation we set $\overline{p}(s) = u'(r(s, \overline{\alpha}), \overline{\alpha})$ and $p(s) = u'(r(s, \alpha), \alpha)$. Accordingly $p(u_*) < \overline{p}(u_*)$.

We now argue by contradiction, and assume $\alpha \notin \mathcal{N}$. In this case there exists $s_1 \in [0, u_*)$ such that

$$p(s) < \overline{p}(s) \ \text{for all} \ s \in (s_1, u_*)$$

and

$$p(s_1) = \overline{p}(s_1).$$

By direct differentiation we get

$$\frac{\partial W}{\partial s}(s, \alpha) = \frac{2F(s) - (N-2)p^2(s)}{p(s)(p^2(s) + 2F(s))^{\frac{1}{2}}}, \tag{2.7}$$

and

$$\frac{\partial W}{\partial s}(s, \overline{\alpha}) = \frac{2F(s) - (N-2)\overline{p}^2(s)}{\overline{p}(s)(\overline{p}^2(s) + 2F(s))^{\frac{1}{2}}}. \tag{2.8}$$

Integrating (2.7) and (2.8) between s_1 and u_* and then subtracting, we obtain

$$0 > (W(u_*, \overline{\alpha}) - W(u_*, \alpha)) - (W(s_1, \overline{\alpha}) - W(s_1, \alpha))$$

$$= \int_{s_1}^{u_*} 2F(s) \left(\frac{1}{\overline{p}(s)(\overline{p}^2(s) + 2F(s))^{\frac{1}{2}}} - \frac{1}{p(s)(p^2(s) + 2F(s))^{\frac{1}{2}}} \right) ds$$

$$-(N-2) \int_{s_1}^{u_*} \left(\frac{\overline{p}^2(s)}{\overline{p}(s)(\overline{p}^2(s) + 2F(s))^{\frac{1}{2}}} - \frac{p^2(s)}{p(s)(p^2(s) + 2F(s))^{\frac{1}{2}}} \right) ds > 0.$$

since $F(s) \leq 0$ for $s \in (0, u_*)$. This contradiction proves that $(\overline{\alpha}, \overline{\alpha} + \delta) \subset \mathcal{N}$. The proof that $(\overline{\alpha} - \delta, \overline{\alpha}) \subset \mathcal{P}$ is similar. This ends the proof of Proposition 2.1.

We will also need the following

PROPOSITION 2.2 *Under the hypotheses (f1)-(f4'), if $\overline{\alpha} \in \mathcal{G}$ then there exist $u_* \in [u_0, \beta]$ and $\delta > 0$ such that $r_*(\alpha)$ is decreasing and $W(u_*, \alpha)$ is strictly increasing in α on $(\overline{\alpha} - \delta, \overline{\alpha} + \delta)$.*

From now on we will denote the functions $u(r, \overline{\alpha})$ and $\varphi(r, \overline{\alpha})$ simply by $u(r)$ and $\varphi(r)$, without making $\overline{\alpha}$ explicit. We will also write R instead of $R(\overline{\alpha})$, and $r_0 = r(u_0, \overline{\alpha})$.

We prove Proposition 2.2 by studying the derivatives of $r_*(\alpha)$ and $W(u_*, \alpha)$ with respect to α. We have the following formulae, valid for $u_* \geq u_0$,

$$\frac{\partial r_*}{\partial \alpha}(\alpha) = -\frac{\varphi(r_*(\alpha), \alpha)}{u'(r_*(\alpha), \alpha)}, \tag{2.9}$$

$$\frac{\partial W}{\partial \alpha}(u_*, \alpha) = (I)^{-\frac{1}{2}} \left\{ \frac{(r^{N-2}\varphi(r, \alpha))'}{r^{N-3}} u'(r, \alpha) \right.$$
$$\left. + r f(u_*)\varphi(r, \alpha) - 2F(u_*)\frac{\varphi(r, \alpha)}{u'(r, \alpha)} \right\}_{r=r_*(\alpha)}, \tag{2.10}$$

where $I = I(r, \alpha)$ was defined by (2.2).

We note that Proposition 2.2 follows from an analysis of the signs of φ and $(r^{N-2}\varphi(r))'$.

PROOF OF PROPOSITION 2.2. First assume that $\varphi(r) > 0$ on $[0, r_0)$. Using (2.4) and integrating by parts, we get

$$0 = \int_0^{r_0} (\varphi(r)u'(r) + f'(u(r))\varphi(r)(u(r) - u_0))r^{N-1}dr. \tag{2.11}$$

Integrating by parts again and using (2.1), we obtain

$$0 = -r_0^{N-1}\varphi(r_0)u'(r_0) + \int_0^{r_0} \varphi(r)\left(f'(u(r))(u(r) - u_0) - f(u(r))\right)r^{N-1}dr,$$

and then using (f3) we obtain

$$r_0^{N-1}\varphi(r_0)u'(r_0) \geq 0. \tag{2.12}$$

Therefore, since $u'(r_0) < 0$, we must have $\varphi(r_0) = 0$. The uniqueness of the solution of (2.4) implies that also $\varphi'(r_0) < 0$.

From (2.10) we see that $\partial W(u_0, \overline{\alpha})/\partial \alpha > 0$, and then, since φ and φ' are continuous, it follows that $W(u_*, \alpha)$ is strictly increasing near $\overline{\alpha}$ with $u_* = u_0$.

We still have to show that r_* is decreasing. In order to do this, from (2.12), (2.11) and (f3) we get

$$f'(s)(s - u_0) - f(s) = 0 \quad \text{on } (u_0, \overline{\alpha}).$$

Hence

$$\int_0^{r(\overline{\alpha},\alpha)} \varphi(r,\alpha) \left(f'(u(r,\alpha))(u(r,\alpha) - u_0) - f(u(r,\alpha)) \right) r^{N-1} dr =$$

$$\text{(2.13)}$$

$$= r_0^{N-1}(\alpha)\varphi(r_0(\alpha),\alpha)u'(r_0(\alpha),\alpha).$$

Let $a > 0$ be a small number. If $f'(s)(s - u_0) - f(s) = 0$ on $(\overline{\alpha}, \overline{\alpha} + a)$, then $\varphi(r_0(\alpha), \alpha) \equiv 0$ in a neighborhood of $\overline{\alpha}$. Otherwise, from (2.13), the continuity of φ, and the fact that $\varphi(0, \alpha) \equiv 1$ we find $\varphi(r_0(\alpha), \alpha) \leq 0$ in a neighborhood of $\overline{\alpha}$.

It then follows from (2.9) that $r_*(\alpha) = r_0(\alpha)$ is decreasing near $\overline{\alpha}$. This finishes the proof of Proposition 2.2 under the assumption $\varphi > 0$ on $[0, r_0)$.

Next we consider the case when φ vanishes at least once in $(0, r_0)$. Let r_1 be the first zero of φ. We distinguish two cases.

First assume $u(r_1) \in (u_0, \beta]$. In this case we have $\varphi(r_1) = 0$ and, by uniqueness, $\varphi'(r_1) < 0$. Since $u(r_1) > u_0$, by using continuity, we see that there exists $r_2 > r_1$ such that $u(r_2) > u_0$, $\varphi(r_2) < 0$ and $\varphi'(r_2) < 0$. Then, from (2.9) and (2.10),

$$\frac{\partial r_*}{\partial \alpha}(\overline{\alpha}) < 0 \quad \text{and} \quad \frac{\partial W}{\partial \alpha}(u_*, \overline{\alpha}) > 0 \quad \text{with} \quad u_* = u(r_2).$$

As before, we again obtain Proposition 2.2.

To deal with the case when $u(r_1) > \beta$ we need the following lemma.

LEMMA 2.1 *If* $\overline{\alpha} \in \mathcal{G}$ *and* $u(r_1) > \beta$, *then* $\varphi(r) < 0$ *for* $r \in (r_1, r_0]$ *and* $(r^{N-2}\varphi(r))'(r_0) \leq 0$.

PROOF. Let $v(r) = ru'(r) + cu(r)$, where c is a constant to be fixed later. A direct computation shows that

$$v''(r) + \frac{N-1}{r}v'(r) + f'(u(r))v(r) = \phi(u(r)),$$

where

$$\phi(u) = cf'(u)u - (c+2)f(u). \qquad \text{(2.14)}$$

Since we are assuming that $u(r_1) > \beta$, it follows from the hypotheses (f3), (f4') and the fact $u'(r) < 0$ for $r \in (0, R)$, that we can choose $c \geq N - 2$ such that $\phi(u(r)) \geq 0$ for $r \in (r_1, r_0)$, and $\phi(u(r)) \leq 0$ for $r \in (0, r_1)$. In order to prove the lemma, assume there exists a first value r_2 in $(r_1, r_0]$ such that $\varphi(r_2) = 0$. We have

$$0 \geq \int_0^r \varphi(t)\phi(u(t))t^{N-1}dt = \int_0^r (\varphi(t)\Delta v(t) - v(t)\Delta\varphi(t))t^{N-1}dt$$

$$= r^{N-1}[\varphi(r)v'(r) - \varphi'(r)v(r)]. \qquad \text{(2.15)}$$

Therefore

$$r^{N-1}[\varphi(r)v'(r) - \varphi'(r)v(r)] \leq 0 \qquad \text{(2.16)}$$

for any $r \in (0, r_2]$. Evaluating (2.16) at $r = r_1$ gives

$$-r_1^{N-1}\varphi'(r_1)v(r_1) \leq 0.$$

Since $\varphi(r_1) = 0$ and φ is a solution of (2.4), we have $\varphi'(r_1) < 0$. Hence $v(r_1) \leq 0$. As

$$v'(r) = ru''(r) + (c+1)u'(r) \leq ru''(r) + (N-1)u'(r) = -rf(u(r)) < 0 \quad (2.17)$$

we deduce that $v(r) < 0$ for $r_1 < r \leq r_0$. In particular $v(r_2) < 0$ and $v(r_0) < 0$. Evaluating (2.16) at r_2, we obtain next

$$-r_2^{N-1}\varphi'(r_2)v(r_2) \leq 0,$$

a contradiction since $\varphi'(r_2) > 0$. (Here we have used again that φ is a solution of (2.4)). Now evaluating (2.15) at r_0 gives

$$0 \geq r_0^{N-1}(\varphi(r_0)v'(r_0) - \varphi'(r_0)v(r_0)).$$

It follows that $\varphi'(r_0) \leq 0$ and hence $(r^{N-2}\varphi(r))'(r_0) \leq 0$. This ends the proof of the lemma.

We recall that we still have to finish the proof of Proposition 2.2 in the case when $u(r_1) > \beta$. We note $u'(r_0) < 0$, $F(u_0) < 0$ and $f(u_0) = 0$. Moreover by Lemma 2.1 we have $\varphi(r_0) < 0$ and $(r^{N-2}\varphi(r))'(r_0) \leq 0$. It follows now from (2.10) at $u_* = u_0$, that $\partial W(u_*, \overline{\alpha})/\partial\alpha(u_*, \overline{\alpha}) > 0$. By continuity, the same holds for α in a neighborhood of $\overline{\alpha}$. Hence $W(u_*, \alpha)$ is strictly increasing in α on $(\overline{\alpha} - \delta, \overline{\alpha} + \delta)$ for some $\delta > 0$. Similarly, using Lemma 2.1 and (2.9) we see that $r_*(\alpha)$ is decreasing in α near $\overline{\alpha}$. This completes the proof of Proposition 2.2.

We can now prove Theorem 1.2. We argue by contradiction, and assume there exist α_1 and α_2 in $\mathcal{G}$ with $\alpha_1 < \alpha_2$. By Proposition 2.1 and Proposition 2.2 we can assume that $(\alpha_1, \alpha_2) \subset \mathcal{N} \cup \mathcal{P}$. Moreover, again by Proposition 2.1 and Proposition 2.2, it follows that $(\alpha_1, \alpha_2) \cap \mathcal{N} \neq \emptyset$ and $(\alpha_1, \alpha_2) \cap \mathcal{P} \neq \emptyset$. This contradicts the connectedness of (α_1, α_2), since $\mathcal{N}$ and $\mathcal{P}$ are open. The proof of Theorem 1.2 is thus complete.

3 REMARKS ABOUT SYSTEMS

We end this note with some observations about elliptic systems. We will see that for a certain class of systems, uniqueness results for a single equation implies precise multiplicity results for systems.

Let $h : \mathbf{R} \to \mathbf{R}$ be an even function and $p : R \times R \to R$ be an homogeneous polynomial of degree q. Radial solutions of the elliptic system

$$\Delta u + uh(p(u, v)) = 0$$
$$\Delta v + vh(p(u, v)) = 0$$

in R^N satisfy the ODE system

$$u''(r) + \frac{N-1}{r}u'(r) + u(r)h(p(u(r), v(r))) = 0 \qquad (3.1)$$

$$v''(r) + \frac{N-1}{r}v'(r) + v(r)h(p(u(r), v(r))) = 0 \qquad (3.2)$$

in $[0, +\infty)$ with $u'(0) = v'(0) = 0$. We also assume that $\lim_{r \to \infty} u(r) = 0$ and $\lim_{r \to \infty} v(r) = 0$.

Let us also consider the equation

$$w''(r) + \frac{N-1}{r} w'(r) + u(r) h(w^q r)) = 0 \qquad (3.3)$$

in $[0, +\infty)$ with $w'(0) = 0$ and $\lim_{r \to \infty} w(r) = 0$.

If (u, v) is a non negative solution of (3.1)-(3.2), then setting $H(r) = u(r)v'(r) - u'(r)v(r)$ we have

$$H'(r) = \frac{N-1}{r} H(r)$$

and, since $H(0) = 0$, we must have $H(r) \equiv 0$. This implies

$$v(r) = \lambda u(r)$$

for some constant $\lambda \geq 0$. Next we substitute $v(r) = \lambda u(r)$ in equation (3.1) and we obtain

$$u''(r) + \frac{N-1}{r} u'(r) + u(r) h(z(\lambda) u^q(r)) = 0 \qquad (3.4)$$

where

$$z(\lambda) = p(1, \lambda).$$

Multiplying (3.4) by $|z(\lambda)|^{\frac{1}{q}}$ we get that the function $|z(\lambda)|^{\frac{1}{q}} u(r)$ is a solution of (3.3). Therefore

$$u(r) = \frac{1}{|z(\lambda)|^{\frac{1}{q}}} w(r)$$

where w is a solution of (3.3). In an analogous way we obtain that

$$v(r) = \frac{\lambda}{|z(\lambda)|^{\frac{1}{q}}} w(r).$$

In this fashion, we have that any non-negative solution of (3.1)-(3.2) is of the form

$$(u(r), v(r)) = (\frac{1}{|z(\lambda)|^{\frac{1}{q}}} w(r), \frac{\lambda}{|z(\lambda)|^{\frac{1}{q}}} w(r)), \qquad (3.5)$$

where w is a solution of (3.3), and $\lambda \geq 0$. In the case that (3.3) has a unique non-negative solution, then the set of solutions of (3.2) is a one parameter family of pairs of functions of the form (3.5). It is easy now to give examples where the set of solutions of (3.2) is bounded, connected, unbounded or disconnected.

One particular case where these ideas apply corresponds to a Lagrangean system where the nonlinearity is given by a potential of the form

$$V(u, v) = G(u^2 + v^2),$$

with $G : \mathbf{R} \to \mathbf{R}$ smooth.

An interesting problem suggested by the discussion above is to study the structure of the set of solutions for more general systems of the form

$$u''(r) + \frac{N-1}{r} u'(r) + u(r) g(u(r), v(r)) = 0$$

$$v''(r) + \frac{N-1}{r} v'(r) + v(r) g(u(r), v(r)) = 0$$

in $[0, +\infty)$ with $u'(0) = v'(0) = 0$ under suitable hypotheses on g.

REMARK   For sublinear systems, in bounded domains, uniqueness results have been obtained by Felmer and Martínez in [5]. In the simplest case, it is shown that the system

$$\Delta u + v^p = 0$$
$$\Delta v + u^q = 0$$

in Ω, with Dirichlet boundary condition, possesses exactly one positive solution. This is in contrast with the one parameter family obtained above.

REFERENCES

[1] C. V. Coffman, Uniqueness of the ground state solution for $\Delta u - u + u^3 = 0$ and a variational characterization of other solutions, *Arch. Rational Mech. Anal.* **46**, 1972, 81-95.

[2] C. C. Chen & C. S. Lin, Uniqueness of the ground state solutions of $\Delta u + f(u) = 0$ in $\mathbb{R}^N$, $N \geq 3$, *Comm. in Partial Diff. Equations* **16**, 1991, 1549-1572.

[3] C. Cortázar, M. Elgueta & P. Felmer, On a semilinear elliptic problem in $\mathbb{R}^N$ with a non-Lipschitzian non-linearity, *Advances in Differential Equations* **1**, 1996, 199-218.

[4] C. Cortázar, M. Elgueta & P. Felmer, Uniqueness of positive solutions of $\Delta u + f(u) = 0$ in $\mathbb{R}^N$, $N \geq 3$. To appear in Arch. Rat. Mech. Anal.

[5] P. Felmer & S. Martínez, Existence and uniqueness of positive solutions to certain differential systems. To appear in Differential and Integral Equations.

[6] B. Franchi, E. Lanconelli & J. Serrin, Existence and uniqueness of nonnegative solutions of quasilinear equations in $\mathbb{R}^N$. To appear in *Advances in Mathematics* 1996.

[7] B. Gidas, W. M. Ni & L. Nirenberg, Symmetry and related properties via the Maximum Principle, *Commun. Math. Phys.* **68**, 1979, 209-243

[8] B. Gidas, W. M. Ni & L. Nirenberg, Symmetry of positive solutions of nonlinear elliptic equations in $\mathbb{R}^N$, *Advances in Math. Studies* **7 A** 1979, 209-243.

[9] I. I. Kolodner, Heavy rotating string-A nonlinear eigenvalue problem, *Comm. Pure Appl. Math.* **8**, 1955, 395-408.

[10] M. K. Kwong, Uniqueness of positive solutions of $\Delta u - u + u^p = 0$ in $\mathbb{R}^N$. *Arch. Rational Mech. Anal.* **105**, 1989, 243-266.

[11] M. K. Kwong & L. Zhang, Uniqueness of the positive solution of $\Delta u + f(u) = 0$ in an annulus, *Differential and Integral Equations* **4**, 1991, 583-599.

[12] K. McLeod, Uniqueness of positive radial solutions of $\Delta u + f(u) = 0$ in $\mathbb{R}^N$, II *Trans. AMS.* **339**, 1993, 495-505.

[13] K. McLeod & J. Serrin, Uniqueness of positive radial solutions of $\triangle u + f(u) = 0$, *Arch. Rational Mech. Anal.* **99**, 1987, 115-145.

[14] L. A. Peletier & J. Serrin, Uniqueness of positive solutions of semilinear equations in $\mathbf{R}^N$ *Arch. Rational Mech. Anal.* **81**, 1983, 181-197.

[15] L. A. Peletier & J. Serrin, Uniqueness of non-negative solutions of semilinear equations in $\mathbf{R}^N$, *J. Diff. Eq.* **61**, 1986, 380-397.

[16] E. Yanagida, Uniqueness of positive radial solutions of $\triangle u + g(r)u + h(r)u^p = 0$ in $\mathbf{R}^N$, *Arch. Rational Mech. Anal.* **115**, 1991, 257-274.

A Strong Comparison Principle for the Dirichlet p-Laplacian

MABEL CUESTA   Département de Mathématique, Université Libre de Bruxelles, Campus Plaine C.P. 214, B–1050 Bruxelles, Belgique;
e-mail: `mcuesta@ulb.ac.be`

PETER TAKÁČ   Fachbereich Mathematik, Universität Rostock, Universitätsplatz 1, D–18055 Rostock, Germany;
e-mail: `peter.takac@mathematik.uni-rostock.de`

1   INTRODUCTION

We consider the following elliptic boundary value problem,

$$(1) \qquad -\Delta_p u = \lambda \psi_p(u) + f(x) \text{ in } \Omega; \qquad u = 0 \text{ on } \partial\Omega.$$

Here, $\Omega \subset \mathbb{R}^N$ is a bounded domain whose boundary $\partial\Omega$ is a connected $C^{2,\alpha}$-manifold, for some $\alpha \in (0,1)$, Δ_p denotes the p-Laplacian defined by $\Delta_p u \stackrel{\text{def}}{=} \operatorname{div}(|\nabla u|^{p-2}\nabla u)$ for $p \in (1,\infty)$, and $\psi_p(u) \stackrel{\text{def}}{=} |u|^{p-2}u$. We denote by $\nu \equiv \nu(x_0)$ the exterior unit normal to $\partial\Omega$ at $x_0 \in \partial\Omega$. Finally, $\lambda \in \mathbb{R}$ is a real parameter, and $f \in L^\infty(\Omega)$ is a given function.

We define

$$\lambda_1 \equiv \lambda_1(\Omega) \stackrel{\text{def}}{=} \inf \left\{ \int_\Omega |\nabla u|^p \, dx : u \in W_0^{1,p}(\Omega) \text{ with } \int_\Omega |u|^p \, dx = 1 \right\}.$$

79

It is well-known, cf. Anane [1, Théorème 1, p. 727], that $\lambda_1 > 0$, and λ_1 is the first eigenvalue of the negative Dirichlet p-Laplacian $-\Delta_p$ in Ω.

In this paper we assume that $-\infty < \lambda < \lambda_1$ and $f \geq 0$ in Ω. We investigate the validity of the *strong comparison principle* for positive weak solutions $u \in W_0^{1,p}(\Omega)$ to Problem (1). That is to say, let f and g be two functions from $L^\infty(\Omega)$ satisfying $0 \leq f \leq g$ and $f \not\equiv g$ in Ω. Assume that $u, v \in W_0^{1,p}(\Omega)$ are any weak solutions to the following problems, respectively,

$$(2) \qquad -\Delta_p u = \lambda \psi_p(u) + f \ \text{in} \ \Omega; \qquad u = 0 \ \text{on} \ \partial\Omega,$$

$$(3) \qquad -\Delta_p v = \lambda \psi_p(v) + g \ \text{in} \ \Omega; \qquad v = 0 \ \text{on} \ \partial\Omega.$$

PROBLEM   *Are the following inequalities valid for u and v,*

$$(4) \qquad 0 \leq u < v \ \text{in} \ \Omega \quad \text{and} \quad \frac{\partial v}{\partial \nu} < \frac{\partial u}{\partial \nu} \leq 0 \ \text{on} \ \partial\Omega?$$

For $0 \leq \lambda < \lambda_1$ and $\Omega = (-a, a) \subset \mathbb{R}$, an open interval in the space dimension one, the inequalities (4) are proved in Fleckinger et al. [4, Prop. 4.1]. In the case of the regular Laplace operator Δ (i.e. $p = 2$), the inequalities (4) follow from the classical strong maximum and boundary point principles, whenever $-\infty < \lambda < \lambda_1$. A number of uniqueness results for strong solutions of linear and semilinear boundary value problems involving the operator Δ follow (directly or indirectly) from (4), cf. Gilbarg and Trudinger [8, Chapt. 3] and Takáč [12].

In the case $p \neq 2$ and $N \geq 2$, the validity of (4) is still an open question, except for a few special cases mentioned above. In this paper we give an affarmative answer to this question for $0 \leq \lambda < \lambda_1$, see Theorem 1 below. We essentially extend the methods employed in [4] for $N = 1$. In Remark 1 we also briefly discuss the case when $0 < \lambda < \lambda_1$ and the function $f(x)$ in Problem (1) has indefinite sign. In analogy with the case $p = 2$, also for $p \neq 2$, the inequalities (4) imply uniqueness and nonexistence results for strong solutions of quasilinear boundary value problems involving the operator Δ_p, cf. Fleckinger et al. [3] or [4, Theorem 2.1].

2   THE MAIN RESULT

Let N be a positive integer, $p \in (1, \infty)$, and Ω a bounded domain in $\mathbb{R}^N$. We denote by $\overline{\Omega}$ the closure of Ω in $\mathbb{R}^N$. For a pair of Lebesgue measurable functions $f, g : \Omega \to \mathbb{R}$, we write $f \leq g$ ($f \not\equiv g$, respectively) in Ω if and only if $f(x) \leq g(x)$ ($f(x) \neq g(x)$) holds for all $x \in \Omega'$ from some set $\Omega' \subset \Omega$ of positive Lebesgue measure.

$W^{1,p}(\Omega)$ denotes the Sobolev space of all functions from $L^p(\Omega)$ whose all first-order partial derivatives (in the sense of distributions) also belong to $L^p(\Omega)$. We set $W_0^{1,p}(\Omega) = \{u \in W^{1,p}(\Omega) : u = 0 \text{ on } \partial\Omega\}$. We refer to Nečas [10, Sect. 2.4, pp. 30–32] for a definition of the trace $u|_\Omega \in W^{1-(1/p),p}(\partial\Omega)$ of a function $u \in W^{1,p}(\Omega)$.

We have the following *strong comparison principle* for the weak solutions $u, v \in W_0^{1,p}(\Omega)$ of the partial differential equations (2) and (3), respectively. This theorem is our main result in this paper.

THEOREM 1   *Let $\Omega \subset \mathbb{R}^N$ be a bounded domain whose boundary $\partial\Omega$ is a connected $C^{2,\alpha}$-manifold, for some $\alpha \in (0,1)$, $1 < p < \infty$ and $0 \leq \lambda < \lambda_1$. Let $f, g \in L^\infty(\Omega)$ satisfy $0 \leq f \leq g$ with $f \not\equiv g$ in Ω. Assume that $u, v \in W_0^{1,p}(\Omega)$ are any weak solutions of Eqs. (2) and (3). Then the strong comparison principle (4) is valid.*

For $0 \equiv f \leq g \not\equiv 0$ and $0 \equiv u \leq v \not\equiv 0$ in Ω, this result is the strong maximum principle due to Tolksdorf [13, Prop. 3.2.2, p. 801] and Vázquez [15, Theorem 5, p. 200]. The following version of the strong comparison principle, which is considerably weaker than our Theorem 1, was shown in Tolksdorf [13, Prop. 3.3.2, p. 803]: Let Ω' be a subdomain (i.e., an open connected subset) of Ω, such that $|\nabla u(x)| \geq \delta > 0$ for every $x \in \Omega'$, where δ is a constant. Then $u < v$ in Ω'. It is obvious that $x_0 \notin \overline{\Omega}'$ whenever the function u attains a local minimum or maximum at $x_0 \in \Omega$. Consequently $\Omega' \neq \Omega$.

In our *proof* of Theorem 1, we will make use of the following three results, see Propositions 2 and 3, and Lemma 4 below, respectively: (a) a *weak comparison principle* due to Tolksdorf [13, Lemma 3.1, p. 800] for $\lambda \leq 0$, and to Fleckinger et al. [5, Theorem 2] for $0 < \lambda < \lambda_1$; (b) a strong comparison principle *near the boundary* shown in Fleckinger and Takáč [6, Prop. 2, p. 448] or [7, Prop. 5.1, p. 1238]; and (c) a *regular approximation* result due to Lieberman [9, Theorem 1, p. 1203].

These three results can be stated as folows:

First, consider the Dirichlet problems

(5) $-\Delta_p u = \lambda \psi_p(u) + f$ in Ω; $u = f'$ on $\partial\Omega$,

(6) $-\Delta_p v = \lambda \psi_p(v) + g$ in Ω; $v = g'$ on $\partial\Omega$.

PROPOSITION 2   *Let Ω be a bounded domain in $\mathbb{R}^N$ with a $C^{1,\alpha}$-boundary $\partial\Omega$, for some $\alpha \in (0,1)$, $1 < p < \infty$ and $-\infty < \lambda < \lambda_1$. Assume that $f \leq g$ in $L^{p/(p-1)}(\Omega)$, $f' \leq g'$ in $W^{1-(1/p),p}(\partial\Omega)$, and $u, v \in W^{1,p}(\Omega)$ are any weak solutions of the Dirichlet problems (5) and (6), respectively.*

If also $\lambda \leq 0$, then $u \leq v$ holds almost everywhere in Ω.

If $\lambda < \lambda_1$, $0 \leq f \leq g$ in $L^\infty(\Omega)$, and $f' \equiv g' \equiv 0$ on $\partial\Omega$, then $0 \leq u \leq v$ holds almost everywhere in Ω with $u, v \in L^\infty(\Omega)$.

The $L^\infty(\Omega)$-regularity result is proved in Anane [2, Théorème A.1, p. 96].

Next we consider the open δ-neighborhood $\Omega_\delta \subset \Omega$ of the boundary $\partial\Omega$,

(7) $\Omega_\delta = \{x \in \Omega : \text{dist}(x, \partial\Omega) < \delta\}$ for $\delta > 0$ small enough.

Notice that, if $\partial\Omega$ is a compact manifold of class $C^{2,\alpha}$, then $\overline{\Omega_\delta}$ is $C^{2,\alpha}$-diffeomorphic to $\partial\Omega \times [0,1]$ with $x \mapsto (x,0)$ for all $x \in \partial\Omega$, and $\Omega \setminus \Omega_\delta$ is $C^{2,\alpha}$-diffeomorphic to $\overline{\Omega}$. Both diffeomorphisms are considered between manifolds with boundary of class $C^{2,\alpha}$.

PROPOSITION 3   *Let $\Omega \subset \mathbb{R}^N$ be a bounded domain with a $C^{2,\alpha}$-boundary $\partial\Omega$, for some $\alpha \in (0,1)$, $1 < p < \infty$ and $-\infty < \lambda < \lambda_1$. Let $f, g \in L^\infty(\Omega)$ satisfy $0 \leq f \leq g$ in Ω. Assume that $u, v \in W_0^{1,p}(\Omega)$ are any weak solutions of Eqs. (2) and (3). Then, for every $\delta > 0$ small enough and for every connected component Σ of Ω_δ, we have either $0 \leq u \equiv v$ in Σ or else*

(8) $0 \leq u < v$ *in* $\overline{\Sigma} \setminus \partial\Omega$ *and* $0 \geq \dfrac{\partial u}{\partial \nu} > \dfrac{\partial v}{\partial \nu}$ *on* $\partial\Omega \cap \overline{\Sigma}$.

Here, the solutions u and v belong to the Hölder space $C^{1,\beta}(\overline{\Omega})$, for some $\beta \in (0,1)$, see Lieberman [9, Theorem 1, p. 1203] (and Tolksdorf [14] for interior regularity) combined with the $L^\infty(\Omega)$-regularity from Anane [2, Théorème A.1, p. 96].

The proof of Proposition 3 in Fleckinger and Takáč [7, Prop. 5.1, p. 1238] uses, among other results, the following regular approximation of a weak solution $u \in W_0^{1,p}(\Omega)$ to Problem (1), which is due to Lieberman [9, Theorem 1, p. 1203]. Let us replace Problem (1) by the following more general Dirichlet problem, for any fixed $\varepsilon \in [0, 1]$,

(9)
$$\begin{cases} -\operatorname{div}((\varepsilon^2 + |\nabla u^\varepsilon|^2)^{(p-2)/2} \nabla u^\varepsilon) = \lambda \psi_p(u^\varepsilon) + f & \text{in } \Omega; \\ u^\varepsilon = g & \text{on } \partial\Omega, \end{cases}$$

where $u^\varepsilon \in W^{1,p}(\Omega)$ is a weak solution of Problem (9). We assume that λ is a constant, $-\infty < \lambda \leq 0$, $0 \leq f \in L^\infty(\Omega)$ with the norm $\|f\|_{\infty;\Omega}$ and $0 \leq g \in C^{1,\alpha}(\partial\Omega)$ with the Hölder norm $|g|_{1+\alpha;\partial\Omega}$. The Dirichlet problem (9) has a unique weak solution $u^\varepsilon \in W^{1,p}(\Omega)$ by variational methods, cf. Nečas [10, Sect. 3.4, pp. 55–59]. The uniform boundedness of u^ε in $L^\infty(\Omega)$ follows from Anane [2, Théorème A.1, p. 96].

LEMMA 4   *Assume that $u^\varepsilon \in W^{1,p}(\Omega)$ is a weak solution of Problem (9), for any fixed $\varepsilon \in [0, 1]$. Let $C' \geq 0$ be any constant such that*

$$\|u^\varepsilon\|_{\infty;\Omega} \leq C', \quad \|f\|_{\infty;\Omega} \leq C' \quad and \quad |g|_{1+\alpha;\partial\Omega} \leq C'.$$

Then there exist a constant $\beta \equiv \beta(\alpha, p, \lambda, N)$, $0 < \beta < 1$, depending solely upon α, p, λ and N, and another constant $C \equiv C(\alpha, p, \lambda, N, \Omega, C')$, $0 \leq C < \infty$, depending solely upon α, p, λ, N, Ω and C' such that $u^\varepsilon \in C^{1,\beta}(\overline{\Omega})$ with the Hölder norm

$$|u^\varepsilon|_{1+\beta;\overline{\Omega}} \leq C(\alpha, p, \lambda, N, \Omega, C').$$

Proof of Theorem 1. To begin with, the weak comparison principle (Proposition 2 above) forces $0 \leq u \leq v$ in Ω. So we may assume $\lambda = 0$. Indeed, if $0 < \lambda < \lambda_1$, then we may simply replace $\lambda \psi_p(u) + f$ by f and $\lambda \psi_p(v) + f$ by g. Moreover, we have $u, v \in C^{1,\beta}(\overline{\Omega})$, for some $\beta \in (0, 1)$, see Lieberman [9, Theorem 1, p. 1203] (and Tolksdorf [14] for interior regularity).

Next, by Proposition 3 with $\Sigma = \Omega_\delta$, precisely one of the following two mutually exclusive alternatives must be valid:

(a1) $u \equiv v$ in Ω_δ, for some $\delta > 0$;

(a2) the strong comparison principle (8) holds in Ω_δ, for some $\delta > 0$.

In the remaining part of this proof, we rule out the first alternative and show that the second one implies $u < v$ throughout Ω.

Alt. (a1): First, we claim that, for $0 < \eta < \delta$, η small enough, the divergence theorem may be applied to Eqs. (2) and (3) over the domain $\Omega'_\eta = \Omega \setminus \overline{\Omega}_\eta$. That is to say, we will show that

$$(10) \qquad -\int_{\partial\Omega'_\eta} \psi_p(\nabla u(x)) \cdot \nu(x)\, d\sigma(x) \;=\; \int_{\Omega'_\eta} f(x)\, dx,$$

$$(11) \qquad -\int_{\partial\Omega'_\eta} \psi_p(\nabla v(x)) \cdot \nu(x)\, d\sigma(x) \;=\; \int_{\Omega'_\eta} g(x)\, dx.$$

Indeed, for $0 < \varepsilon \le 1$, every weak solution $u^\varepsilon \in W_0^{1,p}(\Omega)$ of the Dirichlet problem (9), with $g \equiv 0$ on $\partial\Omega$, satisfies also $u^\varepsilon \in W^{2,q}(\Omega)$, for any $q \in (1,\infty)$, since this problem is regular, see Gilbarg and Trudinger [8, Chapt. 9]. Choose η small enough, $0 < \eta < \delta$, so that $\partial\Omega_\eta$ be C^1-diffeomorphic to $\partial\Omega$. Notice that $\partial\Omega'_\eta = \partial\Omega_\eta \subset \Omega_\delta$.

Now we may apply the divergence theorem to Eq. (9) over the domain Ω'_η, thus obtaining

$$(12) \qquad -\int_{\partial\Omega'_\eta} (\varepsilon^2 + |\nabla u^\varepsilon|^2)^{(p-2)/2} \nabla u^\varepsilon \cdot \nu(x)\, d\sigma(x) = \int_{\Omega'_\eta} f(x)\, dx.$$

Next, we apply Lemma 4 to conclude that the set of weak solutions $\{u^\varepsilon : \varepsilon \in [0,1]\}$ to Eq. (9) is bounded in the Hölder space $C^{1,\beta}(\overline{\Omega})$, and hence, relatively compact in $C^{1,\beta^*}(\overline{\Omega})$ for any fixed $\beta^* \in (0,\beta)$, by Arzelà-Ascoli's theorem. Hence, given any sequence $\{\varepsilon_n\} \subset (0,1)$, $\varepsilon_n \to 0$ as $n \to \infty$, we can extract a subsequence $\{\varepsilon_{n_k}\}$, $n_k \nearrow \infty$ as $k \to \infty$, such that $u^{\varepsilon_{n_k}} \to u^*$ in $C^{1,\beta^*}(\overline{\Omega})$ as $k \to \infty$. In particular, since the function $\psi_p(\mathbf{u}) = |\mathbf{u}|^{p-2}\mathbf{u}$ is locally α-Hölder continuous in $\mathbf{u} \in \mathbb{R}^N$, whenever $\alpha \in (0,1)$ and $\alpha \le p-1$, we obtain

$$(13) \qquad (\varepsilon_{n_k}^2 + |\nabla u^{\varepsilon_{n_k}}|^2)^{(p-2)/2} \nabla u^{\varepsilon_{n_k}} \overset{k\to\infty}{\longrightarrow} |\nabla u^*|^{p-2}\nabla u^* \quad \text{in } C(\overline{\Omega}).$$

Combining (9) with (13), we conclude that $u^* \in C^{1,\beta^*}(\overline{\Omega})$ is a weak solution of Problem (1), whence $u^* = u^0 = u \in C^{1,\beta}(\overline{\Omega})$ by uniqueness.

We deduce from the convergence $u^{\varepsilon_{n_k}} \to u^0$ in $C^{1,\beta^*}(\overline{\Omega})$ as $k \to \infty$ that Eq. (12) is valid also for $\varepsilon = 0$, which proves Eq. (10).

Since $u \equiv v$ in Ω_δ and $\partial\Omega'_\eta \subset \Omega_\delta$, the two surface integrals on the right-hand side in Eqs. (10) and (11), respectively, are equal. Therefore, we obtain

$$\int_{\Omega'_\eta} f(x)\, dx = \int_{\Omega'_\eta} g(x)\, dx.$$

Combined with $f \leq g$ in Ω, this equality forces $f \equiv g$ in Ω'_η. From $u \equiv v$ in Ω_δ, we have also $f \equiv g$ in Ω_δ. Thus, we arrive at $f \equiv g$ throughout $\Omega = \Omega'_\eta \cup \Omega_\delta$, a contradiction to our hypothesis $f \not\equiv g$ in Ω. We have ruled out Alt. (a1).

Alt. (a2): Again, choose η small enough, $0 < \eta < \delta$, so that $\partial\Omega_\eta$ be $C^{1,\alpha}$-diffeomorphic to $\partial\Omega$. Set $\Omega'_\eta = \Omega \setminus \overline{\Omega}_\eta$. We have $0 \leq u < v$ on $\partial\Omega'_\eta$, by the strong comparison principle (8) in Ω_δ. Since also $u, v \in C^{1,\beta}(\overline{\Omega})$, for some $\beta \in (0, 1)$, there exists a constant $c > 0$ such that $u + c \leq v$ on $\partial\Omega'_\eta$. Furthermore, we have

$$-\Delta_p(u + c) = -\Delta_p u = f \leq g = -\Delta_p v \quad \text{in } \Omega'_\eta.$$

Hence, we may apply the weak comparison principle (Proposition 2) to the pair $u + c$ and v in Ω'_η, thus arriving at $u + c \leq v$ throughout Ω'_η. Thus, the strong comparison principle (4) is valid in the entire domain Ω.

Theorem 1 is proved. ∎

3   DISCUSSION

If $0 < \lambda < \lambda_1$ and the functions $f \leq g$ in $L^\infty(\Omega)$ have indefinite sign, then even the weak comparison principle stated in Proposition 2 cannot be valid. This is an easy consequence of the following remark about the nonuniqueness of a weak solution to Problem (1) in the case when $0 < \lambda < \lambda_1$ and the function $f(x)$ has indefinite sign.

REMARK 1   For $0 < \lambda < \lambda_1$ and $p \neq 2$, it is possible to construct simple examples of the domain Ω and the function $f \in L^\infty(\Omega)$ (with indefinite sign) such that Problem (1) exhibits multiple solutions. In fact, for $2 < p < \infty$, this nonuniqueness was shown in del Pino, Elgueta and Manasevich [11, Eq. (5.26), p. 12]. For $1 < p < 2$, it was shown in Fleckinger et al. [5, Example 2].

Acknowledgments   The research of M. Cuesta was supported in part by the European Community Contract ERBCHRXCT940555. The research of P. Takáč was supported in part by the U.S. National Science Foundation Grant DMS-9401418 to Washington State University, Pullman, WA 99164–3113, U.S.A.

References

1. Anane, A. (1987). Simplicité et isolation de la première valeur propre du
   p-laplacien avec poids, *Comptes Rendus Acad. Sc. Paris, Série I,* **305**:
   725–728.

2. Anane, A. (1988). "Etude des valeurs propres et de la résonance pour
   l'opérateur *p*-Laplacien", *Thèse de doctorat,* Université Libre de Bruxelles, Brussels.

3. Fleckinger, J., Gossez, J.-P., Takáč, P., and de Thélin, F. (1995). Existence, nonexistence et principe de l'antimaximum pour le *p*-laplacien,
   Comptes Rendus Acad. Sc. Paris, Série I, **321**: 731–734.

4. Fleckinger, J., Gossez, J.-P., Takáč, P., and de Thélin, F. (1996). Nonexistence of solutions and an anti-maximum principle for cooperative systems with the *p*-Laplacian, *Math. Nachrichten,* submitted.

5. Fleckinger, J., Hernández, J., Takáč, P., and de Thélin, F. (1996).
   "Uniqueness and Positivity for Solutions of Equations with the *p*-
   Laplacian", Proceedings of the Conference on Reaction-Diffusion Equa-
   tions 1995, Trieste, Italy, submitted.

6. Fleckinger, J. and Takáč, P. (1994). Unicité de la solution d'un systéme
   non linéaire strictement coopératif, *Comptes Rendus Acad. Sc. Paris,
   Série I,* **319**: 447–450.

7. Fleckinger, J. and Takáč, P. (1994). Uniqueness of positive solutions
   for nonlinear cooperative systems with the *p*-Laplacian, *Indiana Univ.
   Math. J.,* **43**(4): 1227–1253.

8. Gilbarg, D. and Trudinger, N. S. (1977). "Elliptic Partial Differential Equations of Second Order", Springer-Verlag, New York–Berlin–
   Heidelberg.

9. Lieberman, G. (1988). Boundary regularity for solutions of degenerate
   elliptic equations, *Nonlinear Anal.,* **12**(11): 1203–1219.

10. Nečas, J. (1986). "Introduction to the Theory of Nonlinear Elliptic Equa-
    tions", John Wiley & Sons, New York.

11. del Pino, M., Elgueta, M., and Manasevich, R. (1989). A homotopic deformation along p of a Leray-Schauder degree result and existence for $(|u'|^{p-2}u')' + f(t,u) = 0$, $u(0) = u(T) = 0$, $p > 1$, *J. Differential Equations*, **80**(1): 1–13.

12. Takáč, P. (1996). Convergence in the part metric for discrete dynamical systems in ordered topological cones, *Nonlinear Anal.*, **26**(11): 1753–1777.

13. Tolksdorf, P. (1983). On the Dirichlet problem for quasilinear equations in domains with conical boundary points, *Comm. P.D.E.*, **8**(7): 773–817.

14. Tolksdorf, P. (1984). Regularity for a more general class of quasilinear elliptic equations, *J. Differential Equations*, **51**: 126–150.

15. Vázquez, J. L. (1984). A strong maximum principle for some quasilinear elliptic equations, *Appl. Math. Optim.*, **12**: 191–202.

Geometric Properties of Solutions to Subelliptic Equations in Nilpotent Lie Groups

DONATELLA DANIELLI and NICOLA GAROFALO, Dipartimento di Metodi e Modelli Matematici per le Scienze Applicate, Università di Padova, via Belzoni 7, 35131 Padova, Italy

1. Introduction.

The aim of this paper is to establish some geometric properties of the level sets of solutions to sub-Laplacians in stratified, nilpotent Lie groups. Such properties are reminiscent of classical ones for harmonic functions, but the fact that they hold in the complex subelliptic geometry is a perhaps unexpected and interesting phenomenon. For the sake of simplicity we will focus on the *capacitary problem*, but similar ideas apply to more general situations and also, we hope, to nonlinear equations. Capacitary estimates play an important role in the study of the asymptotic behavior (both local and at infinity) of solutions to second order partial differential equations. This is well witnessed in the famous paper of Littman, Stampacchia and Weinberger [LSW] on uniformly elliptic equations in divergence form. In [CDG2] L. Capogna and the authors established sharp capacitary estimates for metric rings in the Carnot-Carathéodory space generated by a system $X = \{X_1, \ldots, X_m\}$ of smooth vector fields in $\mathbb{R}^n$ satisfying Hörmander's finite rank condition [H]: *rank Lie*$[X_1, \ldots, X_m] \equiv n$. Following the ideas in the celebrated works of Serrin [Se1], [Se2], such estimates were used to establish the local asymptotic behavior of singular solutions to large classes of nonlinear subelliptic equations which arise in CR, or in conformal geometry. To give a flavor of the results involved and provide some background for this paper we recall a fundamental theorem obtained independently by A. Sanchez-Calle [SC] and by Nagel, Stein and Wainger [NSW]. Let $X = \{X_1, \ldots, X_m\}$ be a system of smooth vector fields satisfying Hörmander's condition and consider the sub-Laplacian $\mathcal{L} = \sum_{j=1}^{m} X_j^* X_j$ associated to X. Then, the fundamental solution $\Gamma(x, y)$ of the operator $\mathcal{L}$ satisfies the estimate

$$C \, \frac{d(x,y)^2}{|B(x,d(x,y))|} \leq \Gamma(x,y) \leq C^{-1} \, \frac{d(x,y)^2}{|B(x,d(x,y))|}. \tag{1.1}$$

Here, $d(x,y)$ is the Carnot-Carathéodory distance defined via the *sub-unit curves* generated by the system X, $B(x,r)$ is the metric ball centered at x with radius r, x runs in an arbitrarily fixed compact set K and y has to be uniformly close to x. The positive constant C only depends on K and X. Consider now a quasi-linear equation of the type

$$\sum_{j=1}^{m} X_j^* A_j(x,u,X_1u,\ldots,X_mu) = B(x,u,X_1u,\ldots,X_mu),$$

where the structural assumptions on the functions $A = (A_1, \ldots, A_m)$ and B are tailored on the *subelliptic p-Laplacian*

$$\mathcal{L}_p(u) = \sum_{j=1}^{m} X_j^*(|Xu|^{p-2} X_j) = 0, \quad 1 < p < \infty. \tag{1.2}$$

Then, it was proved in [CDG2] that the fundamental solution $\Gamma_p(x,y)$ satisfies the estimate

$$C\left(\frac{d(x,y)^p}{|B(x,d(x,y))|}\right)^{1/(p-1)} \le \Gamma_p(x,y) \le C^{-1}\left(\frac{d(x,y)^p}{|B(x,d(x,y))|}\right)^{1/(p-1)}. \tag{1.3}$$

It is clear that when $p=2$, and the equation is linear and with a zero right-hand side B, then (1.3) gives back (1.1). Estimate (1.3) was conjectured on the basis of some remarkable explicit fundamental solutions for groups of *Heisenberg type*, or *H-type* groups, found in [CDG2]. Such explicit fundamental solutions constitute the background motivation of the present paper.

To make this point precise, we introduce the type of questions we are interested in. Consider a stratified, nilpotent Lie group of step r, G, and let $\mathfrak{g} = V_1 \oplus \ldots \oplus V_r$ be a stratification of its Lie algebra $\mathfrak{g}$. If $X = \{X_1, \ldots, X_m\}$ denotes a basis of the first layer V_1 of the stratification, let $\mathcal{L}_p$ be the *p-sub-Laplacian* (1.2) generated by X. Let $\Omega \subset G$ be a smooth domain containing the identity $e \in G$ and denote by u the *p-capacitary potential* of Ω, i.e., u is the solution to the problem

$$\begin{cases} \mathcal{L}_p(u) = 0 & in \ G \setminus \overline{\Omega} \\ u|_{\partial\Omega} = 1, & \lim_{d(x,e)\to\infty} u(x) = 0. \end{cases} \tag{1.4}$$

In (1.4) we have denoted by $d(x,y)$ the Carnot-Carathéodory distance on G generated by the system X. We recall that the *homogeneous dimension* of G is given by the number $Q = \sum_{i=1}^{r} i(\dim V_i)$. In the special case in which G is a group of Heisenberg type, and $\rho = \rho(x)$ denotes Kaplan's gauge function on G (see [K] and (2.8) below), it was proved in [CDG2] that when Ω is a gauge ball of radius R, i.e., $\Omega = \{x \in G \mid \rho(x) < R\}$, then the *p*-capacitary potential of Ω is given by

$$u(x) = \begin{cases} \left(\dfrac{R}{\rho(x)}\right)^{(Q-p)/(p-1)} & when \ p \ne Q, \\[3mm] \log\left(\dfrac{R}{\rho(x)}\right) & when \ p = Q. \end{cases} \tag{1.5}$$

Now the gauge balls enjoy a special geometric property: They are *starlike* with respect to the group identity e and to the anisotropic dilations of the group. The notion of starlikeness will be made precise below. We have thus come to one of the questions of interest for us. By analyzing the special situation of groups of Heisenberg type we were led to formulate the following

Conjecture: *Let G be a stratified, nilpotent Lie group and $\Omega \subset G$ be a sufficiently smooth domain. If Ω is starlike with respect to the identity $e \in G$, then every level set of the p-capacitary potential of Ω is starlike with respect to e.*

In this paper we will prove the conjecture true when **G** is a group of Heisenberg type (see section two for the relevant definition) and $p=2$.

Theorem 1.1. *Let* **G** *be a group of Heisenberg type. Let* $\Omega \subset \mathbf{G}$ *be a* C^2 *domain which is starlike with respect to the identity* $e \in \mathbf{G}$, *and let u be a solution to* (1.4) *with* $p=2$. *Then every level set* $E_t = \{x \in \mathbf{G} \setminus \overline{\Omega} \mid u(x) > t\}$, $0<t<1$, *is strictly starlike.*

The case of a general stratified, nilpotent Lie group is technically more involved and will be the object of a forthcoming article [DG]. The nonlinear case $p \neq 2$ is much harder and still open. We plan to come back to it in a future study. Among the stratified, nilpotent Lie groups the Heisenberg group $\mathbb{H}^n$ occupies a special position because of its ubiquitous role in CR geometry, harmonic analysis and pde's. When $\mathbf{G} = \mathbb{H}^n$ one can prove that the solution to (1.4) possesses additional symmetry properties provided that so does the ground domain Ω. We emphasize that a positive answer to the above conjecture constitutes only a first step toward the general problem of understanding completely the symmetry properties of harmonic functions in the Heisenberg group or, in more general stratified, nilpotent Lie groups. This question is very hard since complex geometry is involved and the similarities with the Euclidean setting most of the time fail miserably. For instance, in connection with the isoperimetric inequalities established in [CDG3], [GN], it would be extremely important to characterize the extremal sets for the subelliptic capacity, i.e., sets of least capacity among all those having fixed volume.

2. Background material.

In this section we collect some known background material. Let **G** be a stratified, nilpotent Lie group of step r with Lie algebra $\mathfrak{g} = V_1 \oplus ... \oplus V_r$. We assume that $[V_1, V_j] \subset V_{j+1}$, $j=1,...,r-1$, $[V_1, V_r] = \{0\}$, and that, furthermore, V_1 generates $\mathfrak{g}$ as an algebra. A natural family of dilations on $\mathfrak{g}$ is given as follows: For $\lambda > 0$ and $X \in \mathfrak{g}$, with $X = \sum_{j=1}^{r} X_j$, $X_j \in V_j$, let

$$\Delta_\lambda(X) = \lambda X_1 + \lambda^2 X_2 + ... + \lambda^r X_r. \qquad (2.1)$$

Since exp: $\mathfrak{g} \to \mathbf{G}$ is a diffeomorphism, a family of dilations on **G** is given by

$$\delta_\lambda(x) = \exp \circ \Delta_\lambda \circ \exp^{-1}(x). \qquad (2.2)$$

We recall that $\{\Delta_\lambda\}_{\lambda>0}$ and $\{\delta_\lambda\}_{\lambda>0}$ are respectively algebra and group automorphisms. Identifying the elements $X \in \mathfrak{g}$ with the corresponding left-invariant vector fields on **G**, one easily recognizes that $X \in V_j$ if and only if X is homogeneous of degree j with respect to (2.1), i.e., for any $f \in C^\infty(\mathbf{G})$ one has

$$X(f \circ \delta_\lambda) = \lambda^j \delta_\lambda \circ (Xf). \qquad (2.3)$$

Let now ∥ ∥ indicate a Euclidean norm on $\mathfrak{g}$ with respect to which the $\mathbb{V}_j$'s are mutually orthogonal and define for $X = \sum_{j=1}^{r} X_j$

$$|X|_{\mathfrak{g}} = \left(\sum_{j=1}^{r} \|X_j\|^{\frac{2r!}{j}} \right)^{\frac{1}{2r!}} .$$

It is clear that $|\ |_{\mathfrak{g}}$ is homogeneous of degree one with respect to (2.1), i.e., $|\Delta_\lambda X|_{\mathfrak{g}} = \lambda |X|_{\mathfrak{g}}$ for $X \in \mathfrak{g}$ and $\lambda > 0$. A homogeneous norm, or *gauge*, on $\mathbf{G}$ is given by $|x| = |X|_{\mathfrak{g}}$, if $x = \exp(X)$. We observe that

$$|\delta_\lambda(x)| = |\delta_\lambda(\exp X)| = |\exp \Delta_\lambda(X)| = |\Delta_\lambda(X)|_{\mathfrak{g}} = \lambda |X|_{\mathfrak{g}} = \lambda |x|,$$

so that $|\ |$ is homogeneous of degree one with respect to (2.2).

Since $\mathbb{V}_1$ generates $\mathfrak{g}$ as an algebra, if we choose a basis $X = \{X_1, \ldots, X_m\}$ of $\mathbb{V}_1$, one sees that X satisfies Hörmander's finite rank condition. Denote by $d(x,y)$ the Carnot-Carathéodory distance associated to X, see, e.g., [NSW], [VSC], [Gr], [G]. The left-invariance of $d(x,y)$ gives $d(gx,gy) = d(x,y)$ for every x, y, $g \in \mathbf{G}$. We now consider the sub-Laplacian $\mathcal{L} = \mathcal{L}_2$ associated to X obtained from (1.2) when $p=2$ (we note, here, that in any stratified, nilpotent Lie group one has $X_j^* = -X_j$). It was proved by Folland [F2] that the fundamental solution $\Gamma(x) = \Gamma(x,e)$ of $\mathcal{L}$, with singularity at the identity $e \in \mathbf{G}$, is a distribution in $L_{loc}^1(\mathbf{G})$ which is homogeneous of degree $2-Q$ with respect to the dilations (2.2). Here, Q is the homogeneous dimension of the group defined in section one. Clearly, by Hörmander's theorem we have $\Gamma \in C^\infty(\mathbf{G} \setminus \{e\})$. By left-invariance $\Gamma(x,y) = \Gamma(x^{-1}y)$. These observations and the compactness of the gauge sphere $S = \{x \in \mathbf{G} \mid |x| = 1\}$, see [F2], allow to conclude that there exist $C_1, C_2 > 0$ such that for all $x, y \in \mathbf{G}$

$$\frac{C_1}{|x^{-1}y|^{Q-2}} \le \Gamma(x,y) \le \frac{C_2}{|x^{-1}y|^{Q-2}} . \tag{2.4}$$

The following well-known self-similarity property of the Carnot-Carathéodory distance is often useful, see, e.g., [G] for a proof: For any $x, y \in \mathbf{G}$ and $\lambda > 0$ one has

$$d(\delta_\lambda(x), \delta_\lambda(y)) = \lambda d(x,y). \tag{2.5}$$

It is clear that (2.4), (2.5) give for all $x, y \in \mathbf{G}$

$$\frac{C_1}{d(x,y)^{Q-2}} \le \Gamma(x,y) \le \frac{C_2}{d(x,y)^{Q-2}} . \tag{2.6}$$

Since we will be primarily working with a subclass of that of all stratified nilpotent Lie groups we next introduce this class. Groups of Heisenberg type (also known as *Kaplan's*, or *H*-groups) were invented by Kaplan in [K] in connection with hypoellipticity questions. They constitute a generalization of the Heisenberg group $\mathbb{H}^n$. There are infinitely many isomorphism classes of such groups which include the nilpotent component in the Iwasawa decomposition of simple groups of rank one. The best way to describe a group of *H*-type is through its Lie algebra $\mathfrak{g}$. One assumes that a positive definite inner product $< , >$ is given on $\mathfrak{g}$ and an orthogonal decomposition $\mathfrak{g} =

$V_1 \oplus V_2$ (stratification of step 2). Here, V_2 is the center of the group and for every unit vector $t \in V_2$ the mapping $J_t : V_1 \to V_1$, defined by

$$< J_t(z), z' > = < t, [z, z'] >,$$

is assumed orthogonal. It is important to note that for every $z \in V_1$, and $t \in V_2$ one has

$$< J_t(z), z > = 0, \qquad\qquad < J_t(z), J_t(z) > = |t|^2 |z|^2. \qquad (2.7)$$

For every positive integer n there exists a group of H-type having center of dimension n. When $n=1$ one obtains the Heisenberg groups. Next, we choose an orthogonal basis $X = \{X_1, \ldots, X_m\}$ of V_1, and we assume that the Riemannian metric on $\mathbf{G}$ is such that $< X_i, X_j > = \delta_{ij}$. Using the exponential map we can define two analytic diffeomorphisms $z: \mathbf{G} \to V_1$ and $t: \mathbf{G} \to V_2$ by $x = \exp[z(x) + t(x)]$. Following [K] we introduce the *gauge*

$$\rho(x) = \left[|z(x)|^4 + 16|t(x)|^2 \right]^{\frac{1}{4}}. \qquad (2.8)$$

The remarkable fact about H-type groups is that, in contrast with more general groups for which one only has an estimate such as (2.4), the fundamental solution of the sub-Laplacian $\mathcal{L} = \sum_{j=1}^{m} X_j^* X_j$ is exactly given by a power of the gauge. Precisely, let $m = \dim V_1$, $k = \dim V_2$, so that the dimension of $\mathbf{G}$ is $N = m + k$. The homogeneous dimension of $\mathbf{G}$ is $Q = m + 2k$. Then, the fundamental solution of $\mathcal{L}$ is given by

$$\Gamma(x, y) = C_Q \rho(x^{-1}y)^{2-Q}. \qquad (2.9)$$

This result was generalized to quasilinear equations such as (1.2) in [CDG2] (cfr. also (1.5)). Equation (2.9) extends to groups of H-type a basic discovery made by Folland [F1] for the Heisenberg group $\mathbb{H}^n$.

For a group of H-type $\mathbf{G}$ the dilations (2.2) are given by $\delta_\lambda(x) = \exp[\lambda z(x) + \lambda^2 t(x)]$. Let us denote by Z the infinitesimal generator of the one-parameter group $\{\delta_\lambda\}_{\lambda > 0}$. Then, a function u is homogeneous of degree k with respect to $\{\delta_\lambda\}_{\lambda > 0}$ if and only if $Zu(x) = ku(x)$ for every $x \in \mathbf{G}$. We will need the following

Lemma 2.1. *For any* $X_j \in X = \{X_1, \ldots, X_m\}$ *one has* $[X_j, Z] = X_j$.

Proof. Let $\theta(s, x)$ be the one-parameter group action associated to Z, i.e., the solution to the system

$$\begin{cases} \frac{d}{ds} \theta(s, x) = Z(\theta(s, x)) \\ \theta(0, x) = x. \end{cases} \qquad (2.10)$$

If $u \in C_0^\infty(\mathbf{G})$ from the definition of $\{\delta_\lambda\}_{\lambda > 0}$ and the fact that each X_j is homogeneous of degree one (see (2.3)), we obtain

$$X_j(u(\theta(s, \cdot)))(x) = e^s \theta_s(X_j u)(x) \qquad (2.11)$$

where for a given function f we have denoted with $\theta_s(f)$ the action defined by $\theta_s(f)(x) = f(\theta(s,x))$. Differentiating (2.11) with respect to s, and using (2.10), gives

$$\tfrac{d}{ds} X_j(u(\theta(s,\cdot)))(x)|_{s=0} = X_j u(x) + Z(X_j u)(x). \qquad (2.12)$$

On the other hand we can write

$$u(\theta(s,x)) = u(x) + \int_0^s \tfrac{d}{d\tau} u(\theta(\tau,x)) d\tau = u(x) + \int_0^s Zu(\theta(\tau,x)) d\tau = u(x) + sg(s,x)$$

where $g(s,\cdot)$ is a smooth function such that $g(0,x) = Zu(x)$. Therefore,

$$X_j(u(\theta(s,\cdot)))(x) = X_j u(x) + sX_j(g(s,\cdot))(x),$$

and we obtain

$$\tfrac{d}{ds} X_j(u(\theta(s,\cdot)))(x)|_{s=0} = X_j(Zu)(x). \qquad (2.13)$$

From (2.12), (2.13) we reach the conclusion.

                                                                                                              o

Finally, we will need the following mean value theorem for homogeneous functions. Its proof can be found in [F2, Proposition 1.15].

Lemma 2.2. *Let* **G** *be a stratified, nilpotent Lie group and* u *a function which is* $C^2(\mathbf{G} \setminus \{e\})$ *and homogeneous of degree* $k \in \mathbb{R}$. *Then, there exists* $C > 0$ *such that*

$$|u(xy) - u(x)| \le C|y||x|^{k-1} \qquad\qquad \text{whenever} \quad |y| \le \tfrac{1}{2}|x|.$$

3. Proofs.

In this section we prove the main result of the paper. We begin by recalling a basic pointwise a priori estimate which is a special case of Theorem 2.1 in [DG]. Let **G** be a stratified, nilpotent Lie group with homogeneous dimension Q and gauge $|\ |$. Denote $F(t) = t^{Y_{(Q-2)}}$. We choose a nonnegative function $f \in C_0^\infty(1,2)$, such that $\int_{\mathbb{R}} f(s)ds = 1$, and let $f_R(s) = R^{-1} f(R^{-1}s)$. Following [CDG4] we introduce the kernel

$$K_R(x,y) = f_R(F(\Gamma(x,y)^{-1}) \frac{|X_y\Gamma(x,y)|^2}{\Gamma(x,y)^2} F'(\Gamma(x,y)^{-1}),$$

where $\Gamma(x,y)$ is the fundamental solution of $\mathcal{L} = \sum_{j=1}^m X_j^* X_j$. It is important to observe that for any fixed $x \in \mathbf{G}$, we have $supp\, K_R(x,\cdot) \subset B(x,2R) \setminus B(x,R)$. Here, we have let

$$B(x,R) = \{y \in \mathbf{G}|\ |x^{-1}y| < R\}.$$

Given a function $u \in L^1_{loc}(\mathbb{R}^n)$ we define a family of mollifiers of u as follows

$$J_R u(x) = \int_G u(y) K_R(x,y) dy, \qquad R>0,$$

where dy denotes the bi-invariant Haar measure on **G** obtained by pushing forward the Lebesgue measure on $\mathfrak{g}$ through the exponential mapping. Our main a priori estimate is contained in the following

Theorem 3.1. *There exists a constant* C>0, *depending only on* **G** *and on the system* X, *such that for any* $u \in L^1_{loc}(\mathbf{G})$, $x \in \mathbf{G}, R > 0$ *one has*

$$|X J_R u(x)| \le \frac{C}{R} \frac{1}{|B(x,2R)|} \int_{B(x,R)} |u(y)| dy.$$

Hereafter, the notation Xu indicates the horizontal gradient $(X_1 u, \ldots, X_m u)$ of a function u with respect to X, whereas $|Xu| = \left(\sum_{j=1}^m (X_j u)^2 \right)^{1/2}$. For the proof of Theorem 3.1 we refer the reader to [DG], where the same result is established for a general system of Hörmander type. A remarkable consequence of Theorem 3.1 is the following

Theorem 3.2. *Let* u *be a harmonic function in* $B(x,4R) \subset \mathbf{G}$, *i.e., a solution to* $\mathcal{L} u = \sum_{j=1}^m X_j^* X_j u = 0$. *Then, there exists* C>0, *depending only on* **G** *and* X, *such that*

$$|Xu(x)| \le \frac{C}{R} \frac{1}{|B(x,2R)|} \int_{B(x,2R)} |u(y)| dy.$$

We emphasize that Theorem 3.2 cannot be obtained similarly to its classical ancestor, where one uses the mean value formula, coupled with the trivial observation that any derivative of a harmonic function is still harmonic. In the present non abelian context if u is harmonic, then it is not true that so is $X_j u$, $j = 1, \ldots, m$. We note in passing the following important corollary of Theorem 3.2.

Corollary 3.1. (Liouville theorem) *Let* u *be harmonic on* **G** *and having sub-linear growth at infinity, i.e., there exist constants* $C > 0, 0 \le \varepsilon < 1$, *such that* $|u(x)| \le C(1 + d(x,e)^\varepsilon)$. *Then,* u *is constant.*

We now turn our attention to the capacitary problem. In what follows, **G** is a group of Heisenberg type with gauge ρ. Given an open set $\Omega \subset \mathbf{G}$ we denote by $\mathcal{L}^{1,2}(\Omega)$ the Folland-Stein Sobolev space of those functions $u \in L^2(\Omega)$ such that $X_j u \in L^2(\Omega), j = 1, \ldots, m$. Endowed with the norm $\|u\|_{\mathcal{L}^{1,2}(\Omega)} = \|u\|_{L^2(\Omega)} + \|Xu\|_{L^2(\Omega)}$, $\mathcal{L}^{1,2}(\Omega)$ is a Banach space. One also defines $\mathcal{L}^{1,2}_0(\Omega) = \overline{C_0^\infty(\Omega)}^{\mathcal{L}^{1,2}(\Omega)}$, while the local space $\mathcal{L}^{1,2}_{loc}(\Omega)$ is defined as for the classical Sobolev spaces.

Given a connected, bounded open set $\Omega \subset \mathbf{G}$, with $e \in \Omega$, by a solution to the *capacitary problem* for Ω we mean a function $u \in \mathcal{L}^{1,2}_{loc}(\mathbf{G} \setminus \overline{\Omega})$ such that $u - 1 \in \mathcal{L}^{1,2}_0(\mathbf{G} \setminus \overline{\Omega})$,

$\lim_{\rho(x)\to\infty} u(x) = 0$, and u is harmonic in the weak sense in $G\backslash\overline{\Omega}$, i.e., for any function $\varphi \in$ $\mathcal{L}_0^{1,2}(G\backslash\overline{\Omega})$, with $\text{supp}\,\varphi$ compact, one has

$$\int_{G\backslash\overline{\Omega}} < Xu, X\varphi > dx \; = \; 0.$$

We will indicate the capacitary problem with the following notation

$$\begin{cases} \mathcal{L}u = \sum_{j=1}^m X_j^* X_j u = 0 & in \; G\backslash\overline{\Omega}, \\ u|_{\partial\Omega} = 1, & \lim_{\rho(x)\to\infty} u(x) = 0. \end{cases} \tag{3.1}$$

In the sequel we will not concern ourselves with the question of the existence of a unique solution to (3.1). This can be proved following classical ideas using, e.g., the results in [CDG1], [D]. The solution to (3.1) will be called the (subelliptic) *capacitary potential* of Ω. Hereafter, we assume that Ω be sufficiently smooth, let us say C^2. We stress (see [CG]) that from the point of view of the sub-Riemannian geometry of G, Euclidean smoothness does not necessarily mean favorable geometric properties of the domain. For instance, there exist $C^{1,\alpha}$ domains, in sufficiently high Heisenberg groups $\mathbb{H}^n$, which are not regular for the Dirichlet problem, see [HaH]. We need to introduce a relevant definition.

Definition 3.1. Given an open set $\Omega \subset G$ we say that Ω admits an interior corkscrew at $x_0 \in \partial\Omega$ if for some $K>0$ and $R_0 > 0$, and any $0 < r < R_0$, one can find $A_r(x_0) \in \Omega$ such that

$$\frac{r}{K} \leq d(A_r(x_0), x_0) \leq r, \qquad dist(A_r(x_0), \partial\Omega) \geq \frac{r}{K}.$$

If the same K and R_0 work for every $x_0 \in \partial\Omega$, then we say that Ω has the *uniform interior corkscrew condition*. Finally, Ω is said to satisfy the *uniform corkscrew condition* if both Ω and $\overline{\Omega}^c$ fulfill the uniform interior corkscrew condition.

It has been proved in [CG] that in a stratified, nilpotent Lie group of step 2, and therefore in particular in any H-type group, any bounded $C^{1,1}$ domain satisfies the uniform corkscrew condition. In particular, thanks to the results in [C] or [D], the existence on an interior corkscrew implies that the weak solution to (3.1) be continuous up to the boundary. In fact, there exists $0 < \alpha < 1$ such that $u \in \Gamma^\alpha(U\backslash\Omega)$, where U is a bounded open set containing $\overline{\Omega}$. It is clear that, because of Hörmander's theorem, $u \in C^\infty(G\backslash\overline{\Omega})$. We next introduce the notion of subelliptic capacity of a condenser, see [D]. Let K be a compact set and $\Omega \subset G$ be an open set such that $K \subset \Omega$. The couple (K, Ω) is called a condenser. Consider the set $\mathcal{F}(K,\Omega) = \{u \in C_0^\infty(\Omega) | u \geq 0, u \geq 1 \text{ on } K\}$. The capacity of the condenser is defined as follows.

$$Cap(K,\Omega) \; = \; \inf_{u \in \mathcal{F}(K,\Omega)} \int_\Omega |Xu|^2 dx \; .$$

The capacity of an open set Ω is defined as the limit $Cap(\Omega) = \lim_{R \to \infty} Cap(\overline{\Omega}, B(e, R))$. One can show that if u is the unique solution to (3.1), then

$$Cap(\Omega) = \int_{G \setminus \Omega} |Xu|^2 dx. \tag{3.2}$$

Our next task is to prove a basic representation formula for the capacitary potential of Ω. Before we do so we establish a simple, but useful, lemma.

Lemma 3.1. *Let u be the solution to (3.1). There exist constants $\alpha, \beta > 0$, depending on Ω, such that for every $x \in G \setminus \overline{\Omega}$ one has*

$$\frac{\alpha}{\rho(x)^{Q-2}} \leq u(x) \leq \frac{\beta}{\rho(x)^{Q-2}} .$$

Proof. Thanks to (2.9) the function $x \to \rho(x)^{2-Q}$ is harmonic in $G \setminus \{e\}$. The result then follows from an elementary application of Bony's maximum principle [B] since, by hypothesis, u decays at infinity.

Theorem 3.3. *Let u be the solution to (3.1). Then the following asymptotic representation holds*

$$u(x) = Cap(\Omega) \, \Gamma(x, e)(1 + \omega(x))$$

where $|\omega(x)| \leq \dfrac{C}{\rho(x)}$ *as $\rho(x) \to \infty$. Furthermore, if Z denotes the infinitesimal generator of the group dilations, then we also have*

$$Zu(x) = (2 - Q) Cap(\Omega) \Gamma(x, e) \, (1 + \overline{\omega}(x))$$

with $|\overline{\omega}(x)| \leq \dfrac{C}{\rho(x)}$ *as $\rho(x) \to \infty$.*

Proof. To prove the theorem we fix a point $x \in G \setminus \overline{\Omega}$ and consider $R > 0$ so large that $x \in B(e, R) = B_R$. Next, we choose $\varepsilon > 0$ sufficiently small so that $\overline{B(x, \varepsilon)} \subset B_R \setminus \overline{\Omega}$. Letting $v(y) = \Gamma(y, x)$, where $\Gamma(x, y) = \Gamma(y, x)$ is as in (2.9), an application of the divergence theorem gives

$$0 = \int_{(B_R \setminus \overline{\Omega}) \setminus B(x, \varepsilon)} (v\mathcal{L}u - u\mathcal{L}v) dy = -\sum_{j=1}^{m} \int_{(B_R \setminus \overline{\Omega}) \setminus B(x, \varepsilon)} div_G \Big[(vX_j u - uX_j v) X_j \Big] dy \tag{3.3}$$

$$= -\int_{\partial B_R} v \sum_{j=1}^{m} X_j u < X_j, \eta > dH_{N-1} + \int_{\partial B(x, \varepsilon)} v \sum_{j=1}^{m} X_j u < X_j, \eta > dH_{N-1}$$

$$+ \int_{\partial\Omega} v \sum_{j=1}^{m} X_j u < X_j, \eta > dH_{N-1} \ + \ \int_{\partial B_R} u \sum_{j=1}^{m} X_j v < X_j, \eta > dH_{N-1}$$

$$- \int_{\partial B(x,\varepsilon)} u \sum_{j=1}^{m} X_j v < X_j, \eta > dH_{N-1} - \int_{\partial\Omega} \sum_{j=1}^{m} X_j v < X_j, \eta > dH_{N-1} \ .$$

In (3.3) we have used the fact that $\mathcal{L} v(y) = 0$ for $y \in \mathbf{G} \setminus \{x\}$. Furthermore, in any stratified, nilpotent Lie group one has $X_j^* = -X_j$, and consequently $div_{\mathbf{G}} X_j = 0$. The symbol η denotes the Riemannian unit normal pointing outward $B_R, \overline{\Omega}$, and $B(x,\varepsilon)$, whereas we have indicated with dH_{N-1} the $(N\text{-}1)$-dimensional Riemannian measure (here, N indicates the topological dimension of the group $\mathbf{G}$). One of the above integrals, namely $\int_{\partial\Omega} v \sum_{j=1}^{m} X_j u < X_j, \eta > dH_{N-1}$, requires an explanation . As we pointed out above, the function u is in a Hölder class near $\partial\Omega$, i.e., $u \in \Gamma^{\alpha}(U \setminus \Omega)$, however this does not guarantee the existence of the above integral. The latter must therefore be interpreted as a limit of integrals performed on the boundaries of a sequence of smooth domains which shrink monotonically to Ω. This being said, we return to (3.3). A computation using (2.9) shows that

$$\int_{\partial B(x,\varepsilon)} u \sum_{j=1}^{m} X_j v < X_j, \eta > dH_{N-1} \ = \ - \frac{C_Q(Q-2)}{\varepsilon^{Q-1}} \int_{\partial B(x,\varepsilon)} u \frac{|X\rho|^2}{|\nabla_{\mathbf{G}}\rho|} dH_{Q-1} \ ,$$

where $|\nabla_{\mathbf{G}}\rho|$ denotes the length of the Riemannian gradient of $y \rightarrow \rho(x^{-1}y)$. Taking into account that the constant C_Q (see also [CDG2]) is chosen so that for any $\varepsilon > 0$ and $x \in \mathbf{G}$

$$\frac{C_Q(Q-2)}{\varepsilon^{Q-1}} \int_{\partial B(x,\varepsilon)} \frac{|X\rho|^2}{|\nabla_{\mathbf{G}}\rho|} dH_{Q-1} = 1,$$

we infer

$$\lim_{\varepsilon \to 0} \int_{\partial B(x,\varepsilon)} u \sum_{j=1}^{m} X_j v < X_j, \eta > dH_{N-1} \ = - u(x). \tag{3.4}$$

Next, we analyze the sixth integral in the right-hand side of (3.3). By the harmonicity of v in $\mathbf{G} \setminus \{x\}$ we infer

$$\int_{\partial\Omega} \sum_{j=1}^{m} X_j v < X_j, \eta > dH_{N-1} = - \int_{\Omega} \mathcal{L} v \, dy = 0. \tag{3.5}$$

If we pass to the limit as $\varepsilon \to 0$ in (3.3), and we use (3.4), (3.5), we obtain

$$u(x) = \int_{\partial\Omega} \Gamma(x,y) \sum_{j=1}^{m} X_j u(y) < X_j(y), \eta(y) > dH_{N-1} + \tag{3.6}$$

$$- \int_{\partial B_R} v \sum_{j=1}^{m} X_j u < X_j, \eta > dH_{N-1} + \int_{\partial B_R} u \sum_{j=1}^{m} X_j v < X_j, \eta > dH_{N-1} \ .$$

Next, we estimate the two integrals on ∂B_R in the right hand side of (3.6). In the analysis of the former one we use the delicate pointwise control of Xu provided by Theorem 3.2. We consider the pseudo-distance $d^*(x,y) = \rho(x^{-1}y)$ generated by the gauge. We will need the following fact: There exists a constant $C>0$, depending only on **G**, such that for every $x \in \mathbf{G} \setminus \overline{\Omega}$ for which $\rho(x) \geq 1$, and every $R = R(x) = 2Cd^*(x,e) = 2C\rho(x)$, we have

$$\sup_{y \in \partial B_R} \frac{1}{d^*(y,x)} \leq \frac{2C}{d^*(y,e)}. \tag{3.7}$$

To prove (3.7) we observe that there exists $C>0$ such that
$$d^*(x,y) \leq C(d^*(x,z) + d^*(z,y)).$$

Therefore, for $y \in \partial B_R$ we have: $d^*(y,e) \leq C(d^*(y,x) + d^*(x,e))$. It is then enough to show that for any $y \in \partial B_R$ one has: $C d^*(x,e) \leq \frac{1}{2}d^*(y,e)$. Since by assumption

$d^*(y,e) = R = 2Cd^*(x,e) \geq 2Cd^*(x,e)$, the latter inequality is true, hence so is (3.7). If we

take into account that on ∂B_R the outward unit normal is given by $\eta = \dfrac{\nabla_G \rho}{|\nabla_G \rho|}$, we now have

$$\left| \int_{\partial B_R} v \sum_{j=1}^{m} X_j u < X_j, \eta > dH_{N-1} \right| \leq \int_{\partial B_R} \frac{C_Q}{d^*(y,x)^{Q-2}} |Xu(y)| \frac{|X\rho|}{|\nabla_G \rho|} dH_{N-1}. \tag{3.8}$$

By Bony's maximum principle [B], the solution to (3.1) is nonnegative. By the Harnack inequality, see [B] or also [CDG1], there exists a constant $C>0$ such that for every $y \in \mathbf{G} \setminus \overline{\Omega}$ for which $B\left(y, 2^{-1}d^*(y,e)\right) \subset \mathbf{G} \setminus \overline{\Omega}$, one has

$$\max_{B\left(y, \frac{d^*(y,e)}{8}\right)} u \leq C \min_{B\left(y, \frac{d^*(y,e)}{8}\right)} u \leq C\, u(y). \tag{3.9}$$

Suppose now that $\rho(x) > 100 diam(\Omega)$. Then, for $y \in \partial B_R$ we have $B(y, 2^{-1}d^*(x,e)) \subset \mathbf{G} \setminus \overline{\Omega}$. Applying Theorem 3.2 in combination with (3.9) to such a ball we infer

$$|Xu(y)| \leq \frac{C}{d^*(y,e)} \frac{1}{|B(y,d^*(y,e))|} \int_{B(y,\frac{d^*(y,e)}{8})} |u| dz \leq \frac{C}{d^*(y,e)} u(y). \tag{3.10}$$

This important estimate is now used in (3.8), together with (3.7), to obtain

$$\left| \int_{\partial B_R} v \sum_{j=1}^{m} X_j u < X_j, \eta > dH_{N-1} \right| \leq \frac{C}{R^{Q-2}} \int_{\partial B_R} \frac{1}{d^*(y,e)^{Q-1}} \frac{|X\rho|}{|\nabla_G \rho|} dH_{N-1}$$

$$\leq (by \ Lemma \ 3.1) \ \frac{C}{R^{2Q-3}} \int_{\partial B_R} \frac{|X\rho|}{|\nabla_G \rho|} dH_{N-1}.$$

We now observe that the function $|X\rho| = \dfrac{|z|}{\rho}$ is homogeneous of degree zero with respect to the group dilations. Therefore an easy calculation gives

$$\int_{B_R} |X\rho| dx = \omega_Q R^Q$$

for some positive constant ω_Q depending only on **G**. On the other hand we have from the coarea formula [Ch]

$$\int_{B_R} |X\rho| dx = \int_0^R \int_{\partial B_s} \frac{|X\rho|}{|\nabla_G \rho|} dH_{N-1} \, ds.$$

This formula and the previous one give

$$\int_{\partial B_R} \frac{|X\rho|}{|\nabla_G \rho|} dH_{N-1} = Q\omega_Q R^{Q-1}. \tag{3.11}$$

Thanks to (3.11), and recalling that $R = R(x) = 2Cd^*(x,e)$, we conclude

$$\left| \int_{\partial B_R} v \sum_{j=1}^{m} X_j u < X_j, \eta > dH_{N-1} \right| \leq \frac{C}{R^{Q-2}} = \frac{C}{d^*(x,e)^{Q-1}} \tag{3.12}$$

$$= \Gamma(x,e)\left(1 + 0\left(\frac{1}{d^*(x,e)}\right)\right).$$

In the right hand side of (3.12) we have denoted with $0\left(\dfrac{1}{d^*(x,e)}\right)$ a function which in absolute value is bounded by $d^*(x,e)^{-1}$ as $\rho(x) \to \infty$.

We now estimate the third integral in the right hand side of (3.6). Lemma 3.1 allows to infer

$$\left| \int_{\partial B_R} u \sum_{j=1}^{m} X_j v < X_j, \eta > dH_{N-1} \right| \leq \frac{C}{R^{Q-2}} \int_{\partial B_R} |Xv(y)| \frac{|X\rho|}{|\nabla_G \rho|} dH_{N-1}.$$

Using the estimate

$$|Xv(y)| = |X\Gamma(x,y)| \leq \frac{C}{d^*(x,y)^{Q-1}}$$

we obtain

$$\left| \int_{\partial B_R} u \sum_{j=1}^{m} X_j v < X_j, \eta > dH_{N-1} \right| \leq \frac{C}{R^{Q-2}} \int_{\partial B_R} \frac{1}{d^*(x,y)^{Q-1}} \frac{|X\rho|}{|\nabla_G \rho|} dH_{N-1} .$$

From (3.7) and (3.11) one concludes

$$\left| \int_{\partial B_R} u \sum_{j=1}^{m} X_j v < X_j, \eta > dH_{N-1} \right| \leq \frac{C}{R^{Q-2}} = Cap(\Omega)\Gamma(x,e)\left(1 + 0\left(\frac{1}{d^*(x,e)}\right)\right). \tag{3.13}$$

Using (3.12), (3.13) in (3.6) we conclude

$$u(x) = \int_{\partial \Omega} \Gamma(x,y) \sum_{j=1}^{m} X_j u(y) < X_j(y), \eta(y) > dH_{Q-1} . \tag{3.14}$$

We now analyze the integral on in the right-hand side of (3.14). To this end, we use Lemma 2.2 applied to the function $w(x) = \Gamma(x,e)$, which is homogeneous of degree $2-Q$. This gives for any $x \in \mathbf{G} \setminus \overline{\Omega}$ such that $\rho(x)$ is large enough

$$|\Gamma(x,y) - \Gamma(x,e)| \leq C d^*(y,e) d^*(x,e)^{1-Q} \qquad\qquad y \in \partial\Omega. \tag{3.15}$$

Then

$$\int_{\partial\Omega} \Gamma(x,y) \sum_{j=1}^{m} X_j u(y) < X_j(y), \eta(y) > dH_{Q-1} = \Gamma(x,e) \int_{\partial\Omega} \sum_{j=1}^{m} X_j u(y) < X_j(y), \eta(y) > dH_{Q-1}$$

$$+ \int_{\partial\Omega} [\Gamma(x,y) - \Gamma(x,e)] \sum_{j=1}^{m} X_j u(y) < X_j(y), \eta(y) > dH_{Q-1} = (I) + (II). \tag{3.16}$$

Using the observation $\mathcal{L}(u^2) = -2|Xu|^2$, we obtain from the divergence theorem

$$\int_{B_R \setminus \Omega} |Xu|^2 \, dy = \int_{\partial B_R} u \sum_{j=1}^{m} X_j u < X_j, \eta > dH_{N-1} - \int_{\partial\Omega} \sum_{j=1}^{m} X_j u < X_j, \eta > dH_{N-1} ,$$

where, again, the last integral has to be interpreted as a limit of boundary integrals over a sequence of smooth regions shrinking to Ω. With arguments similar to those which led to (3.2) we infer

$$\lim_{R \to \infty} \int_{\partial B_R} u \sum_{j=1}^{m} X_j u < X_j, \eta > dH_{N-1} = 0.$$

This observation and (3.2) imply

$$Cap(\Omega) = - \int_{\partial\Omega} \sum_{j=1}^{m} X_j u < X_j, \eta > dH_{N-1} , \tag{3.17}$$

which gives

$$(I) = Cap(\Omega)\Gamma(x,e).$$

As for the second integral in the right-hand side of (3.16), we obtain from (3.15)

$$(II) \leq \frac{C}{d^*(x,e)^{Q-1}} \int_{\partial\Omega} \sum_{j=1}^{m} |X_j u||< X_j, \eta >| dH_{N-1} = Cap(\Omega)\Gamma(x,e) \, 0\left(\frac{1}{d^*(x,e)}\right).$$

We conclude

$$\int_{\partial\Omega} \Gamma(x,y) \sum_{j=1}^{m} X_j u(y) < X_j(y), \eta(y) > dH_{Q-1} = Cap(\Omega)\Gamma(x,e)\left(1 + 0\left(\frac{1}{d^*(x,e)}\right)\right). \tag{3.18}$$

From (3.16), and (3.18) we finally obtain the estimate

$$u(x) = Cap(\Omega)\Gamma(x,e)\left(1 + 0\left(\frac{1}{d^*(x,e)}\right)\right),$$

which is what we wanted to prove. To complete the proof of the theorem we are left with proving that a similar estimate holds for Zu. This is achieved by differentiating both sides of (3.14) along the vector field Z. Indicating with τ_y the group left-translations, i.e., $\tau_y(x) = yx$, we have

$$Zu(x) = \int_{\partial\Omega} Z_x(\Gamma \circ \tau_{y^{-1}})(x) \sum_{j=1}^{m} X_j u(y) < X_j(y), \eta(y) > dH_{Q-1} ,$$

where we have denoted $\Gamma(x) = \Gamma(x,e)$. Next, we write

$$Zu(x) = Z\Gamma(x) \int_{\partial\Omega} \sum_{j=1}^{m} X_j u(y) < X_j(y), \eta(y) > dH_{Q-1} = \tag{3.19}$$

$$\int_{\partial\Omega} [Z_x(\Gamma \circ \tau_{y^{-1}})(x) - Z\Gamma(x)] \sum_{j=1}^{m} X_j u(y) < X_j(y), \eta(y) > dH_{Q-1} .$$

Since the function $\Gamma(x)$ is homogeneous of degree $2 - Q$ we have $Z\Gamma(x) = (2 - Q)\Gamma(x)$. Using exponential coordinates one can recognize that for any $y \in \partial\Omega$ one has for $\rho(x) \to \infty$

$$|Z_x(\Gamma \circ \tau_{y^{-1}})(x) - Z\Gamma(x)| = \Gamma(x)\left(1 + 0\left(\frac{1}{d^*(x,e)}\right)\right).$$

Inserting this estimate, along with (3.17), into (3.19) we finally conclude

$$Zu(x) = (Q - 2)Cap(\Omega)\Gamma(x,e)\left(1 + 0\left(\frac{1}{d^*(x,e)}\right)\right).$$

This completes the proof of the theorem.

In the proof of the main result we will need the notion of starlikeness.

Definition 3.2. Let Ω be a C^1 domain containing the identity e. We say that Ω is *starlike* with respect to e if for every $x \in \partial\Omega$ one has

$$< Z(x), \eta(x) >_G \geq 0.$$

When the strict inequality holds, then we say that Ω is *strictly starlike* with respect to e.

We are now ready to give the

Proof of Theorem 1.1. Let u be a solution of (3.1) and suppose that Ω be a C^2 connected, bounded open set which is starlike with respect to the group identity e. We consider the function $w = -Zu$. Since $u \in C^\infty(G \setminus \overline{\Omega})$, then $Zu \in C^\infty(G \setminus \overline{\Omega})$. At this point we use Lemma 2.1. The latter gives

$$\mathcal{L}w = \sum_{j=1}^m X_j(X_j Zu) = \sum_{j=1}^m X_j(ZX_j u + X_j u) = \sum_{j=1}^m (ZX_j + X_j)(X_j u) + \sum_{j=1}^m X_j(X_j u)$$

$$= -Z(\mathcal{L}u) - 2\mathcal{L}u = 0.$$

Therefore, w is a harmonic function in $G \setminus \overline{\Omega}$ which, according to Theorem 3.3, vanishes at infinity. By the assumption of starlikeness on Ω we see that $w|_{\partial\Omega} \geq 0$. By the maximum principle, see [CDG2, Theorem 5.1], we infer that $w \geq 0$ in $G \setminus \overline{\Omega}$. Once we know this, we have in fact much more. As a consequence of the Harnack inequality, the nonnegative harmonic function w must be strictly positive in $G \setminus \overline{\Omega}$. This implies that the standard gradient of u, $\nabla_G u$, never vanishes since, if it did, we would have $Zu = < Z, \nabla_G u >_G = 0$. We conclude that each level set $\partial E_\alpha = \{x \in \mathbf{G} \setminus \overline{\Omega} \,|\, u(x) = \alpha\}$ is a C^∞ submanifold of $\mathbf{G}$ which is the boundary of a domain strictly starlike with respect to e. This finishes the proof.

BIBLIOGRAPHY

[B]   J.-M. Bony, *Principe du maximum, inégalité de Harnack et problème de Cauchy pour les operateurs elliptique degeneres*, Ann. Inst. Fourier, Grenoble **19** (1969), 277-304.

[CDG1] L. Capogna, D. Danielli and N. Garofalo, *An embedding theorem and the Harnack inequality for nonlinear subelliptic equations*, Comm P.D.E., 18: 9,**10** (1993), 1765-1794.

[CDG2] L. Capogna, D. Danielli, and N. Garofalo, *Capacitary estimates and the local behavior of solutions of nonlinear subelliptic equations*, Amer. Jour. of Math. (6) **118** (1996), 1153-1196.

[CDG3] L. Capogna, D. Danielli, and N. Garofalo, *The geometric Sobolev embedding for vector fields and the isoperimetric inequality*, Comm. in Anal. and Geometry (2) **2** (1994), 203-215.

[CDG4] L. Capogna, D. Danielli, and N. Garofalo, *Subelliptic mollifiers and a basic pointwise estimate of Poincaré type*, Math. Zeitschrift, to appear.

[CG]   L. Capogna and N. Garofalo, *NTA domains for Carnot-Carathéodory spaces and a Fatou type theorem*, Preprint.

[Ch]   I. Chavel, Eigenvalues in Riemannian Geometry, Academic Press, Orlando, 1984.

[C]   G. Citti, *Wiener estimates at boundary points for Hörmander operators*, Boll. U.M.I 2-B (1988), 667-681.

[D]   D. Danielli, *Regularity at the boundary for solutions of nonlinear subelliptic equations*, Indiana Univ. Math. Jour. 44 (1995), 269-286.

[DG]   D. Danielli and N. Garofalo, *A priori estimates and geometric properties of level sets of solutions to subelliptic equations in nilpotent Lie groups*, preprint.

[F1]   G. B. Folland, *A fundamental solution for a subelliptic operator*, Bull. Amer. Math Soc. **79** (1973), 373-376.

[F2]   G. B. Folland , *Subelliptic estimates and function spaces on nilpotent Lie groups*, Arkiv.für Math., **3** (1975), 161-207.

[G]   N. Garofalo, *Recent Developments in the Theory of Subelliptic Equations and Its Geometric Aspects*, Birkhäuser, book in preparation.

[GN]   N. Garofalo and D. M. Nhieu, *Isoperimetric and Sobolev inequalities for Carnot-Carathéodory metrics and the existence of minimal surfaces*, Comm. in Pure and Appl. Math., **49** (1996), 1081-1144.

[Gr]   M. Gromov, *Carnot-Carathéodory spaces seen from whithin*, IHES, preprint.

[HaH]   W. Hansen and H. Hüber, *The Dirichlet problem for sub-Laplacians on nilpotent groups. Geometric criteria for regularity*, Math. Ann. **246** (1984), 537-547.

[H]   L. Hörmander, *Hypoelliptic second order differential equations*, Acta Math. **19** (1967), 147—171.

[K]   A. Kaplan, *Fundamental solutions for a class of hypoelliptic PDE generated by composition of quadratic forms*, Trans. Amer. Math. Soc. **258** (1980), 147-153.

[LSW]   W. Littman, G. Stampacchia, and H. F. Weinberger, *Regular points for elliptic equations with discontinuous coefficients*, Ann. Scuola Norm. Sup. Pisa (3) **18** (1963), 43-77.

[NSW]   A. Nagel, E. M. Stein, and S. Wainger, *Balls and metrics defined by vector fields I: basic properties*, Acta Math. **55** (1985), 103—147.

[SC]   A. Sanchez-Calle, *Fundamental solutions and geometry of sum of squares of vector fields*, Invent. Math. **78** (1984), 143-160.

[Se1]   J. Serrin, *Local behavior of solutions of quasilinear equations*, Acta Math. **111** (1964), 243-302.

[Se2]   J. Serrin, *Isolated singularities of solutions to quasilinear equations*, Acta Math. **113** (1965), 219-240.

[VSC]   N. Th. Varopoulos, L. Saloff-Coste, and T. Coulhon, *Analysis and Geometry on Groups*, Cambridge Univ. Press, Cambridge 1992.

Recent Progress on Super-Diffusive Nonlinear Parabolic Problems

PANAGIOTA DASKALOPOULOS   Department of Mathematics, University of California, Irvine, CA 92717, USA. [1]

MANUEL DEL PINO   Departamento de Ingeniería Matemática F.C.F.M. Universidad de Chile, Casilla 170 Correo 3, Santiago, CHILE. [2]

The aim of this note is to review some recent results concerning solvability and well posedness of the Cauchy problem

$$\begin{cases} \partial u/\partial t = \operatorname{div}\left(u^{m-1}\nabla u\right) & \text{in } \mathbf{R}^N \times (0,T) \\ \\ u(x,0) = f(x) & x \in R^N, \end{cases} \tag{1}$$

in the range of exponents $m \leq 0$. Here $T > 0$ is a given constant and f a nonnegative, locally integrable function. We shall refer to (1) in this case as *super-diffusive*, in opposition to $m > 1$ and $0 < m < 1$, called in standard terminology *slow* and *fast* diffusion cases respectively.

This equation can also be written as

$$\frac{\partial u}{\partial t} = \Delta\phi(u) \quad \text{in } Q_T, \tag{2}$$

---

[1]Partially supported by NSF grant DMS-9596195

[2]Partially supported by grants Fondecyt 1950-303, CI1*CT93-0323 CCE and by Cátedra Presidencial.

where $\phi(u) = u^m/m$ if $m \neq 0$, $= \log u$ if $m = 0$, and $Q_T = \mathbf{R}^N \times (0, T)$. By a solution of (1) we understand a nonnegative function u in $C([0, T); L^1_{\mathrm{loc}}(\mathbf{R}^N))$ such that $\phi(u)$ belongs to $L^1_{\mathrm{loc}}(\mathbf{R}^N \times [0, T))$ and which satisfies the equation in the distributional sense. Note that the assumption $\phi(u)$ locally integrable implies that u is nonzero almost everywhere in Q_T when $m \leq 0$.

If $m > 0$ it is customary to scale out the factor $1/m$ in the equation and write it in the form $\partial u/\partial t = \Delta u^m$. We recover for $m = 1$ the classical heat equation. [26]) and for $m > 1$ the well studied *porous medium equation*, which models the flow of an isotropic gas through a porous medium, see [1].

The case $m \leq 0$ arises in a number of different contexts. A known example is that of *diffusion in plasma*, see Lonngren & Hirose [20], Berrymann & Holland [5]. The case $m = 0$ arises in the modelling of *spreading of microscopic droplets*, proposed by de Gennes [15], [14] as a model for long range Van der Waals interactions in thin films spreading on solid surfaces, see also López, Miller & Ruckenstein, [21]. Negative values of m have also been considered in this context, see Stratov [23]. In a different setting, Chayes, Osher and Ralston [6] have studied this equation in connection with interacting particle systems with *self-organized criticality*.

When $m = 0$, $N = 2$, this equation arises in Geometry, cf. Wu [27] and Hamilton [16],[17], in the following setting. On $\mathbf{R}^2$ any metric can be expressed as $ds^2 = e^v (dx^2 + dy^2)$, where x, y are rectangular coordinates in $\mathbf{R}^2$ and v is a function. Let $\mathcal{R}$ denote the scalar curvature. The *Ricci flow* on $\mathbf{R}^2$ is the equation

$$\frac{\partial}{\partial t} ds^2 = -\mathcal{R} \, ds^2,$$

which may be rewritten as

$$\frac{\partial v}{\partial t} = e^{-v} \Delta v$$

or, setting $u = e^v$,

$$\frac{\partial u}{\partial t} = \Delta \log u. \tag{3}$$

where Δ denotes the Euclidean Laplacian.

It is natural to ask under which conditions on f the initial value problem (1) is solvable, and our purpose here is to present some recent results in this direction. It is illustrative to establish comparisons with the essentially settled existence theory for the cases $m > 1$ and $0 < m < 1$ which exhibit very different phenomena.

Observe that in the porous medium equation where $m > 1$ the *diffusion coefficient* u^{m-1} is small for small u, which means that if we think of u as a quantity like the density of a gas or temperature, it will spread out at a slow rate. In mathematical terms, this translates into the known phenomenon of *finite speed of propagation*, namely if f is, say, compactly supported, a solution of (1). will remain compactly supported in the space variable at all times. For this reason, the case $m > 1$ is also called *slow diffusion*. This phenomenon is in strong opposition with the heat equation $m = 1$ or the *fast diffusion* case $0 < m < 1$ in which the solution becomes instantly positive.

It is natural to ask what happens with an initial data compactly supported when $m \leq 0$, where the speed of propagation should be even higher. Vázquez showed in [24] that if $m < 0$ and $N \geq 2$ or if $m = 0$ and $N \geq 3$ and $f \in L^1(R^N)$, the solution *vanishes instantaneously* so that the high diffusion takes all mass to infinity in no time, therefore yielding nonexistence of a local solution for the Cauchy problem. On the other hand, there exist solutions to (1) which decay at infinity (at a non-integrable rate), for instance

$$v^T(x,t) = (2\alpha(T-t)_+|x|^{-2})^{1/(1-m)}, \qquad \alpha = (N - 2/(1-m)) > 0. \quad (4)$$

On the other hand, when $m = 0$, $N = 2$ one has the explicit solution

$$u(x,t) = \frac{8(T-t)_+}{(1+|x|^2)^2},$$

which have integrable initial data.

The term *slow diffusion* for $m > 1$ is somewhat misleading since u^{m-1} is large for large u. Thus it is natural to expect that if one starts with an initial data too large at infinity, the diffusion brings infinite mass in short time from infinity, so that the solution blows up. This phenomenon is well known in the case of the heat equation $m = 1$, where with the assistance of the explicit representation formula for the solution it is shown that (1) has a solution defined up to time T if and only if

$$\int e^{\frac{-|x|^2}{4T}} f(x)dx < +\infty,$$

In particular, if f grows faster than $e^{\frac{|x|^2}{4T}}$, then the solution must blow up at some time $\leq T$, see Widder [26]. After the works by Aronson and Caffarelli [3], Benilan, Crandall and Pierre [4] and Dahlberg and Kenig [7], the existence theory for the porous medium equation, $m > 1$ has been essentially settled to a level of knowledge similar to that for the heat equation. The nonlinear case is of course much harder because of the lack of a representation formula for the solutions. In [3] a Harnack estimate for the solutions was proven which in particular implies that if f grows faster than $|x|^{\frac{2}{m-1}}$ then no local solution exists. More precisely, it is established the existence of a constant $C_1 = C_1(m, N) > 0$ such that if (1) is solvable, then

$$\limsup_{R \to \infty} \frac{1}{R^{N+2/(m-1)}} \int_{B_R(0)} f(x)\, dx \le C_1\, T^{-1/(m-1)}. \qquad (5)$$

In [4], it was shown that the reciprocal also holds true, namely condition (5) with C_1 possibly replaced by another constant $C_2(m, N)$, implies solvability of (1). In particular, if $f(x) = o(|x|^{\frac{2}{m-1}})$, then existence of a solution globally defined in time holds. In [7] uniqueness was established, as well as exact asymptotic behavior in the space variable.

Concerning the *fast diffusion case* $0 < m < 1$, a key work is the paper by Herrero and Pierre [18]. In fact this case for which more properly speaking, diffusion is not too slow nor too fast, existence of a unique distributional solution which is global in time holds, whenever f is locally integrable, no matter how fast or slow it grows at infinity. If $m > \frac{N-2}{N}$ the solution becomes instantaneously smooth and positive at all times, while this is not necessarily the case if $m < \frac{N-2}{N}$, as the explicit solutions in (4) show, which are defined not only for $m < 0$ but for for all $m < \frac{N-2}{N}$. However, in the case $0 < m < \frac{N-2}{N}$ they are globally defined in time, although they vanish after $t = T$.

In light of the results above it is natural to seek for a similar existence theory for the *super-diffusive case* $m \le 0$. We have already mentioned that if f is in L^1, $m < 0$, $N \ge 2$, no solution exists. Thus f cannot decay too fast at infinity if local existence holds. The obvious question is: what is the fastest possible decay allowed on f if there is a local solution (i.e. not instant vanishing) ?

It turns out that a notable symmetry between the cases $m < 0$ and $m > 1$ appears. As suggested from our discussion above, blow-up for $m > 1$ plays a similar role as vanishing when $m < 0$. It may not be too surprising that the answer to the above question is that f should not decay faster than $|x|^{\frac{-2}{1-m}}$. More precisely we have the following result, which is shown in [8].

THEOREM 1 *Assume that there is a solution to (1). Then the inital data f*

must satisfy the growth condition

$$\limsup_{R \to \infty} \frac{1}{R^{N-2/(1-m)}} \int_{B_R} f \geq C^* T^{1/(1-m)}. \tag{6}$$

with C^* the precise constant

$$C^* = [2(N - \frac{2}{1-m})]^{1/(1-m)} \omega_N. \tag{7}$$

Here ω_N denotes the surface area of the unit sphere.

In other words, if the limit above is strictly less than $C^* T^{1/(1-m)}$, then any local solution must die by vanishing before time T. This result, symmetric with condition (5) for $m > 1$ is optimal, since for the explicit solution v^T in (4), which vanishes exactly at time T one has

$$\lim_{R \to \infty} \frac{1}{R^{N-2/(1-m)}} \int_{B_R} f = C^* T^{1/(1-m)},$$

where $f(x) = v^T(x, 0)$.

It is tempting to guess that a condition of the form (6), possibly replacing the *limsup* by *liminf*, is sufficient for existence when $m < 0$. The answer is affirmative in the radially symmetric case. In fact the following result has been established in [9]

THEOREM 2 *Assume that $f(x) \geq g(|x|)$, where g is radially symmetric and*

$$\liminf_{R \to \infty} \frac{1}{R^{N-2/(1-m)}} \int_{B_R} g(|x|) dx \geq C^* T^{1/(1-m)}. \tag{8}$$

where C^* is given by (7). Then Problem (1) is solvable.

This result basically settles the issue of existence-nonexistence in the radially symmetric case, except for the possible gap between the liminf and the limsup in conditions (6) and (8). The analogy with the conditions for $m > 1$ is obvious. The latter ones are better in the sense that both of them carry lim sup, but they are not as precise, since the constants C_1 and C_2 in the corresponding proofs may differ and not be optimal as C^* is.

Since the results of existence and nonexistence for $m > 1$ do not require radiality, one may think that the condition in Theorem 2 is just technical. Strikingly enough, this is not the case, as shown via an example in [9]. Roughly speaking it is shown that if f vanishes on a "sufficiently wide" logarithmic spiral region, then no local solution exists. In such case the value of the limit in (3) can be made arbitrarily large. Actually this example can be simplified when $m \leq -1/3$, to yield that if f vanishes on the sector $0 < \theta < l^*$ where $l^* = (m-1)\pi/2m > 0$, then no local solution exists, see Theorem 4 below. These facts show that the super-diffusive case may hide a rich and possibly very complex nonradial structure behind the existence question, not present in the slower diffusion case.

Of course a natural direction to investigate is that of finding sufficient conditions for existence in cases where radial symmetry is violated. The following condition for existence has been found in [9].

For a number $\rho > 0$ we denote by G_ρ the Green's function of the ball B_ρ. For a locally bounded function h, we set

$$G_\rho^*(h)(x) = \int_{B_\rho} [G_\rho(0,y) - G_\rho(x,y)]h(y)dy, \qquad x \in \overline{B}_\rho.$$

THEOREM 3 *Let $E^* = -2mC^*/(1-m)$. Assume that there exists a nonnegative, locally bounded function $\tilde{f}$ for which $f \geq \tilde{f}$ and a sequence $\rho_n \uparrow +\infty$ such that*

$$|x|^{2m/(1-m)}G_{\rho_n}^*(\tilde{f})(x) \geq E^*T^{1/(1-m)} + \theta(x), \tag{9}$$

for all $|x| < \rho_n$. Here $\theta(x)$ is a function such that $\theta(x)|x|^{-2m/(1-m)}$ is locally bounded and $\theta(x) \to 0$ as $|x| \to \infty$. Then, problem (1) is solvable.

This result roughly asserts that if

$$\int (\Gamma|x-y| - \Gamma|y|)f(y)dy \geq E^*T^{1/1-m}|x|^{-2m/1-m},$$

where Γ denotes the fundamental solution of the Laplacian, and the integral is understood in a certain principal value sense, then (2) is solvable. It is not hard to check that this result implies Theorem 2 in case that f is radially symmetric. However, it does not answer a basic question: If f is, say, the characteristic function of the a sector of angle wider than $2\pi - l^*$, is problem (1) solvable? Recently in [11] we have shown that the answer to this question is indeed affirmative.

Assume that $N = 2$ and $m < 0$. We consider, for $0 < l \leq 2\pi$, the sector

$$C_l = \{(r\cos\theta, r\sin\theta) \mid r > 0, \ 0 < \theta < l\}, \tag{10}$$

and denote by χ_l its characteristic function. We define the *critical lenght* (or aperture) l_* as

$$l^* = \min\{\frac{(m-1)}{2m}\pi, 2\pi\} > 0.$$

Note that $l^* = 2\pi$ iff $m \geq -1/3$. The following result holds.

THEOREM 4 *(i) If $f \equiv 0$ a.e. on C_{l^*}, then (1) has no local solution.*
   (ii) Let $l < l^*$ and $f = \chi_{2\pi - l}$. Then (1) admits a solution for $T = +\infty$.

(ii) actually follows from the following more general statement, proved in [11].

Let $L_{2\pi}^\infty$ denote the space of 2π-periodic bounded functions on the real line. Let us define, for $0 < l \leq 2\pi$, and $g \in L_{2\pi}^\infty$ the number $K_l(g)$ as

$$K_l(g) = \inf \left\{ \int_I g(\theta)d\theta \mid I \text{ interval, } |I| = l \right\}. \tag{11}$$

THEOREM 5 *Assume that $N = 2$, $m < 0$, and that $f \in L_{loc}^1(\mathbf{R}^2)$ satisfies, for some $R > 0$, that*

$$f(x) \geq r^{-2/(1-m)} g(\theta), \tag{12}$$

for $|x| > R$, where $g \in L_{2\pi}$ is nonnegative. Assume also that for some $l < l^*$ one has $K_l(g) > 0$. Then (2) has a local solution defined at least up to time T given by

$$T = \left(\frac{K_l(g)\,(l^* - l)^4}{C^*} \right)^{1-m}, \tag{13}$$

where C^* is a positive constant dependent only on m.

Note that combining Theorems 4 and 5, we have that if $f(x) = r^{\frac{-2}{1-m}} g(\theta)$, then (1) admits a local solution if and only if $K_{l^*}(g) > 0$.

Our approach in proving this result is based on the connection between the local solvability of the parabolic problem (1) and the existence of solutions for the elliptic problem

$$\Delta v + v^{-\nu} = f(x) \qquad \text{in } \mathbf{R}^2, \tag{14}$$

where $\nu = -1/m > 0$. The relationship between the two problems comes at the formal level from discretization in time of problem (1) which amounts to solving elliptic problems of the form (14), see [24]. Existence and nonexistence results for (14), which are indeed analogous to Theorems 1 and 2, are contained in the paper [8].

A key fact which allows us to connect the elliptic and parabolic problems is that if u is a positive solution of (1) then the function $\Phi(x) = |m|^{-1} \int_0^T u^m(x, s)\, ds$ satisfies

$$\Delta \Phi + (\gamma T)^{-1/m} \Phi^{1/m} \geq f, \tag{15}$$

with $\gamma = (1 - m)/|m|$. Therefore the solvability of (14) provides through (15) a control from bellow of the positive approximations of the maximal solution of (1).

For the purpose of proving existence of solutions for the elliptic problem (14) we study first the special case of an f of the form

$$f(x) = r^{-2\nu/(1+\nu)} g(\theta),$$

with $g \in L_{2\pi}^\infty(\mathbf{R})$, nonnegative. In this case we can try a solution of the form

$$v(x) = r^{2/(1+\nu)} w(\theta),$$

with w a function in $L^\infty 2\pi$. A direct computation shows that w must solve the one dimensional equation

$$w_{\theta\theta} + \beta^2 w + w^{-\nu} = g(\theta), \tag{16}$$

where $\beta = \pi/l^* = 2/(\nu + 1)$, since $\nu = -1/m$. This reduces our problem into solving (16). Our main result for existence of (16) asserts essentially that if $K_l(g)$ is sufficiently large, for some $l < l^*$, then (16) is solvable. Its proof relies on a somewhat technical construction of a supersolution for this equation. Having shown the solvability of (16) we extend it to general elliptic problem and show the analogue of Theorem 5 for it. This is done through a comparison argument. Finally for the proof of the parabolic result we first solve the problem (1) with initial data $f_\epsilon = f + \epsilon$ and then we combine the elliptic existence result together with (15) to show that actually the sequence of solutions $\{u_\epsilon\}$ converges as $\epsilon \to 0$ to a strictly positive function which is the desired solution.

We conclude this expository note by mentioning some results on the initial value problem (1) in the critical case $m = 0$, $N = 2$. In [10] we characterized the solvability of problem (1) in terms of the the initial condition f. The nonradial structure mentioned above is not present here, however a strong *non-uniqueness phenomenon* takes place, which does not occur when $m > 0$.

Our first result provides a necessary and sufficient condition for solvability: it states in particular that there exists a solution defined up to time $T \le +\infty$, if and only if $\int_{\mathbf{R}^2} f \, dx \ge 4\pi T$.

THEOREM 6 *Assume $m = 0$, $N = 2$. Let f be a nonnegative, locally integrable function. Then, there exists a solution u to problem (1) with*

$$T = \frac{1}{4\pi} \int_{\mathbf{R}^2} f(x) \, dx \le +\infty.$$

In particular, if $\int_{\mathbf{R}^2} f(x)dx = +\infty$, then (1) admits a globally defined solution. Reciprocally, if there is a solution to problem (1), then

$$T \le \frac{1}{4\pi} \int_{\mathbf{R}^2} f\, dx. \tag{17}$$

Moreover,

$$\int_{\mathbf{R}^2} u(x,t)\, dx \le \int_{\mathbf{R}^2} f\, dx - 4\pi t, \tag{18}$$

for all $t < T$.

Thus, in particular all solutions to problem (1) must cease to exist by vanishing before time $(1/4\pi) \int_{\mathbf{R}^2} f$. We note that the fact that $\int_{\mathbf{R}^2} f < +\infty$ implies finite extinction time was already established in [8].

On the other hand our second result shows that given any integrable initial data f, we may find solutions which vanish at any given time less than or equal to $(1/4\pi) \int_{\mathbf{R}^2} f$. This shows in particular that there exists infinitely many solutions of (1) for any given initial data.

THEOREM 7 *Assume $m = 0$, $N = 2$ and $\int_{\mathbf{R}^2} f(x)dx < +\infty$. Then for every $\mu > 0$ there exists a solution u_μ to problem (1) for*

$$T = T_\mu = \frac{1}{2\pi(2 + \mu)} \int_{\mathbf{R}^2} f\, dx,$$

with the property that

$$\int_{\mathbf{R}^2} u_\mu(x,t)dx = \int_{\mathbf{R}^2} f(x)\, dx - 2\pi(2 + \mu)t, \tag{19}$$

for all $t < T_\mu$.

It should be observed that, at least at a formal level, the number $\mu > 0$ in the above constructed solutions is related to their decay rate at infty. In fact, in the radially symmetric case condition (19) roughly tells us that at any time $u(|x|, t)$ decays as the power $|x|^{-(2+\mu)}$. As for the maximal solutions constructed in Theorem 1, the expected decay in x is $(|x| \log |x|)^{-2}$.

In the same way, if for a solution of (1) with integrable initial data we define $\phi(t)$ by the relation

$$\int_{\mathbf{R}^2} u(x,t)\, dx = \int_{\mathbf{R}^2} f\, dx - 2\pi\phi(t), \tag{20}$$

then one has from Theorem 6 that $\phi(t) \geq 2t$ and, formally, one expects u to decay roughly as $|x|^{-\phi'(t)}$ at infinity, in case that ϕ is differentiable. Since u must remain integrable in x at all times, $\phi'(t)$ should not be less than two. Our next result shows rigorously this fact. Moreover, we are also able to establish partially the reciprocal assertion, which generalizes Theorem 7. given any function ϕ of class C^1 with $\phi'(t) \geq 2$, there exists a solution u so that (20) holds.

THEOREM 8 *Assume that $\int_{\mathbf{R}^2} f \, dx < \infty$ and that u solves (1). Let $\phi(t)$ be the function defined by the relation (6). Then, for all $t \in [0, T)$ and all $h > 0$ so that $t + h < T$, we have*

$$\frac{\phi(t + h) - \phi(t)}{h} \geq 2. \tag{21}$$

Reciprocally, if $\phi : [0, T] \to [0, \infty)$ is any continuously differentiable function such that $\phi'(t) \geq 2$, $\phi(0) = 0$ and $\phi(T) = (1/2\pi) \int_{\mathbf{R}^2} f$, then there exists a solution u to (1) such that (19) holds.

Even though smoothness of ϕ can be relaxed in the construction of these solutions, see the Remark at the end of §4, some regularity of the given ϕ seems to be needed.

We mention that independent existence results in the case $m = 0$, $N = 2$ similar to those exposed above have been obtained independently by Esteban, Rodríguez and Vázquez in [25] for the radial case, and by DiBeneddeto and Diller in [12]. In the first reference, characterization of the different solutions is given in terms of the *flux* of the solutions at infinity and implies precise pointwise asymptoic behavior. The method relies on a transformation that takes this radial equation into the corresponding one-dimensional case. In [12], solvability of problem (1) is studied via the method of discretization in time. Also, existence and long time behavior of solutions of this equation under different assumptions on the initial data, including slower, nonintegrable decay, had been previously obtained by Wu in [27]. The case $m = 0$, $N = 1$ has been treated in [13], [22]. In [22] it is shown that for any initial data f with infinite mass, there is a family of solutions of (1) with finite mass for all $t > 0$. This result, which shows non-uniqueness, can be extended to the 2-dimensional radially symmetric case. However the corresponding question in the non-radial setting with initial data of infinite mass remains open.

References

[1] Aronson, D.G., The Porous Medium Equation, in Nonlinear Diffusion problems, em Lecture Notes in Maths, 1224, Springer Verlag, 1986.

[2] Aronson, D.G., Bénilan P., Régularité des solutions de l'équation de milieux poreux dans $\mathbf{R}^n$, *C.R. Acad. Sci. Paris, 288*, 1979, pp 103-105.

[3] Aronson, D.G. and Caffarelli, L.A., The initial trace of a solution of the porous medium equation, *Trans. Amer. Math. Soc., 280*, 1983, pp. 351-366.

[4] Bénilan, P., Crandall, M.G. and Pierre, M., Solutions of the porous medium equation under optimal conditions on the initial values, *Indiana Univ. Math.J., 33*, 1984, pp. 51-87.

[5] Berryman and J.G., Holland, C.J., Asymptotic behavior of the nonlinear differential equation $n_t = (n^{-1}n_x)_x$, *J. Math Phys.*, 23, 1982, pp. 983-987.

[6] Chayes, J.T., Osher, S.J. and Ralston, J.V., On singular diffusion equations with applications to self-organized criticality, *Comm. Pure Appl. Math.* 46, 1993, pp. 1363-1377.

[7] Dahlberg, B.E.G., Kenig, C.E., Nonnegative solutions of the Porous Medium equation, *Comm. Part. Diff. Eq., 9*, 1984, pp. 409-437.

[8] Daskalopoulos, P., del Pino M.A., On fast diffusion nonlinear heat equations and a related singular elliptic problem, *Indiana Univ. Math. J., 43*, 1994, pp. 703-728.

[9] Daskalopoulos,P., del Pino M.A., On nonlinear parabolic equations of very fast diffusion, *Arch. Rational Mech. Analysis, 103*, to appear.

[10] Daskalopoulos,P., del Pino M.A., On a Singular Diffusion Equation, *Comm. in Analysis and Geometry, Vol. 3*, 1995, pp 523-542.

[11] Daskalopoulos,P., del Pino M.A., Nonradial solvability structure of super-diffusive nonlinear parabolic equations, *preprint*.

[12] DiBenedetto, E., Diller, D.,J. About a singular parabolic equations arising in thin film dynamics and in the Ricci flow for complete $\mathbf{R}^2$, *Lecture Notes in Pure and Applied Mathematics, 177*, Marcel Dekker, 1996.

[13] Esteban, J.R., Rodríguez, A., Vazquez, J.L., A nonlinear heat equation with singular diffusivity, *Arch. Rational Mech. Analysis, 103*, 1988, pp. 985-1039.

[14] de Gennes, P.G., Spreading laws for microscopic doplets, *C.R. Acad. Sci., Paris II, 298*, pp. 475-478.

[15] de Gennes, P.G., Wetting: statics and dynamics, *Reviews of Modern Physics, 57 No 3*, 1985, pp. 827-863.

[16] Hamilton, R., The Ricci flow on surfaces, *Contemp. Math., 71, Amer. Math. Soc., Providence, RI*, 1988, pp. 237-262.

[17] Hamilton, R., The Harnack estimate for the Ricci flow, *J. Differential Geometry, 37*, 1993, p.p. 225-243.

[18] Herrero, M. and Pierre, M., The Cauchy problem for $u_t = \Delta u^m$ when $0 < m < 1$, *Trans. Amer. Math. Soc., 291*, 1985, pp. 145-158.

[19] Ladyzenskaja, O.A., Solonnikov, V.A., and Ural'ceva, N.N., Linear and quasilinear equations of parabolic type *Translations of Mathematical Monogrphs, 23*, 1968.

[20] Lonngren, K.E., Hirose, A., Expansion of an electron cloud *Phys. Lett. A, 59*, 1976, pp 285-286.

[21] López, J., Miller, C.A., Ruckenstein, E., Spreading kinetics of liquid drops on solids *J. Coll. Int. Sci. 56*, 1976, pp 460-46

[22] Rodríguez, A., Vázquez, J.L., A Well Posed Problem in Singular Fickian Diffusion, *Arch. rational Mech. Anal., 110*, 1990, pp 141-163.

[23] Stratov, V.M., Speading of droplets of nonvotiable liquids over flat solid surface em Coll. J. USSR 45, 1983, 1009-1014.

[24] Vázquez, J.L., Nonexistence of solutions for nonlinear heat equations of fast-diffusion type *J. Math. Pures Appl., 71*, 1992, pp. 503-526.

[25] Vázquez, J.L., Esteban, J.R., Rodriguez, R., The fast diffusion with logarithmic nonlinearity and the evolution of conformal metrics in the plane, *Advances in Diff. Equations, 1*, 1996.

[26] Widder, D.G., The Heat Equation, *Academic Press , New York*, 1975.

[27] Wu, L.-F., A new result for the porous medium equation derived from the Ricci flow, *Bull. Amer. Math. Soc., 28*, 1993, pp 90-94.

On Linear Perturbations of
Superquadratic Elliptic Systems

D. G. DE FIGUEIREDO[(*)] IMECC-UNICAMP, Caixa Postal 6065, 13081-970 Campinas S.P., Brazil

M. RAMOS[(†)] CMAF, Faculdade de Ciencias, Universidade de Lisboa, Av. Prof. Gama Pinto 2, 1699 Lisboa Codex, Portugal

1 INTRODUCTION

We are concerned with existence of nonzero solution for elliptic systems of the form

$$-\Delta U = MU + [H_v \ H_u]^T \ \text{ in } \ \Omega, \ U = 0 \ \text{ on } \ \partial\Omega, \qquad (1.1)$$

where M is a 2×2 real matrix $M = \begin{bmatrix} a & b \\ c & a \end{bmatrix}$, $U = [u \ v]^T$ and $H :$ $\overline{\Omega} \times I\!\!R^2 \to I\!\!R$ is a C^1 function. Here Ω is a bounded open subset of $I\!\!R^N (N \geq 3)$ with smooth boundary $\partial\Omega$. We denote $H_u(x, u, v) = \dfrac{\partial H}{\partial u}(x, u, v)$ and similarly for H_v. With these notations, the system in (1.1) reads

$$-\Delta u = au + bv + H_v(x, u, v), \quad -\Delta v = cu + av + H_v(x, u, v). \qquad (1.2)$$

We refer to the survey paper [4] for an extensive account of recent results concerning (1.1), as well as for some relevant bibliography. Here we

[(*)] Supported partially by CNPq
[(†)] Supported by JNICT (Program PRAXIS/2/2.1/MAT/125/94) and CNPq.

apply variational methods to (1.1) assuming the Hamiltonian H to be superquadratic both at zero and at infinity. We recall that if

$$uH_u(x, u, v) + vH_v(x, u, v) \geq \mu H(x, u, v) > 0 \tag{1.3}$$

for some $\mu > 2$ and every (x, u, v), and moreover H satisfies (1.13) below and it has the subcritical growth of scalar equations, then system (1.1) with $M = 0$ admits a nonzero solution (see e.g. Benci and Rabirowitz [1]). This includes the Hamiltonian

$$H(u, v) = \frac{|u|^p}{p} + \frac{|v|^q}{q} \tag{1.4}$$

with both $p, q > 2$. The subcritical condition in [1] is max $\{p, q\} < 2N/(N2)$. More recently, attention has been drawn to systems including Hamiltonian as in (1.4), but with p and q satisfying the condition

$$1 - \frac{2}{N} < \frac{1}{p} + \frac{1}{q} < 1, \tag{1.5}$$

which is more natural when dealing with systems.

Again, we refer the reader to [4] for a discussion on the subject.

In particular, still with $M = 0$, variational methods suitable to deal with such kind of nonlinearities were introduced in de Figueiredo and Felmer [5]. In [6], de Figueiredo and Magalhães considered linear pertubations of systems which are nonquadratic at infinity. For a system as (1.10), the main theorem in [6], which considers $M = \begin{bmatrix} a & b \\ c & a \end{bmatrix}$ with both b and c positive and the product bc small includes (1.4) under the assumptions

$$\min\{p, q\} > 2 \quad \text{and} \quad 1 - \frac{2}{N} < \frac{1}{p} + \frac{1}{q}. \tag{1.6}$$

In this note we extend the result in [6] by allowing M to be any matrix while heeping (1.4), (1.6) as a model problems. Observe that for H given by (1.4) the general assuption (1.5) is not sufficient to provide the existence of nonzero solutions for any given linear perturbation: just take $q = 2$ and $M = \begin{bmatrix} 0 & -1 \\ 0 & 0 \end{bmatrix}$.

We now describe our main result. Let us consider real constants $p \geq \alpha > p - 1$, $q \geq \beta > q - 1$ with $\alpha, \beta > 2$. Moreover, as in [5] we impose the conditions

$$[2 - (\frac{1}{p} + \frac{1}{q})] \max\{\frac{p}{\alpha}, \frac{q}{\beta}\} < 1 + \frac{2}{N}, \tag{1.7}$$

$$\frac{p-1}{p}\frac{q}{\beta} < 1 \quad \text{and} \quad \frac{q-1}{q}\frac{p}{\alpha} < 1. \tag{1.8}$$

Furthermore, in the case of $N \geq 5$, we take

$$\frac{p-1}{q} \max\{\frac{p}{\alpha}, \frac{q}{\beta}\} < \frac{N+4}{2N} \quad \text{and} \quad \frac{q-1}{q} \max\{\frac{p}{\alpha}, \frac{q}{\beta}\} < \frac{N+4}{2N}. \quad (1.9)$$

Observe that (1.8) trivializes in case $p = \alpha, q = \beta$. Also, in this case (1.7) reduces to the second inequality in (1.5) and (1.9) reads $\max\{p, q\} < 2N/(N-4)$ (see [4] for some comments on this restriction).

As in [5] we take the more general framework given by (1.7)-(1.9) in order to allow a growth of the type

$$c_1(|u|^\alpha + |v|^\beta - 1) \leq H(x, u, v) \leq c_2(|u|^p + |v|^q + 1) \quad (1.10)$$

for every x, u, v, where c_1, c_2 are positive constants. In fact, we suppose that H satisfies the second inequality above in a stronger form, namely

$$|H_u(x, u, v)| \leq c_3(|u|^{p-1} + |v|^{(p-1)\frac{q}{p}} + 1), \quad (1.11)$$
$$|H_v(x, u, v)| \leq c_3(|v|^{q-1} + |u|^{(q-1)\frac{p}{q}} + 1).$$

The analogous of (1.3) will be: there exists $R > 0$ such that, for any $x \in \overline{\Omega}$ and $|u| + |v| \geq R$,

$$\frac{1}{2}(uH_u(x, u, v) + vH_v(x, u, v)) - H(x, u, v) \geq c_4(|u|^\alpha ||v|^\beta) \quad (1.12)$$

The behaviour of H at the origin is given by

$$\lim_{|u|+|v| \to 0} \frac{H(x, u, v)}{|u|^2 + |v|^2} = 0 , \text{ uniformly in } x \in \overline{\Omega}. \quad (1.13)$$

Finally, instead of assuming $H \geq 0$ everywhere, as in [5,6] we merely suppose that there exists $r > 0$ such that, either

$$H(x, u, v) \geq 0, \quad \forall x \in \overline{\Omega}, \ \forall |u| + |v| \leq r \quad \text{or} \quad (1.14)$$
$$H(x, u, v) \leq 0, \quad \forall x \in \overline{\Omega}, \ \forall |u| + |v| \leq r \quad .$$

Now we con state the following.

THEOREM 1 Let a, b, c be any real constants. For α, β, p, q as in (1.7)-(1.9), suppose that H satisfies (1.10)-(1.14). Then system (1.1) admits a nonzero solution (u, v) with $u \in W^{2,p/(p-1)}(\Omega) \cap W_0^{1,p/(p-1)}(\Omega), v \in W^{2,q(q-1)}(\Omega) \cap W_0^{1,q(q-1)}(\Omega)$

The proof will be given in the next section. It combines some ideas from [5] and [9]. We finish with the following remarks.

REMARK 1 Denote by $\lambda_1 < \lambda_2 \leq \lambda_3...$ the sequence of eigenvalues of $(-\Delta, H_0^1(\Omega))$. In case $(a - \lambda_j)^2 \neq bc$ for every j, then Theorem 1 holds true without assumption (1.14) (see the proof of Proposition 2).

REMARK 2 In the case that Ω is a ball, H is as in (1.4), $a = 0$ and $bc < \lambda_1^2$, this result has been proved by L.A. Peletier and R.C.A.M. van der Vorst [10] using ODE methods. In [7] a similar result is proved for the case of $a = 0$ and less general Hamiltonians H. Also in [2] and [3] similar results are obtained for less general Hamiltonians.

PROOF OF THEOREM 1

We first set up some functional analytic framework. Denote by (φ_j) the sequence of eigenfunctions associated to (λ_j), with $\int_\Omega \varphi_j^2 = 1$. If $u \in L^2(\Omega)$ we write $u = \sum a_j \varphi_j$ as its Fourier series with respect to (φ_j). Following [5], we introduce the fractional Sobolev spaces $E^s, 0 \leq s \leq 2$, as

$$E^s = \{u = \sum a_j \varphi_j \in L^2(\Omega) \quad \sum \lambda_j^s |a_j|^2 < \infty\}$$

and the linear operator

$$A^s u = \sum \lambda_j^{s/2} a_j \varphi_j \quad , \quad u \in E^s.$$

The spaces E^s are Hilbert spaces with inner product

$$\langle u, v \rangle_{E^s} = \int_\Omega A^s u A^t v = \sum \lambda_j^s a_j b_j$$

for $u = \sum a_j \varphi_j,\ v = \sum b_j \varphi_j$, and with associated norm

$$\|u\|_{E^s}^2 = \int_\Omega |A^s u|^2.$$

The Sobolev imbedding theorem for E^s states that $E^s \hookrightarrow L^r(\Omega)$ continuously for $r \geq 1$, such that $\dfrac{1}{r} \geq \dfrac{1}{2} - \dfrac{s}{N}$ and compactly in case of strict inequality. For this and other properties of E^s we refer to [5] and the references therein. Now, for given $s > 0, t > 0$ with $s + t = 2$, consider the Cartesian product $E = E^s \times E^t$ with inner product

$$\langle (u, v), (\phi, \psi) \rangle_E = \langle u, \phi \rangle_{E^s} + \langle v, \psi \rangle_{E^t}.$$

We introduce the symmetric and continuous bilinear form $B : E \times E \to \mathbb{R}$, given by

$$B((u, v), (\phi, \psi)) = \int_\Omega (A^s u A^t \psi + A^s \phi A^t v - au\psi - av\phi - bv\psi - cu\phi)$$

and the associated quadratic form

$$Q(z) = \frac{1}{2}B(z,z) = \int_\Omega (A^s u A^t v - auv - \frac{b}{2}v^2 - \frac{c}{2}u^2)$$

for all $z = (u, z) \in E$. We now particularize the constants s and t as in [5]. It follows easily from (1.7) - (1.9) that there exist $s > 0, t > 0, s + t = 2$ such that

$$\frac{p-1}{p}\max\{\frac{p}{\alpha}, \frac{q}{\beta}\} < \frac{1}{2} + \frac{s}{N} \quad \text{and} \quad \frac{q-1}{q}\max\{\frac{p}{\alpha}, \frac{q}{\beta}\} < \frac{1}{2} + \frac{t}{N}. \quad (2.1)$$

In particular, $\frac{1}{p} > \frac{1}{2} - \frac{s}{N}$ and $\frac{1}{q} > \frac{1}{2} - \frac{t}{N}$. Finally, let $\Phi : E \to I\!R$ be given by

$$\Phi(z) = Q(z) - \int_\Omega H(x, u, v), \quad z = (u, v) \in E.$$

Conditions (1.7) and (1.9) imply that Φ is C^1. Then $\Phi'(z) = 0$, for $z = (u, v)$ means that

$$B((u, v), (\phi, \psi)) - \int_\Omega (\phi H_u(x, u, v) + \psi H_v(x, u, v)) = 0$$

for all $(\phi, \psi) \in E$. It follows then from L^p-regularity theory that (u, v) is a solution of (1.1) in the sense of the statement in Theorem 1 (see [5, Prop. 1.2]). In order to find a nonzero critical point for Φ we shall employ a linking theorem due essentially to Li and Liu [8], in the improved version of Li and Willem in [9]. At first, we use the spectral decomposition of B to split E in an orthogonal direct sum

$$E = E^- \oplus E^0 \oplus E^+ \quad (2.2)$$

in such a way that, for some $\delta > 0$,

$$Q(z) \le -\delta\|z\|_E^2, \ \forall z \in E^- \quad \text{and} \quad Q(z) \ge \delta\|z\|_E^2, \ \forall z \in E^+ \quad (2.3)$$

and $Q|_{E^0} = 0$. To be precise, we consider the eigenvalue problem

$$B(z, \eta) = \mu\langle z, \eta\rangle_E, \quad \forall \eta = (\phi, \psi) \in E. \quad (2.4)$$

If we write $z = (u, v)$, $u = \sum a_j\varphi_j$, $v = \sum b_j\varphi_j$ then, by definition,

$$\int_\Omega A^s\varphi_j A^s u = \lambda_j^s a_j \quad \text{and} \quad \int_\Omega A^s\varphi_j A^t v = \lambda_j b_j.$$

By testing (2.4) with $(\varphi_j, 0)$ and $(0, \varphi_j)$, we thus see that (2.4) reads as

$$(\lambda_j - a)a_j = (b + \lambda_j^t\mu)b_j \quad \text{and} \quad (c + \mu\lambda_j^s)a_j = (\lambda_j - a)b_j \quad (2.5)$$

for every $j \in I\!N$. It follows from (2.5) that there is a double sequence of eigenvalues given by

$$\mu_j^{\pm} = \frac{1}{2\lambda_j^2}[-(b\lambda_j^s + c\lambda_j^t) \pm \sqrt{(b\lambda_j^s - c\lambda_j^t)^2 + 4\lambda_j^2(\lambda_j - a)^2}] \ .$$

In particular, we see that 0 is an eigenvalue of (2.4) if and only if there exists $j \in I\!N$ such that $(\lambda_j - a)^2 = bc$. (see the remark following Theorem 1). Using (2.5), the eigenspaces can be easily described. For definiteness, consider the more involved case where $b > 0$ and $c > 0$. For any j, such that $\lambda_j \neq a$, let

$$\alpha_j = \frac{c + \mu_j^+ \lambda_j^s}{\lambda_j - a} \quad \text{and} \quad \beta_j = \frac{c + \mu_j^- \lambda_j^s}{\lambda_j - a}$$

Observe that $\alpha_j \neq \beta_j$. Then

$$
\begin{aligned}
E^0 &= sp\{(\varphi_j, \frac{\lambda_j - a}{b}\varphi_j), \ j : (\lambda_j - a)^2 = bc\}, \\
E^+ &= \overline{sp}\{(\varphi_j, \alpha_j\varphi_j), \ j : bc < (\lambda_j - a)^2\}, \\
E^- &= sp\{(\varphi_j, \alpha_j\varphi_j), \ j : bc > (\lambda_j - a)^2 > 0\} \oplus \overline{sp}\{(\varphi_j, \beta_j\varphi_j), \ j : a \neq \lambda_j\} \\
&\quad \oplus \{(M\varphi_j, N\varphi_j), \ j : a = \lambda_j, M; N \in I\!R\},
\end{aligned}
$$

correspond respectively to the zero, positive and negative eigenvalues of (2.4). It is easily seen that there are orthogonal spaces both with respect to B and to the inner product in E. The direct sum in (2.2) is then an easy consequence of the identities

$$(\varphi_j, 0) = \frac{\alpha_j}{\alpha_j - \beta_j}(\varphi_j, \beta_j\varphi_j) - \frac{\beta_j}{\alpha_j - \beta_j}(\varphi_j, \alpha_j\varphi_j)$$

and

$$(0, \varphi_j) = \frac{1}{\alpha_j - \beta_j}(\varphi_j, \alpha_j\varphi_j) - \frac{1}{\alpha_j - \beta_j}(\varphi_j, \beta_j\varphi_j)$$

for $\lambda_j \neq a$. Observing that

$$\lim_{j \to \infty} \mu_j^{\pm} = \pm 1,$$

this proves (2.3). Now we can state the following

PROPOSITION 2 Φ has a local linking at the origin.

PROOF   The statement means that we can split

$$E = E_1 \oplus E_2$$

in such a way that, for some $r > 0$,

$$\Phi(z) \leq 0, \quad \forall z \in E_1, \ ||z||_E \leq r \quad \text{and} \quad \Phi(z) \geq 0 \ \forall z \in E_2, ||z||_E \leq r \ . \quad (2.6)$$

Indeed, it follows from (1.10) and (1.13) that for any $\varepsilon > 0$ there exists $c_\varepsilon > 0$ such that, for any $x \in \overline{\Omega}$, $(u, v) \in \mathbb{R}^2$,

$$|H(x, u, v)| \leq \varepsilon(|u|^2 + |v|^2) + c_\varepsilon(|u|^p + |v|^q),$$

Thus, for any $z = (u, v) \in E$,

$$|\int_\Omega H(x, u, v)| \leq \varepsilon ||z||_E^2 + C_\varepsilon(||u||_{E^s}^p + ||v||_{E^t}^q). \quad (2.7)$$

If $E_0 = 0$, we choose $E_1 = E^-$ and $E_2 = E^+$.

Since $p > 2$ and $q > 2$, (2.6) follows from (2.3) and (2.7). In case $E^0 \neq 0$ we use assumption (1.14) and choose $E_1 = E^- \oplus E^0$, $E_2 = E^+$ or else $E_1 = E^-, E_2 = E^0 \oplus E^+$ according to whether (1.14) holds with $\geq$ or $\leq$ respectively. Indeed, since E^0 is finite dimensional, this follows by the argument in [6, p. 24]. ∎

Our next result shows that Φ has the required compactness property in order to apply the quoted linking theorem. Let $(\tilde{z}_n), (\hat{z}_n)$, be orthogonal bases for E^- and E^+, respectively. Let E_n^+ be the finite dimensional subspace of E^+ generated by $(\hat{z}_j), j = 1, \ldots, n$. A similar definition for E_n^-. Let $E_n = E_n^- \oplus E^0 \oplus E_n^+$

PROPOSITION 3 Φ satisfies the (PS)* condition with respect to (E_n).

PROOF Let $z_n = (u_n, v_n) \in E_n$ and $\varepsilon_n \to 0$ be such that $\sup_n \Phi(z_n) < \infty$ and

$$|\Phi'(z_n)\eta| \leq \varepsilon_n ||\eta||_E, \ \forall \eta \in E_n. \quad (2.8)$$

We must prove that (z_n) contains a subsequence which converges to a critical point of Φ. We denote by $c_1, c_2, \ldots$ various positive constants whose values do not depend on the sequence (z_n). For simplicity, we write $H(x, u_n, v_n) = H(u_n, v_n)$. Observe first that

$$\begin{aligned}
c_1(1 + ||z_n||_E) &\geq \Phi(z_n) - \frac{1}{2}\Phi'(z_n)z_n \\
&= \int_\Omega (\frac{1}{2}\nabla H(z_n) \cdot z_n - H(z_n))
\end{aligned}$$

so that (1.12) implies

$$||u_n||^{\alpha}_{L^{\alpha}(\Omega)} + ||v_n||^{\beta}_{L^{\beta}(\Omega)} \leq c_2(||z_n||_E + 1). \tag{2.9}$$

Write $z_n = z_n^- + z_n^0 + z_n^+$ according to the decomposition (2.2). It follows from (2.3) and (2.8) that

$$2\delta||z_n^+||^2_E \leq \Phi'(z_n)z_n^+ + \int_{\Omega} \nabla H(z_n) \cdot z_n^+$$

$$\leq \varepsilon_n||z_n^+||_E + \int_{\Omega} \nabla H(z_n) \cdot z_n^+. \tag{2.10}$$

Conditions (1.8) and (2.1) imply the continuous embedding

$$E^s \hookrightarrow L^{\frac{\alpha}{\alpha-p+1}}(\Omega) \text{ and } E^s \subset L^{\frac{p\beta}{p\beta-(p-1)q}}(\Omega\cap)$$

and similarly with t, β, q, p in place of s, α, p, q respectively. Estimating $\nabla H(z_n) \cdot z_n^+$ with the help of (1.11), and then using Hölder inequality to estimate the several integrals we obtain

$$|\int_{\Omega} \nabla H(z_n) \cdot z_n^+| \leq c_3(||u_n||^{\mu}_{L^{\alpha}(\Omega)} + ||v_n||^{\nu}_{L^{\beta}(\Omega)} + 1)||z_n^+||_E, \tag{2.11}$$

where $\mu := \max\{p-1, (q-1)\frac{p}{q}\} < \alpha$, $\nu := \max\{q-1, (p-1)\frac{q}{p}\} < \beta$.
Using (2.10), (2.11) and proceeding similarly with z_n^-, we deduce that

$$||z_n^+||_E + ||z_n^-||_E \leq c_3(||u_n||^{\mu}_{L^{\alpha}(\Omega)} + ||v_n||^{\nu}_{L^{\beta}(\Omega)} + 1). \tag{2.12}$$

Using (2.9), (2.12) and the fact that E^0 is finite dimensional we conclude that (z_n) is a bounded sequence in E. Thus, up to a subsequence, $z_n \rightharpoonup z$ in E and $z_n^0 \to z^0$. Since

$$\int_{\Omega}(\nabla H(z_n) - \nabla H(z)) \cdot (z_n^+ - z^+) = 2Q(z_n^+ - z^+) - (\Phi'(z_n) - \Phi'(z))(z_n^+ - z^+)$$

$$= 2Q(z_n^+ - z^+) + o(1).$$

Using the compactness of the first member (see [5, Prop. 1.2]) we deduce that $z_n^+ \to z^+$. Similarly $z_n^- \to z^-$ and it follows easily that $\Phi'(z) = 0$. ∎

PROOF OF THEOREM 1 COMPLETED The theorem is a direct consequence of Theorem 2 in [9] once we check that, for every $n \in I\!N$,

$$\Phi(z) \to -\infty \text{ as } ||z||_E \to \infty, \quad z \in E^- \oplus E^0 \oplus E_n^+. \tag{2.13}$$

To see this, write $z = z^- + z^0 + z^+$ according to (2.2). Since $z^0 + z^+$ lies in a finite dimensional space, we see from the expression of the spaces in

(2.2) that we can still decompose orthogonally $z^- = z_1^- + z_2^-$ where z_2^- lies in a finite dimensional space and z_1^- is orthogonal to $\overline{z} := z_2^- + z^0 + z^+$ in $L^2(\Omega) \times L^2(\Omega)$. Using Hölder inequality and obvious notations, we have

$$\int_\Omega |u|^\alpha \geq c_1 (\int_\Omega |u|^2)^{\alpha/2} \geq c_2 (\int_\Omega |\overline{u}|^2)^{\alpha/2} \geq c_3 \|\overline{u}\|_{E^s}^\alpha$$

and similarly for $\int_\Omega |v|^\beta$. Thus, for $\gamma := \min\{\alpha, \beta\} > 2$,

$$\int_\Omega |u|^\alpha + \int_\Omega |v|^\beta \geq c_4 (\|\overline{z}\|_E^\gamma - 1) \geq c_5 (\|z^0\|_E^\gamma + \|z^+\|_E^\gamma - 1).$$

Using (1.10) we deduce

$$\Phi(z) \leq c_6 (\|z^+\|_E^2 - \|z^-\|_E^2 - \|z^0\|_E^\gamma - \|z^+\|_E^\gamma + 1)$$

and this proves (2.13) and finishes the proof. ∎

REFERENCES

[1]   V. Benci and P. Rabinowitz, *Critical point theorems for indefinite functionals*, Invent. Math. **52** (1979), 241-273.

[2]   D.G. Costa ad C. Magalhães, *A variational approach to noncooperative ellyptic systems*, Nonlinear Analysis TMA 25(1995) p. 699-715.

[3]   ——————————————, *A unified approach to a class of strongly indefinitc functionals*, J. Diff Eq. 125(1996).

[4]   D.G. de Figueiredo, *Semilinear elliptic systems: a survey of superlinear problems.* Relatório de Pesquisa Univ. Campinas, 1996. to appear in Resenhas.

[5]   D.G. de Figueiredo and P. Felmer, *On superquadratic elliptic systems*, Trans. Amer.Math. Soc. **343** (1996), 99-116.

[6]   D.G. de Figueiredo and C.A. Magalhães, *On nonquadratic Hamiltonian elliptic systems*, Advances in Differential Equations **1** (1996), 881-898.

[7]   J. Hulshof and R. van der Vorst, *Differential Systems with Strongly Indefinite Variational Sctruture*, J. Fctl. Anal. 114(1993) p. 32-58.

[8]   J.Q. Liu and S. Li, *Some existence theorems on multiple critical points and their applications*, Kexue Tongbao **17** (1984).

[9]   S. Li and M. Willem, *Application of local linking to critical point theory*, J. Math. Anal. Appl. **189** (1995), 6-32.

[10]   L.A. Peletier and R.C.A.M van der Vorst. *Existence and non-existence of positive solutions of non-linear elliptic systems and the biharmonic equation.* Diff. Int. Eq. 5(1992) p. 747-767.

[11]   H. Rabinowitz, *Critical Point Theory and Applications to Differential Equations: a survey*, Topological Nonlinear Analysis: Degree, Singularity and Variations. Colection Progress in Nonlinear Diff. Eq. and their applications, Birkhäuser (1994) vol. 15 p. 465-513.

[12]   E.A.B. Silva, *Critical point theorems and applications to differential equations*, thesis University of Wisconsin-Madson (1988).

[13]   E.A.B. Silva, *Linking theorems and applications to nonlinear elliptic equations at resonance.* Nonlinear Anal., TMA 16(1991) p. 455-477.

An Age Dependent Regularization of Martin's Problem

W. E. FITZGIBBON Department of Mathematics, University of Houston, Houston, Texas 77204-3476

J.J. MORGAN Department of Mathematics, Texas A& M University, College Station, Texas 77843

M.E. PARROTT Department of Mathematics, University of South Florida, Tampa, Florida 33620

G.F. WEBB Department of Mathematics, Vanderbilt University, Nashville, Tennessee 37240

1 INTRODUCTION

Several recent papers have concerned the diffusive spread of epidemics through age structured populations [3], [4], [5], [6], [7], [11], [12], [21]. Basically the models have equations of reaction diffusion type with one or more of the time derivatives replaced by an "age transport" term of the form $\partial/\partial t + \partial/\partial a$. In this case the state variable also depends on "a", the age of the disease within an individual as well as spatial location, x, and time, t. At first glance one might be led to think that the hybrid parabolic-hyperbolic nature of the system would complicate matters and thereby render its analysis more difficult. However, this is not always the case. Indeed, as we can see in [3], [4], [5], [6], that the presence of an age structure can make the requisite process of obtaining a priori estimates much more straightforward and simple. Speaking in rough terms the presence of age structure introduces a delay which will allow the diffusion operator to regularize the system.

Rather than overview diffusive, age dependent, epidemic models we choose to illustrate the final remark of the previous paragraph by demonstrating how the introduction of a fictive age structure can regularize Martin's problem. Martin's problem, named after its popularizer R.H. Martin Jr. has the nature of a Turing instability in that it illustrates how the addition of diffusion can complicate the dynamics of systems of ordinary differential equations. It concerns the global well-posedness and boundedness of balanced, positivity preserving reaction diffusion systems. The problem remains unsolved, however, recently we have witnessed some dramatic progress [10], [20].

2 MARTIN'S PROBLEM

In this section we describe the current state of affairs regarding Martin's problem. We begin with the introduction of the following system of ordinary differential equations

$$du/dt = -u\varphi(v) \tag{2.1a}$$
$$dv/dt = u\varphi(v) \tag{2.1b}$$

with

$$u(0) = u_0 > 0 \tag{2.1c}$$
$$v(0) = v_0 > 0 \tag{2.1d}$$

where $\varphi(\ ) \in C()$ with $\varphi(v) > 0$ for $v > 0$. Elementary analysis yields a complete description of the behavior of solutions to (2.1a-d). We know that the nonnegative quadrant $\frac{2}{+}$ is a flow invariant region and that solutions eminating from (u_0, v_0) flow along the line $u + v = k_0$ (with $u(t)$ decreasing and $v(t)$ increasing). Moreover,

$$\lim_{t \to \infty} u(t) = 0 \tag{2.2a}$$

and

$$\lim_{t \to \infty} v(t) = k_0 \tag{2.2b}$$

We now "add diffusion" to the system and consider the semilinear parabolic system with Neumann boundary conditions,

$$\partial u / \partial t = d_1 \Delta u - u\varphi(v), \tag{2.3a}$$
$$\partial v / \partial t = d_2 \Delta v + u\varphi(u), \tag{2.3b}$$

for $x \in \Omega$ and $t > 0$ with boundary conditions,

$$\partial u / \partial n = \partial v / \partial u = 0, \tag{2.3c}$$

for $x \in \partial\Omega$ and $t > 0$ and having initial data,

$$u(x, 0) = u_0(x), \tag{2.3d}$$
$$v(x, 0) = v_0(x). \tag{2.3e}$$

for $x \in \Omega$. We require that the diffusivites d_1, d_2 be positive and allow for the possibility that $d_1 \neq d_2$. The spatial domain Ω is assumed to be a bounded region of ^{n} and the boundary $\partial\Omega$ is assumed to be a $C^{2+\epsilon} (\epsilon > 0)$ $(n-1)$-dimensional manifold such that, locally, Ω lies on one side of $\partial\Omega$. Finally we assume that the initial data is positive and continuous, i.e.,

$$u_0(\ ), v_0(\ ) \in C(\overline{\Omega}) \text{ with } u_0(x), v_0(x) > 0 \text{ for } x \in \overline{\Omega}. \tag{2.4}$$

If u_0, v_0 are constant functions on $\overline{\Omega}$ then a moment's reflection leads to the observation that if $u(t), v(t)$ satisfy (2.1 a-d) then they are spatially homogeneous solutions to (2.3 a-e) and are the unique solutions to the initial boundary value problem (2.3 a-e). Solutions to (2.3 a-e) are locally well-posed and the global continuation theory is contingent upon obtaining continuous a prior estimates for $\|v(\ , t)\|_{\infty, \Omega}$. The Maximum Principle guarantees that

$$\|u(\ , t)\|_{\infty, \Omega} \leq \|u_0(\ )\|_{\infty, \Omega} \tag{2.5}$$

and we may observe that the nonnegative orthant is an invariant rectangle , cf. [19], for solutions. Thus, solutions with nonnegative initial data remain nonnegative on their interval of existence. If we formally integrate (2.3a) and (2.3b) on the space time cylinder $Q(t) = \Omega \times (0, t)$ and sum these results we obtain the a priori $L_1(\Omega)$ estimate

$$\|u(\ , t)\|_{1, \Omega} + \|v(\ , t)\|_{1, \Omega} = \|u_0\|_{1, \Omega} + \|v_0\|_{1, \Omega} = K_0 \tag{2.6}$$

This equality is sometimes called conservation of mass. The basic conundrum which has thus far stymied researchers is whether this a priori $L_1(\Omega)$ estimate can be bootstrapped to an $L_\infty(\Omega)$ estimate. Many recent papers [1], [10], [12], [11], [15], treat Martin's problem and its variants.

By use of a clever construction, cf. [9], one may observe that for sufficiently small $\delta, \epsilon > 0$

$$d/dt \int_\Omega \{1 + \delta(u(x,(\ ) + u^2(x,(\ ))e^{\epsilon v(x,t)}\} \, dx \leq 0. \tag{2.7}$$

This will produce uniform estimates for $\|v(\ ,t)\|_{p,\Omega}$ for $p > 1$. If the nonlinearity $\varphi(v)$ is polynomially bounded we can then apply either the parabolic regularity estimates of [14] or the semigroup methods in [18] to produce a priori estimates for $\iota\|v(\ ,t)\|_{\infty,\Omega}$. We state the following theorem whose proof can be found in [9].

THEOREM 2.7. *Let (2.8) and all the foregoing hypotheses hold. If there exist constants $k_1, k_2 > 0$ and $r \in {}^+$ so that for all $v \geq 0$*

$$\varphi(v) \leq k_1 v^r + k_2 \tag{2.9}$$

then (2.3a-e) has globally defined unique solutions. Moreover there exists a $\mathcal{K}_1 > 0$ so that

$$\sup_{t>0}\{\max[\|u(\ ,t)\|_{\infty,\Omega} \|v(\ ,t)\|_{\infty,\Omega}\} \leq \mathcal{K}_1. \tag{2.10}$$

We remark that the constant k_1 depends upon both initial functions and both diffusion constants. Moreover, a simple example in [11] shows that this constant cannot be obtained via a maximum principle argument. The presence of uniform L_∞ a priori estimates allows the complete determination of the longtime behavior of solutions to (2.3a-e). Toward this end we introduce the spatial averages

$$\bar{u}(t) = |\Omega|^{-1} \int_\Omega u(x,t) \, dx$$

$$\bar{v}(t) = |\Omega|^{-1} \int_\Omega v(x,t) \, dx \tag{2.11}$$

where $|\Omega|$ denotes the m-dimensional Lebesque measure of Ω. We have the following result.

PROPOSITION 2.12. *If the hypotheses of Theorem 2.8 hold and $v^* = K_0(\Omega)$ when K_0 is the constant given by (2.6) we have*

$$\lim_{t\to\infty} \|u(\ ,t)\|_{\infty,\Omega} = 0 \tag{2.13a}$$

and

$$\lim_{t\to\infty} \|v(\ ,t) - v^*\|_{\infty,\Omega} = 0. \tag{2.13b}$$

Proof. We begin by working with equation (2.3a). If we integrate on $Q(t)$ we may observe that,

$$\|(u,\ ,t)\|_{1,\Omega} = \|u_0\|_{1,\Omega} - \int_0^t \int_\Omega u(x,s)\phi(v,(x,s)) \, dx.$$

Because the integrand is positive we are assure the convergence of the improper integral

$$\int_0^\infty \|u\varphi(v)\|_{1,\Omega} \, dt = \int_0^\infty \int_\Omega u(x,t)\varphi(v(x,t)) \, dx \, dt. \tag{2.14}$$

By virtue of the fact that $u\varphi(v) > 0$ for $u > 0$, $v > 0$ we may use the estimates on the derivatives of u and v, cf. the comments following (2.20) to observe that

$$\lim_{t\to\infty} \int_\Omega u(x,t) \, dx = \lim_{t\to\infty} \|u(\,,t)\|_{1,\Omega} = 0 \tag{2.15}$$

and hence that

$$\lim_{t\to\infty} \overline{u}(t) = 0. \tag{2.16}$$

Because (2.6) holds we have

$$\lim_{t\to\infty} \int_\Omega v(x,t) \, dx = \lim_{t\to\infty} \|v(\,,t)\|_{1,\Omega} = K_0, \tag{2.17}$$

and

$$\lim_{t\to\infty} v(t) = K_0/|\Omega| = v^*. \tag{2.18}$$

Therefore our proof will be complete if we can establish that

$$\lim_{t\to\infty} \|u(\,,t) - \overline{u}(t)\|_{\infty,\Omega} = 0, \tag{2.19a}$$

and

$$\lim_{t\to\infty} \|v(\,,t) - \overline{v}(t)\| = 0.$$

We will limit ourselves to the second assertion and merely observe that the arguments which establish this assertion immediately adapt to the first case. We begin by multiplying equation (2.36) by v and integrating on $Q(t)$ to observe that,

$$\frac{1}{2} \|v(\,,t)\|_{2,\Omega}^2 + \int_0^t \int_\Omega |\nabla v|^2 \, dx \, dt = \frac{1}{2} \|v_0\|_{2,\Omega}^2$$

The uniform a priori estimates on $\|v(\,,t)\|_{\infty,\Omega}$ together with the convergence of $\int_0^t \int u\varphi(v) \, dx \, dt$ (cf. 2.14) guarantee that

$$\int_0^\infty \int_\Omega |\nabla v|^2 \, dx \, dt < \infty. \tag{2.20}$$

We now recall that invariance arguments (cf. [19]) can be extended to insure that $|\partial u/\partial t|$, $|\partial v/\partial t|$, $|\nabla u|$, $|\nabla u|$, $|\partial/\partial t \nabla u|$ and $|\partial/\partial t \nabla v|$ are all uniformly bounded and hence we may deduce from (2.20) that,

$$\lim_{t\to\infty} \||\nabla v|\|_{2,\Omega} = \int_\Omega |\nabla v|^2 \, dx = 0, \tag{2.21}$$

and by virtue of the uniform boundedness of $|\nabla v|$ we have,

$$\lim_{t\to\infty} \||\nabla v|\|_{p,\Omega} = 0, \tag{2.22}$$

for any $p > 0$. We further know, cf [19], that there exists a $C_2 > 0$ so that

$$\|v(\ ,t) - \overline{v}(t)\|_{2,\Omega} \leq C_2 \|\|\nabla v\|\|_{2,\Omega} \tag{2.23}$$

and hence a $C_p > 0$ so that

$$\|v(\ ,t) - \overline{v}(t)\|_{p,\Omega} \leq C_p \|\|\nabla v\|\|_{p,\Omega}. \tag{2.24}$$

Thus

$$\|v(,t) - \overline{v}(t)\|_{p,\Omega} + \|\|\nabla v\|\|_{p,\Omega} \leq (1 + C_p) \|\|\nabla v\|\|_{p,\Omega}$$

for any $p > 1$ and we invoke the Sobolov imbedding theorem to observe that,

$$\lim_{t \to \infty} \|v(\ ,t) - \overline{v}(t)\|_{\infty,\Omega} = 0, \tag{2.25}$$

and reach our desired conclusion.

We would like to point out that Condition 2.9 is certainly not the optimal restriction upon the growth of $\varphi(v)$. Indeed the argument of [9] covers the case of $\varphi(v) = e^{v\alpha}$ with $0 < \alpha < 1$. A yet unpublished manuscript [10] covers the case of $\varphi(v) = e^v$ for the pure initial valued problem with initial data continuous and bounded in ^{n}.

3 AN "AGE DEPENDENT" MARTIN'S PROBLEM

In this section we introduce a variant of Martin's problem with an artifical age dependence. We let $a \in [0, \infty)$ be another independent variable and select $\theta(\ ) \in C^1[0, \infty)$ so that

$$\int_0^\infty \theta(a)\ da = 1 \tag{3.1}$$

$$\theta(a) \geq 0 \ \text{ for } a \geq 0 \tag{3.2}$$

$$\lim_{a \to \infty} \theta(a) = 0 \tag{3.3}$$

and define $\ell_0(x, a)$ by

$$\ell_0(x, a) = v_0(x)\theta(a) \text{ for } x \in \Omega, \ a \geq 0$$

whose $v_0(\ )$ is the initial function introduced in the previous sections. We consider the following system of partial differential equations with state variables u and ℓ, and $\tau > 0$.

$$\partial u / \partial t = d_1 \Delta u - u\phi\left(\int_\tau^\infty \ell(x, a, t)\ da\right) \tag{3.4a}$$

$$\partial \ell / \partial t + \partial \ell / \partial a = d_2 \Delta \ell \tag{3.4b}$$

for $x \in \Omega, t > 0$ and $a \geq 0$. We impose homogeneous Neumann boundary conditions,

$$\partial u / \partial n = 0, \tag{3.4c}$$

$$\partial \ell / \partial n = 0, \tag{3.4d}$$

for $x \in \partial\Omega, \ t > 0$ and $a \geq 0$ and prescribe initial data

$$u(x, 0) = u_0(x) \qquad \text{for } x \in \overline{\Omega}. \tag{3.4e}$$

$$\ell(x, a, 0) = \ell_0(x, a) = v_0(x)\theta(a) \qquad \text{for } x \in \overline{\Omega}, a > 0. \tag{3.4f}$$

We also require that

$$\ell(x, 0, t) = B(x, t) = u(x, t)\varphi \int_C^\infty \ell(x, a, t) \, da \qquad \text{for } x \in \overline{\Omega}, \ t \geq 0. \tag{3.4g}$$

If we define

$$v(x, t) = \int_\tau^\infty \ell(x, a, t) \, da \tag{3.5a}$$

we may formally integrate (3.4b) in (τ, ∞) to produce

$$\partial v/\partial t = d_2 \Delta v + \ell(x, \tau, t) \tag{3.5b}$$

with

$$\partial v/\partial n = 0 \qquad \text{for } x \in \partial\Omega, \ t > 0 \tag{3.5c}$$

and

$$v(x, 0) = v_0(x) \qquad \text{for } x \in \overline{\Omega}. \tag{3.5d}$$

Here we have made the assumption that $\lim_{a \to \infty} \ell(x, a, t) = 0$.

The well-posedness theory for (3.4a-g) is well in hand. Essentially the fact that $\tau > 0$ allows us to effectively decouple the system with delay and apply the method of steps to obtain globally defined solutions. The following result can be obtained by a minor modification of arguments in [3], [5] and we state it without proof.

THEOREM 3.6. *Let $u_0(\ )$, $v_0(\ )$ satisfy the hypotheses of the proceeding section, assume that $\theta(\ )$ satisfies (3.1-3.3) and set*

$$\ell_0(x, a) = v_0(x)\theta(a) \text{ for } x \in \Omega, \ a \geq 0. \tag{3.7}$$

Then there exists a unique pair $u(x, t), \ell(x, a, t)$ which is continuously differentiable if $a \neq t$, nonnegative, and satisfies (3.4 a-g).

Our result allows for the possibility of a jump discontinuity across the characteristic line $a = t$. The reason for this jump will become apparent. The well-posedness result incorporates a representation result. We let $\{T(t)|t \geq 0\}$ be the analytic semigroup and $\rho(x, t)$ be the Green's function associated with

$$\partial z/\partial t = d_2 \Delta z \qquad x \in \Omega, \ t > 0 \tag{3.8a}$$
$$\partial z/\partial n = 0 \qquad x \in \partial\Omega, \ t > 0. \tag{3.8b}$$

Then it may be seen that solutions to (3.4b) have the form

$$\ell(x, a, t) = \begin{cases} (T(t)\ell_0(\ , a - t))(x) & \text{if } x \in \Omega, \quad a > t \\ (T(a)B(\ , t - a))(x) = \ (T(a)(u(\ , t - a)\varphi(v(\ , t - a))(x) & \text{if } x \in \Omega, t > a \end{cases} \tag{3.9}$$

or using Green's functions we have

$$\ell(x, a, t) = \begin{cases} \int_\Omega \rho(x - y, t)\ell_0(y, a - t) \, dy & \text{if } x \in \Omega, \ a > t \\ \int_\Omega \rho(x - y, a)u(y, t - a)\varphi(v(y, t - a)) \, dy & \text{if } x \in \Omega, \ t \geq a \end{cases} \tag{3.10}$$

We keep this in mind. We may observe that $\lim_{a \to \infty} \ell(x, a, t) = 0$ and we may write (3.4b) for $t > \tau$ as

$$\partial v / \partial t = d_2 \Delta v + T(\tau)(u(\ , t - \tau)\varphi(v(\cdot,\ t - \tau)))(x) \tag{3.11}$$

or

$$\partial v / \partial t = d_2 \Delta v + \int_\Omega \rho(x - y, \tau)u(y, t - \tau)\varphi(v(y, t - \tau)) \, dy. \tag{3.12}$$

Thus, the system consisting of (3.4a) and (3.11 or 3.12) with initial conditions $u(x, 0) = u_0(x)$, $v(x, 0) = v_0(x)$ and boundary conditions $\partial u / \partial n = 0$ for $x \in \partial\Omega, t > 0$ may be considered as a parabolic regularization of the Martin's problem of previous section.

In this case it becomes relatively easily to obtain uniform a priori estimates for $\|v(\ , t)\|_{\infty,\Omega}$.

PROPOSITION 3.13. *If $t > \tau$ and $v(x, t)$ is defined via (3.5a) then there exists $\hat{K}_1 > 0$ so that*

$$\sup_{t>0} \|v(\ , t)\|_{\infty,\Omega} \leq \hat{K}_1 \tag{3.14}$$

Proof. We begin with equation (3.4a) and integrate on $Q(t)$ to observe that,

$$\int_0^t \|u(\ , s)\varphi(v(\ , s))\|_{1,\Omega} \, ds < \infty \tag{3.15}$$

If we apply a variation of parameters formula to (3.5b) we may observe that

$$\|v(\ , t)\|_{\infty,\Omega} \leq \|v(\ , \tau)\|_{\infty,\Omega} + \int_\tau^t \|\ell(\ , \tau, s)\|_{\infty,\Omega} \, ds. \tag{3.16}$$

However, applying Hölder's inequality we have,

$$\|\ell(\ , \tau, s)\|_{\infty,\Omega} \leq \|\rho(\ , \tau)\|_{\infty,\Omega} \|u(\ , s - \tau)\varphi(v(\ , s - \tau))\|_{1,\Omega}. \tag{3.17}$$

We let $K(\tau) = \|\rho(\ , \tau)\|_{\infty,\Omega}$ and observe that

$$\|v(\ , t)\|_{\infty,\Omega} \leq \|v(\ , \tau)\|_{\infty,\Omega} + K(\tau) \int_0^\infty \|u(\ , s)\varphi(v(\ , s))\|_{1,\Omega} \, ds \tag{3.18}$$

and (3.14) now follows from 3.15.

It is also possible to characterize the long term behavior of solutions.

THEOREM 3.19. *If the foregoing hypotheses hold and $u(x, t)$ and $v(x, t)$ are solutions to (3.4a) and (3.5b) respectively, then*

$$\lim_{t \to \infty} \|u(\ , t)\|_{\infty,\Omega} = 0 \tag{3.20}$$

and there exists a $v^\sharp \leq v^* = K_0/|\Omega|$ so that

$$\lim_{t \to \infty} \|v(\ , t) - v^\sharp\|_{\infty,\Omega} = 0 \tag{3.21}$$

Proof. The arguments that lim

follow in essentially the same manner as Proposition (2.12). If we integrate (3.5b) on Ω we may use the positivity of $\ell(x, \tau, t)$ to see that

$$d/dt \int_\Omega v(x,t) \, dt \geq 0 \qquad \text{for } t > 0 \tag{3.22}$$

and hence there is a $v^\sharp$ so that

$$\lim_{t \to \infty} \overline{v}(t) = v^\sharp. \tag{3.23}$$

We let $B(x,s) = u(x,s)\varphi(v(x,s))$ and introduce

$$w(x,t) = \int_0^\infty \ell(x,a,t) \, da. \tag{3.24}$$

It should be clear that

$$v(x.t) \leq w(x,t) \qquad \text{for } x \in \Omega, \ t > 0. \tag{3.25}$$

Moreover, if we integrate (3.4b) on $(0,\infty)$ we obtain:

$$\partial w/\partial t = d_2 \Delta w + B(x,t). \tag{3.26}$$

Furthermore, $\partial u/\partial t$ satisfies,

$$\partial u/\partial t = d_1 \Delta u - B(x,t). \tag{3.27}$$

The uniform estimates on $\|v(\ ,t)\|_{\infty,\Omega}$ together with the representation formula (3.9) and (3.10) will insure uniform estimates on $\|w(\ ,t)\|_{\infty,\Omega}$. Consequently, we can apply the arguments of Proposition (2.12) to see that

$$\lim_{t \to \infty} \|w(\ ,t) = v^*\|_{\infty,\Omega} = 0 \tag{3.28}$$

where $v^* = K_0/|\Omega|$ and K_0 is the constant of (2.6). By virtue of (3.25) we have

$$v^\sharp \leq v^* \tag{3.29}$$

and conclude our proof.

If $\tau > 0$ is small one can see that $v^* - v^\sharp$ is small and expect solution pairs satisfying (3.4a) and (3.5b) to be good approximations to solutions to (2.3a) and (2.3b). Nevertheless we cannot, in general, let $\tau \downarrow 0$ and guarantee that solutions to (3.4a) and (3.5b) will converge to solutions to (2.3a-b). The difficulty lies in the fact that $\rho(\ ,\tau)_{\infty,\Omega}$ blows up as $\tau \downarrow 0$.

In the parlance of mathematical epidimiology the term $\ell(x,a,t)$ is called a time dependent density of infection with respect to space and the age of the disease in a given individual. The independent variable a representing the age of the disease is the time elapsed since contracting the infection. The spatial density of individuals who have had the disease between a_1 and a_2 units of time would be computed by integration over the age interval, i.e., we evaluate $\int_{a_1}^{a_2} \ell(x,a,t) \, dt$. For example, one may wish to consider an infected population subdivided into two classes, exposed individuals and infectuous individuals, whose spatial densities will be denoted by $y(x,t)$ and $v(x,t)$. Here one might assume that individuals who contract the disease do not immediately become infected but enter a period of latency or incubation which lasts a fixed period of time $\tau > 0$. After the latent period individuals enter the infectious state. Here we would have

$$y(x,t) = \int_0^\tau \ell(x,a,t) \, da$$

and

$$v(x,t) = \int_\tau^\infty \ell(x,a,t) \, da.$$

The term $f(x,t) = u(x,t)\varphi(v(x,t))$, called an incidence term, describes the transfer from the susceptible to the infectious class. The model provided by 3.4a-g, which is admittedly both simple and unrealistic, would be called an SE (Susceptible-Infective) Model.

Heretofore, one of the standard methods of modelling the spread of infectious disease has involved reaction diffusion equations. We feel that more complete descriptions are given by models such as (3.4a-g) which involve an age dependence. We also feel that the introduction of an age variable or similar structural variable has potential for playing a role in other situations where reaction diffusion equations are applied.

REFERENCES

1. J. Bebernes and A. Lacey, *Finite blow up for a parabolic system*, preprint.
2. S. Bonafede and D. Schmidt, *Triangular reaction diffusion systems with integrable data*, Preprint, Institute Henri Cartan, 1996.
3. W. Fitzgibbon, M. Langlais, M. Parrott, G. Webb, *A diffusive system with age dependence modelling FIV*, Nonlinear Anal: TMA **25** (1995), 975-989.
4. W.E. Fitzgibbon, J. Morgan, and M. Parrott, *Periodicity in diffusive age structured SEIR models*, preprint.
5. W.E. Fitzgibbon, M. Parrott, G. Webb, *Diffusive epidemic models with spatial and age dependent heterogeneity*, Discrete and Cont. Systems **1** (1995), 35-57.
6. _____ , *A diffusive age structured SEIRS model*, preprint.
7. _____ , *Diffusion epidemic models with incubation and criss-cross dynamics*, Math. Bioscience (to appear).
8. A. Haraux and M. Kirane, *Extimation C' des problemes paraboliques semilinéaires*, Ann. Fac. Sci. Toulouse **5** (1983), 265-280.
9. N. Haraux and A. Youkana, *On a result of K. Masuda concerning reaction diffusion equations*, Tokuku J. Math., SIAM J. Math. Anal **40** (1988), 159-163.
10. M. Herrero, A. Lacey and J. Velazquez, *Global existence for systems modelling ignition* (to appear).
11. S. Hollis, R. Martin, and M. Pierre, *Global existence and boundedness for reaction diffusion systems*, SIAM J. Math. Anal. **8** (1987), 747-762.
12. Y.I. Kanel, *Cauchy's problem for semilinear parabolic equations with balance conditions*, Trans. Diff. Uray. **20** (1984), 1753-1760.
13. M. Kubo and M. Langlais, *Periodic solutions for a population dynamics problem with age dependence and spatial structure*, J. Math. Biology **26** (1991), 363-378.
14. O. Ladyshenskaya, V. Solonnikov and N. Ural'ceva, *Linear and Quasilinear Equations of Parabolic Type*, AMS Trans., Vol 23, Amer. Math. Soc., Providence, 1968.
15. M. Langlais, *Remarks on an epidemic model with age structure*, Mathematical Population Dynamics, Lecture Notes in Pure and Applied Mathematics **131** (1991), Marcel Dekker, New York, 319-346.
16. R. Martin and M. Pierre, *Nonlinear reaction diffusion systems*, Nonlinear Equations in the Applied Sciences (ed. W. Ames), Academic Press, San Diego, 1992.
17. K. Masuda, *On the global existence and asymptotic behavior of solutions of reaction diffusive systems*, Hokkaido Math. J. **12** (1982), 360-370.
18. J. Morgan, *Boundedness and decay for reaction diffusion systems*, SIAM J. Math. Anal. **20** (1989), 1128-1149.
19. J. Smoller, *Shock Waves and Reactive Diffusion Systems*, Springer Verlag, New York, 1984.
20. M. Pierre and D. Schmitt, *Blow-up in reaction diffusion systems with dissipation*, to appear in SIAM J. Math. Anal.
21. G.F. Webb, *An age dependent epidemic model with spatial diffusion*, Arch. Rat. Mech. Anal. **75** (1981), 150-161.

Uniqueness and Positivity for Solutions of Equations with the p-Laplacian

JACQUELINE FLECKINGER-PELLÉ   ICM et CEREMATH, Université des Sciences Sociales, 21 Allee de Brienne, F–31042 Toulouse Cedex, France; *e-mail:* `jfleck@cict.fr`

JESÚS HERNÁNDEZ   Departamento de Matemáticas, Universidad Autónoma de Madrid, E–28049 Madrid, Spain; *e-mail:* `hernande@ccuam3.sdi.uam.es`

PETER TAKÁČ   Fachbereich Mathematik, Universität Rostock, Universitätsplatz 1, D–18055 Rostock, Germany; *e-mail:* `peter.takac@mathematik.uni-rostock.de`

FRANÇOIS DE THÉLIN   UMR MIP et UFR MIG, Université Paul Sabatier, 118 Route de Narbonne, F–31062 Toulouse Cedex, France; *e-mail:* `dethelin@cict.fr`

The problem of *existence*, *uniqueness* and *positivity* for weak solutions u to the following elliptic boundary value problem is investigated,

$$-\operatorname{div}(|\nabla u|^{p-2}\nabla u) = \mu a(x)|u|^{p-2}u + h(x) \ \text{ in } \Omega; \quad u = 0 \ \text{ on } \partial\Omega. \qquad (P)$$

Here, $p \in (1, \infty)$ is a given number, $\Omega \subset \mathbb{R}^N$ is a bounded domain with a $C^{1,\alpha}$-boundary $\partial\Omega$, for some $\alpha \in (0, 1)$, μ is a constant, and $a, h \in L^\infty(\Omega)$.

It is assumed that $0 \leq a \not\equiv 0$ and $h \not\equiv 0$ in Ω, and $-\infty < \mu < \mu_1(a)$, where $\mu_1(a)$ denotes the first eigenvalue associated with the corresponding eigenvalue problem for $\mu \in \mathbb{R}$ (with $h \equiv 0$ in (P)). For $h \geq 0$ throughout Ω, it is shown that Problem (P) has a unique weak solution $u \in W_0^{1,p}(\Omega)$. This solution satisfies also $u > 0$ in Ω and $\partial u/\partial \nu < 0$ on $\partial\Omega$. For $p \neq 2$, $a \equiv 1$ and $0 < \mu < \mu_1(1)$, a function h (with indefinite sign) is constructed such that Problem (P) has at least two weak solutions in $W_0^{1,p}(\Omega)$. Also analogous problems with $f(x, u(x))$ in place of $h(x)$ are addressed. The proofs of these results make use of standard methods of variational calculus.

1   INTRODUCTION

We consider the following elliptic boundary value problem,

$$- \Delta_p u = \mu a(x)\psi_p(u) + h(x) \ \text{ in } \Omega; \qquad u = 0 \ \text{ on } \partial\Omega. \tag{1}$$

Here, $\Omega \subset \mathbb{R}^N$ is a bounded domain with a $C^{1,\alpha}$-boundary $\partial\Omega$, for some $\alpha \in (0,1)$, Δ_p denotes the p-Laplacian defined by $\Delta_p u \overset{\text{def}}{=} \operatorname{div}(|\nabla u|^{p-2}\nabla u)$ for $p \in (1, \infty)$, and $\psi_p(u) \overset{\text{def}}{=} |u|^{p-2}u$. We denote by $\nu \equiv \nu(x_0)$ the exterior unit normal to $\partial\Omega$ at $x_0 \in \partial\Omega$. Finally, $\mu \in \mathbb{R}$ is a constant, and $a, h : \Omega \to \mathbb{R}$ are given functions such that $a, h \in L^\infty(\Omega)$, $0 \leq a \not\equiv 0$ and $h \not\equiv 0$ in Ω. The function $a(x)$ plays the role of a weight. Until present, most of the work concerning the p-Laplacian has considered only a being a constant. In this note we have decided to treat the more general case of $a(x)$ being a function of $x \in \Omega$, as the same mathematical techniques apply.

We define

$$\lambda_1 \equiv \lambda_1(\Omega) \overset{\text{def}}{=} \inf \left\{ \int_\Omega |\nabla u|^p \, dx : u \in W_0^{1,p}(\Omega) \ \text{ with } \ \int_\Omega |u|^p \, dx = 1 \right\}.$$

It is well-known, cf. Anane [1, Théorème 1, p. 727], that $\lambda_1 > 0$, and λ_1 is the first eigenvalue of the negative Dirichlet p-Laplacian $-\Delta_p$ in Ω. In analogy with the definition of λ_1, we introduce the number

$$\mu_1(a) \equiv \mu_1(a, \Omega)$$
$$\overset{\text{def}}{=} \inf \left\{ \int_\Omega |\nabla u|^p \, dx : u \in W_0^{1,p}(\Omega) \ \text{ with } \ \int_\Omega a|u|^p \, dx = 1 \right\}.$$

Again, cf. Anane [1], one can show that $\mu_1(a) > 0$, and $\mu_1(a)$ is the first eigenvalue associated with the corresponding eigenvalue problem for $\mu \in \mathbb{R}$,

$$- \Delta_p u = \mu a(x)\psi_p(u) \ \text{ in } \Omega; \qquad u = 0 \ \text{ on } \partial\Omega. \tag{2}$$

Notice that if a is a positive constant, then $\mu_1(a)a = \lambda_1$.

For $-\infty < \mu < \mu_1(a)$, the existence of a weak solution $u \in W_0^{1,p}(\Omega)$ to Problem (1) follows easily (see J.-L. Lions [11]) from the weak lower semicontinuity and coercivity of the functional

$$\mathcal{J}(u) \stackrel{\text{def}}{=} \frac{1}{p}\int_\Omega |\nabla u|^p \, dx - \frac{\mu}{p}\int_\Omega a|u|^p \, dx - \int_\Omega hu \, dx, \quad u \in W_0^{1,p}(\Omega). \tag{3}$$

Indeed, every local minimizer for the functional $\mathcal{J}$ over $W_0^{1,p}(\Omega)$ is a weak solution to Problem (1). Furthermore, for $-\infty < \mu \leq 0$, the uniqueness of such a solution u follows from the strict convexity of the functional $\mathcal{J}$. However, the uniqueness of u, which holds for $p = 2$ and μ not an eigenvalue by the Fredholm alternative in $L^2(\Omega)$, is false for $p \neq 2$ and $0 < \mu < \mu_1(a)$ unless an additional restriction is imposed on h. For $p \neq 2$, $\Omega = (-1, 1)$, $a \equiv 1$ in Ω, and any constant μ with $0 < \mu < \lambda_1$, it is possible to construct a function $h \in L^\infty(\Omega)$ (with indefinite sign) such that Problem (P) has at least two weak solutions in $W_0^{1,p}(\Omega)$. In fact, for $2 < p < \infty$, this nonuniqueness was shown in del Pino, Elgueta and Manasevich [12, Eq. (5.26), p. 12], cf. Example 1 below. For $1 < p < 2$, we show it in our Example 2.

Therefore, in this note we focus our attention on the special case $h \geq 0$ in Ω. We extend the uniqueness result for u to $0 < \mu < \mu_1(a)$, see Theorem 1 below. The existence of a (strictly) positive weak solution $u \in W_0^{1,p}(\Omega)$ to (1) has been established in Fleckinger, Hernández and de Thélin [4, Théorème 2, p. 666] or [5, Theorem 6]. In fact, our methods apply to any $\mu \in (-\infty, \mu_1(a))$. Our proof makes essentiall use of the variational method due to Díaz and Saa [3, Lemme 1 et 2, p. 522] who treated the case when the function $h(x)$ is replaced by a function $f(x, u)$ with $u \mapsto f(x, u)/u^{p-1}$ strictly decreasing in $(0, \infty)$, for almost every $x \in \Omega$ ([3, Théorème 1 et 2, p. 521]). Notice that $u \mapsto h(x)/u^{p-1}$ is not strictly decreasing for such $x \in \Omega$ at which $h(x) = 0$. ¿From Theorem 1 we deduce a weak comparison principle (Theorem 2) which extends a result for $\mu \leq 0$ due to Tolksdorf [13, Lemma 3.1, p. 800] (see also Fleckinger and Takáč [6, Prop. 1, p. 448] or [7, Prop. 4.1, p. 1235]) also to any constant $\mu < \mu_1(a)$.

Finally, in Theorem 3 we present a generalization of Theorem 1 with $f(x, u(x))$ in place of $h(x)$. In Díaz and Saa [3, Théorème 1 et 2, p. 521], an inequality [3, Lemme 2, p. 522] was employed to investigate the existence and uniqueness in the case when $u \mapsto f(x, u)/u^{p-1}$ is strictly decreasing in $(0, \infty)$, for almost every $x \in \Omega$. This condition on f is satisfied if, in particular, the function $v \mapsto f(x, v^{1/p})v^{1/p}$ is strictly concave in $[0, \infty)$, for almost every $x \in \Omega$. In our proof of Theorem 3, which is given in the Appendix, we use an approach different from [3]. We take advantage of the strict convexity of the functional $v \mapsto \mathcal{J}(v^{1/p})$ for $v > 0$ in Ω with $v^{1/p} \in W_0^{1,p}(\Omega)$, cf. [3, Lemme 1, p. 522]. This is the case provided the function f satisfies the latter (stronger) condition.

2   THE MAIN RESULT

The following theorem is our main result; the uniqueness is new even if a is a positive constant and $0 < \mu a < \lambda_1 = \mu_1(a)a$:

THEOREM 1 *Let $\Omega \subset \mathbb{R}^N$ be a bounded domain with a $C^{1,\alpha}$-boundary $\partial\Omega$, for some $\alpha \in (0,1)$, and $p \in (1, \infty)$. Assume that $a, h \in L^\infty(\Omega)$ are given functions satisfying $0 \leq a \not\equiv 0$ and $0 \leq h \not\equiv 0$ in Ω. Let μ be a number with $-\infty < \mu < \mu_1(a)$. Then the boundary value problem (1) has a unique weak solution $u \in W_0^{1,p}(\Omega)$. Moreover, this solution satisfies $u \in C^{1,\beta}(\overline{\Omega})$, for some $\beta \in (0, \infty)$, and also the strong maximum and boundary point principles, respectively,*

$$u > 0 \ \ in \ \Omega \qquad and \qquad \frac{\partial u}{\partial \nu} < 0 \ \ on \ \partial\Omega. \tag{4}$$

Proof. First we prove the uniqueness.

As the first step of our proof, we show that every weak solution $u \in W_0^{1,p}(\Omega)$ of Problem (1) must be nonnegative, that is, $u \geq 0$ almost everywhere in Ω. We recall that if $a(x) > 0$ for a.e. $x \in \Omega$, then this claim can be proved in the same way as in the case $a = \text{const}$ in Fleckinger, Hernández and de Thélin [4, Théorème 1, p. 666] or [5, Theorem 5]. If a is not positive almost everywhere in Ω, we need to modify their proof as follows.

We set $u^+ = \max\{u, 0\}$ and $u^- = \max\{-u, 0\}$; thus $u = u^+ - u^-$ and $|u| = u^+ + u^-$. Then also $u^+, u^- \in W_0^{1,p}(\Omega)$ with $\nabla u^+ = 0$ a.e. in $\Omega_- =$

$\{x \in \Omega : u(x) \le 0\}$ and $\nabla u^- = 0$ a.e. in $\Omega_+ = \{x \in \Omega : u(x) \ge 0\}$, see Gilbarg and Trudinger [8, Theorem 7.8, p. 153]. Now we multiply Eq. (1) by u^- and integrate over Ω, thus obtaining

$$-\int_\Omega |\nabla u^-|^p \, dx = -\mu \int_\Omega a(x)(u^-)^p \, dx + \int_\Omega h(x)u^- \, dx. \tag{5}$$

Next we notice that, by our definition of the number $\mu_1(a)$, we have

$$\int_\Omega |\nabla u^-|^p \, dx \ge \mu_1(a) \int_\Omega a(x)(u^-)^p \, dx \tag{6}$$

where the equality holds if and only if u^- satisfies

$$-\Delta_p(u^-) = \mu_1(a)a(x)(u^-)^{p-1} \text{ in } \Omega; \qquad u^- = 0 \text{ on } \partial\Omega. \tag{7}$$

Inserting (6) into Eq. (5) we arrive at

$$(\mu_1(a) - \mu)\int_\Omega a(x)(u^-)^p \, dx + \int_\Omega h(x)u^- \, dx \le 0. \tag{8}$$

As $\mu < \mu_1(a)$, this inequality forces

$$\int_\Omega a(x)(u^-)^p \, dx = 0 \quad \text{and} \quad \int_\Omega h(x)u^- \, dx = 0, \tag{9}$$

together with Eq. (7). The strong maximum principle from Tolksdorf [13, Prop. 3.2.2, p. 801] or Vázquez [15, Theorem 5, p. 200] applied to Eq. (7) yields that either $u^- \equiv 0$ or else $u^- > 0$ throughout Ω. However, the latter case contradicts Eqs. (9). We have proved that $u \ge 0$ a.e. in Ω.

Obviously $u \not\equiv 0$ in Ω, since $h \not\equiv 0$ in Ω. Furthermore, $a, h \in L^\infty(\Omega)$ forces $u \in L^\infty(\Omega)$, by Anane [2, Théorème A.1, p. 96]. ¿From the regularity result of Lieberman [10, Theorem 1, p. 1203] (and Tolksdorf [14]) we then obtain even $u \in C^{1,\beta}(\overline{\Omega})$ with a constant $\beta \equiv \beta(\alpha, p, a, N)$, $0 < \beta < 1$. The strong maximum and boundary point principles from Tolksdorf [13, Prop. 3.2.2, p. 801] or Vázquez [15, Theorem 5, p. 200] then yield the desired inequalities (4). Consequently, it suffices to consider only those weak solutions u of Problem (1) which belong to the set

$$\overset{\circ}{V}_+ \overset{\text{def}}{=} \{f \in C^1(\overline{\Omega}) : f > 0 \text{ in } \Omega \text{ and } \frac{\partial f}{\partial \nu} < 0 \text{ on } \partial\Omega\}.$$

Finally, let us consider two weak solutions $u_1, u_2 \in \overset{\circ}{V}_+$ of Problem (1). An inequality due to Díaz and Saa [3, Lemme 2, p. 522] reads

$$\int_\Omega \left(\frac{-\Delta_p u_1}{u_1^{p-1}} - \frac{-\Delta_p u_2}{u_2^{p-1}} \right) (u_1^p - u_2^p) \, dx \ge 0 \tag{10}$$

where the equality holds if and only if u_1 and u_2 are colinear (i.e., $u_2 = \xi u_1$ in Ω for some $\xi \in (0, \infty)$). We insert Eq. (1) for u_1 and u_2 into (10), which then becomes

$$\int_\Omega h(x) \left(\frac{1}{u_1^{p-1}} - \frac{1}{u_2^{p-1}} \right) (u_1^p - u_2^p)\, dx \geq 0. \tag{11}$$

However, the integrand on the left-hand side is nonpositive, and thus, it must vanish almost everywhere in Ω. It follows from (10) that $u_2 = \xi u_1$ in Ω for some $\xi \in (0, \infty)$. Inserting this relation into Eq. (1) for u_1 and u_2, we conclude that $(\xi - 1)h \equiv 0$ in Ω. So $\xi = 1$ and $u_2 = u_1$ must be valid.

The existence of a weak solution $u \in W_0^{1,p}(\Omega) \cap \overset{\circ}{V}_+$ to (1) has been established in Fleckinger, Hernández and de Thélin [4, Théorème 2, p. 666] or [5, Theorem 6]. We have finished our proof of Theorem 1. ∎

3 A WEAK COMPARISON PRINCIPLE

In this section we extend the standard weak comparison principle for the Dirichlet problem (1) for $\mu \leq 0$ due to Tolksdorf [13, Lemma 3.1, p. 800] (see also Fleckinger and Takáč [6, Prop. 1, p. 448] or [7, Prop. 4.1, p. 1235]) also to any constant $\mu < \mu_1(a)$.

THEOREM 2 *Let Ω and a be as in Theorem 1, $1 < p < \infty$ and $-\infty < \mu < \mu_1(a)$. Assume that $u, u' \in W_0^{1,p}(\Omega)$, respectively, are weak solutions of*

$$-\Delta_p u = \mu a(x)\psi_p(u) + h \quad \text{in } \Omega; \qquad u = 0 \quad \text{on } \partial\Omega, \tag{12}$$

$$-\Delta_p u' = \mu a(x)\psi_p(u') + h' \quad \text{in } \Omega; \qquad u' = 0 \quad \text{on } \partial\Omega, \tag{13}$$

where $0 \leq h \leq h'$ in $L^\infty(\Omega)$. Then $0 \leq u \leq u'$ throughout Ω with $u, u' \in C^{1,\beta}(\overline{\Omega})$, for some $\beta \in (0, 1)$.

Proof. For $\mu \leq 0$, see [13] (or [6, 7]). ¿From now on we assume $0 \leq \mu < \mu_1(a)$. Set $L_+^\infty(\Omega) \overset{\text{def}}{=} \{f \in L^\infty(\Omega) : f \geq 0 \text{ a.e. in } \Omega\}$. Given any $h \in L_+^\infty(\Omega)$, define the mapping $T_h : L_+^\infty(\Omega) \to L_+^\infty(\Omega)$ by $v \mapsto T_h v \overset{\text{def}}{=} w$, where $w \in W_0^{1,p}(\Omega)$ is a weak solution of

$$-\Delta_p w = a(x)\psi_p(v) + h \quad \text{in } \Omega; \quad w = 0 \quad \text{on } \partial\Omega. \tag{14}$$

The following properties of the mapping T_h are proved in Fleckinger and Takáč [7, Lemma 6.1, p. 1243]:

(i) $T_h : L_+^\infty(\Omega) \to L_+^\infty(\Omega)$ is continuous and maps bounded sets into relatively compact sets in $C^{1,\beta}(\overline{\Omega})$, for some $\beta \in (0,1)$.

(ii) The mapping $(v,h) \mapsto T_h v$ is monotone, i.e., $v \le v'$ and $h \le h'$ in $L_+^\infty(\Omega)$ implies $T_h u \le T_{h'} u'$.

By Theorem 1, Problem (1) has a unique weak solution $u \in W_0^{1,p}(\Omega)$. Consequently, the sequence $0 \le T_h 0 \le (T_h)^2 0 \le \cdots \le (T_h)^n 0 \le \cdots \le u$ converges in $C^{1,\beta}(\overline{\Omega})$ to the smallest weak solution $\underline{u} \in W_0^{1,p}(\Omega)$ of Problem (1). Hence $\underline{u} = u$ and $(T_h)^n 0 \nearrow u$ in $C^{1,\beta}(\overline{\Omega})$ as $n \to \infty$. Similarly $(T_{h'})^n 0 \nearrow u'$. Since also $(T_h)^n 0 \le (T_{h'})^n 0$ for all $n \ge 1$, by $0 \le h \le h'$, we arrive at $0 \le u \le u'$ in Ω. ■

4 NONUNIQUENESS

We present two examples of nonuniqueness of a weak solution to Problem (1) for $0 < \mu < \mu_1(a)$. The first one, for $2 < p < \infty$, is due to del Pino, Elgueta and Manasevich [12, Eq. (5.26), p. 12], cf. Example 1 below. In the second one, for $1 < p < 2$, we need to modify their variational method, see Example 2. The main difference between the two examples is the degeneracy/singularity of the p-Laplacian at $\nabla u = 0$: Δ_p is degenerate for $2 < p < \infty$ and singular for $1 < p < 2$. In both these examples, the nonuniqueness for $0 < \mu < \mu_1(a)$ is caused by the fact that the functional $\mathcal{J} : W_0^{1,p}(\Omega) \to \mathbb{R}$ defined in (3) is not convex, although it is still coercive. Consequently, $\mathcal{J}$ being also weakly lower semicontinuous, it possesses a global minimizer u^* in $W_0^{1,p}(\Omega)$. Then u^* is a weak solution to Problem (1). Hence, in order to show nonuniqueness, it suffices to construct a function $u_0 \in W_0^{1,p}(\Omega)$ with the following properties:

$$
\begin{cases}
u_0 \in C^2(\overline{\Omega}), \quad -\Delta_p u_0 \in C^0(\overline{\Omega}), \quad \text{and for} \\
h(x) \stackrel{\text{def}}{=} -\Delta_p u_0 - \mu a(x) \psi_p(u_0), \quad x \in \Omega, \\
u_0 \text{ is } not \text{ a local minimizer for } \mathcal{J}.
\end{cases}
\tag{15}
$$

So u_0 is another weak solution to Problem (1) with $u_0 \not\equiv u^*$.

As $1 < p < \infty$, we set $p' = p/(p-1)$. In both examples below we take the domain $\Omega = (-1,1)$, $a \equiv 1$ in Ω, and $0 < \mu < \mu_1(1) = \lambda_1$.

EXAMPLE 1 ([12, Eq. (5.26), p. 12]) Let $2 < p < \infty$. Fix any number $\varepsilon \in (0,1)$. Let $u_0 \in C^2([-1,1])$ be any function such that $u_0(1) = u_0(-1) = 0$ and $u_0(x) = \text{const} \neq 0$ for $|x| \leq \varepsilon$. Define $h \in C^0([-1,1])$ by

$$h(x) \stackrel{\text{def}}{=} -\Delta_p u_0 - \mu a(x)\psi_p(u_0) \quad \text{for} \ \ x \in [-1,1]. \tag{16}$$

We claim that at u_0, $\mathcal{J}$ does not attain its minimum in $W_0^{1,p}([-1,1])$, and therefore, there are two distinct solutions to Problem (1), namely, u_0 and u^*.

To prove this claim, we note that for $2 < p < \infty$, $\mathcal{J}$ is twice Fréchet differentiable in $W_0^{1,p}([-1,1])$. If $\mathcal{J}''(u_0)$ denotes the second derivative of $\mathcal{J}$ at u_0, then

$$\langle \mathcal{J}''(u_0)v, w \rangle = (p-1)\left(\int_{-1}^{1} |u_0'|^{p-2} v' w' \, dx - \mu \int_{-1}^{1} |u_0|^{p-2} vw \, dx \right) \tag{17}$$

for all $v, w \in W_0^{1,p}([-1,1])$. Here, $\langle \bullet, \bullet \rangle$ denotes the duality pairing between $W_0^{1,p}([-1,1])$ and its dual space $W^{-1,p'}([-1,1])$. Next let $z \in C^1([-1,1])$ be any nontrivial function such that $z(x) = 0$ for $\varepsilon \leq |x| \leq 1$. We find from (17) and the definition of u_0 that

$$\langle \mathcal{J}''(u_0)z, z \rangle = -(p-1)\mu \int_{-1}^{1} |u_0|^{p-2} z^2 \, dx < 0,$$

which shows that at u_0, the functional $\mathcal{J}$ does not attain its minimum.

EXAMPLE 2 Let $1 < p < 2$. Fix any numbers ε, ε_1 and ε_2 such that $0 < \varepsilon < \varepsilon_1 < \varepsilon_2 < 1/2$. For a constant m to be determined later, $p' \leq m < \infty$, let $u_0 \in C^2([-1,1])$ be any function such that

$$\begin{cases} u_0(x) = |x|^m & \text{for} \ \ |x| \leq \varepsilon_1; \\ x u_0'(x) > 0 & \text{for} \ \ \varepsilon_1 \leq |x| \leq \varepsilon_2; \\ u_0(x) = \frac{1}{2^m} - \left| \frac{1}{2} - |x| \right|^m & \text{for} \ \ \varepsilon_2 \leq |x| \leq 1. \end{cases} \tag{18}$$

Notice that the condition $x u_0'(x) > 0$ for $\varepsilon_1 \leq |x| \leq \varepsilon_2$ can be satisfied, because

$$u_0(\pm\varepsilon_1) < u_0(\pm\varepsilon_2) \quad \text{follows from} \quad \varepsilon_1^m + \left| \frac{1}{2} - \varepsilon_2 \right|^m < \frac{1}{2^m}.$$

Clearly, we have $u_0(1) = u_0(-1) = 0$ and $\Delta_p u_0 \in C^0([-1,1])$. Again, define $h \in C^0([-1,1])$ by (16). We claim that at u_0, $\mathcal{J}$ does not attain its minimum in $W_0^{1,p}([-1,1])$.

To prove this claim, we note that for $1 < p < \infty$, $\mathcal{J}$ is once Fréchet differentiable in $W_0^{1,p}([-1,1])$. If $\mathcal{J}'(u)$ denotes the first derivative of $\mathcal{J}$ at $u \in W_0^{1,p}([-1,1])$, then

$$\langle \mathcal{J}'(u), v \rangle = \int_{-1}^{1} |u'|^{p-2} u' v' \, dx - \mu \int_{-1}^{1} |u|^{p-2} uv \, dx - \int_{-1}^{1} hv \, dx \qquad (19)$$

for all $v \in W_0^{1,p}([-1,1])$. Next let $z \in C^1([-1,1])$ be any function such that

$$\begin{cases} z(x) = 1 & \text{for } |x| \leq \varepsilon; \\ 0 \leq z(x) \leq 1 & \text{for } \varepsilon \leq |x| \leq \varepsilon_1; \\ z(x) = 0 & \text{for } \varepsilon_1 \leq |x| \leq 1. \end{cases} \qquad (20)$$

We find from (19) and the definition of z that

$$\langle \mathcal{J}'(u), z \rangle = \int_{-\varepsilon_1}^{\varepsilon_1} |u'|^{p-2} u' z' \, dx - \mu \int_{-\varepsilon_1}^{\varepsilon_1} |u|^{p-2} uz \, dx - \int_{-\varepsilon_1}^{\varepsilon_1} hz \, dx \qquad (21)$$

for all $u \in W_0^{1,p}([-1,1])$.

Now we investigate the function

$$\zeta(t) \overset{\text{def}}{=} \frac{1}{t} \left(\langle \mathcal{J}'(u_0 + tz), z \rangle - \langle \mathcal{J}'(u_0), z \rangle \right) \qquad \text{for } 0 < |t| < 1. \qquad (22)$$

Observe that $\zeta(t) = J(t) + J_1(t)$ for $0 < |t| < 1$, where

$$J(t) = -\frac{\mu}{t} \int_{-\varepsilon}^{\varepsilon} \left(|u_0 + t|^{p-2}(u_0 + t) - |u_0|^{p-2} u_0 \right) dx \qquad (23)$$

and

$$\begin{aligned} J_1(t) &= \tfrac{1}{t} \int_{\varepsilon \leq |x| \leq \varepsilon_1} \left(|u_0' + tz'|^{p-2}(u_0' + tz') - |u_0'|^{p-2} u_0' \right) z' \, dx \\ &\quad - \tfrac{\mu}{t} \int_{\varepsilon \leq |x| \leq \varepsilon_1} \left(|u_0 + tz|^{p-2}(u_0 + tz) - |u_0|^{p-2} u_0 \right) z \, dx. \end{aligned} \qquad (24)$$

Since $u_0(x) = |x|^m$ for $|x| \leq \varepsilon_1$, we obtain the (possibly infinite) limit

$$\begin{aligned} J(0) &= \lim_{t \to 0} J(t) = \\ &\quad -(p-1)\mu \int_{-\varepsilon}^{\varepsilon} |u_0|^{p-2} \, dx = -(p-1)\mu \int_{-\varepsilon}^{\varepsilon} |x|^{m(p-2)} \, dx, \end{aligned} \qquad (25)$$

whence $-\infty \leq J(0) < 0$, and the finite limit

$$J_1(0) = \lim_{t \to 0} J_1(t) = (p-1) \int_{\varepsilon \leq |x| \leq \varepsilon_1} \left(|u_0'|^{p-2} |z'|^2 - \mu |u_0|^{p-2} z^2 \right) dx. \qquad (26)$$

Combining (22) with (25) and (26), we arrive at

$$\zeta(t) = \frac{1}{t}\left(\langle \mathcal{J}'(u_0 + tz), z\rangle - \langle \mathcal{J}'(u_0), z\rangle\right)$$

$$\xrightarrow{t\to 0} -(p-1)\mu \int_{-\varepsilon}^{\varepsilon} |x|^{m(p-2)}\, dx + J_1(0).$$

We find that $\zeta(0) = \lim_{t\to 0} \zeta(t) = -\infty < 0$ provided $m(p-2) \le -1$, which shows that at u_0, the functional $\mathcal{J}$ does not attain its minimum.

Last but not least, notice that the constant m in (18) can be any number satisfying

$$\max\{p/(p-1), 1/(2-p)\} \le m < \infty.$$

This means $m \ge p' = p/(p-1)$ for $1 < p \le (1+\sqrt{5})/2$, and $m \ge 1/(2-p)$ for $(1+\sqrt{5})/2 \le p < 2$.

5   A GENERALIZATION OF THE MAIN RESULT

Finally, we present an obvious generalization of our Theorem 1 to the following Dirichlet boundary value problem which includes (1), cf. [4, Théorème 2, p. 666] or [5, Theorem 6]:

$$-\Delta_p u = \mu a(x)\psi_p(u) + f(x, u(x)) \quad \text{in } \Omega; \qquad u = 0 \quad \text{on } \partial\Omega. \quad (27)$$

We assume that $a \in L^\infty(\Omega)$ is a given function satisfying $0 \le a \not\equiv 0$ in Ω, μ is a number with $-\infty < \mu < \mu_1(a)$, and the function $f : \Omega \times \mathbb{R} \to \mathbb{R}$, which is a Carathéodory function, satisfies

$$0 \le f(x, u) \le C(1 + |u|^{\sigma(p-1)}) \quad \text{for all } u \in \mathbb{R} \text{ and a.e. } x \in \Omega, \quad (28)$$

where $C \ge 0$ and $0 \le \sigma < 1$ are some constants, and

(H) for a.e. $x \in \Omega$, the function $v \in \mathbb{R}_+ \mapsto f(x, v^{1/p})v^{1/p}$ is concave, and for all $x \in \Omega'$ from some set $\Omega' \subset \Omega$ of positive Lebesgue measure, this function is strictly concave.

We have denoted $\mathbb{R}_+ = [0, \infty)$.

Then a similar proof as for Theorem 1 yields the following analogous result:

THEOREM 3 *Let Ω and a be as in* Theorem 1, $1 < p < \infty$, *and assume that both hypotheses (28) and (H) are satisfied. Then every weak solution $u \in W_0^{1,p}(\Omega)$ of the boundary value problem (27) satisfies $u \in C^{1,\beta}(\overline{\Omega})$, for some $\beta \in (0, \infty)$. Furthermore, we have the following two alternatives:*

(i) *If $f(\bullet, 0) \not\equiv 0$ in Ω, then* Problem (27) *has a unique weak solution $u \in W_0^{1,p}(\Omega)$. Moreover, the strong maximum and boundary point principles (4) hold.*

(ii) *If $f(\bullet, 0) \equiv 0$ in Ω, then, besides the trivial solution $u_0 \equiv 0$ in Ω,* Problem (27) *has at most one other weak solution $u \in W_0^{1,p}(\Omega)$. The nontrivial solution obeys (4).*

The complete *proof* of Theorem 3 is given in the Appendix below.

To investigate the existence and uniqueness for Problem (27), Díaz and Saa [3, Théorème 1 et 2, p. 521] employed an inequality [3, Lemme 2, p. 522], which required the following hypothesis to be imposed on $f(x, u)$:

(DS) the function $u \mapsto f(x, u)/u^{p-1}$ of $u \in (0, \infty)$ is nonincreasing for a.e. $x \in \Omega$ and strictly decreasing for all $x \in \Omega'$ from some set $\Omega' \subset \Omega$ of positive Lebesgue measure.

To see that our hypothesis (H) implies (DS), notice that, for a.e. $x \in \Omega$, if the function $v \in \mathbb{R}_+ \mapsto f(x, v^{1/p})v^{1/p}$ is concave (strictly concave, respectively), then

$$f(x, (0 + \lambda v)^{1/p})(0 + \lambda v)^{1/p} \geq (>) \, 0 + \lambda f(x, v^{1/p})v^{1/p} \quad \text{for each } \lambda \in (0, 1).$$

Taking $u = v^{1/p}$ and $u' = (\lambda v)^{1/p}$, we arrive at $f(x, u')/(u')^{p-1} \geq (>)$ $f(x, u)/u^{p-1}$ whenever $0 < u' < u$.

REMARK 1 A result similar to Díaz and Saa [3, Théorème 1 et 2, p. 521] and to our Theorem 3, as well, has been obtained recently by Idogawa and Ôtani [9, Theorem 4, p. 8]. They work in a more general setting which uses abstract convex functionals and their subdifferentials.

6   APPENDIX

Here, we *prove* Theorem 3.

As the first step of this proof, in Problem (27) we replace the unknown function $u \in W_0^{1,p}(\Omega)$ by $v = |u|^{p-1}u \equiv \psi_{p+1}(u)$ with

$$|v|^{(1/p)-1}v \equiv \psi_{(1/p)+1}(v) = u \in W_0^{1,p}(\Omega).$$

Notice that $\psi_p(\psi_{(1/p)+1}(v)) \equiv \psi_{2-(1/p)}(v)$. Then Problem (27) is equivalent to

$$\begin{cases} -\Delta_p(\psi_{(1/p)+1}(v)) = \mu a(x)\psi_{2-(1/p)}(v) \\ \qquad\qquad +f(x, \psi_{(1/p)+1}(v(x))) \quad \text{in } \Omega; \\ \qquad v = 0 \quad \text{on } \partial\Omega; \quad \psi_{(1/p)+1}(v) \in W_0^{1,p}(\Omega). \end{cases} \tag{29}$$

It is a basic fact from the calculus of variations (see J.-L. Lions [11]) that a function v with $\psi_{(1/p)+1}(v) \in W_0^{1,p}(\Omega)$ is a weak solution of Problem (29) if and only if v is a critical point of the functional $\mathcal{K}(v) = \mathcal{J}(\psi_{(1/p)+1}(v))$ defined by

$$\begin{aligned} \mathcal{K}(v) \stackrel{\text{def}}{=} & \tfrac{1}{p}\int_\Omega |\nabla(\psi_{(1/p)+1}(v))|^p\, dx - \tfrac{\mu}{p}\int_\Omega a|v|\, dx \\ & - \int_\Omega f(x, \psi_{(1/p)+1}(v(x)))\psi_{(1/p)+1}(v)\, dx \\ \equiv & \tfrac{1}{p}\mathcal{K}_1(v) - \tfrac{\mu}{p}\mathcal{K}_2(v) - \mathcal{K}_3(v), \quad \psi_{(1/p)+1}(v) \in W_0^{1,p}(\Omega). \end{aligned} \tag{30}$$

In the second step we recall that every weak solution $u = |v|^{(1/p)-1}v \in W_0^{1,p}(\Omega)$ of Problem (27) must be nonnegative, $u \geq 0$ almost everywhere in Ω, by Fleckinger, Hernández and de Thélin [4, Théorème 1, p. 666] or [5, Theorem 5]. If $f(\bullet, 0) \not\equiv 0$ in Ω, then obviously $u \not\equiv 0$ in Ω. Furthermore, $h \in L^\infty(\Omega)$ forces $u \in L^\infty(\Omega)$, by Anane [2, Théorème A.1, p. 96]. ¿From the regularity result of Lieberman [10, Theorem 1, p. 1203] (and Tolksdorf [14]) we then obtain even $u \in C^{1,\beta}(\overline{\Omega})$ with a constant $\beta \equiv \beta(\alpha, p, a, N)$, $0 < \beta < 1$. If $u \not\equiv 0$, then the strong maximum and boundary point principles from Tolksdorf [13, Prop. 3.2.2, p. 801] or Vázquez [15, Theorem 5, p. 200] yield the desired inequalities (4). Consequently, it suffices to consider the restriction

$$\begin{aligned} \mathcal{K}(v) &= \tfrac{1}{p}\int_\Omega |\nabla(v^{1/p})|^p\, dx - \tfrac{\mu}{p}\int_\Omega av\, dx - \int_\Omega f(x, v^{1/p})v^{1/p}\, dx, \\ v \in \overset{\bullet}{V}_+ &\stackrel{\text{def}}{=} \{v : \Omega \to (0, \infty) \mid v^{1/p} \in W_0^{1,p}(\Omega) \cap C_0(\overline{\Omega})\}. \end{aligned} \tag{31}$$

In the third step we prove the strict convexity of $\mathcal{K}$ restricted to the set $\overset{\bullet}{V}_+$. By a result due to Díaz and Saa [3, Lemme 1, p. 522] (and its proof), the first functional $\mathcal{K}_1 : \overset{\bullet}{V}_+ \to \mathbb{R}$ in (30) is *ray-strictly convex*: for all $\theta \in (0, 1)$ and $v_1, v_2 \in \overset{\bullet}{V}_+$, we have

$$\mathcal{K}_1((1 - \theta)v_1 + \theta v_2) \leq (1 - \theta)\mathcal{K}_1(v_1) + \theta\mathcal{K}_1(v_2)$$

where the equality holds if and only if v_1 and v_2 are colinear (i.e., $v_2 = \xi v_1$ or $v_1 = \xi v_2$ for some $\xi \in \mathbb{R}_+$).

To give a brief idea of the proof of this result, one first denotes $a_1 = (1-\theta)|\nabla v_1(x)|$, $b_1 = (1-\theta)v_1(x)$ and $a_2 = \theta|\nabla v_2(x)|$, $b_2 = \theta v_2(x)$ for almost every fixed $x \in \Omega$, then proves the following lemma, and finally integrates the inequality (32) (below) over Ω.

LEMMA 4 *Let* $a_i \in [0, \infty)$ *and* $b_i \in (0, \infty)$ *for* $i = 1, 2$, *and* $p \in (1, \infty)$. *Then we have*

$$\frac{(a_1 + a_2)^p}{(b_1 + b_2)^{p-1}} \leq \frac{a_1^p}{b_1^{p-1}} + \frac{a_2^p}{b_2^{p-1}} \tag{32}$$

where the equality holds if and only if $a_1/b_1 = a_2/b_2$.

Proof. The function $t \mapsto t^p$ of $t \in \mathbb{R}_+$ is strictly convex. Thus, we obtain

$$\left(\frac{b_1}{b_1 + b_2} \frac{a_1}{b_1} + \frac{b_2}{b_1 + b_2} \frac{a_2}{b_2} \right)^p \leq \frac{b_1}{b_1 + b_2} \left(\frac{a_1}{b_1} \right)^p + \frac{b_2}{b_1 + b_2} \left(\frac{a_2}{b_2} \right)^p,$$

where the equality holds if and only if $a_1/b_1 = a_2/b_2$. Simplifying all expressions we arrive at (32). ∎

The second functional $\mathcal{K}_2 : \overset{\bullet}{V}_+ \to \mathbb{R}$ in (30) is linear.

Finally, by the hypothesis (H), the function $v \in \mathbb{R}_+ \mapsto f(x, v^{1/p})v^{1/p}$ is concave for a.e. $x \in \Omega$, and strictly concave for all $x \in \Omega'$ from some set $\Omega' \subset \Omega$ of positive Lebesgue measure. Hence, the third functional $\mathcal{K}_3 : \overset{\bullet}{V}_+ \to \mathbb{R}$ in (30) is concave and, moreover, if $v_i \in L^1_+(\Omega)$ ($i = 1, 2$) and $0 < \theta < 1$, then

$$\mathcal{K}_3((1 - \theta)v_1 + \theta v_2) \geq (1 - \theta)\mathcal{K}_3(v_1) + \theta\mathcal{K}_3(v_2)$$

where the equality holds only if $v_1 = v_2$ almost everywhere in Ω'. In particular, if v_1 and v_2 are colinear, say $v_2 = \xi v_1$ for some $\xi \in \mathbb{R}_+$, and also $v_1 > 0$ in Ω, then $(\xi - 1)v_1 = 0$ a.e. in Ω' forces $\xi = 1$, i.e., $v_1 = v_2$ in Ω.

Thus, we have just proved

PROPOSITION 5 *The functional $\mathcal{K}$ defined in (30) is strictly convex in $\overset{\bullet}{V}_+$.*

Using the well-known characterization of a critical point of a convex functional (which is assumed to be Gâteaux-differentiable), see J.-L. Lions [11], namely, that every critical point is a global minimizer (with the set of all global minimizers being convex), we conclude that $\mathcal{K}$ has at most one critical point in $\overset{\bullet}{V}_+$, namely, its unique minimizer $v = u^p$ in $\overset{\bullet}{V}_+$. Knowing (4) for $u = v^{1/p} \not\equiv 0$, we conclude that Problem (29) has at most one nontrivial weak solution v with $|v|^{(1/p)-1}v \in W_0^{1,p}(\Omega)$ or, equivalently, our original Problem (27) has at most one nontrivial weak solution $u = |v|^{(1/p)-1}v \in W_0^{1,p}(\Omega)$.

The existence of a nonnegative weak solution $u \in W_0^{1,p}(\Omega)$ to Problem (29) has been established in Fleckinger, Hernández and de Thélin [4, Théorème 2, p. 666] or [5, Theorem 6]. We have finished our proof of Theorem 3.

Acknowledgments The research of J. Fleckinger-Pellé was supported in part by the European Community Contract ERBCHRXCT930409. The research of J. Hernández was supported in part by the European Community Contract ERBCHRXCT930409, and by DGICYT (Spain) Project PB93/0443. The research of P. Takáč was supported in part by the U.S. National Science Foundation Grant DMS-9401418 to Washington State University, Pullman, WA 99164–3113, U.S.A. The research of F. de Thélin was supported in part by the European Community Contract ERBCHRXCT930409.

References

1. Anane, A. (1987). Simplicité et isolation de la première valeur propre du p-laplacien avec poids, *Comptes Rendus Acad. Sc. Paris, Série I*, **305**: 725–728.

2. Anane, A. (1988). "Etude des valeurs propres et de la résonance pour l'opérateur p-Laplacien", *Thèse de doctorat*, Université Libre de Bruxelles, Brussels.

3. Díaz, J. I., and Saa, J. E. (1987). Existence et unicité de solutions positives pour certaines équations elliptiques quasilinéaires, *Comptes Rendus Acad. Sc. Paris, Série I*, **305**: 521–524.

4. Fleckinger, J., Hernández, and de Thélin, F. (1992). Principe du maximum pour un système elliptique non linéaire, *Comptes Rendus Acad. Sc. Paris, Série I*, **314**: 665–668.

5. Fleckinger, J., Hernández, and de Thélin, F. (1995). On maximum principles and existence of positive solutions for some cooperative elliptic systems, *Diff. and Int. Equations*, **8**: 69–85.

6. Fleckinger, J. and Takáč, P. (1994). Unicité de la solution d'un systéme non linéaire strictement coopératif, *Comptes Rendus Acad. Sc. Paris, Série I*, **319**: 447–450.

7. Fleckinger, J. and Takáč, P. (1994). Uniqueness of positive solutions for nonlinear cooperative systems with the *p*-Laplacian, *Indiana Univ. Math. J.*, **43**(4): 1227–1253.

8. Gilbarg, D. and Trudinger, N. S. (1977). "Elliptic Partial Differential Equations of Second Order", Springer-Verlag, New York–Berlin–Heidelberg.

9. Idogawa, T. and Ôtani, M. (1995). The first eigenvalues of some abstract elliptic operators, *Funkcialaj Ekvacioj*, **38**: 1–9.

10. Lieberman, G. (1988). Boundary regularity for solutions of degenerate elliptic equations, *Nonlinear Anal.*, **12**(11): 1203–1219.

11. Lions, J.-L. (1969). "Quelques méthodes de résolution des problèmes aux limites non linéaires", Dunod et Gauthier Villars, Paris.

12. del Pino, M., Elgueta, M., and Manasevich, R. (1989). A homotopic deformation along p of a Leray-Schauder degree result and existence for $(|u'|^{p-2}u')' + f(t,u) = 0$, $u(0) = u(T) = 0$, $p > 1$, *J. Differential Equations*, **80**(1): 1–13.

13. Tolksdorf, P. (1983). On the Dirichlet problem for quasilinear equations in domains with conical boundary points, *Comm. P.D.E.*, **8**(7): 773–817.

14. Tolksdorf, P. (1984). Regularity for a more general class of quasilinear elliptic equations, *J. Differential Equations*, **51**: 126–150.

15. Vázquez, J. L. (1984). A strong maximum principle for some quasilinear elliptic equations, *Appl. Math. Optim.*, **12**: 191–202.

On the First Curve in the
Fučik Spectrum of a Mixed Problem

J.-P. GOSSEZ [1] Département de Mathématique, C.P. 214, Université Libre de Bruxelles, 1050 Bruxelles, Belgique.

A. MARCOS [2] Institut de Mathématiques et de Sciences Physiques, B.P. 613, Porto Novo, Bénin.

1 INTRODUCTION

The Fučik spectrum of the Laplacian under the Dirichlet boundary conditions is defined as the set Σ of those $(\alpha, \beta) \in \mathbb{R}^2$ such that the problem

$$\begin{cases} -\Delta u = \alpha u^+ - \beta u^- \text{ in } \Omega, \\ u = 0 \text{ on } \partial\Omega \end{cases} \tag{1}$$

has a nontrivial solution u. Here $u^+ = \max(u, 0)$, $u = u^+ - u^-$, and Ω is a smooth bounded domain in R^N, $N \geq 1$. One recovers the usual spectrum by taking $\alpha = \beta$ in (1). This generalized spectrum plays a central role in the study of asymmetric semilinear problems, as was already observed in the pioneering papers of Fučik [8] and Dancer [3]. Recent works dealing with Σ include [4],[10], [1], [7], ...

Of special interest for our present purposes is [7] where the existence of a first non trivial curve in Σ is derived and some of its properties established. In particular it is shown there that this first curve is asymptotic to the lines $\{\lambda_1\} \times \mathbb{R}$ and $\mathbb{R} \times \{\lambda_1\}$. The case of the Neumann boundary is also considered in [7]. In this case the first curve is still asymptotic to the lines $\{\tilde{\lambda}_1\} \times \mathbb{R}$ and $\mathbb{R} \times \{\tilde{\lambda}_1\}$ when $N \geq 2$ but exhibits a gap at infinity with respect to these lines when $N = 1$. (We have denoted by $\lambda_1 > 0$ and $\tilde{\lambda}_1 = 0$ the principal eigenvalues of $-\Delta$ under respectively the Dirichlet and the Neumann boundary conditions). This difference in the asymptotic behaviour of the first curve has several interpretations and consequences. In particular, as shown

---

[1] Partially supported by E.C. project ERBCHRXCT 940555.

[2] This work was carried out during a stay of this author at U.L.B., which is here acknowledged for its support.

in [7], the existence or non existence of a gap at infinity turns out to be closely related to the validity of an uniform form of the antimaximum principle of [2].

The question now naturally arises of the study of the beginning of the Fučik spectrum and in particular of the asymptotic behaviour of the first curve under other boundary conditions. It is our purpose in this paper to investigate a case which is somehow intermediate between Dirichlet's and Neumann's, i.e. the case of the classical "mixed boundary conditions". We show that the situation here turns out to be similar to that in the Dirichlet case, i.e. that the first curve is always asymptotic to the vertical and horizontal lines through the corresponding principal eigenvalue. As a consequence, the "uniform antimaximum principle" does not hold under these boundary conditions.

For simplicity we limit ourselves below to the Laplacian operator, but our results can clearly be extended to a second order symmetric elliptic operator in divergence form with suitable regularity assumptions on the coefficients.

2   STATEMENT

Let Ω be a smooth bounded domain in $\mathbb{R}^N$, $N \geq 1$, whose boundary $\partial\Omega$ is made of two closed disjoint non empty pieces Γ_1 and Γ_2, with Γ_1 and Γ_2 smooth manifolds of dimension $N - 1$.

The Fučik spectrum of the Laplacian under mixed boundary conditions is defined as the set Σ' of those $(\alpha, \beta) \in \mathbb{R}^2$ such that

$$\begin{cases} -\Delta u = \alpha u^+ - \beta u^- \text{ in } \Omega, \\ u = 0 \text{ on } \Gamma_1, \partial u/\partial n = 0 \text{ on } \Gamma_2 \end{cases} \tag{1}$$

has a non-trivial solution u. Here $\partial/\partial n$ denotes the exterior normal derivative. We recall that u is a (weak) solution of (1) if $u \in E(\Omega)$ and

$$\int_\Omega \nabla u \cdot \nabla v = \alpha \int_\Omega u^+ v - \beta \int_\Omega u^- v \tag{2}$$

for all $v \in E(\Omega)$, where the space $E(\Omega)$ is defined as

$$E(\Omega) = \{v \in H^1(\Omega) \, ; \, v = 0 \text{ on } \Gamma_1 \text{ in the sense of traces }\}$$

(cf. e.g. [5]). Using the regularity of Ω, one can show that $(\int_\Omega |\nabla v|^2)^{\frac{1}{2}})$ is a norm on $E(\Omega)$ which is equivalent to the $H^1(\Omega)$ norm (cf. [9, 11]).

We will denote by

$$\mu_1 = \inf\{\int_\Omega |\nabla u|^2 / \int_\Omega u^2 \, ; \, u \in E(\Omega), u \neq 0\}$$

the principal eigenvalue of $-\Delta$ under the above mixed boundary conditions. Arguing as in [6], one can show that μ_1 is > 0, simple, and that the associated L^2

normalized eigenfunction Ψ_1 can be taken > 0 in Ω. Clearly the lines $\{\mu_1\} \times \mathbb{R}$ and $\mathbb{R} \times \{\mu_1\}$ belong to the spectrum Σ'.

To motivate our construction of a first curve in Σ', we observe that if u solves (1), then, by taking $v = \Psi_1$ in (2), one has

$$\int_\Omega ((\alpha - \mu_1)u^+\Psi_1 - (\beta - \mu_1)u^-\Psi_1) = 0.$$

This is some sort of asymmetric orthogonality relation between u and Ψ_1 which is going to be crucial in our approach.

Define, for $r > 0$,

$$\mathcal{M}_r = \{u \in E(\Omega)\,;\, \int_\Omega (u^+\Psi_1 - ru^-\Psi_1) = 0\},$$

$$\mathcal{N}_r = \{u \in E(\Omega)\,;\, \int_\Omega ((u^+)^2 - r(u^-)^2) = 1\},$$

and let

$$\begin{cases} \alpha = \alpha(r) = \mu_1 + \inf\{\int_\Omega(|\nabla u|^2 - \mu_1 u^2)\,;\, u \in \mathcal{M}_r \cap \mathcal{N}_r\}, \\ \beta = \beta(r) = \mu_1 + r(\alpha(r) - \mu_1). \end{cases} \quad (3)$$

The following facts can be proved by a simple adaptation of the arguments in [7]:

- α and β defined by (3) are $> \mu_1$ and (α, β) is the first intersection point with Σ' of the line of slope r emanating from (μ_1, μ_1),

- the infimum in (3) is achieved precisely by the solutions of (1) which are normalized so as to belong to $\mathcal{N}_r$,

- the set $\mathcal{C}_1 = \{(\alpha(r), \beta(r))\,;\, r > 0\} \subset \mathbb{R}^2$ is a continuous strictly decreasing curve, which is symmetric with respect to the diagonal,

- $\alpha(r) \to +\infty$ as $r \to 0$ and $\beta(r) \to +\infty$ as $r \to +\infty$.

With respect to the asymptotic behaviour of the first curve $\mathcal{C}_1$, we have the following

PROPOSITION 2.1. $\alpha(r) \to \mu_1$ as $r \to +\infty$ and $\beta(r) \to \mu_1$ as $r \to 0$, i.e. the curve $\mathcal{C}_1$ is asymptotic to the lines $\{\mu_1\} \times \mathbb{R}$ and $\mathbb{R} \times \{\mu_1\}$.

The proof of this proposition is given in the following section. It borrows some ideas from [7] but new difficulties arise in connexion with the use of the space $E(\Omega)$.

3   PROOF OF PROPOSITION 2.1

By symmetry it suffices to prove that $\alpha(r) \to \mu_1$ as $r \to +\infty$, and we already know that $\alpha(r)$ decreases as r increases to infinity. Assume by contradiction that there exists $\eta > 0$ such that $\alpha(r) \geq \mu_1 + \eta$ for all $r > 0$. This means that

$$\int_\Omega (|\nabla u|^2 - \mu_1 u^2) \geq \eta \quad (1)$$

for all $u \in E(\Omega)$ such that, for some $r > 0$, $u \in \mathcal{M}_r \cap \mathcal{N}_r$.

For $\varepsilon > 0$ sufficiently small let us take a nonempty open ball $\tilde{\Omega} = \tilde{\Omega}_\varepsilon$ with

$$\tilde{\Omega} = \tilde{\Omega}_\varepsilon \subset \{x \in \Omega \, ; \, \text{dist}\,(x, \Gamma_1) < \varepsilon\}$$

and such that

$$\hat{\Omega} = \hat{\Omega}_\varepsilon = \text{int}\,(\Omega \setminus \tilde{\Omega}_\varepsilon)$$

is a smooth bounded domain. On $\hat{\Omega}$ we consider the boundary conditions of Neumann on Γ_2 and of Dirichlet on $\partial\hat{\Omega} \setminus \Gamma_2$; we denote by $\hat{\mu}_1 = \hat{\mu}_{1,\varepsilon}$ the corresponding principal eigenvalue of $-\Delta$ and by $\hat{\Psi}_1 = \hat{\Psi}_{1,\varepsilon}$ the associated L^2 normalized positive eigenfunction. By the regularity of $\hat{\Omega}$, extending $\hat{\Psi}_1$ by zero on $\Omega \setminus \hat{\Omega}$ yields a function in $E(\Omega)$, which we still denote by $\hat{\Psi}_1$. On $\tilde{\Omega}$ we consider the Dirichlet boundary conditions; we denote by $\tilde{\mu}_1 = \tilde{\mu}_{1,\varepsilon}$ the corresponding principal eigenvalue of $-\Delta$ and by $\tilde{\Psi}_1 = \tilde{\Psi}_{1,\varepsilon}$ the associated L^2 normalized positive eigenvalue. Extending $\tilde{\Psi}_1$ by zero on $\Omega \setminus \tilde{\Omega}$ yields a function in $H^1_0(\Omega) \subset E(\Omega)$, which we still denote by $\tilde{\Psi}_1$.

Define, for $\delta > 0$,

$$u = u_{\varepsilon,\delta} = \begin{cases} \hat{\Psi}_1 & \text{ou } \hat{\Omega}, \\ -\delta\tilde{\Psi}_1 & \text{ou } \tilde{\Omega} \end{cases}$$

Clearly $u = \hat{\Psi}_1 - \delta\tilde{\Psi}_1 \in E(\Omega)$ and we have

$$\begin{aligned} \int_\Omega \left(|\nabla u|^2 - \mu_1 u^2\right) &= \int_{\hat{\Omega}} \left(|\nabla\hat{\Psi}_1|^2 - \mu_1\hat{\Psi}_1^2\right) + \delta^2 \int_{\tilde{\Omega}} \left(|\nabla\tilde{\Psi}_1|^2 - \mu_1\tilde{\Psi}_1^2\right) \\ &= (\hat{\mu}_1 - \mu_1) + \delta^2(\tilde{\mu}_1 - \mu_1). \end{aligned}$$

We first choose $\varepsilon > 0$ such that $\hat{\mu}_1 - \mu_1 < \eta/4$, which is possible by lemma 3.1 below. We then choose $\delta > 0$ such that $\delta^2(\tilde{\mu}_1 - \mu_1) < \eta/4$. With this choice of ε, δ, one has

$$\int_\Omega \left(|\nabla u|^2 - \mu_1 u^2\right) < \eta/2. \tag{2}$$

Let us now look at the constraint $\mathcal{M}_r$. Since u changes sign, we have that $u \in \mathcal{M}_r$ for a suitable $r = r(\varepsilon, \delta)$. For the other constraint $\mathcal{N}_r$, we have

$$\begin{aligned} \int_\Omega \left((u^+)^2 + r(U^-)^2\right) &= \int_{\hat{\Omega}} \hat{\Psi}_1^2 + r\delta^2 \int_{\tilde{\Omega}} \tilde{\Psi}_1^{\,2} \\ &= 1 + r\delta^2 > 1. \end{aligned}$$

Consequently, by (2), the function $u/(1 + r\delta^2)^{1/2}$ leads to a contradiction with (1). Q.E.D.

LEMMA 3.1. With the notations of the above proof, $\hat{\mu}_{1,\varepsilon} \to \mu_1$ and $\hat{\Psi}_{1,\varepsilon} \to \Psi_1$ in $H^1(\Omega)$ when $\varepsilon \to 0$.

PROOF. From the definitions and the fact that extending a function by zero on $\Omega \setminus \hat{\Omega}$ carries $E(\hat{\Omega})$ into $E(\Omega)$, one clearly has $\hat{\mu}_{1,\varepsilon} \geq \mu_1$.

Take $\nu > 0$ with $\nu < 1$. By the regularity of Ω, the $C^\infty(\overline{\Omega})$ functions which vanish near Γ_1 are dense in $E(\Omega)$ with respect to the $H^1(\Omega)$ norm. Consequently there exists $\varphi \in C^\infty(\overline{\Omega})$ with $\varphi = 0$ in a neighbourhood of Γ_1 such that $|\int_\Omega |\nabla\Psi_1|^2 -$

$\int_\Omega |\nabla\varphi|^2| < \nu$ and $|\int_\Omega \Psi_1^2 - \int_\Omega \varphi^2| < \nu$. Choose $\varepsilon_0 > 0$ such that supp $\varphi \subset \hat\Omega_\varepsilon$ for $\varepsilon < \varepsilon_0$. We have

$$\hat\mu_{1,\varepsilon} \leq \int_\Omega |\nabla\varphi|^2 / \int_\Omega \varphi^2 \leq \left(\int_\Omega |\nabla\Psi_1|^2 + \nu\right) \Big/ \left(\int_\Omega \Psi_1^2 - \nu\right)$$

and so

$$\limsup_{\varepsilon\to 0} \hat\mu_{1,\varepsilon} \leq \left(\int_\Omega |\nabla\Psi_1|^2 + \nu\right) \Big/ \left(\int_\Omega \Psi_1^2 - \nu\right).$$

Since this holds for all $\nu > 0$ sufficiently small, we deduce

$$\limsup \hat\mu_{1,\varepsilon} \leq \mu_1,$$

and the first assertion of the lemma follows.

To prove that $\hat\Psi_{1,\varepsilon}$ converges to Ψ_1 in $H^1(\Omega)$, we first observe that since $\hat u_{1,\varepsilon} = \int_\Omega |\nabla\hat\Psi_{1,\varepsilon}|^2$ converges (cf. above), $\hat\Psi_{1,\varepsilon}$ remains bounded in $E(\Omega)$. Consequently, for a subsequence, $\hat\Psi_{1,\varepsilon} \to v$ weakly in $E(\Omega)$ and strongly in $L^2(\Omega)$. Clearly $\int_\Omega v^2 = 1$, and

$$\begin{aligned}\int_\Omega |\nabla v|^2 &\leq \liminf \int_\Omega |\nabla\hat\Psi_{1,\varepsilon}|^2 \\ &= \liminf \hat\mu_{1,\varepsilon} = \mu_1,\end{aligned}$$

which implies $v = \Psi_1$. The strong convergence in $E(\Omega)$ now follows from the weak convergence combined with the convergence of the equivalent Hilbert space norm $\left(\int_\Omega |\nabla\hat\Psi_{1,\varepsilon}|^2\right)^{1/2}$. Q.E.D.

References

[1] M. ARIAS and J. CAMPOS, Radial Fučik spectrum of the Laplace operator, to appear.

[2] P. CLEMENT and L. PELETIER, An antimaximum principle for second order elliptic operators, J. Diff. Eq., 34 (1979), 218-229.

[3] E. DANCER, On the Dirichlet problem for weakly non linear elliptic partial differential equations, Proc. Royal. Soc. Edimb., 76 (1977), 283-300.

[4] E. DANCER, Generic domain dependence for nonsmooth equations and the open set problem for jumping nonlinearities, Proc. Royal Soc. Edimb., 1995.

[5] R. DAUTRAY et J.L. LIONS, Analyse Mathématique et calcul numérique pour les sciences et les techniques, Masson, Paris, 1984.

[6] D. DE FIGUEIREDO, Positive solutions of semi-linear elliptic equations, Lect. Notes Math., Springer, 957 (1982), 34-87.

[7] D. DE FIGUEIREDO and J.P. GOSSEZ, On the first curve of the Fučik spectrum of an elliptic operator, Diff. Int. Eq., 7 (1994), 1285-1302.

[8] S. FUČIK, Solvability of nonlinear equations and boundary value problems, Reidel, Dordrecht, 1980.

[9] D. KINDERLEHRER and G. STAMPACCHIA, An introduction to variational inequalities and their applications, Academic Press, New York, 1980.

[10] A.M. MICHELETTI, A remark on the resonance set for a semilinear elliptic equation, Proc. Royal. Soc. Edim., 124 (1994), 803-809.

[11] W. ZIEMER, Weakly differentiable functions, Springer, New York, 1989.

The Maximum Principle and Positive Principal Eigenfunctions for Polyharmonic Equations

HANS-CHRISTOPH GRUNAU Mathematisches Institut, Universität Bayreuth, D-95440 Bayreuth, Germany

GUIDO SWEERS Vakgroep Algemene Wiskunde, Technische Universiteit Delft, Postbus 5031, 2600 GA Delft, Netherlands

1 SECOND ORDER ELLIPTIC EQUATIONS

It is well known that in solving second order elliptic boundary value problems such as

$$\begin{cases} -\Delta u & = & \lambda u + f & \text{in } \Omega, \\ u & = & 0 & \text{on } \partial\Omega, \end{cases}$$

for quite general domains $\Omega \subset R^n$ and functions f, one has the following sign results.

- *There is λ_1 such that for all $0 \neq f \geq 0$*

  i. *if $\lambda < \lambda_1$, then there is a solution u and $u > 0$;*

  ii. *if $\lambda = \lambda_1$, then there is no solution u;*

  iii. *if $\lambda > \lambda_1$, then $u \neq 0$; that is, either no solution or if there is a solution there exists $\tilde{x} \in \Omega$ such that $u(\tilde{x}) < 0$.*

For bounded smooth domains the number λ_1 is the so-called principal eigenvalue. It has a unique eigenfunction, which is positive, and this eigenfunction is the only positive one (up to normalization).

Two main references for this type of results, which are usually called maximum principles, are the books by Walter (1964) and by Protter and Weinberger

(1967). Extensions to general bounded non-smooth domains are studied by Beresty-cki, Nirenberg and Varadhan (1994).

Due to Clément and Peletier (1979) there is even a property which is called the anti-maximum principle.

- *For every $0 \neq f \geq 0$ there is $\delta_f > 0$ such that*

    4. if $\lambda_1 < \lambda < \lambda_1 + \delta_f$, then there is a solution u and $u < 0$.

A crucial difference with the maximum principle is the fact that the constant δ_f depends on f in general. This result cannot be made uniform in general: that is $\inf \left\{ \delta_f; f \in C\left(\bar{\Omega}; R_0^+\right) \right\} = 0$. It is shown (Sweers, 1996b) that, even on smooth domains, $f \in L^p\left(\Omega\right)$ with $p > n$ is a necessary condition for the anti-maximum principle to hold. The only uniform result that is known holds for non-Dirichlet boundary conditions in one dimension. See (Clément and Peletier, 1979). The anti-maximum principle is extended to non-smooth domains by Birindelli (1996) but only for right hand sides f that have its support outside of the non-smooth boundary. Behaviour at cone shaped boundary points is studied in (Sweers, 1996a).

Similar results hold for cooperative, weakly coupled systems of second order elliptic equations. For more details see (Walter, 1964), (Mitidieri and De Figueiredo, 1990), (Cosner and Schaefer, 1989), (Pao, 1992), (Mitidieri and Sweers, 1995) and (Sweers, 1992). The last paper also contains an anti-maximum principle for systems.

Systems that are coupled by derivatives or where the coupling is noncooperative in general do not have the sign results mentioned above. Except some higher order operators that can be rewritten as a cooperative system, elliptic operators of order larger than two do not have such features. Remaining positivity preserving properties will be subject of the present paper.

2   POLYHARMONICS ON BALLS

For higher order equations with Dirichlet boundary conditions, such as the poly-harmonic, only a very restricted result seems to remain. A basic result goes back 91 years to (Boggio, 1905). Boggio gave an explicit formula for the Green functions of all polyharmonic equations with Dirichlet boundary conditions on the unit ball $B \subset R^n$ for any n. Dirichlet boundary condition for $(-\Delta)^m$ means that all 0 to $m - 1$ derivatives are zero at the boundary. His formula immediately shows one that the Green function is positive:

$$G_{m,n}\left(x, y\right) = k_{m,n} \, |x - y|^{2m-n} \int_1^{A(x,y)} \frac{\left(v^2 - 1\right)^{m-1}}{v^{n-1}} dv, \tag{1}$$

with

$$A\left(x, y\right) = \frac{[XY]}{|x - y|}.$$

The constants $k_{m,n}$ are positive and the expression $[XY]$ denotes the 'Kelvin-transformed' distance of x and y :

$$[XY] = \left| x\,|y| - \frac{y}{|y|} \right| = \sqrt{|x|^2\,|y|^2 - 2x \cdot y + 1}.$$

Positivity of $G_{m,n}\,(\cdot,\cdot)$ follows since $[XY] > |x - y|$ implies $A\,(\cdot,\cdot) > 1$ on B^2. Indeed

$$[XY]^2 - |x - y|^2 = \left(1 - |x|^2\right)\left(1 - |y|^2\right) > 0 \text{ on } B^2.$$

The solution of

$$\begin{cases} (-\Delta)^m\,u &=& f &\text{in } B, \\ \mathcal{D}_m u &=& 0 &\text{on } \partial B, \end{cases} \tag{2}$$

where $\mathcal{D}_m$ is the Dirichlet boundary condition

$$\mathcal{D}_m u = (D^\alpha u)_{\alpha \in N^n,\,|\alpha| \leq m-1},$$

is given by

$$u\,(x) = \int_{y \in B} G_{m,n}\,(x,y)\ f\,(y)\ dy.$$

By a rescaling argument one recovers from (1) a similar Green function on the half-space $R_+^n = \left\{(x',x_n)\,;x' \in R^{n-1}, x_n > 0\right\}$. Replacing $[XY]$ in $A\,(x,y)$ of (1) by

$$[XY] = |x - y^*|$$

where $y^* = (y_1, \ldots, y_{n-1}, -y_n)$ one finds a solution of

$$\begin{cases} (-\Delta)^m\,u &=& f &\text{in } R_+^n, \\ \mathcal{D}_m u &=& 0 &\text{on } \partial R_+^n, \end{cases} \tag{3}$$

for suitable f. Since we still have that $A\,(x,y) > 1$ for $x,y \in R_+^n$ this Green function on R_+^n is positive.

3   OTHER HIGHER ORDER EQUATIONS AND OTHER DOMAINS

Before Boggio showed that the polyharmonic Green function on the ball is positive, he (1901), and also Hadamard (1908a), conjectured that in arbitrary reasonable domains Ω, $f \geq 0$ implies $u \geq 0$. After Hadamard (1908b) showed that such a result does not hold in annuli with small inner radius, the conjecture remained for convex domains.

For $m = n = 2$ there is a physical interpretation. For Ω being the plate, f the load and u the displacement one obtains the so-called *clamped plate equation*:

$$\begin{cases} (-\Delta)^2\,u &=& f &\text{in } \Omega, \\ u = \frac{\partial}{\partial \nu} u &=& 0 &\text{on } \partial \Omega. \end{cases} \tag{4}$$

For (4) the conjecture can be formulated as:

Pushing upwards implies bending upwards.

3.1 Examples for non-positivity

The conjecture of Boggio and Hadamard proved to be wrong. In 1949 a counter example appeared by Duffin, soon to be followed by Garabedian (1951), see also (Garabedian, 1986,p. 275), Loewner (1953) and Szegö (1953). Garabedian showed that already in nice domains such as an ellipse in R^2 with the ratio of the half axes $\simeq 2$, the Green function for the biharmonic operator Δ^2 changes sign. Hedenmalm (1994) has numerical evidence that for the ratio of the axes between 1.5933... and 2.4716... the Green function $G(\cdot,\cdot)$ at x,y, with x,y close to the two opposite extreme points, is negative. For ratio larger then 2.4716... the sign of $G(x,\cdot)$, with x fixed close to the end of the longer axis, is expected to change at least twice and for growing ratio we expect an oscillatory behaviour. An elementary proof that an eccentric ellipse gives a counter example has recently been published by Shapiro and Tegmark (1994).

A renewed interest in sign properties for the biharmonic started in the seventies. Osher (1973) studied the Green function for the biharmonic in a wedge. In the eighties Coffman (1982) and Coffman and Duffin (1980) studied the Green function for the biharmonic on rectangles and obtained that the Green function has infinitely many sign-changes near a corner. Also Kozlov, Kondrat'ev and Maz'ya (1990) should be mentioned.

Altogether we may conclude that neither arbitrary smoothness, nor uniform convexity or symmetry of domains yields a positive Green function. The question that comes to ones mind is the following.

Is the polyharmonic on the unit ball the only higher order elliptic operator for which the inverse for the Dirichlet problem is sign preserving?

A trivial answer no is obtained by using the same transformation both for the operator and the domain. But it has been shown that the Green function for the Dirichlet problem is positive for a more general class of elliptic operators than the ones obtained by this trivial transformation.

3.2 Examples for positivity

The perturbation of the polyharmonic that has been considered in (Grunau and Sweers, 1996a) adds small lower order derivatives to the operator. Which means that $(-\Delta)^m$ is replaced by

$$L = (-\Delta)^m + \sum_{|\alpha| \leq 2m-1} a_\alpha(\cdot)\, D^\alpha, \tag{5}$$

with $\alpha \in R^n$ a multi-index, $D^\alpha = \prod_{i=1}^n \left(\frac{\partial}{\partial x_i}\right)^{\alpha_i}$ and $|\alpha| = \sum_{i=1}^n \alpha_i$. We were able to show that for $\|a_\alpha\|_\infty$ sufficiently small the corresponding Green function $G_L(\cdot,\cdot)$ remains positive. The proof uses a power series expansion of the Green operator $\mathcal{G}_L$ in terms of $\mathcal{G}_{(-\Delta)^m}$:

$$\mathcal{G}_L = \mathcal{G}_{(-\Delta)^m} \left(\mathcal{I} + \sum_{k=1}^\infty \left(- \sum_{|\alpha| < 2m-1} M_{a_\alpha} D^\alpha \mathcal{G}_{(-\Delta)^m} \right)^k \right),$$

where

$$(\mathcal{G}_L f)(x) = \int_{y \in \Omega} G_L(x,y) f(y) \, dy,$$

$$(M_a f)(x) = a(x) f(x).$$

The crucial step is the pointwise estimate

$$G_{(-\Delta)^m}(x,y) \geq c \left| \int_{z \in B} G_{(-\Delta)^m}(x,z) \, D_z^\alpha G_{(-\Delta)^m}(z,y) \, dz \right| \qquad (6)$$

(for some $c > 0$) which is a polyharmonic equivalent of the so-called 3G-Theorem of Cranston, Fabes and Zhao (1988). Their theorem holds for bounded Lipschitz domains. Our estimates, valid only on B, are proved by pointwise estimates for the Green function and its derivatives. For these estimates see the appendix.

THEOREM 3.1 (Grunau and Sweers, 1996a) *Let L be defined in (5). There exists $\varepsilon_{m,n} > 0$ such that if $\|a_\alpha\|_\infty \leq \varepsilon_{m,n}$ for $|\alpha| < 2m$ then for all $f \in L^p(B)$, with B the unit ball in R^n and $p \in (1, \infty)$, there is a unique solution $u \in W^{2m,p}(B) \cap W_0^{m,p}(B)$ of*

$$\begin{cases} Lu = f & \text{in } B, \\ \mathcal{D}_m u = 0 & \text{on } \partial B. \end{cases} \qquad (7)$$

Moreover, if $0 \neq f \geq 0$ in Ω then $u(x) > 0$ for all $x \in B$.

Remark 3.1: For $mp > n$ one finds that $u \in C^m(\bar{B})$ and for $0 \neq f \geq 0$ an equivalent of Hopf's boundary point Lemma follows for $\|a_\alpha\|_\infty < \varepsilon_{m,n}$; namely $\left(-\frac{\partial}{\partial \nu}\right)^m u > 0$ on ∂B where ν is the outward normal. See (Grunau and Sweers, 1996d).

Domain perturbation yields a more complicated problem. In two dimensions, by using the link with conformal mappings (see Courant, 1950), the following is proven.

THEOREM 3.2 (Grunau and Sweers, 1996c) *Fix $m \in N$ and let $\Omega \subset R^2$ with $\partial\Omega \in C^{2m,\gamma}$ be such that there exists a mapping $g \in C^{2m}(\bar{\Omega}; R^2)$ with $\|g - Id\|_{C^{2m}}$ small enough and $g(\bar{\Omega}) = \bar{B}$. Then the Green function for*

$$\begin{cases} (-\Delta)^m u = f & \text{in } \Omega, \\ \mathcal{D}_m u = 0 & \text{on } \partial\Omega, \end{cases} \qquad (8)$$

is positive.

The similar question for higher dimensional domains ($n > 2$) remains open.

Hedenmalm (1994) exploited the relation with conformal mappings and studied positivity preserving properties on the disk for the operator $\Delta |z|^{-2\alpha} \Delta$ with $\alpha > -1$ and Dirichlet boundary conditions.

Using pseudoconformal mappings one can even allow small perturbations in the leading order terms of the polyharmonic equation on the ball in R^2 and still have positivity of the solution operator. This type of result can also be found in (Grunau and Sweers, 1996c).

4   POSITIVE RESOLVENTS FOR THE POLYHARMONIC

We consider the elliptic problem

$$\begin{cases} (-\Delta)^m u = \lambda u + f & \text{in } \Omega, \\ \mathcal{D}_m u = 0 & \text{on } \partial\Omega, \end{cases} \tag{9}$$

where Ω is a bounded C^{2m}-smooth domain in R^n. Let us define

$$\lambda_1 = \inf \left\{ \frac{\int_\Omega u \left((-\Delta)^m u\right) dx}{\int_\Omega u^2 \, dx}; u \in W^{2m,2}(\Omega) \cap W_0^{m,2}(\Omega) \right\}, \tag{10}$$

which is positive by an integration by parts of $\int_\Omega u \left((-\Delta)^m u\right) dx$ and a repeated use of the Poincaré-Friedrichs inequality. The number λ_1 is the first eigenvalue and we let ϕ_1 denote a corresponding eigenfunction. Assume that ϕ_1 is unique up to normalization. We want to remark that for $m > 1$ in general one doesn't have uniqueness, or positivity, of the first eigenfunction. Both uniqueness and positivity are lost on annuli with very small inner radius (Coffman, Duffin, Shaffer, 1979).

PROPOSITION 4.1 (Grunau and Sweers, 1996a, Theorem 6.1, Lemma 6.2) *Suppose that for some $\tilde{\lambda} < \lambda_1$ the Green function $G_{\tilde{\lambda}}(x, y)$ for (9) is positive:*

$$G_{\tilde{\lambda}}(x, y) > 0 \text{ for all } x \neq y \in \Omega.$$

Then for all $\lambda \in \left(\tilde{\lambda}, \lambda_1\right)$ the Green function $G_\lambda(x, y)$ is positive and moreover, the first eigenfunction is of fixed sign.

Remark 4.1: A fixed sign implies uniqueness: if an eigenvalue doesn't have a unique eigenfunction then obviously there exists a sign-changing one.

Proof: Since $(-\Delta)^m : W^{2m,2}(\Omega) \cap W_0^{m,2}(\Omega) \to L^2(\Omega)$ is self-adjoint all eigenvalues are real and the geometric multiplicity equals the algebraic multiplicity. Because of (10) we find $\lambda_i \geq \lambda_1$. Let us denote the solution operator of (9) for $\lambda \notin \{\lambda_i\}_{i=1}^\infty$ by $\mathcal{G}_\lambda$:

$$(\mathcal{G}_\lambda f)(x) := \int_{y \in \Omega} G_\lambda(x, y) f(y) \, dy.$$

The eigenvalues of $\mathcal{G}_{\tilde{\lambda}}$ we denote by $\{\mu_i\}_{i=1}^\infty$. The eigenvalues $\{\lambda_i\}_{i=1}^\infty$ of (9) and $\{\mu_i\}_{i=1}^\infty$ are related through $\mu_i = \left(\lambda_i - \tilde{\lambda}\right)^{-1}$. For $\left|\lambda - \tilde{\lambda}\right| < \nu\left(\mathcal{G}_{\tilde{\lambda}}\right)^{-1}$ the following series converges and we find

$$\mathcal{G}_\lambda = \sum_{k=0}^\infty \left(\lambda - \tilde{\lambda}\right)^k \mathcal{G}_{\tilde{\lambda}}^{k+1}.$$

Since $\mathcal{G}_{\tilde{\lambda}} : C(\bar{\Omega}) \to C(\bar{\Omega})$ is a compact integral operator with a strictly positive kernel a theorem of Jentzsch (1912) (a predecessor of the Krein-Rutman Theorem) implies that the spectral radius $\mu = \nu\left(\mathcal{G}_{\tilde{\lambda}}\right)$ is the largest eigenvalue of $\mathcal{G}_{\tilde{\lambda}}$ and that the corresponding eigenfunction is positive. Hence $\nu\left(\mathcal{G}_{\tilde{\lambda}}\right) = \left(\lambda_1 - \tilde{\lambda}\right)^{-1}$ and the series converges for λ with $\left|\lambda - \tilde{\lambda}\right| < \lambda_1 - \tilde{\lambda}$. For $\lambda \in \left[\tilde{\lambda}, \lambda_1\right)$ we do not only find convergence but also that $\mathcal{G}_\lambda$ is positive.

COROLLARY 4.2 (Grunau and Sweers, 1996a, Corollary 6.4) *Let λ_1 be the first eigenvalue and assume that $\phi_1\left(x\right) > 0$ for all $x \in \Omega$. Then there exists $\lambda_c \in [-\infty, \lambda_1]$ such that for λ, u and f as in (9) we have*

$$
\begin{array}{llllll}
i) & \textit{if} & \lambda \in (\lambda_1, \infty) & \textit{then} & 0 \neq f \geq 0 & \textit{implies} & u0 \textit{ or no solution } u; \\
ii) & \textit{if} & \lambda = \lambda_1 & \textit{then} & 0 \neq f \geq 0 & \textit{implies} & \textit{no solution } u; \\
iii) & \textit{if} & \lambda \in (\lambda_c, \lambda_1) & \textit{then} & 0 \neq f \geq 0 & \textit{implies} & u > 0; \\
iv) & \textit{if} & \lambda = \lambda_c < \lambda_1 & \textit{then} & 0 \neq f \geq 0 & \textit{implies} & u \geq 0; \\
v) & \textit{if} & \lambda \in (-\infty, \lambda_c) & \textit{then} & 0 \neq f \geq 0 & \textit{implies} & u0.
\end{array}
$$

Remark 4.2: For second order elliptic operators $\lambda_c = -\infty$; for higher order elliptic operators one finds that $\lambda_c > -\infty$. For the polyharmonic Dirichlet problem we find that $\lambda_c < 0$ if $\Omega = B$. See respectively the counterexample and Theorem 3.1 in (Grunau and Sweers, 1996a).

Remark 4.3: For $\lambda = \lambda_c$ one finds $f > 0$ implies $u \geq 0$. We expect the positivity preserving property to break down at the boundary first. That is, for $\lambda < \lambda_c$ and $|\lambda - \lambda_c|$ small enough the Green function satisfies $G_\lambda\left(x, y\right) > 0$ on Ω^2 except for some x, y near $\partial\Omega \times \partial\Omega$. Some numerical evidence is mentioned by Hedenmalm (1994).

Proof of *i)* and *ii)*. The usual multiplication with the eigenfunction for a weight yields after integrating by parts:

$$
0 < \int_\Omega \phi_1 f \, dx = \int_\Omega \phi_1 \left((-\Delta)^m - \lambda\right) u \, dx =
$$

$$
= \int_\Omega \left((-\Delta)^m - \lambda\right) \phi_1 u \, dx = (\lambda_1 - \lambda) \int_\Omega \phi_1 u \, dx. \tag{11}
$$

For $\lambda_1 < \lambda$ one finds $u0$ and for $\lambda_1 = \lambda$ one gets a contradiction.

iii). Set $\lambda_c = \inf \left\{\lambda \in [-\infty, \lambda_1)\,; G_\lambda\left(x, y\right) > 0 \text{ for all } x \neq y \in \Omega\right\}$ if the infimum exists; otherwise set $\lambda_c = \lambda_1$. Proposition 4.1 shows that for all $\lambda \in (\lambda_c, \lambda_1)$ the operator $\mathcal{G}_\lambda$ is positive.

iv). A continuity argument for $\lambda \downarrow \lambda_c$ and *iii)* imply *iv)*.

v). With (11) one finds a contradiction whenever $0 \neq u \leq 0$.

5   A POSITIVE PRINCIPAL EIGENFUNCTION

As mentioned before we could not solve in dimensions $n \geq 3$ the question whether or not the resolvent remains positive under small smooth perturbations of the domain. We can show however that such small perturbations do not change the positivity of the first eigenfunction. This result will be the consequence of a lemma for more general elliptic operators. Let w be a smooth, strictly positive function on Ω and let L be a self-adjoint elliptic operator on $W_0^{m,2}(\Omega, wdx)$ as follows

$$L = w(x)^{-1} M^*(x, D) \cdot w(x) M(x, D) \tag{12}$$

where

$$M(x, D) = \sum_{|\alpha| \leq m} b_\alpha(x) D^\alpha \quad \text{and} \quad M^*(x, D) = \sum_{|\alpha| \leq m} (-1)^{|\alpha|} D^\alpha b_\alpha(x)$$

and with the functions $b_\beta(\cdot)$, possibly vectorvalued, having appropriate regularity and satisfying for some $c > 0$ and for all $x \in \bar{\Omega}$ and $\xi \in R^n$

$$\sum_{|\alpha|=m} \sum_{|\beta|=m} b_\alpha(x) \cdot b_\beta(x) \xi^{\alpha+\beta} \geq c_e |\xi|^{2m} . \tag{13}$$

Condition (13) shows that for some $c_0 = C(c_e, \|b_\alpha\|_\infty, \Omega) \in R$ large enough the operator $L + c_0$ is coercive with constant $c'_e = c'_e(c_e, \inf w, \sup w)$:

$$\int_\Omega u(L + c_0) u \, wdx \geq c'_e \|u\|_{W_0^{m,2}(\Omega, wdx)} .$$

The first eigenvalue of L is then well defined by

$$\lambda_1 = \inf \left\{ \frac{\int_\Omega (Mu \cdot Mu) \, wdx}{\int_\Omega u^2 \, wdx} ; u \in W_0^{m,2}(\Omega) \right\} . \tag{14}$$

Note that $\lambda_1 \geq C(c_0, c_e)$.

We assume that a corresponding eigenfunction ϕ is normalized by

$$\int_\Omega \phi^2 \, dx = 1.$$

LEMMA 5.1 *Let $\gamma \in (0, 1)$, let the operators L and $\tilde{L}$ be as in (12) and assume that $\|b_\alpha\|_{C^{m,\gamma}(\bar{\Omega})}$, $\|\tilde{b}_\alpha\|_{C^{m,\gamma}(\bar{\Omega})}$, $\|w\|_{C^{m,\gamma}(\bar{\Omega})}$, $\|\tilde{w}\|_{C^{m,\gamma}(\bar{\Omega})}$, $\|w^{-1}\|_\infty$ and $\|\tilde{w}^{-1}\|_\infty$ are bounded, say by κ. Suppose that the multiplicity of λ_1, the principal eigenvalue of L, is 1.*
Then for all $\varepsilon > 0$ there exists $\delta > 0$ such that if

$$\|b_\alpha - \tilde{b}_\alpha\|_{C^{m,\gamma}(\bar{\Omega})} \leq \delta \quad \text{and} \quad \|w - \tilde{w}\|_{C^{m,\gamma}(\bar{\Omega})} \leq \delta$$

then the multiplicity of $\tilde{\lambda}_1$ is 1 and the eigenfunction $\tilde{\phi}_1$ (or $-\tilde{\phi}_1$) satisfies

$$\left\| \phi_1 - \tilde{\phi}_1 \right\|_{C^{2m}(\bar{\Omega})} \leq \varepsilon. \tag{15}$$

Proof: For short notation we use $W_w^{m,2} := W^{m,2}(\Omega, wdx)$ and $W^{m,2} := W^{m,2}(\Omega)$. First we will estimate the difference in the eigenvalues. Testing with an auxiliary function in (14) one finds that $\lambda_1 \leq C(\|b_\alpha\|_\infty, \Omega, \min w, \max w)$ and the analogous result for $\tilde\lambda_1$. Hence we have a uniform bound for $\|\phi_1\|_{W_w^{m,2}}$, $\|\tilde\phi_1\|_{W_{\tilde w}^{m,2}}$ and also for $\|\phi_1\|_{W^{m,2}}$, $\|\tilde\phi_1\|_{W^{m,2}}$, $\|\phi_1\|_{W_{\tilde w}^{m,2}}$ and $\|\tilde\phi_1\|_{W_w^{m,2}}$. Let us denote this bound by $C_1 = C(\kappa, c_e, \Omega)$.

Writing $\mathcal{F}_w^M(u) = \int_\Omega (Mu \cdot Mu) \, wdx$ for $u \in W_0^{m,2}(\Omega)$, we find that

$$\tilde\lambda_1 - \lambda_1 = \frac{\mathcal{F}_{\tilde w}^{\tilde M}(\tilde\phi_1)}{\|\tilde\phi_1\|_{L_{\tilde w}^2}^2} - \frac{\mathcal{F}_w^M(\phi_1)}{\|\phi_1\|_{L_w^2}^2} \leq \frac{\mathcal{F}_{\tilde w}^{\tilde M}(\phi_1)}{\|\phi_1\|_{L_{\tilde w}^2}^2} - \frac{\mathcal{F}_w^M(\phi_1)}{\|\phi_1\|_{L_w^2}^2} =$$

$$= \frac{\left(\mathcal{F}_{\tilde w}^{\tilde M}(\phi_1) - \mathcal{F}_w^{\tilde M}(\phi_1)\right)\|\phi_1\|_{L_w^2}^2 + \left(\mathcal{F}_w^{\tilde M}(\phi_1) - \mathcal{F}_w^M(\phi_1)\right)\|\phi_1\|_{L_w^2}^2 + \mathcal{F}_w^M(\phi_1)\left(\|\phi_1\|_{L_w^2}^2 - \|\phi_1\|_{L_{\tilde w}^2}^2\right)}{\|\phi_1\|_{L_{\tilde w}^2}^2 \|\phi_1\|_{L_w^2}^2} \leq$$

$$\leq C_1^2 \frac{\|\tilde b_\alpha\|_{L^\infty}^2 \|w - \tilde w\|_{L^\infty}\|w\|_{L^\infty} + \|\tilde b_\alpha - b_\alpha\|_{L^\infty}\|\tilde b_\alpha + b_\alpha\|_{L^\infty}\|w\|_{L^\infty}^2 + \|b_\alpha\|_{L^\infty}^2 \|w\|_{L^\infty}\|w - \tilde w\|_{L^\infty}}{\|\phi_1\|_{L_{\tilde w}^2}^2 \|\phi_1\|_{L_w^2}^2} \leq$$

$$\leq 2\kappa^5 C_1^2 \left(\|w - \tilde w\|_{L^\infty} + \left\|\tilde b_\alpha - b_\alpha\right\|_{L^\infty}\right). \tag{16}$$

By a similar argument one estimates $\lambda_1 - \tilde\lambda_1$ to find with $C^* = 2\kappa^5 C_1^2$ that

$$\left|\tilde\lambda_1 - \lambda_1\right| \leq C^* \delta. \tag{17}$$

Next we will estimate the L_w^2–difference of the eigenfunctions. Let P_1 denote the w-weighted projection on ϕ_1:

$$P_1(u)(x) = \frac{\int_\Omega \phi_1 u \, wdx}{\|\phi_1\|_{L_w^2}^2} \phi_1(x).$$

We have for $u \in W_0^{m,2}(\Omega)$ that

$$\mathcal{F}_w^M((I - P_1)u) \geq \lambda_2 \|(I - P_1)u\|_{L_w^2}^2.$$

Then it follows that

$$\mathcal{F}_w^M(u) = \int_\Omega (MP_1u + M(I - P_1)u)^2 \, wdx \geq \lambda_1 \|P_1u\|_{L_w^2}^2 + \lambda_2 \|(I - P_1)u\|_{L_w^2}^2$$

implying

$$\mathcal{F}_w^M(\tilde\phi_1) \geq \lambda_1 \left\|P_1\tilde\phi_1\right\|_{L_w^2}^2 + \lambda_2 \left\|(I - P_1)\tilde\phi_1\right\|_{L_w^2}^2. \tag{18}$$

Using (16) and the similar estimate with L and $\tilde L$ interchanged, we find

$$\frac{\mathcal{F}_{\tilde w}^{\tilde M}(\phi_1)}{\|\phi_1\|_{L_{\tilde w}^2}^2} \leq \frac{\mathcal{F}_w^M(\phi_1)}{\|\phi_1\|_{L_w^2}^2} + C^*\delta \quad \text{and} \quad \frac{\mathcal{F}_w^M(\tilde\phi_1)}{\|\tilde\phi_1\|_{L_w^2}^2} \leq \frac{\mathcal{F}_{\tilde w}^{\tilde M}(\tilde\phi_1)}{\|\tilde\phi_1\|_{L_{\tilde w}^2}^2} + C^*\delta$$

implying that

$$\mathcal{F}_w^M(\tilde\phi_1) \leq \left\|\tilde\phi_1\right\|_{L_w^2}^2 \left(\tilde\lambda_1 + C^*\delta\right) \leq \left\|\tilde\phi_1\right\|_{L_w^2}^2 (\lambda_1 + 2C^*\delta) =$$

$$= \lambda_1 \left\| P_1 \tilde{\phi}_1 \right\|_{L_w^2}^2 + \lambda_1 \left\| (I - P_1) \tilde{\phi}_1 \right\|_{L_w^2}^2 + 2C^* \delta \left\| \tilde{\phi}_1 \right\|_{L_w^2}^2 . \tag{19}$$

Set $\delta_0 = \frac{\lambda_2 - \lambda_1}{2C^*}$ which is positive since $\lambda_2 > \lambda_1$. Combining (18)-(19) it follows for $\delta < \delta_0$ that

$$\left\| (I - P_1) \tilde{\phi}_1 \right\|_{L_w^2}^2 \leq \frac{2C^* \delta}{\lambda_2 - \lambda_1} \left\| \tilde{\phi}_1 \right\|_{L_w^2}^2$$

and hence

$$\left\| P_1 \tilde{\phi}_1 \right\|_{L_w^2}^2 \geq \left(1 - \frac{2C^* \delta}{\lambda_2 - \lambda_1} \right) \left\| \tilde{\phi}_1 \right\|_{L_w^2}^2 .$$

One has $\phi_1 = \pm \frac{\|\phi_1\|_{L_w^2}}{\|P_1 \tilde{\phi}_1\|_{L_w^2}} P_1 \tilde{\phi}_1$. Assuming a +-sign we first estimate

$$\left\| \frac{\|\tilde{\phi}_1\|_{L_w^2}}{\|\phi_1\|_{L_w^2}} \phi_1 - \tilde{\phi}_1 \right\|_{L^2} =$$

$$= \left\| \left(\frac{\|\tilde{\phi}_1\|_{L_w^2}}{\|P_1 \tilde{\phi}_1\|_{L_w^2}} - 1 \right) P_1 \tilde{\phi}_1 - (I - P_1) \tilde{\phi}_1 \right\|_{L^2} \leq$$

$$\leq \frac{\|\tilde{\phi}_1\|_{L_w^2} - \|P_1 \tilde{\phi}_1\|_{L_w^2}}{\|P_1 \tilde{\phi}_1\|_{L_w^2}} \left\| P_1 \tilde{\phi}_1 \right\|_{L^2} + \left\| (I - P_1) \tilde{\phi}_1 \right\|_{L^2} \leq$$

$$\leq \left(1 - \sqrt{1 - \frac{2C^* \delta}{\lambda_2 - \lambda_1}} \right) \frac{\|P_1 \tilde{\phi}_1\|_{L^2}}{\|P_1 \tilde{\phi}_1\|_{L_w^2}} \left\| \tilde{\phi}_1 \right\|_{L_w^2} + \sqrt{\frac{2C^* \delta}{\lambda_2 - \lambda_1}} \frac{\|(I - P_1)\tilde{\phi}_1\|_{L^2}}{\|(I - P_1)\tilde{\phi}_1\|_{L_w^2}} \left\| \tilde{\phi}_1 \right\|_{L_w^2} \leq$$

$$\leq 2\kappa \sqrt{\frac{2C^* \delta}{\lambda_2 - \lambda_1}}. \tag{20}$$

By this estimate and the normalisation $\|\phi_1\|_{L^2} = \left\| \tilde{\phi}_1 \right\|_{L^2} = 1$, we also have

$$\left\| \phi_1 - \tilde{\phi}_1 \right\|_{L^2} \leq$$

$$\leq \left| 1 - \frac{\|\tilde{\phi}_1\|_{L_w^2}}{\|\phi_1\|_{L_w^2}} \right| \|\phi_1\|_{L^2} + \left\| \frac{\|\tilde{\phi}_1\|_{L_w^2}}{\|\phi_1\|_{L_w^2}} \phi_1 - \tilde{\phi}_1 \right\|_{L^2} \leq$$

$$\leq \left| \left\| \tilde{\phi}_1 \right\|_{L^2} - \left\| \frac{\|\tilde{\phi}_1\|_{L_w^2}}{\|\phi_1\|_{L_w^2}} \phi_1 \right\| \right|_{L^2} + 2\kappa \sqrt{\frac{2C^* \delta}{\lambda_2 - \lambda_1}} \leq$$

$$\leq \left\| \tilde{\phi}_1 - \frac{\|\tilde{\phi}_1\|_{L_w^2}}{\|\phi_1\|_{L_w^2}} \phi_1 \right\|_{L^2} + 2\kappa \sqrt{\frac{2C^* \delta}{\lambda_2 - \lambda_1}} \leq$$

$$\leq 4\kappa \sqrt{\frac{2C^* \delta}{\lambda_2 - \lambda_1}}.$$

The C^{2m}-estimates for $\phi_1 - \tilde{\phi}_1$ follow from the regularity theory (Agmon, Douglis, Nirenberg, 1959) for the boundary value problem

$$\begin{cases} (L + c_0) \left(\phi_1 - \tilde{\phi}_1 \right) = f & \text{in } \Omega, \\ \mathcal{D}_m \left(\phi_1 - \tilde{\phi}_1 \right) = 0 & \text{on } \partial\Omega, \end{cases} \tag{21}$$

with $f = \left(\lambda_1 - \tilde{\lambda}_1\right)\phi_1 + \left(\tilde{\lambda}_1 + c_0\right)\left(\phi_1 - \tilde{\phi}_1\right) + \left(\tilde{L} - L\right)\tilde{\phi}_1$. Note that the $C^{m,\gamma}$-bounds on $b_\alpha, \tilde{b}_\alpha, w$, and $\tilde{w}$ will be used in $\left(\tilde{L} - L\right)\tilde{\phi}_1$ as well as for the Schauder-type regularity.

Let Ω be a domain in R^n, $k \in N$, $\gamma \in [0,1)$ and $\varepsilon > 0$. We call Ω ε-close in $C^{k,\gamma}$-sense to the unit ball B if there exists a surjective mapping $g : C^{k,\gamma}\left(\bar{B}; \bar{\Omega}\right)$ such that

$$\|g - Id\|_{C^{k,\gamma}(\bar{B})} \leq \varepsilon. \tag{22}$$

THEOREM 5.2 *There is $\varepsilon_{m,n} > 0$ such that if Ω is ε-close in $C^{2m,\gamma}$-sense to B with $\varepsilon < \varepsilon_{m,n}$, then the eigenfunction $\phi_{1,\Omega}$ for the first eigenvalue of*

$$\begin{cases} (-\Delta)^m \phi &= \lambda\phi & \text{in } \Omega, \\ \mathcal{D}_m \phi &= 0 & \text{on } \partial\Omega, \end{cases} \tag{23}$$

is unique (up to normalization) and there exists $c > 0$ such that $\phi_{1,\Omega}(x) \geq c\, d(x)^m$ for all $x \in \Omega$.

Here we denote by $d(x)$ the distance of x to the boundary of $\partial\Omega$:

$$d(x) = \inf_{y \in \partial\Omega} |x - y|. \tag{24}$$

Proof: Let $g : \bar{B} \to \bar{\Omega}$ be as in (22) and denote the inverse by h. For $\varepsilon \in (0, \varepsilon_0)$ with ε_0 small we find that the inverse h of g exists and satisfies

$$\|h - Id\|_{C^{2m,\gamma}(\bar{\Omega})} = \mathcal{O}\left(\|g - Id\|_{C^{2m,\gamma}(\bar{B})}\right).$$

For $u \in W_0^m(\Omega)$ define $\tilde{u} \in W_0^m(B)$ by

$$\tilde{u}(x) = u(g(x)).$$

For m even one finds that

$$\int_\Omega \left(\Delta^{\frac{m}{2}} u(y)\right)^2 dy = \int_B \left(\tilde{M}\tilde{u}(x)\right)^2 J_g(x)\, dx$$

with the Jacobian $J_g(\cdot) \in C^{2m-1,\gamma}$ ε-close to 1, and $\tilde{M}$ defined by

$$\left(\tilde{M}(\tilde{u})\right)(x) = \left(\Delta^{\frac{m}{2}} u\right)(g(x)) = A^{\frac{m}{2}}\tilde{u}(x)$$

where

$$A = \sum_{k=1}^n \sum_{\ell=1}^n \left(\left(\nabla h_k \cdot \nabla h_\ell\right) \circ g(x)\right)\frac{\partial}{\partial x_k}\frac{\partial}{\partial x_\ell} + \sum_{\ell=1}^n \left(\left(\Delta h_\ell\right) \circ g(x)\right)\frac{\partial}{\partial x_\ell}.$$

For m odd

$$\int_\Omega \left(\nabla\Delta^{\frac{m-1}{2}} u(y)\right)^2 dy = \int_B \left(\tilde{M}\tilde{u}(x)\right)^2 J_g(x)\, dx$$

with $\tilde{M} = \left(\tilde{M}_1, \dots, \tilde{M}_n \right)$ defined by

$$\left(\tilde{M}_i \left(\tilde{u} \right) \right) (x) = \left(\frac{\partial}{\partial y_i} \Delta^{\frac{m-1}{2}} u \right) (g(x)) = \sum_{p=1}^{n} \frac{\partial h_p}{\partial y_i} \frac{\partial}{\partial x_p} A^{\frac{m-1}{2}} \tilde{u}(x).$$

Using this transformation and $\tilde{w} = J_g$ we find that the eigenvalues and eigenfunctions of

$$\begin{cases} (-\Delta)^m \phi = \lambda \phi & \text{in } \Omega, \\ \mathcal{D}_m \phi = 0 & \text{on } \partial\Omega, \end{cases} \quad \text{and} \quad \begin{cases} \tilde{M}^* \tilde{w} \tilde{M} \tilde{\phi} = \tilde{\lambda} \tilde{w} \tilde{\phi} & \text{in } B, \\ \mathcal{D}_m \tilde{\phi} = 0 & \text{on } \partial B, \end{cases}$$

are corresponding through

$$\lambda_{i,\Omega} = \tilde{\lambda}_i \quad \text{and} \quad \phi_{i,\Omega} \circ g = \tilde{\phi}_i.$$

In the next step one shows that $\tilde{L} = \tilde{w}^{-1} \tilde{M}^* \tilde{w} \tilde{M}$ with $\tilde{w} = J_g$ and $L = (-\Delta)^m$ with $w = 1$ satisfy the conditions of Lemma 5.1 whenever $\|g - Id\|_{C^{2m,\gamma}}$ is sufficiently small. Indeed for m even, writing the mth-order terms of $\tilde{M}$ as in (12), we find

$$\sum_{\substack{|\beta| = \frac{1}{2}m \\ \beta \in N^{n \times n}}} \binom{\frac{1}{2}m}{\beta} \left(\prod_{k,\ell=1}^{n} \left((\nabla h_k \cdot \nabla h_\ell) \circ g(x) \right)^{\beta_{k\ell}} \right) \times$$

$$\times \left(\prod_{\ell=1}^{n} \left(\frac{\partial}{\partial x_\ell} \right)^{\sum_{k=1}^{n} \beta_{k\ell}} \right) \left(\prod_{k=1}^{n} \left(\frac{\partial}{\partial x_k} \right)^{\sum_{\ell=1}^{n} \beta_{k\ell}} \right)$$

and

$$\left\| (\nabla h_k \cdot \nabla h_\ell) \circ g - \delta_{kl} \right\|_{C^{m,\gamma}} = \mathcal{O} \left(\|g - Id\|_{C^{m+1,\gamma}} \right).$$

The lower order terms of $\tilde{M}$ each contain at least one derivative of $(\nabla h_k \cdot \nabla h_\ell) \circ g$ of at least order 1 and at most a derivative of $(\nabla h_k \cdot \nabla h_\ell) \circ g$ of order $m - 1$. For $\beta \in N^n$ with $|\beta| = m - 1$ we have

$$\left\| D^\beta (\nabla h_k \cdot \nabla h_\ell) \circ g \right\|_{C^{m,\gamma}} = \mathcal{O} \left(\|g - Id\|_{C^{2m,\gamma}} \right).$$

Similar results hold for m odd. We also find that

$$\left\| J_g - 1 \right\|_{C^{m,\gamma}} = \mathcal{O} \left(\|g - Id\|_{C^{m+1,\gamma}} \right).$$

Since $\phi_{1,B}$ has the property above, namely it is the unique principal eigenfunction satisfying the estimate from below by Proposition 4.1 and Remark 3.1, we are done for $\|g - Id\|_{C^{2m,\gamma}}$ sufficiently small by comparing with Lemma 5.1 the first eigenvalues/functions of

$$\begin{cases} (-\Delta)^m \phi = \lambda \phi & \text{in } B, \\ \mathcal{D}_m \phi = 0 & \text{on } \partial B, \end{cases} \quad \text{and} \quad \begin{cases} \tilde{L} \phi = \lambda \phi & \text{in } B, \\ \mathcal{D}_m \phi = 0 & \text{on } \partial B. \end{cases} \tag{25}$$

PROPOSITION 5.3 (An anti-maximum principle) *We consider* (9). *Suppose that the principal eigenfunction ϕ_1 is unique and that for some $C > 0$ one has*

$$\phi_1(x) > C\,d(x)^m \text{ for all } x \in \Omega.$$

Then for all $f \in L^p(\Omega)$, with $p > \frac{n}{m}$ and $p \geq 2$, and $f > 0$ there exists $\delta_f > 0$ such that the solution u of (9) *satisfies*

$$
\begin{array}{llcccll}
vi) & if & \lambda_1 - \delta_f & < & \lambda & < & \lambda_1 & then & u > 0, \\
vii) & if & & \lambda_1 & < & \lambda & < & \lambda_1 + \delta_f & then & u < 0.
\end{array}
$$

See (24) *for $d(x)$.*

Remark 5.1: By the previous theorem the assumption $\phi_1(x) > C\,d(x)^m$ holds for $\Omega \subset R^n$ that is ε-close to a ball with ε sufficiently small.

Proof: The proof uses similar steps as Clément and Peletier (1979). Some steps we can simplify because of the special form of our operator.

We write $L^2(\Omega) = [\![\phi_1]\!] \oplus E$ where $E = \left\{ u \in L^2(\Omega); \int_\Omega u\phi_1\, dx = 0 \right\}$. The operator $A : \left(W^{2m}(\Omega) \cap W_0^m(\Omega) \right) \to L^2(\Omega)$ defined by $A = (-\Delta)^m$ is self-adjoint. By assumption λ_1 has a unique eigenfunction ϕ_1 and all other eigenvalues are real. Consequently we have $\lambda_1 < \lambda_2 \leq \lambda_3 \leq \ldots$. Moreover by using the eigenfunction expansion the operator

$$T_2 = A - \lambda_1 I : \left(W^{2m,2}(\Omega) \cap W_0^{m,2}(\Omega) \cap E \right) \to E$$

has a well defined inverse $T_2^{-1} f = \sum_{i=2}^\infty (\lambda_i - \lambda_1)^{-1} \langle f, \phi_i \rangle \phi_i$ and T_2 is an isomorphism. Since one finds for $f_e \in L^p(\Omega) \cap E$ with $2 \leq p \leq \frac{2n}{n-4m}$ that

$$\left\| (A - \lambda_1 I)^{-1} f_e \right\|_{W^{2m,p}} \leq c \left(\left\| A(A - \lambda_1 I)^{-1} f_e \right\|_{L^p} + \left\| (A - \lambda_1 I)^{-1} f_e \right\|_{L^p} \right) \leq$$

$$\leq c \left(\left\| (A - \lambda_1 I)(A - \lambda_1 I)^{-1} f_e \right\|_{L^p} + (1 + |\lambda_1|) \left\| (A - \lambda_1 I)^{-1} f_e \right\|_{L^p} \right) \leq$$

$$\leq c \left\| f_e \right\|_{L^p} + c' \left\| (A - \lambda_1 I)^{-1} f_e \right\|_{W^{2m,2}} \leq$$

$$\leq c \left\| f_e \right\|_{L^p} + c'' \left\| f_e \right\|_{L^2} \leq \left(c + c'' \right) \left\| f_e \right\|_{L^p}$$

also

$$T_p = A - \lambda_1 I : \left(W^{2m,p}(\Omega) \cap W_0^{m,p}(\Omega) \cap E \right) \to \left(L^p(\Omega) \cap E \right)$$

is an isomorphism for such p. A bootstrapping argument shows that T_p is an isomorphism for all $p \in [2, \infty)$. Note that since $W^{2m,p}(\Omega) \hookrightarrow C^m(\bar{\Omega})$ for $p > \frac{n}{m}$ (see Gilbarg and Trudinger 1983, Theorem 7.26), the boundary conditions are satisfied in the classical sense, implying $(A - \lambda_1 I)^{-1} f_e \in W_0^{m,p}(\Omega)$.

Let $0 < \theta < \lambda_2 - \lambda_1$. Note that $\lambda_2 > \lambda_1$ follows from the assumption and (10). Then for $|\lambda - \lambda_1| < \theta$ the operators

$$A - \lambda I : \left(W^{2m,p}(\Omega) \cap W_0^{m,p}(\Omega) \cap E \right) \to \left(L^p(\Omega) \cap E \right)$$

are isomorphisms. For $f_e \in W^{2m,p}(\Omega) \cap W_0^{m,p}(\Omega) \cap E$ we have

$$\left\| (A - \lambda I)^{-1} f_e \right\|_{W^{2m,p}} =$$

$$= \left\| \sum_{k=0}^{\infty} \left((\lambda - \lambda_1)(A - \lambda_1 I)^{-1} \right)^k (A - \lambda_1 I)^{-1} f_e \right\|_{W^{2m,p}} \leq$$

$$\leq C_\theta \left\| (A - \lambda_1 I)^{-1} f_e \right\|_{W^{2m,p}} \leq C_\theta' \left\| f_e \right\|_{L^p}.$$

Then the solution of $(A - \lambda) u = f$ with $f = c_f \phi_1 + f_e$ and $f_e \in L^p(\Omega) \cap E$ can be written as

$$u = \frac{c_f}{\lambda_1 - \lambda} \phi_1 + (A - \lambda)^{-1} f_e.$$

The continuous imbedding $W^{2m,p}(\Omega) \hookrightarrow C^m(\bar{\Omega})$ for $p > \frac{n}{m}$ shows that

$$\left\| (A - \lambda)^{-1} f_e \right\|_{C^m(\bar{\Omega})} \leq c_{p,m,n,\theta} \left\| f_e \right\|_{L^p}$$

and hence we obtain from the boundary condition that

$$\left| \left((A - \lambda)^{-1} f_e \right)(x) \right| \leq c_{p,m,n,\theta}' \left\| f_e \right\|_{L^p} (d(x))^m.$$

For $f > 0$ it follows that $c_f > 0$ and since $\phi_1(x) > C\, d(x)^m$ we find by

$$u(x) = \frac{c_f}{\lambda_1 - \lambda} \phi_1(x) + \left((A - \lambda)^{-1} f_e \right)(x)$$

$$\begin{cases} \geq \left(\frac{c_f}{\lambda_1 - \lambda} - \frac{c_{p,m,n,\theta}'}{C} \left\| f_e \right\|_{L^p} \right) \phi_1(x) \\[2mm] \leq \left(\frac{c_f}{\lambda_1 - \lambda} + \frac{c_{p,m,n,\theta}'}{C} \left\| f_e \right\|_{L^p} \right) \phi_1(x) \end{cases}$$

that for $|\lambda_1 - \lambda| \leq \frac{C\, c_f}{c_{p,m,n,\theta}' \|f_e\|_{L^p}}$ the sign of u equals the sign of $\lambda_1 - \lambda$.

6   AN APPLICATION TO SEMILINEAR EQUATIONS

The first author (Grunau, 1990 and 1991) studied growth conditions that imply the existence of a strong solution for the following type of problems:

$$\begin{cases} Lu + g(\cdot, u) = f & \text{in } \Omega, \\ \mathcal{D}_m u = 0 & \text{on } \partial\Omega, \end{cases} \tag{26}$$

with $L = (-\Delta)^m$ and where $u \mapsto g(\cdot, u)$ exceeds the controllable growth rate $u^{\frac{n+2m}{n-2m}}$. Recently in (Grunau and Sweers, 1996b) results have been extended to the following

$$L = \left(-\sum_{i,j=1}^{n} a_{ij} \frac{\partial^2}{\partial x_i \partial x_j} \right)^m + \sum_{|\alpha| \leq 2m-1} b_\alpha(x) D^\alpha \tag{27}$$

with $a_{ij} \in R$, $\sum_{i,j=1}^{n} a_{ij} \xi_i \xi_j \geq c |\xi|^2$, $b_\alpha \in C^{|\alpha|,\gamma}(\bar{\Omega})$ and L assumed to be coercive, which means that for some $c > 0$ and all $u \in W_0^{m,2}(\Omega) \cap C^{2m}(\bar{\Omega})$ one has

$$\int_\Omega u L u \, dx \geq c \|u\|_{W^{m,2}(\Omega)}^2 .$$

In this section we will briefly explain the arguments necessary in proving a result as follows. For the sake of argument we assume that g and Ω are sufficiently smooth.

THEOREM 6.1 (Grunau and Sweers, 1996b) *Fix $n \geq 2m$ and suppose that g satisfies the sign condition $u g(\cdot, u) \geq 0$ for all $u \in R$ and the one-sided growth condition*

$$g(x, u) \geq -c(1 + |u|^\sigma) \tag{28}$$

with

$$
\begin{array}{llll}
\sigma & = & 1 & \text{if} \quad 6m \leq n, \\
\sigma & < & \frac{4m}{n-2m} & \text{if} \quad 2m < n < 6m, \\
\sigma & < & \infty & \text{if} \quad n = 2m.
\end{array}
\tag{29}
$$

Then for every $f \in C^\alpha(\bar{\Omega})$ problem (26) has a solution $u \in C^{2m,\alpha}(\Omega) \cap W_0^{m,2}(\Omega)$.

Remark 6.1: We do not consider $n < 2m$ since there is no critical growth rate.

For the following two-sided growth condition, instead of (28),

$$
\begin{array}{llll}
g(x, u) & \geq & -c(1 + |u|^\tau) & \text{for} \quad u \leq 0 \\
g(x, u) & \leq & c(1 + |u|^\tau) & \text{for} \quad u \geq 0
\end{array}
\tag{30}
$$

with $\tau \leq \frac{n+2m}{n-2m}$ existence of a weak solution $u \in W_0^{m,2}(\Omega)$ follows from the coercivity of (26). Moreover, for $\tau < \frac{n+2m}{n-2m}$ a linear argument, bootstrapping between Sobolev imbedding and regularity theory (see (Agmon, Douglis, Nirenberg 1959)), shows existence of a strong solution $u \in C^{2m}(\bar{\Omega})$ as well as regularity of any weak solution. Luckhaus (1979) proved for general elliptic operators that the solutions for (26) are classical, meaning $u \in C^{2m}(\bar{\Omega})$, whenever (30) holds with $\tau \leq \frac{n+2m}{n-2m}$.

For $m = 1$ no controllable growth conditions are needed. Here the maximum principle together with the sign condition for g give an L^∞-bound to start the bootstrapping. For $m = 2$ Tomi in 1976 obtains a classical solution by using the maximum principle for an auxiliary function like $a(\Delta u)^2 + G(u)$ where $G' = g$ and $a \in R$.

These approaches do not work for general higher order elliptic equations with zero Dirichlet boundary conditions. Not only no maximum principle on general domains exists but also the restriction to a level set defines a new non-zero Dirichlet problem. By exploiting the Green function estimates on balls a local maximum principle can however be proven.

THEOREM 6.2 (A local maximum principle; see (Grunau and Sweers, 1996b)) *Let* $\Omega \subset R^n$ *be open and* $K \subset \Omega$ *be compact, and suppose that* L *is as in (27).* *Let* $q \in R$ *be such that* $q > \frac{n}{2m}$ *and* $q \geq 1$. *Then there exists* $c \in R$ *such that for every* $u \in C^{2m}(\bar{\Omega})$, $f \in C^0(\bar{\Omega})$ *that satisfy the differential inequality*

$$Lu \leq f \quad in \ \Omega$$

it follows that

$$\sup_{x \in K} u(x) \leq c \left(\|f^+\|_{L^q(\Omega)} + \|u\|_{W^{m-1,1}(\Omega)} \right).$$

This local maximum principle is proven by taking the Green function on the unit ball, rescaling it for small balls in Ω and using it for a test function on these balls.

We end with the explanation how the growth rates for $u < 0$ in (29) appear. By the coercivity of L and the sign condition for g one finds

$$\|u\|_{W^{m,2}(\Omega)}^2 \leq c \int_{\Omega} u L u \, dx \leq c \int_{\Omega} u \left(Lu + g(x,u) \right) \, dx \leq c \|u\|_{L^2(\Omega)} \|f\|_{L^2(\Omega)}$$

implying with Poincaré-Friedrichs that

$$\|u\|_{W^{m,2}(\Omega)} \leq c' \|f\|_{L^2(\Omega)}.$$

We have

$$Lu(x) \leq \|f^+\|_{C^0(\bar{\Omega})} - g(x, u(x)) \leq$$

$$\leq \|f^+\|_{C^0(\bar{\Omega})} + c\chi_{[u<0]} \left(1 + |u(x)|^\sigma \right).$$

In order to apply the local maximum principle we need the right hand side to be in $L^q(\Omega)$ for some q satisfying $q > \frac{n}{2m}$ and $q \geq 1$. If $2m < n < 6m$ one may take $q = \frac{2n}{\sigma(n-2m)}$, hence $q > \frac{n}{2m} > 1$, to find by the imbedding of $W^{m,2}(\Omega)$ in $L^{\frac{2n}{n-2m}}(\Omega)$ that

$$\|1 + |u|^\sigma\|_{L^q(\Omega)} \leq c \left(1 + \|u\|_{L^{\sigma q}(\Omega)}^\sigma \right) \leq$$

$$\leq c' \left(1 + \|u\|_{W^{m,2}(\Omega)}^\sigma \right) \leq c'' \left(1 + \|f\|_{L^2(\Omega)}^\sigma \right).$$

If $n = 2m$ then any $q > 1$ will do.

For $n \geq 6m$ the number $\frac{4m}{n-2m}$ is less or equal 1. In this case one replaces the operator L by $L + b_0(x)$ for an appropriately chosen function b_0. See (Grunau and Sweers, 1996c). One can show that the dependence of the constant in the local maximum principle on the function b_0 can be controlled.

A   ESTIMATES FOR THE POLYHARMONIC GREEN FUNCTION

DEFINITION A.1 *Let* $g, h : \Omega \to R$ *with* $g, h \geq 0$. *We say that*

$$g(x) \preceq h(x) \quad on \ \Omega$$

if there exists $c > 0$ *such that* $g(x) \leq c\, h(x)$ *for all* $x \in \bar{\Omega}$. *We say that*

$$g(x) \sim h(x) \quad on \ \Omega$$

if $g(x) \preceq h(x)$ *on* Ω *and* $h(x) \preceq g(x)$ *on* Ω.

Let $G_{m,n}(x, y)$ denote the Green function for

$$\begin{cases} (-\Delta)^m u &= \ f \quad in \ B, \\ D_m u &= \ 0 \quad on \ \partial B. \end{cases}$$

The following estimates are proven in (Grunau and Sweers, 1996a). These estimates are the crucial tools in most of the results mentioned in this paper.

The distance $d(x)$ of $x \in B$ to the boundary satisfies $d(x) = 1 - |x|$.

PROPOSITION A.2 *On* B^2 *(that is* $(x, y) \in B^2$*) we have the following.*

i. For $2m < n$ *:*
$$G_{m,n}(x, y) \sim |x - y|^{2m-n} \left(1 \wedge \frac{d(x)^m\, d(y)^m}{|x - y|^{2m}} \right).$$

ii. For $2m = n$ *:*
$$G_{m,n}(x, y) \sim \log \left(1 + \frac{d(x)^m\, d(y)^m}{|x - y|^{2m}} \right).$$

iii. For $2m > n$ *:*
$$G_{m,n}(x, y) \sim (d(x)\, d(y))^{m-\frac{1}{2}n} \left(1 \wedge \frac{d(x)^{\frac{1}{2}n}\, d(y)^{\frac{1}{2}n}}{|x - y|^n} \right).$$

PROPOSITION A.3 *Let* $\alpha \in N^n$. *Then on* B^2 *we have the following.*

i. For $|\alpha| \geq 2m - n$ *and* n *odd, or,* $|\alpha| > 2m - n$ *and* n *even:*

(a) if $|\alpha| \leq m$ *then*
$$|D_x^\alpha G_{m,n}(x, y)| \preceq |x - y|^{2m-n-|\alpha|} \left(1 \wedge \frac{d(x)^{m-|\alpha|}\, d(y)^m}{|x - y|^{2m-|\alpha|}} \right) ;$$

(b) if $|\alpha| \geq m$ *then*
$$|D_x^\alpha G_{m,n}(x, y)| \preceq |x - y|^{2m-n-|\alpha|} \left(1 \wedge \frac{d(y)^m}{|x - y|^m} \right).$$

ii. For $|\alpha| = 2m - n$ *and* n *even:*

(a) if $|\alpha| \le m$ *(that is $m \le n$) then*

$$|D_x^\alpha G_{m,n}(x,y)| \preceq \log\left(2 + \frac{d(y)}{|x-y|}\right)\left(1 \wedge \frac{d(x)^{m-\lceil\alpha\rceil}d(y)^m}{|x-y|^{2m-|\alpha|}}\right) ;$$

(b) if $|\alpha| \ge m$ *(that is $m \ge n$) then*

$$|D_x^\alpha G_{m,n}(x,y)| \preceq \log\left(2 + \frac{d(y)}{|x-y|}\right)\left(1 \wedge \frac{d(y)^m}{|x-y|^m}\right).$$

iii. *For $|\alpha| \le 2m - n$ and n odd, or, $|\alpha| < 2m - n$ and n even:*

(a) if $|\alpha| \le m - \frac{1}{2}n$ *then*

$$|D_x^\alpha G_{m,n}(x,y)| \preceq d(x)^{m-\frac{1}{2}n-|\alpha|}d(y)^{m-\frac{1}{2}n}\left(1 \wedge \frac{d(x)^{\frac{1}{2}n}d(y)^{\frac{1}{2}n}}{|x-y|^n}\right) ;$$

(b) if $m - \frac{1}{2}n \le |\alpha| \le m$ *then*

$$|D_x^\alpha G_{m,n}(x,y)| \preceq d(y)^{2m-n-|\alpha|}\left(1 \wedge \frac{d(x)^{m-|\alpha|}d(y)^{n-m+|\alpha|}}{|x-y|^n}\right) ;$$

(c) if $m \le |\alpha|$ *then*

$$|D_x^\alpha G_{m,n}(x,y)| \preceq d(y)^{2m-n-|\alpha|}\left(1 \wedge \frac{d(y)^{n-m+|\alpha|}}{|x-y|^{n-m+|\alpha|}}\right).$$

REFERENCES

1. S. Agmon, A. Douglis, L. Nirenberg (1959), *Estimates near the boundary for so-lutions of elliptic partial differential equations satisfying general boundary con-ditions. I*, Commun. Pure Appl. Math. **12**, 623-727.

2. H. Berestycki, L. Nirenberg and S. Varadhan (1994), *The principal eigenvalue and maximum principle for second-order elliptic operators in general domains*, Commun. Pure Appl. Math. **47**, 47–92.

3. I. Birindelli (1995), *Hopf's lemma and Anti-Maximum Principle in general do-mains*, J. Differ. Equations **119**, 450-472.

4. T. Boggio (1901), *Sull'equilibrio delle piastre elastiche incastrate*, Rend. Acc. Lincei **10**, 197-205.

5. T. Boggio (1905), *Sulle funzioni di Green d'ordine m*, Rend. Circ. Mat. Palermo **20**, 97-135.

6. Ph. Clément and L.A. Peletier (1979), *An anti-maximum principle for second-order elliptic equations*, J. Differ. Equations **34**, 218-229.

7. C. V. Coffman, R. J. Duffin and D.H. Shaffer (1979), *The fundamental mode of vibration of a clamped annular plate is not of one sign*, in: C. V. Coffman and G. J. Fix (eds.), Constructive approaches to mathematical models (Proc. Conf. in honor of R.J. Duffin, Pittsburgh, Pa., 1978), pp. 267–277, Academic Press, New York.

8. C. V. Coffman, R. J. Duffin (1980), *On the structure of biharmonic functions satisfying the clamped plate conditions on a right angle*, Adv. Appl. Math. **1**, 373-389.

9. C. V. Coffman (1982), *On the structure of solutions to* $\Delta^2 u = \lambda u$ *which satisfy the clamped plate condition on a right angle*, SIAM J. Math. Anal. **13**, 746-757.

10. C. Cosner and P. W. Schaefer (1989), *Sign-definite solutions in some linear elliptic systems*, Proc. Roy. Soc. Edinb. Section A **111**, 347-358.

11. R. Courant (1950), *Dirichlet's Principle, Conformal Mapping, and Minimal Surfaces*, Interscience, New York.

12. M. Cranston, E. Fabes and Z. Zhao (1988), *Conditional gauge and potential theory for the Schrödinger operator*, Trans. A.M.S. **307**, 171-194.

13. R. J. Duffin (1949), *On a question of Hadamard concerning super-biharmonic functions*, J. Math. Phys. **27**, 253-258.

14. D.G. De Figueiredo and E. Mitidieri (1990), *Maximum principles for cooperative elliptic systems*, C.R. Acad. Sci. Paris, Sér. I **310**, 49-52.

15. P. R. Garabedian (1951), *A partial differential equation arising in conformal mapping*, Pacific J. Math. **1**, 485-524.

16. P. R. Garabedian (1986), *Partial Differential Equations*, second edition, Chelsea, New York.

17. D. Gilbarg and N. S. Trudinger (1983), *Elliptic Partial Differential Equations of Second Order*, second edition, Springer-Verlag, Berlin etc.

18. H.-Ch. Grunau (1990), *Das Dirichletproblem für semilineare elliptische Gleichungen höherer Ordnung*, Doctoral thesis, Georg-August-Universität Göttingen.

19. H.-Ch. Grunau (1991), *The Dirichlet problem for some semilinear elliptic differential equations of arbitrary order*, Analysis **11**, 83-90.

20. H.-Ch. Grunau and G. Sweers (1996a), *Positivity for equations involving polyharmonic operators with Dirichlet boundary conditions*, to appear in Math. Annal.

21. H.-Ch. Grunau and G. Sweers (1996b), *Classical solutions for some higher order semilinear elliptic equations under weak growth conditions*, to appear in Nonlinear Anal., T.M.A.

22. H.-Ch. Grunau and G. Sweers (1996c), *Positivity for perturbations of polyharmonic operators with Dirichlet boundary conditions in two dimensions*, Math. Nachr. 179, 89-102.

23. H.-Ch. Grunau and G. Sweers (1996d), *Positivity properties of elliptic boundary value problems of higher order*, preprint.

24. J. Hadamard (1908a), *Mémoire sur le problème d'analyse relatif à l'équilibre des plaques élastiques incastrées*, Mèmoires prèsentès par divers savants à l'Acadèmie des Sciences, Vol. **33**, 1-128. Reprinted in: *Œuvres de Jacques Hadamard*, Centre National de la Recherche Scientifique, Paris 1968, Tome II, 515-641.

25. J. Hadamard (1908b), *Sur certains cas intéressants du problème biharmonique*, Atti IVe Congr. Intern. Mat. Rome, 12-14. Reprinted in: *Œuvres de Jacques Hadamard*, Centre National de la Recherche Scientifique, Paris 1968, Tome III, 1297-1299.

26. P. J. H. Hedenmalm (1994), *A computation of Green functions for the weighted biharmonic operators* $\Delta |z|^{-2\alpha} \Delta$ *with* $\alpha > -1$, Duke Math. J. **75**, 51-78.

27. P. Jentzsch (1912), *Über Integralgleichungen mit positivern Kern*, J. Reine Angew. Math. **141**, 235-244.

28. V. A. Kozlov, V. A. Kondrat'ev and V. G. Maz'ya (1990), *On sign variation and the absence of "strong" zeros of solutions of elliptic equations*, Math. USSR Izvestiya **34**, 337-353.

29. Ch. Loewner (1953), *On generation of solutions of the biharmonic equation in the plane by conformal mappings*, Pacific J. Math. **3**, 417-436.

30. St. Luckhaus (1979), *Existence and regularity of weak solutions to the Dirichlet problem for semilinear elliptic systems of higher order*, J. Reine Angew. Math. **306**, 192-207.

31. E. Mitidieri and G. Sweers (1995), *Weakly coupled elliptic systems and positivity*, Math. Nachr. **173**, 259-286.

32. S. Osher (1973), *On Green's function for the biharmonic equation in a right angle wedge*, J. Math. Anal. Appl. **43**, 705-716.

33. C. V. Pao (1992), *Nonlinear parabolic and elliptic equations*, Plenum Press, New York.

34. M. Protter and H.F. Weinberger (1967), *Maximum principles in differential equations*, Prentice Hall, Englewood Cliffs, (reprint by Springer-Verlag 1984).

35. H. S. Shapiro and M. Tegmark (1994), *An elementary proof that the biharmonic Green function of an eccentric ellipse changes sign*, SIAM Rev. **36**, 99-101.

36. G. Sweers (1992), *Strong positivity in $C(\bar{\Omega})$ for elliptic systems*, Math. Z. **209**, 251-271.

37. G. Sweers (1994), *Positivity for a strongly coupled elliptic system by Green function estimates*, J. Geometric Anal. **4**, 121-142.

38. G. Sweers (1996a), *Hopf's lemma and two-dimensional domains with corners*, Report 96-46 TUDelft.

39. G. Sweers (1996b), *L^N is sharp for the antimaximum principle*, to appear in J. Differ. Equations.

40. G. Szegö (1953), *Remark on the preceding paper of Charles Loewner*, Pacific J. Math. **3**, 437-446.

41. F. Tomi (1976), *Über elliptische Differentialgleichunge 4. Ordnung mit einer starken Nichtlinearität*, Nachr. Akad. Wiss. Göttingen II. Math-Phys. Klasse, 33-42.

42. W. Walter (1964), *Differential- und Integral-Ungleichungen*, Springer-Verlag, Berlin.

Positive Solutions for the Logistic Equation with Unbounded Weights

JESUS HERNÁNDEZ, Departamento de Matemáticas, Universidad Autónoma de Madrid, 28049 Madrid, Spain

1 INTRODUCTION

Reaction-diffusion equations have been widely studied in the last twenty years not only by the interest presented by themselves but also because the detailed knowledge of solutions is useful when considering more complicated models involving systems. Very often positive solutions are particularly interesting: this is the case for chemical reactions or population dynamics. See the books (Smoller 1983) or (Pao 1992).

Here we consider the semilinear elliptic equation

$$-\Delta u = \lambda m(x)u - q(x)u^p \text{ in } \Omega$$
$$u = 0 \text{ on } \partial\Omega , \tag{1.1}$$

where Ω is a bounded domain in $\mathbf{R}^N$, λ a real parameter, $p > 1$ and m and q are nonnegative functions. When $m \equiv q \equiv 1$ positive solutions of this equation have been widely studied and these results have been used, in particular, to study Volterra-Lotka systems in mathematical biology. For problems like (1.1) the reader can find results in (Berestycki 1981), (Berestycki and Lions 1980), (Brezis and Oswald 1986), (Cantrell and Cosner 1989), (de Figueiredo 1981), (Hernndez 1986) and the references there.

The coefficient $m(x)$ in (1.1) has the meaning of the intrinsic growth rate for the population and thus it is positive (resp. negative) on the favourable (resp. unfavourable) regions. The effect is measured by $q(x)$, which is supposed to be nonnegative. If m and q are bounded and bounded below from zero, the situation is very similar to the case $m \equiv q \equiv 1$. The case when q vanishes on a subset of Ω is more delicate and we do not deal with it here; there is some interesting work by Ou-yang, Alama and Tarantello, etc. Equation (1.1) with $m \in L^\infty(\Omega)$ and q constant is studied by Cantrell and Cosner (1989) in environmental problems where it is important to allow weights m change sign on the domain and in (Caada and Gmez 1993).

We show in this article that most of the results available for $m \equiv q \equiv 1$ may be extended to the more general situation of m changing sign on Ω and <u>unbounded</u>. We take $q \equiv 1$ in order to simplify the exposition but the general case can be handled under suitable assumptions by the same methods. If m is unbounded we cannot expect (1.1) to have classical solutions but for $r > \frac{N}{2}$ (resp. $r > N$) it is posible to prove under additional assumptions on p and N that weak solutions are in $C(\bar{\Omega})$ (resp. in $C^1(\bar{\Omega})$). Since m is interpreted as providing a limitation for the size of the population, our results show that if the limitation goes to infinity but it is not "too singular" solutions are still bounded.

In Section 2 we prove a general existence result for weak solutions to semilinear elliptic problems when there are ordered (not necessarily bounded) sub and super-solutions. These results are related with previous work, in particular with (Dancer and Sweers 1989), see Remark 1 below. Theorem 6 deals with the regularity of solutions we obtain. Since some of the different methods used to prove uniqueness of positive solutions fail for weak solutions (for example, some require solutions to be in $C^1(\bar{\Omega})$), we treat this point with some care.

However, our main interest is in applying results in Section 2 to the logistic equation. In Section 3 we first apply the method to some sublinear problems studied by Boccardo and Orsina (1994) and give a direct proof avoiding the approximation method used in their paper. Moreover, the case of m changing sign raises two interesting questions; the uniqueness of positive solutions and the (possible) existence of "dead cores". The general results are then applied to equation (1.1). The main point here is to find supersolutions for the corresponding problem; the smoothness of these supersolutions depends on p and r. Finally, we use the same ideas to study the logistic equation, but this time with nonlinear diffusion.

Acknowledgements: This work was partially supported by the European Network <u>Reaction Diffusion Equations</u> (contract) ERBCHRXCT 930409 and Project PB 93/0443 from DGICYT, Spain.

2. EXISTENCE AND UNIQUENESS THEOREMS

In this section we prove the main general existence and uniqueness theorems which will be applied later to the problems mentioned in the introduction.

We consider the semilinear boundary value problem

$$-\Delta\, u = f(x, u) \text{ in } \Omega$$
$$u = 0 \text{ on } \partial\Omega\,. \tag{2.1}$$

Here Ω is a bounded domain in $\mathbf{R}^N (N \geq 3)$ with a very smooth boundary $\partial\Omega$ and $f : \Omega \times \mathbf{R} \longrightarrow \mathbf{R}$ is a Carathodory function (f is continuous in u for a.a. $x \in \Omega$ and measurable in x for all u) such that $f(x,0) = 0$ for any $x \in \Omega$.

DEFINITION. We say that $u \in H_0^1(\Omega)$ is a (weak) solution of (2.1) if $f(x,u)$ is in $L^1(\Omega)$ and for any $\varphi \in H_0^1(\Omega)$ we have

$$\int_\Omega \nabla u \cdot \nabla \varphi - \int_\Omega f(x,u)\ \varphi = 0 .$$

DEFINITION. We say that $u_0 \in H^1(\Omega)$ (resp.u^0) is a subsolution (resp.a supersolution) of (2.1) if

$$\int_\Omega \nabla u_0 \cdot \nabla \varphi - \int_\Omega f(x,u_0)\ \varphi \leq 0 \qquad (2.2)$$

(resp. ≥ 0 for u^0) for all $\varphi \in H_0^1(\Omega)$ such that $\varphi \geq 0$ on Ω and

$$u_0 \leq 0 \leq u^0 \text{ on } \partial\Omega . \qquad (2.3)$$

Here $H^1(\Omega)$ and $H_0^1(\Omega)$ denote the usual Sobolev spaces and (2.3) should be interpreted in the sense of traces (see Lions and Magenes 1968).

Our main existence theorem for solutions of (2.1) is the following.

THEOREM 1 *Assume that the above conditions are satisfied and that there is a Carathodory function* $g : \Omega \times \mathbf{R} \longrightarrow \mathbf{R}$ *such that*

$$g(x,0) = 0 \qquad \forall\ x \in \Omega , \qquad (2.4)$$

$$\sup_{|w| \leq \alpha} |g(x,w)| \leq \emptyset_\alpha(x) \qquad (2.5)$$

with $\emptyset_\alpha \in L^1(\Omega)$ *for all* $\alpha > 0$,

$$g(x,u) \text{ is increasing in } u, \forall\ x \in \Omega \qquad (2.6)$$

$$f(x,u) + g(x,u) \text{ is increasing in } u, \forall\ x \in \Omega \qquad (2.7)$$

$$\sup\{|f(x,w) + g(x,w)| : u_0(x) \leq w \leq u^0(x)\} \in L^\gamma(\Omega) \qquad (2.8)$$

for some $\gamma > \frac{2N}{N+2}$, *where* u_0 *(resp.* u^0*) is a subsolution (resp. a supersolution) such that* $u_0 \leq u^0$. *Then there exists a minimal (resp. maximal) solution* $\underline{u}$ *(resp.* $\overline{u}$*) of (2.1) such that* $u_0 \leq \underline{u} \leq \overline{u} \leq u^0$.

The main auxiliary tool used in the proof is an existence (and uniqueness) theorem which is a very particular case of results by Brezis and Browder (1982).

THEOREM 2 *Let p be a Carathodory function satisfying (2.4)–(2.6). Then for any* $h \in H^{-1}(\Omega)$ *there is a unique solution* $z \in H_0^1(\Omega)$ *to*

$$-\Delta\ z + p(x,z) = h \text{ in } \Omega$$
$$z = 0 \text{ on } \partial\Omega . \qquad (2.9)$$

Now, let P be the solution operator corresponding to equation (2.9). More precisely, if $2^* = \frac{2N}{N-2}$ is the Sobolev exponent, we have the imbedding $H_0^1(\Omega) \hookrightarrow L^{2^*}(\Omega)$ and then $L^{\frac{2N}{N+2}}(\Omega) \hookrightarrow H^{-1}(\Omega)$. (Notice that $(2^*)' = \frac{2N}{N+2}$, where $\frac{1}{s} + \frac{1}{s'} = 1$). The nonlinear operator $P : L^{\frac{2N}{N+2}}(\Omega) \longrightarrow L^2(\Omega)$ is defined for any $h \in L^{\frac{2N}{N+2}}(\Omega)$ given as the unique solution $Ph \in H_0^1(\Omega)$ of (2.9). We collect in a Lemma some useful properties of P.

LEMMA 3 *P is compact. Moreover, if $h \le h'$, then $Ph \le Ph'$. In particular, $Ph \ge 0$ for $h \ge 0$.*

Proof: From (2.9) for h and h' we obtain

$$-\Delta(Ph - Ph') + p(x, Ph) - p(x, Ph') = h - h' \text{ in } \Omega$$
$$Ph - Ph' = 0 \text{ on } \partial\Omega . \tag{2.10}$$

Multiplying (2.10) by $Ph - Ph'$, integrating over Ω and using Hlder's inequality we obtain

$$\int_\Omega -\Delta(Ph - Ph')(Ph - Ph') + \int_\Omega (p(x, Ph) - p(x, Ph'))(Ph - Ph')$$
$$= \int_\Omega (h - h')(Ph - Ph') \le \| h - h' \|_{L^{\frac{2N}{N+2}}(\Omega)} \| Ph - Ph' \|_{L^{2^*}(\Omega)} .$$

Since the second integral is nonnegative by (2.6) we get

$$\int_\Omega |\nabla(Ph - Ph')|^2 = \| Ph - Ph' \|_{H_0^1(\Omega)}^2 \le c \| h - h' \|_{L^{\frac{2N}{N+2}}(\Omega)} \| Ph - Ph' \|_{H_0^1(\Omega)}$$

for some $c > 0$. Hence

$$\| Ph - Ph' \|_{H_0^1(\Omega)} \le c \| h - h' \|_{L^{\frac{2N}{N+2}}(\Omega)} .$$

This gives the continuity of P and, together with Rellich's Lemma, gives its compactness as well.

Suppose that $h \le h'$. If $w^+ = \max(w, 0)$ and $w^- = w^+ - w$ for $w \in H^1(\Omega)$, then it is well-known (see, e.g., Gilbarg and Trudinger 1983) that $w^+ \in H^1(\Omega)$. Multiplying (2.10) by $(Ph - Ph')^+ \in H_0^1(\Omega)$ and integrating again over Ω

$$\int_\Omega -\Delta(Ph - Ph')(Ph - Ph')^+ + \int_\Omega (p(x, Ph) - p(x, Ph'))(Ph - Ph')^+$$
$$= \int_\Omega (h - h')(Ph - Ph')^+ \le 0 .$$

Reasoning as above and using that $\int_\Omega \nabla w^+ . \nabla w^- = 0$ if $w \in H^1(\Omega)$ we obtain $(Ph - Ph')^+ = \text{const.}$ and then $(Ph - Ph')^+ \equiv 0$, which ends the proof.

Proof of Theorem 1: Let us define $K = [u_0, u^0] = \{w \in L^2(\Omega) : u_0(x) \leq w \leq u^0(x)$ a.a. $x \in \Omega\}$; then K is convex, closed and bounded in $L^2(\Omega)$. Let $F + G$ be the Nemitskii operator associated to $f + g$, i.e., $(F + G)\, u(x) = f(x, u(x)) + g(x, u(x))$. It follows easily from (2.8) that $|f(x, \cdot) + g(x, \cdot)| \leq K(x)$ for some $K \in L^\gamma(\Omega)$. By Theorem 2.1 in (Krasnoselski 1964) or Theorem 2.2 in (Ambrosetti and Prodi 1993), $F + G : L^2(\Omega) \longrightarrow L^\gamma(\Omega)$ is continuous and bounded. Since $F(u) + G(u) \in L^{\frac{2N}{N+2}}(\Omega) \hookrightarrow H^{-1}(\Omega)$ for $u \in K$, we can apply Theorem 2 with $p \equiv g$.

If we define $T = P \circ (F + G)$, then $T : K \longrightarrow L^2(\Omega)$ is a well-defined nonlinear operator which is compact by the above considerations and Lemma 3.

Moreover, T satisfies

   i) $u \leq v \Longrightarrow Tu \leq Tv$,
   ii) $u_0 \leq Tu_0$,
   iii) $u^0 \geq Tu^0$.

Property i) follows from (2.7) and Lemma 3. To prove ii) we pick $\varphi = (Tu_0 - u_0)^-$ in (2.2) to obtain

$$0 \leq \int_\Omega \nabla(Tu_0 - u_0) \, \nabla(Tu_0 - u_0)^- + \int_\Omega (g(x, Tu_0) - g(x, u_0))(Tu_0 - u_0)^-$$

$$\leq -\int_\Omega |\nabla(Tu_0 - u_0)^-|^2$$

by (2.7) and hence $(Tu_0 - u_0)^- \equiv 0$. Same proof for iii).

Finally, $T : K \longrightarrow K$ is a compact operator satisfying i) - iii) and the result follows from well-known results by Amann (1976, Thm. 6.1). Since fixed points of T are solutions to (2.1), this ends the proof.

COROLLARY 4. Assume that f is a Carathodory function and that there exists $q \in L^1(\Omega), q \geq 0$ on Ω , such that

$$f(x, u) + q(x)u \text{ is increasing in } u, \forall\, x \in \Omega \tag{2.11}$$

and $f(x, u) + q(x)u$ satisfies (2.8). Then the conclusion of Theorem 1 holds.

Proof: If $g(x, u) = q(x)u$, then (2.6) is trivially satisfied and $\sup_{|w| \leq \alpha} |g(x, w)| \leq \alpha\, q(x) \in L^1(\Omega)$. The rest is trivial.

COROLLARY 5. Assume that f is a Carathodory function and there exists $\beta : \mathrm{R} \longrightarrow \mathrm{R}$ such that $\beta(0) = 0$, continuous, increasing and

$$f(x, u) + \beta(u) \text{ is increasing in } u, \forall\, x \in \Omega \tag{2.12}$$

$$f(x, u) + \beta(u) \text{ satisfies (2.8) .} \tag{2.13}$$

The conclusion of Theorem 1 holds.

Proof: For $g(x, u) = \beta(u)$, $\sup\limits_{|w| \leq \alpha} |g(x, w)|$ is constant and g satisfies (2.6). The rest is trivial.

REMARK 1. Theorem 1 and its corollaries are closely related with some work in the literature. Existence of a weak solution between ordered sub and supersolutions was proved in a more general context (allowing nonlinear differential operators as the p-Laplacian and nonlinearities $f(x, u, \nabla u)$ depending on ∇u as well) by Deuel and Hess (1975) but their proof, which uses in a very elegant way truncation arguments does not provide existence of minimal and maximal solutions; such a result was proved by Dancer and Sweers (1989) without condition (2.7) for $g \equiv 0$ but the proof, which uses Zorn's Lemma, is not constructive. (See Mitidieri and Sweers (1994) for the extension to cooperative systems). The case $\gamma = r = \infty$ was considered by Amann and Crandall (1978) and Berestycki and Lions (1980). Related work can be found in (Berestycki 1981), (Boccardo and Orsina 1994), (Cantrell and Cosner 1989), (Caada and Gmez 1993), (Daz 1985), (de Figueiredo 1980), (Hernndez 1986).

REMARK 2. The differential operator $-\Delta$ can be replaced by a uniformly elliptic second order operator in divergence form. The case of the p-Laplacian can be studied by using results of Brezis and Browder (1982).

The regularity of solutions obtained in Theorem 1 depends on γ in (2.8) for $g \equiv 0$ as stated in the following result.

THEOREM 6. Suppose that all the assumptions in Theorem 1 are satisfied. If

$$\sup\{|f(x, w)| : u_0(x) \leq w \leq u^0(x)\} \in L^\eta(\Omega)$$

with $\eta > \frac{2N}{N+2}$, then solutions of (2.1) are in the Sobolev space $W^{2,\eta}(\Omega) \cap W_0^{1,\eta}(\Omega)$. If $\eta > \frac{N}{2}$, then $u \in C(\bar{\Omega})$ and if, moreover, $\eta > N, u \in C^{1,\alpha}(\bar{\Omega})$ for some $0 < \alpha < 1$.

Proof: The statements follow from the classical L^p-regularity theory (see, e.g., Gilbarg and Trudinger 1983) and Morrey's Lemma.

It is well-known that uniqueness of nontrivial positive solutions follows in one way or another from "concavity" arguments. Most of them are collected in (Hernndez 1986) or (Brezis and Kamin 1992), see also (Berestycki 1981), (Brezis and Oswald 1986), (Cantrell and Cosner 1989), (Cohen and Laetsch 1970). When considering weak solutions, its regularity becomes important and some methods work for C^1 solutions only. If this is not the case the method of Ambrosetti et al. (1994) or Brezis and Kamin (1992) still works. We include a proof for the reader's convenience.

THEOREM 7 Assume that f is a Carathodory function satisfying

$$\frac{f(x, u)}{u} \text{ is decreasing in } u, \forall \, x \in \Omega \tag{2.14}$$

Then there is a unique nontrivial positive solution of (2.1).

Proof: Suppose that $u, v > 0$ are solutions to (2.1). Then we have

$$-v\Delta u + u\Delta v = f(x, u)v - f(x, v) u \qquad (2.15)$$

Now, let θ be a smooth increasing real function such that $\theta(0) = 0, \theta(t) = 1$ if $t \geq 1, \theta(t) = 0$ for $t \leq 0$. Define $\theta_\varepsilon(t) = \theta\left(\frac{t}{\varepsilon}\right)$.

We multiply (2.15) by $\theta_\varepsilon(v - u)$ and integrate by parts over Ω. For the left-hand side this gives

$$\int_\Omega (-v\Delta u + u\Delta v)\,\theta_\varepsilon(v - u) = \int_\Omega v\theta'_\varepsilon(v - u)\,\nabla u \cdot \nabla(v - u)$$

$$-\int_\Omega u\theta'_\varepsilon(v - u)\,\nabla v \cdot \nabla(v - u) = -\int_\Omega v\theta'_\varepsilon(v - u)|\nabla(v - u)|^2$$

$$+\int_\Omega (v - u)\theta'_\varepsilon(v - u)\,\nabla v \cdot \nabla(v - u) \leq \int_\Omega (v - u)\theta'_\varepsilon(v - u)\,\nabla v \cdot \nabla(v - u)$$

$$= \int_\Omega \nabla v \nabla\,[\gamma_\varepsilon(v - u)] = -\int_\Omega \Delta v \cdot \gamma_\varepsilon(v - u)$$

with $\gamma_\varepsilon(s) = \int_0^s t\theta'_\varepsilon(t)dt$. Since $0 \leq \gamma_\varepsilon(s) \leq \varepsilon$ for any $s \in \mathbf{R}$ we obtain

$$\int_\Omega (-v\Delta u + u\Delta v)\,\theta_\varepsilon(v - u) = \int_\Omega uv\left(\frac{f(x, u)}{u} - \frac{f(x, v)}{v}\right)\theta_\varepsilon(v - u) \leq \varepsilon$$

and letting ε go to zero we have

$$\int_{[v > u]} uv\left(\frac{f(x, u)}{u} - \frac{f(x, v)}{v}\right) \leq 0$$

which by (2.14) implies that the set $[v > u]$ has measure 0. Thus $v \leq u$ and this ends the proof.

We add two variants which may be helpful for applications.

PROPOSITION 8. Assume that f satisfies the conditions of Theorem 1 and (2.14) and that, if $u, v > 0$ are solutions to (2.1), there is a supersolution w of (2.1) such that $u, v < w$. Then the nontrivial positive solution of (2.1) is unique.

Proof: First, assume that $0 < u \leq v$ for the solutions u and v. It follows immediately from (2.14) (2.15) that $u \equiv v$. If u and v are not ordered, let w be a supersolution w such that $u < w, v < w$. Taking $u_0 \equiv 0$ as a subsolution, by Theorem 1 there is a maximal solution $\bar{u}$ in the interval $[0, w]$. Then $u \leq \bar{u}, v \leq \bar{u}$ and by the first part $u \equiv v \equiv \bar{u}$.

PROPOSITION 9. Assume that f satisfies the conditions in Theorem 1 and (2.14) and that if $u, v > 0$ are solutions to (2.1), there is a subsolution $z < u, z < v$. Then there is a unique nontrivial positive solution of (2.1).

Proof: We show that ordered solutions coincide as in Proposition 8. If they are not ordered, it was proved in (Dancer and Sweers 1989) that $\inf\{u, v\}$ is a supersolution. By Theorem 1 there is a maximal solution $\bar{u}$ such that $z \leq \bar{u} \leq \inf\{u, v\}$. By the first part $\bar{u} \equiv u \equiv v$.

3. APPLICATIONS

Next we provide some applications for the existence and uniqueness results in Section 2.

3.1 Sublinear elliptic equations.

First, we consider the sublinear elliptic problem

$$-\Delta u = \lambda m(x)u^q \text{ in } \Omega$$
$$u = 0 \text{ on } \partial\Omega , \tag{3.1}$$

where Ω is a bounded smooth domain in $\mathbf{R}^N (N \geq 3)$, as before, and the weight m satisfies

$$m \in L^r(\Omega), \text{ with } r > \frac{N}{2} , m \geq 0 \text{ in } \Omega , \tag{3.2}$$

and

$$0 < q < 1. \tag{3.3}$$

Then we recall that for the associated linear eigenvalue problem

$$-\Delta w = \lambda m(x)w \text{ in } \Omega$$
$$w = 0 \text{ on } \partial\Omega , \tag{3.4}$$

the following auxiliary result holds.

LEMMA 10. Assume that m satisfies (3.2). Then (3.4) posseses a first positive eigenvalue $\lambda_1 > 0$ which is simple and whose corresponding eigenfunction $\varphi_1 \in H_0^1(\Omega)$ can be choosen such that $\varphi_1 > 0$ on Ω . Moreover, $\varphi_1 \in C(\bar{\Omega})$ and, if $r > N$, then $\varphi_1 \in C^1(\bar{\Omega})$ with $\frac{\partial\varphi_1}{\partial n} < 0$ on $\partial\Omega$.

Proof: See de Figueiredo (1981) for the first part. By the results of Brezis and Kato (1979, Thm. 2.3) $\varphi_1 \in L^t(\Omega)$ for all $1 < t < \infty$ and this yields easily $m\varphi_1 \in L^s(\Omega)$ for all $1 < s < r$, giving in turn $\varphi_1 \in C(\bar{\Omega})$ by (3.2). In a similar way we show that, if $r > N, \varphi \in C^1(\bar{\Omega})$ and $\frac{\partial\varphi_1}{\partial n} < 0$ on $\partial\Omega$ (Gilbarg and Trudinger 1983).

From now on we normalize φ_1 by condition $\| \varphi_1 \|_{L^\infty} = 1$.

THEOREM 11. Assume that (3.2) (3.3) are satisfied. Then for every $\lambda > 0$ there is a unique nontrivial positive solution $u > 0$ of (3.1) such that $u \in H_0^1(\Omega) \cap C(\bar{\Omega})$. If, moreover, $r > N$, then $u \in C^1(\bar{\Omega})$.

Proof: By Lemma 10 we have

$$-\Delta\varphi_1 = \lambda_1 m(x)\varphi_1 \, , \varphi_1 > 0 \text{ in } \Omega$$
$$\varphi_1 - 0 \text{ on } \partial\Omega \, . \tag{3.5}$$

If we pick $u_0 \equiv c\varphi_1$ with $c > 0$, then

$$-\Delta u_0 - \lambda m u_0^q = \lambda_1 m c \varphi_1 - \lambda m c^q \varphi_1^q$$
$$= m c^q \varphi_1^q (c^{1-q} \lambda_1 \varphi_1^{1-q} - \lambda)$$

and u_0 will be a subsolution for $c > 0$ small.

We look for a supersolution. Let $\tilde{\Omega}$ be a smooth bounded domain such that $\bar{\Omega} \subset \tilde{\Omega}$ and $\tilde{\lambda}_1 > 0$ and $\tilde{\psi}_1 > 0$ the first eigenvalue and the corresponding eigenfunction to (3.4) for $\tilde{\Omega}$. (We extend m by zero out of Ω , which still satisfies (3.2)). By Lemma 10, $\beta = \min_{\tilde{\Omega}} \tilde{\psi}_1 > 0$ and for $u^0 \equiv C\tilde{\psi}_1$ we obtain

$$-\Delta u^0 - \lambda m (u^0)^q = \tilde{\lambda}_1 m C \tilde{\psi}_1 - \lambda m C^q \tilde{\psi}_1^q$$
$$= m C^q \tilde{\psi}_1^q (\tilde{\lambda}_1 C^{1-q} \tilde{\psi}_1^{1-q} - \lambda)$$

and u^0 will be a supersolution if we choose $C^{1-q} > \frac{\lambda}{\tilde{\lambda}_1 \beta^{1-q}}$.

Since $\varphi_1 \in C(\bar{\Omega})$ it is always possible to pick c and C such that $u_0 \leq u^0$. This together with Theorems 1 (with $\gamma = r > \frac{2N}{N+2}$), 6 and 7 gives the result. (One can also apply Proposition 8 and, if, $r > N$, Proposition 9).

The results in Lemma 10 are still true under the much weaker assumption that the subset $\Omega^+ = \{x \in \Omega \mid m(x) > 0\}$ has positive measure, thus allowing m to change sign in Ω , instead of (3.2). This allows to replace (3.2) by

There exists Ω' open and smooth such that $\overline{\Omega'} \subset \Omega$ and m_0 constant such that

$$m(x) \geq m_0 > 0 \text{ on } \Omega' \, . \tag{3.6}$$

THEOREM 12. Assume that (3.3) and (3.6) are satisfied. Then for every $\lambda > 0$ there is a solution $u \geq 0$ of (3.1) with $u \in H_0^1(\Omega) \cap C(\bar{\Omega})$. If $r > N$, then $u \in C^1(\bar{\Omega})$.

Proof: If (3.6) is satisfied, then it is proved in (Boccardo and Orsina 1994) that $u_0 \equiv c\psi_1$, where $\psi_1 > 0$ is an eigenfunction for the first eigenvalue of (3.4) on Ω' and

$\psi_1 \equiv 0$ on $\Omega - \Omega'$, is a subsolution of (3.1) for $c > 0$ small. (If $r > N$ it is possible to argue as in Berestycki and Lions (1980, pp. 17–19)). We find a supersolution $u^0 \geq u^0$ as in Theorem 11 and use Theorems 1 (with $g(x,u) = m^-(x)u^q$ and $\gamma = r$) and 6.

REMARK 3. Theorem 12 was proved by Boccardo and Orsina (1994) by using an approximation method that we are able to avoid here. There are two interesting open questions concerning Theorem 12: the first one is uniqueness for nonnegative solutions; the second is that in this case solutions could exhibit dead cores, i.e., domains where $u = 0$. See (Daz 1985) for results in this direction.

3.2 The Logistic Equation with Unbounded Weights

We study the generalized logistic equation

$$\Delta u = \lambda m(x)u - u^p \text{ in } \Omega$$
$$u = 0 \text{ on } \partial\Omega , \tag{3.7}$$

with Ω a smooth bounded domain as above, $p > 1, \lambda$ is a real parameter and m satisfies

$$m \in L^r(\Omega), r > \frac{N}{2} \text{ and } m(x) \geq m_0 > 0 \text{ in } \Omega . \tag{3.8}$$

It is well-known that for $m \equiv 1$ there is a unique positive solution to (3.7) for any $\lambda > \mu_1$, where $\mu_1 > 0$ denotes the first eigenvalue of (3.4) for $m \equiv 1$. (See Berestycki (1981), Berestycki and Lions (1980), Brezis and Oswald (1986), Cantrell and Cosner (1989), de Figueiredo (1981), Hernndez (1986) for this and related results). In this case $u^0 \equiv \lambda^{\frac{1}{p-1}}$ is a supersolution to (3.7) and it is easily seen by using the Maximum Principle that is also provides an a priori estimate for positive solutions.

There is an alternative way of finding a supersolution. The real function $f(u) = \lambda u - u^p$ has a maximum for $u = \left(\frac{\lambda}{p}\right)^{\frac{1}{p-1}}$ where it takes the value

$$h(\lambda) = \left(\frac{p-1}{p}\right)\left(\frac{1}{p}\right)^{\frac{1}{p-1}}\lambda^{\frac{p}{p-1}}$$

Let ψ be the solution of the linear equation

$$-\Delta\psi = h(\lambda) \text{ in } \Omega$$
$$\psi = 0 \text{ on } \partial\Omega .$$

Now it easy to see that $\psi > 0$ is a supersolution. If m is not constant $u^0(x) \equiv (\lambda m(x))^{\frac{1}{p-1}}$ is not a supersolution any more but the alternative way still gives a supersolution.

THEOREM 13. Assume that $p > 1, m$ satisfies (3.8) and

$$\frac{pN}{2(p-1)} < r . \tag{3.9}$$

Then for any $\lambda > \lambda_1$ there exists a unique positive solution $u > 0$ of (3.7) with $u \in W^{2,q}(\Omega) \cap W_0^{1,q}(\Omega)$ for some $q > \frac{N}{2}$. Therefore $u \in C(\bar{\Omega})$. If moreover

$$\frac{pN}{p-1} < r \tag{3.10}$$

is satisfied, then $u \in C^{1,\beta}(\bar{\Omega})$ for some $0 < \beta < 1$.

Proof: The key point in order to apply results in Section 2 is to find a supersolution to (3.7). Notice that for almost all $x \in \Omega$ <u>fixed</u> the real function $f(x, u) = \lambda m(x)u - u^p$ is well-defined and attains its maximum at $\bar{u} = \left(\frac{\lambda m(x)}{p}\right)^{\frac{1}{p-1}}$, where it takes the value

$$h(\lambda)(x) = \left(\frac{p-1}{p}\right)\left(\frac{1}{p}\right)^{\frac{1}{p-1}} (\lambda m(x))^{\frac{p}{p-1}} . \tag{3.11}$$

If $h(\lambda) \in L^s(\Omega)$ for some $s > 1$, the linear equation

$$\begin{aligned} -\Delta\psi &= h(\lambda)(x) \text{ in } \Omega \\ \psi &= 0 \text{ on } \partial\Omega \end{aligned} \tag{3.12}$$

has a unique solution $\psi \in W^{2,s}(\Omega) \cap W_0^{1,s}(\Omega)$ and $\psi > 0$ by the Maximum Principle for weak solutions. If, moreover $s > \frac{2N}{N+2}$, $\psi \in H_0^1(\Omega)$ and

$$-\Delta\psi - f(x, \psi) = h(\lambda)(x) - f(x, \psi) \geq 0 \text{ a.e. on } \Omega .$$

Hence ψ is a (pointwise) supersolution and a simple integration by parts shows that it is a weak supersolution as well.

The regularity of ψ depends on s, which in turn depends on p and r. It follows easily from (3.11) that we have

$$s = \frac{(p-1)r}{p} .$$

If (3.9) is satisfied, then $s > \frac{N}{2}$ and $\psi \in C(\bar{\Omega})$. If, moreover, (3.10) is also satisfied, then $\psi \in C^1(\bar{\Omega})$.

Next, we look for a subsolution $u_0 \equiv c\varphi_1$, with φ_1 given by (3.5). For $\lambda > \lambda_1$ we have

$$\begin{aligned} -\Delta u_0 - \lambda m u_0 + u_0^p &= \lambda_1 mc\varphi_1 - \lambda mc\varphi_1 \\ &\quad + c^p\varphi_1^p = c\varphi_1[(\lambda_1 - \lambda)m + c^{p-1}\varphi_1^{p-1}] \end{aligned}$$

and u_0 is a subsolution if $0 < c^{p-1} < m_0(\lambda - \lambda_1)$.

We may choose $u_0 \leq u^0$. Indeed, if (3.10) holds, both are in $C^1(\bar{\Omega})$, the normal derivatives are negative on $\partial\Omega$, and the argument is straightforward. If only (3.9) is satisfied the preceding argument does not work but it is not difficult to see that for arbitrary constants $a, b > 0$, the solution ζ of $-\Delta\zeta = h(\lambda)(x) + a$ in Ω, $\zeta = b$ on $\partial\Omega$ is still a supersolution and $\zeta > b$ by the Maximum Principle. Thus it is enough to pick $u^0 \equiv \zeta$ with $b > c$.

In any case u_0 and u^0 are bounded and we apply Theorem 1 with $f(x, u) = \lambda m(x)u - u^p$ and $g(x, u) = Mu$, where $M > 0$ is such that $Mu - u^p$ is increasing on $[0, \sup u^0]$ and $\gamma = r$ in (2.8). (We can also apply Corollary 4, this time with $g(x, u) = u^p$). The regularity statements follow from (3.9) (3.10) and Theorem 6. Uniqueness follows either from Theorem 7 or from Proposition 8 and the above argument involving ζ. For $r > N$, Proposition 9 is also available.

The case where $\frac{N}{2} < r \leq \frac{pN}{2(p-1)}$ is more complicated. Now u_0 is still bounded but u^0 may be unbounded and the proof of Theorem 13 should be modified accordingly. Theorem 14 below can be proved by using results in Section 2 but the best regularity of solutions is more delicate and will considered elsewhere.

THEOREM 14. Assume that $p > 1, m$ satisfies (3.8) and moreover

$$\frac{2Np}{(p-1)(N+2)} < r \leq \frac{pN}{2(p-1)}$$

and

$$\frac{2N(2p-1)}{(p-1)(N+6)} < r .$$

Then there exists a unique nontrivial positive weak solution of (3.7).

REMARK 4. The more general problem (1.1) can be treated in a very similar way if $q(x)$ also satifies (3.8) with additional conditions. The cases where $m \geq 0, q \geq 0$ but are zero on subsets with positive measure are interesting and will be studied elsewhere.

The Logistic Equation with Nonlinear Diffusion

We consider again a generalized logistic equation, this time with nonlinear diffusion, namely

$$-\Delta w^n = \lambda m(x)w - w^k \text{ in } \Omega$$
$$w = 0 \text{ on } \partial\Omega . \tag{3.13}$$

Here Ω is a above, m satisfies (3.2) and $k > 1, n > 1$.

With the change of variable $u = w^n$ equation (3.13) becomes

$$-\Delta u = \lambda m(x)u^q - u^p \text{ in } \Omega$$
$$u = 0 \text{ on } \partial\Omega , \qquad (3.14)$$

where $q = \frac{1}{n}$, $p = \frac{k}{n}$. Hence we study (3.14) with q and p satisfying

$$0 < q < 1 \qquad q < p . \qquad (3.15)$$

Both cases $p \geq 1$ and $0 < q < p < 1$ are allowed.

THEOREM 15. Assume that $p \geq 1$ and (3.8) and (3.15) are satisfied. Then for any $\lambda > 0$ there is a unique nontrivial positive solution $u \in C(\bar{\Omega})$ of (3.14). If $r > N$, then $u \in C^1(\bar{\Omega})$.

Proof: As before let λ_1 and $\varphi_1 > 0$ be the first eigenvalue and eigenfunction in (3.5). If $u_0 \equiv c\varphi_1$ with $c > 0$, and $\lambda > 0$

$$- \Delta u_0 - \lambda m u_0^q + u_0^p = \lambda_1 m c \varphi_1 \quad \lambda m c^q \varphi_1^q + c^p \varphi_1^p$$
$$= m c^q \varphi_1^q (\lambda_1 c^{1-q} \varphi_1^{1-q} - \lambda) + \frac{1}{m} c^{p-1} \varphi^{p-1}$$

and u_0 is a subsolution for $c > 0$ small.

As a supersolution, one can pick the positive solution of (3.1) given by Theorem 11. One can show as above that they can by choosen satisfying $u_0 \leq u^0$. Now it suffices to apply Theorem 1 with $f(x, u) = \lambda m(x)u^q - u^p$, $g(x, u) = Mu$, where $M > 0$ is such that $Mu - u^p$ is increasing on $[0, \sup u^0]$ and $\gamma = r$. Theorem 6 with $\gamma = r$ gives the regularity and uniqueness follows from Theorem 7 and $p \geq 1$ (or Proposition 8).

The case of m changing sign is very similar to the preceding one, so we only sketch it.

THEOREM 16. Assume that $p > 1$ and (3.2) (3.6) (3.15) are satisfied. Then for any $\lambda > 0$ there is a solution $u \geq 0$ of (3.14) with $u \in C(\bar{\Omega})$. If $r > N$, then $u \in C^1(\bar{\Omega})$.

Proof: Reasoning as for Theorems 13 and 15 one can show that the u_0 in the proof of Theorem 13 is still a subsolution. The rest is as for Theorem 15.

REFERENCES

1. Amann, H. (1976). Fixed point equations and nonlinear eigenvalue problems in ordered Banach spaces, <u>S.I.A.M. Rev. 18</u>: 620–709.

2. Amann, H. and Crandall, M.G. (1978). On some existence theorems for semilinear elliptic equations, <u>Indiana Univ. Math. J. 27</u>: 779–790.

3. Ambrosetti, A., Brezis, H. and Cerami, G. (1994). Combined effects of concave and convex nonlinearities in some elliptic problems, <u>J. Funct. Anal. 122</u>: 519–543.

4. Ambrosetti, A. and Prodi, G. (1993). <u>A primer of Nonlinear Analysis</u>. Cambridge University Press.

5. Berestycki, H. (1981). Le nombre de solutions de certains problmes semilinaires elliptiques, <u>J. Funct. Anal. 40</u>: 1–29.

6. Berestycki, H., and Lions, P.L. (1980). Some applications of the method of sub and supersolutions, <u>Bifurcation and nonlinear eigenvaluie problems</u>, C. Bardos, J.M. Lasry and M. Schatzman (eds.), Springer Lecture Notes 782, pp. 16–41.

7. Boccardo, L., and Orsina, L. (1994) Sublinear elliptic equations in L^s , <u>Houston Math. J. 20</u>: 99–114.

8. Brezis, H., and Browder, F. (1982). Some properties of higher order Sobolev spaces, <u>J. Math. Pures Appl. 61</u>: 245–259.

9. Brezis, H., and Kamin, S. (1992). Sublinear elliptic equations in R^n, <u>Manuscripta Math. 74</u>: 87–106.

10. Brezis, H., and Kato, T. (1979). Remarks on the Schrdinger operator with singular complex potentials, <u>J. Math. Pures Appl. 58</u>: 137–151.

11. Brezis, H., and Oswald, L. (1986). Remarks on sublinear elliptic equations, <u>Nonlinear Anal. 10</u>: 55–64.

12. Cantrell, R.S., and Cosner, C. (1989). Diffusive logistic equations with indefinite weights: population models in disrupted environments, <u>Proc. Roy. Soc. Edinburgh 112 A</u>: 293–318.

13. Caada, A., and Gmez, J.L. (1993). Some new applications of the method of lower and upper solutions to elliptic problems, <u>Appl. Math. Letter 6</u>: 41–45.

14. Cohen, D.S., and Laetsch, T. (1970). Nonlinear boundary value problems suggested by chemical reactor theory, <u>J. Diff. Eqs. 7</u>: 217–226.

15. Dancer, E.N., and Sweers, G. (1989). On the existence of a maximal weak solution for a semilinear elliptic equation, <u>Differ. Int. Equations 2</u>: 533–540.

16. Deuel, J., and Hess, P. (1975). Criterion for the existence of solutions of nonlinear elliptic boundary value problems. <u>Proc. Roy. Soc. Edinburgh 74 A</u>: 49–54.

17. Diaz, J.I. (1985). <u>Nonlinear Partial Differential Equations and Free boundaries</u>, Pitman, Boston.

18. de Figueiredo, D.G. (1981). Positive solutions of semilinear elliptic problems, Lectures Notes, Sao Paulo.

19. Gilbarg, D., and Trudinger, N.S. (1983). <u>Elliptic Partial Differential Equations of Second Order</u>. Springer, Berlin, 2^d ed.

20. Hernndez, J. (1986). Qualitative methods for nonlinear diffusion equations, in <u>Nonlinear Diffusion Problems</u>, A. Fasano, A., and Primicerio, M. (eds.), Springer Lectures Notes 1224, pp. 47–118.

21. Krasnoselski, M. (1964). <u>Topological Methods in the Theory of Nonlinear Integral Equations</u>. Pergamon Press, London.

22. Lions, J.L., and Magenes, E. (1968). <u>Problmes aux limites non homognes et applications</u>, vol. 1. Dunod, Pars.

23. Mitidieri, E., and Sweers, G. (1994). Existence of a maximal solution for quasi-monotone elliptic systems, <u>Differ. Int. Equations</u>, <u>7</u>: 1495–1510.

24. Pao, C.V. (1992). <u>Nonlinear Parabolic and Elliptic Equations</u>. Plenum Press, New York.

25. Smoller, J. (1983). <u>Shock Wawes and Reaction-Diffusion Equations</u>. Springer, Berln.

Recent Results on Selfsimilar Solutions of Degenerate Nonlinear Diffusion Equations

JOSEPHUS HULSHOF Mathematical Institute of the Leiden University, P.O. Box 9512, 2300 RA Leiden, The Netherlands

1 INTRODUCTION

In this survey we collect some of the recent results on the porous medium equation,

$$u_t = \Delta(u^m) = \Delta(|u|^{m-1}u), \tag{1.1}$$

and related problems. Here as usual $u = u(x,t)$ with $x \in \mathbb{R}^N$ and $t > 0$, and m is a fixed positive number. It is also possible to consider $m \leq 0$, but then (1.1) has to be rewritten as, after a scaling of t,

$$u_t = \nabla \cdot (|u|^{m-1}\nabla u). \tag{1.2}$$

As far as the physical background of this equation is concerned, let us just mention that originally it was proposed as a model for gas flow through a homogeneous porous medium. In this model u is the density, and by the equation of state, which relates the pressure and the density, the pressure is given by u^{m-1}. The other two ingredients are the conservation of mass principle and Darcy's law, which postulates that the flux is proportional to (minus) the gradient of the pressure, see e.g. [1].

A first basic observation is that for $m \neq 1$, given any two positive numbers λ and μ, the scaling

$$u_{\lambda\mu}(x,t) = \lambda^{\frac{-2}{m-1}} \mu^{\frac{1}{m-1}} u(\lambda x, \mu t) \tag{1.3}$$

leaves (1.1) invariant. Thus we have a two-parameter scaling group, from which we can extract suitable one-parameter scaling groups, e.g. by setting

$$\lambda = \mu^{\beta}, \tag{1.4}$$

where β is any fixed number. Any solution which is invariant under this particular one-parameter scaling group, satisfies

$$u(x,t) = \mu^\alpha u(\mu^\beta x, \mu t), \quad (m-1)\alpha + 2\beta = 1 \tag{1.5}$$

for all x, t and μ. Substituting $\mu t = 1$, we find

$$u(x,t) = t^{-\alpha} u(xt^{-\beta}, 1) = t^{-\alpha} U(\eta), \quad \eta = xt^{-\beta}. \tag{1.6}$$

For $U(\eta)$ the equation reduces to

$$\Delta(|U|^{m-1}U) + \beta\eta \cdot \nabla U + \alpha U = 0, \tag{1.7}$$

or, for solutions which are also radially invariant,

$$(|U|^{m-1}U)'' + \frac{N-1}{r}(|U|^{m-1}U)' + \beta r U' + \alpha U = 0, \quad r = |\eta|. \tag{1.8}$$

A second basic observation is that integrable (in space) solutions satisfy, after an integration by parts,

$$\frac{d}{dt}\int u(x,t)dx = 0, \tag{1.9}$$

which is a conservation law. For (1.6) this implies, together with (1.5),

$$\alpha = N\beta = \frac{N}{N(m-1)+2}, \tag{1.10}$$

in which case there exists an explicit solution, the Barenblatt-Pattle instantaneous point source solution,

$$u(x,t) = t^{-\frac{N}{N(m-1)+2}} U(\eta), \quad \eta = xt^{-\frac{1}{N(m-1)+2}},$$

$$U(\eta) = (C - \frac{m-1}{2m(N(m-1)+2)}|\eta|^2)_+^{\frac{1}{m-1}}, \tag{1.11}$$

which converges to a multiple of the Dirac measure $\delta(x)$ as $t \downarrow 0$. Here and throughout this paper we restrict ourselves to $m > 1$. This solution, which was first published by Zel'dovich and Kompanyeets [31], initiated the development of an extensive theory for equation (1.1) and its generalizations. Without going into to any detail, let us recall that [1]: the Cauchy problem for nonnegative integrable initial data is wellposed; the intermediate asymptotics of (weak) solutions are given by (1.11) with C determined by $\int u(x,0)dx$; compactly supported solutions have supports with a free boundary moving outwards with finite speed.

In this survey we shall discuss a number of related problems, and investigate to what extent the results above remain valid. All these problems will have in common with (1.1) that they allow a scaling similar to (1.3). Consequently similarity solutions will play an important role. The emphasis will be on those similarity solutions, which are fundamental for the theory of the partial differential equation.

2 COMPACTLY SUPPORTED SOLUTIONS

The instanteneous point source solution (1.11) can be thought of as a fundamental solution for (1.1), in much the same way as the heat kernel

$$u(x,t) = E(x,t) = \frac{1}{(4\pi t)^{N/2}} e^{-\frac{|x|^2}{4t}} \tag{2.1}$$

is the fundamental solution for the heat equation ($m = 1$). Note that (2.1) is also selfsimilar, and that its decay as $|x| \to \infty$ is exponential, and that any derivative of (2.1) is again a selfsimilar solution, having the same exponential decay. For $m > 1$ the role of solutions with exponential decay is taken over by solutions with compact support, so it was natural to ask whether compactly supported selfsimilar solutions are as abundant as exponential decay solutions for the heat equation. For the one-dimensional case ($N = 1$), a complete classification was given in [17], the result being:

THEOREM 2.1. [17] *Let* $m > 1$, $N = 1$, *and let* $(m-1)\alpha + 2\beta = 1$. *Then there exists a strictly increasing sequence*

$$\alpha_1 = \frac{1}{m+1} < \alpha_2 = \frac{1}{m} < \alpha_3 < \alpha_4 < ... \uparrow \frac{1}{m-1}, \tag{2.2}$$

such that (1.7) *has (a one-parameter family of) compactly supported solutions if and only if* $\alpha = \alpha_k$ *for some integer* $k \geq 1$. *Moreover,* $k-1$ *equals exactly the number of sign changes of* $U(\eta)$, *and* $U(\eta)$ *is symmetric (anti-symmetric) if* k *is odd (even).*

Note that every compactly supported solution comes with a one-parameter family in view of the scaling

$$U(\eta; \lambda) = \lambda^{\frac{m-1}{2}} U(\frac{\eta}{\lambda}), \tag{2.3}$$

which (1.7) inherits from (1.3). For k odd, the value of the corresponding eigenfunction U_k in $\eta = 0$ can be used as a parameter, and for k even the value of $(|U_k|^{m-1}U_k)'$. The solutions corresponding to α_1 are the explicit instantaneous point source solutions given by (1.11).

For α_2 the solutions in Theorem 2.1 are also explicit and originally due to Barenblatt and Zel'dovich [8]. In fact they belong to a family of explicit solutions parametrized by the dimension N:

$$\alpha = \frac{1}{m}; \quad \beta = \frac{1}{2m}; \quad U(\eta) = \pm|\eta|^{\frac{2-N}{m}} (C - \frac{m-1}{N(m-1)+2} |\eta|^{N+\frac{2-N}{m}})_+^{\frac{1}{m-1}}. \tag{2.4}$$

For $N = 2$ these solutions coincide with (1.11), whereas for $N > 2$ they are singular in $\eta = 0$. In the one-dimensional case (2.4) gives the dipole profiles for (1.1), which describe the intermediate asymptotics for compactly supported solutions of (1.1) with zero integral. This latter fact was established by Kamin and Vazquez [23] who incidentally also showed that compactly supported solutions of (1.1) with nonzero integral eventually loose all their sign changes, in which case the intermediate asymptotics are again given by (1.11). The values of α and β follow from (1.5) and a generalisation of the mass conservation law (1.9) for (compactly supported) solutions of (1.1): integration by parts yields that for every harmonic function $h(x)$

$$\frac{d}{dt} \int h(x)u(x,t)dx = 0. \tag{2.5}$$

With $N = 1$ and $h(x) = x$ this implies $\alpha = 2\beta$. Consequently each of these dipole solutions converges to a multiple of the distributional derivative of the Dirac measure as $t \downarrow 0$. After setting

$$w(x,t) = \int_{-\infty}^{x} u(s,t)ds, \tag{2.6}$$

we obtain functions which have the the Dirac measure itself as initial data. These are (fundamental) solutions of the equation

$$w_t = \nabla \cdot (|\nabla w|^{m-1} \nabla w) \tag{2.7}$$

in dimension $N = 1$. The right hand side of (2.7) is known in the literature as the p-Laplacian ($p = m+1$). Equation (2.7) was treated in [22]. The theory for compactly supported solutions of this equation is much the same as the theory for (1.1). In particular it allows a family of selfsimilar compactly supported instantaneous point source solutions playing the same role as the Barenblatt-Pattle solutions for the porous medium equation. However in dimension $N > 1$ equation (2.7) cannot be transformed into (1.1) by differentiation.

The first two exponents in Theorem 2.1 are explicit numbers, obtained from combining the relation $(m - 1)\alpha + 2\beta = 1$ with a conservation law. Observe that writing

$$k_n = \frac{\alpha_n}{\beta_n}, \tag{2.8}$$

we have $k_1 = 1$ and $k_2 = 2$. The numbers k_n are now commonly refered to as eigenvalues, and the corresponding solution profiles as eigenfunctions. Note that for $\beta > 0$ we may just as well take as similarity variable

$$\eta = \beta^{\frac{1}{2}} x t^{-\beta}, \tag{2.9}$$

which gives

$$\Delta(|U|^{m-1}U) + \eta \cdot \nabla U + kU = 0, \quad k = \frac{\alpha}{\beta}, \tag{2.10}$$

whence the terminology of eigenvalues and eigenfunctions.

In the original russian literature the selfsimilar solutions corresponding to k_1 and k_2 are called selfsimilar solutions of the first kind. Solutions for which there is no physical or dimensional principle to determine a second relation between the exponents, are called selfsimilar solutions of the second kind, and their exponents are called anomalous. For the newly found higher eigenvalues $k_3, k_4, k_5, \ldots$, it was not immediately clear whether they were of the first or the second kind. In the case of the heat equation, when we look at profiles with exponential decay, $k_n = n$ for all n. The corresponding profiles are easily seen to have the property that their first $n-1$ moments are zero, which is related to the fact that the heat equation conserves all these moments equally well. This gave rise to the conjecture that perhaps also for the porous medium equation $k_n = n$ would be true for all n. However the conjecture turned out to be hard to prove, the reason being:

THEOREM 2.2. [11] *Let* $m > 1$, $N = 1$, *and let* $k_3 = \alpha_3/b_3$, *where* α_3 *and* β_3 *are as in Theorem 4.1. Then*

$$k_3 > 3. \tag{2.11}$$

The proof can be found in [11], which also contains a computation of the limits of all eigenvalues as $m \to \infty$. We have e.g. that

$$k_3 \to 4, \quad k_4 \to 4 + 2\sqrt{2}, \quad k_5 \to 6 + 2\sqrt{5}. \tag{2.12}$$

The limits of the higher order eigenvalues are more complicated algebraic numbers. Numerical experiments indicate that all the eigenvalues are strictly increasing functions of the parameter m, but until now this has not been proved. In particular the inequality $k_n > n$ is still undecided for $n \geq 4$. Another open question is the differentiability of the eigenvalues as functions of m, and, in case of an affirmative answer, what are the explicit values of these derivatives at $m = 1$?

3   THE DUAL POROUS MEDIUM EQUATION

In Section 2 we already saw that (1.1) can be integrated once with respect to x to give a new equation (2.7) for the primitive of $u(x, t)$. A second integration,

$$z(x, t) = \int_{-\infty}^{x} w(s, t) ds, \tag{3.1}$$

yields the equation

$$z_t = |z_{xx}|^{m-1} z_{xx}, \tag{3.2}$$

which is called the dual porous medium equation [11]. Just as (1.1) and (2.7) this is a degenerate nonlinear diffusion equation, but it does not have a conservation law for $\int z(x, t) dx$. The role played by the instanteneous point source solutions for (1.1) and (2.7), is now taken over by the second primitive of the eigenfunction $U_3(\eta)$ corresponding to k_3.

THEOREM 3.1. [11] *There exists a unique family of nonnegative compactly supported similarity solutions of (3.2), given by*

$$z_3(x, t) = t^{-\alpha_3 + 2\beta_3} Z_3(xt^{-\beta_3}) = t^{(-k_3+2)\beta_3} Z_3(xt^{-\beta_3}), \tag{3.3}$$

with

$$Z_3(\eta) = \int_{-\infty}^{\eta} \int_{-\infty}^{\xi} U_3(\zeta) d\zeta d\xi. \tag{3.4}$$

These solutions describe the intermediate asymptotics of all compactly supported nonnegative solutions.

An immediate consequence of the inequality $k_3 > 3$ is that this solution has the property that

$$\int z_3(x, t) dx = t^{(3-k_3)\beta_3} \int Z_3(\eta) d\eta \tag{3.5}$$

goes to infinity if $t \downarrow 0$. Hence this anomalous "fundamental" solution is very singular in the sense that it does not have a distribution as initial value at $t = 0$.

Theorem 2.2 has another peculiar consequence for the dual porous medium equation. If we take the ratio $k = \alpha/\beta = 3$ and look at the symmetric solutions of (1.6) with $N = 1$, we find they have two sign changes and algebraic decay like $|\eta|^{-3}$

as $|\eta| \to \infty$. Integrating twice, the sign changes disappear, and the decay becomes $|\eta|^{-1}$. This implies that the corresponding similarity solutions, which can then be taken strictly positive, converge to multiples of the function $1/|x|$. The convergence is pointwise in every $x \neq 0$. Note that these similarity solutions are not integrable, and that $1/|x|$ is not locally integrable. Also the similarity exponents are such that, had the similarity profile been integrable, the initial data of the similarity solution would have been a multiple of the Dirac δ-measure. In view of the consequences of this fact later on, we state this as a separate theorem.

THEOREM 3.2. *There exists a strictly positive similarity solution*

$$z(x,t) = t^{-\frac{1}{3m-1}} Z(xt^{-\frac{1}{3m-1}}) \tag{3.6}$$

of equation (4.6) with the property that

$$z(x,t) \to \frac{1}{|x|} \tag{3.7}$$

uniformly on every compact $K \subset \mathbf{R} - \{0\}$ as $t \downarrow 0$.

4   THE DIPOLE IN MORE DIMENSIONS

In this section we discuss the more dimensional case, and in particular the difference between $N = 1$ and $N \geq 2$. To begin with, the asymmetric solutions in Theorem 2.1 disappear. Stated in terms of (2.10) rather than (1.7) we have

THEOREM 4.1. [17] *Let $m > 1$ and $N \geq 2$. Then there exists a strictly increasing sequence*

$$k_1 = N < k_3 < k_5 < k_7 ... \uparrow \infty, \tag{4.1}$$

such that (2.10) has (a one-parameter family of) compactly supported radially symmetric solutions U_n if and only if $k = k_n$ for some odd integer $n \geq 1$. Moreover, $(k-1)/2$ equals exactly the number of spheres on which U_n changes sign, and $U_n(0)$ can be used as a parameter.

With the discussion of the one dimensional case in mind, this result raises two questions. First, is there anything in the more dimensional case to replace the number k_2 and the dipole solution? Second, what can be said about k_3? The answers to these two questions are intimately related, as we will point below.

For the heat equation, solutions of dipole type are easily obtained by taking a directional derivative of the heat kernel (2.1). It is well known for instance that

$$u(x,t) = -\frac{\partial}{\partial x_1} E(x,t) = \frac{x_1}{2t} \frac{1}{(4\pi t)^{N/2}} e^{-\frac{x^2}{4t}} \tag{4.2}$$

is a solution of the heat equation with

$$u(x,t) \to -\frac{\partial}{\partial x_1} \delta(x) \tag{4.3}$$

as $t \downarrow 0$. Moreover, this solution decribes the intermediate asymptotics of a large class of solutions on the halfspace $\mathbf{R}^N_+$ with boundary condition $u(0, x_2, \dots, x_N, t) = 0$.

We note that (4.2) is a selfsimilar nonradial solution. The symmetry is lost because of the partial derivative with respect to x_1. Thus we cannot reduce the problem of finding a dipole solution for (1.1) to an ordinary differential equation. In stead a partial differential equations approach has to be pursued. This was done in [19], from which we state

THEOREM 4.2. *There exists a selfsimilar weak solution $u_2(x, t)$ to (1.1) of the form (1.6) with*

$$\alpha = \frac{N+1}{(N+1)m+1-N} = (N+1)\beta, \tag{4.4}$$

which satisfies (4.3) as $t \downarrow 0$. The corresponding profile has compact support and is unique in the class of solutions satisfying $u(0, x_2, \ldots, x_N, t) = 0$.

Note that the uniqueness implies that the solution is radially symmetric in the variables $x_2, \ldots, x_N$. Observe also that the exponents are explicit, since they follow from combining the relation $(m-1)\alpha + 2\beta = 1$ with the conservation law for the first moment with respect to x_1,

$$\frac{d}{dt} \int x_1 u(x, t) dx = 0. \tag{4.5}$$

Thus the dipole solution in more space dimensions is a selfsimilar solution of the first kind, although it is unlikely to have an explicit closed formula representation.

The uniqueness of this dipole follows from a rather standard argument based on a comparison principle for solutions of the porous medium equation on a halfspace and the conservation law (4.5) above. More interesting is the construction of the dipole solution. This is done via the dual porous medium equation.

We have seen in the previous sections that (3.2) can be obtained by twice taking the primitive with respect to x. In more dimensions the dual porous medium equation,

$$z_t = |\Delta z|^{m-1} \Delta z, \tag{4.6}$$

can be related to (1.1) by setting

$$u = -\Delta z. \tag{4.7}$$

Here we have put the minus in front of the Laplacian in view of the positivity properties of this differential operator. To invert the relation between u and z, we use the fundamental solution of Laplace's equation

$$\Gamma(x) = \frac{1}{N(2-N)\omega_N} |x|^{2-N} \quad (N = 2: \quad \Gamma(x) = \frac{1}{2\pi} \log |x|), \tag{4.8}$$

and write

$$z(x, t) = -\int \Gamma(x - y) u(y, t) dy. \tag{4.9}$$

Thus, since $\delta = \Delta\Gamma$, and $u = -\Delta z$, we have , as $t \downarrow 0$,

$$u(x, t) \rightarrow -\frac{\partial}{\partial x_1} \delta(x) \quad \Leftrightarrow \quad z(x, t) \rightarrow z_0(x) = \frac{\partial \Gamma}{\partial x_1}(x) = \frac{1}{N\omega_N} \frac{\cos \varphi_1}{r^{N-1}}, \tag{4.10}$$

where we have used polar coordinats $x_1 = r \cos \varphi_1, x_2 = \sin \varphi_1 \cos \varphi_2, \ldots$.

The dipole solution is then obtained by first solving the dual porous medium equation with initial data $z_0(x)$ as in (4.10), on the half space $\mathbb{R}_+^N$, with zero lateral boundary data, and then putting $u = -\Delta z$. The dual porous medium equation is wellposed for continuous initial data with compact support, and the corresponding semigroup is contractive with respect to the supremum norm. Since $z_0(x)$ is unbounded, a limiting argument is needed to construct a solution of the dual porous medium equation with $z(x,0) = z_0(x)$ as a limit of solutions with bounded initial data. The hard part here is to show that for $t > 0$ this limit is bounded (apriori there is no reason why it should). In particular we need it to be not identically equal to $z_0(x)$, otherwise we would be left with $u = 0$.

For $N = 2$ the proof of this boundedness relies on Theorem 3.2, which is a consequence of the equality $k_3 > 3$. Observing that $z_0(x) < |x|^{1-N}$ we can use the solution in Theorem 3.1 as a supersolution for the limiting problem, provided $N = 2$, forcing the limit solution to be bounded for every positive t.

For $N \geq 3$ the proof relies on a generalization of Theorems 2.2 and 3.2 to the $(N-1)$-dimensional case. Thus information about a specific similarity solution in dimension $N-1$ allows a successful construction of the dipole solution in dimension N.

THEOREM 4.3. [19] *Let $m > 1$, $N > 1$, and let $k_3 = \alpha_3/b_3$, where α_3 and β_3 are as in Theorem 4.1. Then*

$$k_3 > N + 2. \tag{4.11}$$

Moreover, there exists a strictly positive similarity solution

$$z(x,t) = t^{-\frac{1}{m(N+2)-N}} Z(xt^{-\frac{1}{m(N+2)-N}})$$

of equation (3.2) with the property that

$$z(x,t) \to |x|^{2-N}$$

uniformly on every compact $K \subset \mathbb{R}^N - \{0\}$ as $t \downarrow 0$.

5   MORE ANOMALOUS EXPONENTS

Anomalous exponents also appear in the so-called modified heat equation, and in the modified porous medium equation, which read, respectively with $m = 1$ and $m > 1$,

$$u_t + \gamma|u_t| = \Delta u^m. \tag{5.1}$$

Here $\gamma \in (-1,1)$ is a fixed parameter.

With $m = 1$ this is called Barenblatt's equation for elasto-plastic filtration. For $m = N = 2$ it is a model for groundwater flow which takes into account the effect of hysteresis [7]. Denoting the inverse of the function $F : s \to s + \gamma|s|$ by Φ, it can be written as $u_t = \Phi(\Delta u)$, which resembles the dual porous medium equation (4.6), where $F(s) = |s|^{\frac{1}{m}-1}s$ and $\Phi(c) = |c|^{m-1}c$. Kamin, Vazquez and Peletier have observed [25] that the Cauchy problem for this equation is wellposed in the class of nonnegative continuous functions which decay to zero at infinity, and that, with respect to the supremum norm, it generates a contraction semigroup in this class. Moreover they prove

THEOREM 5.1. *Let $m = 1$ and $\gamma \in (-1, 1)$. There exists a unique family of radially symmetric strictly positive similarity solutions*

$$u(x, t) = t^{-k/2} U(x/t^{1/2}) \tag{5.2}$$

of equation (5.1) with the property that $U(\eta)$ decays exponentially to zero as $|\eta| \to \infty$. These solutions are parametrized by $U(0)$. The exponent k is a strictly increasing smooth function $k(\gamma)$ of γ, with $k(\gamma) \to \max(N - 2, 0)$ as $\gamma \to -1$, and $k(\gamma) \to \infty$ as $\gamma \to 1$. Moreover, every solution to the Cauchy problem with nonzero nonnegative initial data

$$u(x, 0) \le M e^{-Ax^2} \tag{5.3}$$

for some positive numbers 1 and M, has intermediate asymptotics described by one of these similarity solutions.

In other words, these similarity solutions can be thought of as fundamental solutions for Barenblatt's equation. Observe that the solution with $U(0) = A \ne 0$ is simply the solution with $U(0) = 1$, multiplied by 1. This reflects the invariance of the equation under scalar multiplication. Just as in the case of the dual porous medium equation, the exponents are anomalous if $\gamma \ne 0$. For $\gamma > 0$ the solution is very singular, just as (3.4), because its integral blows up at $t = 0$. On the other hand, if $\gamma < 0$, the solution is mildly singular, because its integral goes to zero. Note that for $\gamma = 0$ we have the heat equation with $k = N$.

Later it was shown by Aronson and Vazquez [5], using an implicit function type argument in a dynamical systems setting, that

$$\frac{dk}{\delta \gamma}(0) = \frac{4(N/2)^{N/2}}{e^{N/2} \Gamma(N/2)}. \tag{5.4}$$

This number had also been found by Goldenfeld et al. [13] by means of the formal techniques of renormalization. Finally, Peletier [30] showed that

$$\lim_{\gamma \to 1} (1 - \gamma) k(\gamma)^2 = \frac{1}{2} \rho_\nu^2, \tag{5.5}$$

where ρ_ν denotes the first zero of the Bessel function J_ν with $\nu = N/2 - 1$, and that as $\gamma \to 1$, normalizing by $U(0) = 1$,

$$U(\sqrt{k}\eta; \gamma) \to \Gamma(\nu + 1)(\frac{1}{4} |\eta| \rho_\nu)^{-\nu} J_\nu(\frac{1}{2} \rho_\nu | eta|) \chi_{[0,2]}(|\eta|). \tag{5.6}$$

Here $\chi_{[0,2]}$ denotes the characteristic function of the interval $[0, 2]$. Note that this limit profile is a multiple of the first eigenfunction for the radial Laplacian on the ball, extended by zero outside.

Next we consider (5.1) with $m > 1$. Compactly supported similarity solutions were obtained in [20], from which we state

THEOREM 5.2. [20] *Let $m > 1$ and $\gamma \in (-1, 1)$. There exists a unique family of radially symmetric nonnegative similarity solutions*

$$u(x, t) = t^{-\alpha} U(xt^{-\beta}) \tag{5.7}$$

of equation (5.1) with the property that $U(\eta)$ has compact support. These solutions are parametrized by $U(0)$. The exponents satisfy $2\beta + (m-1)\alpha = 1$, and the ratio $k = \alpha/\beta$ is a strictly increasing smooth function $k(\gamma)$ of γ, with $k(\gamma) \to \max(N-2, 0)$ as $\gamma \to -1$, and $k(\gamma) \to \infty$ as $\gamma \to 1$. Moreover

$$\frac{\partial k}{\partial \gamma}(0) = \frac{N^{\frac{N}{2}} \, 2^{\frac{m}{m-1}+1} \, (m-1)^{\frac{N}{2}}}{(2+(m-1)N)^{\frac{1}{m-1}+\frac{N}{2}} \, B\left(\frac{m}{m-1}, \frac{N}{2}\right)}, \tag{5.8}$$

where B is the β-function.

Again the derivative agrees with the formal results of renormalization [10].

The limit behaviour of the similarity solutions as $\gamma \to 1$ was investigated by Ducroo de Jongh, who showed in his graduation work at Leiden University, that

PROPOSITION 5.3. *The behaviour of k as $\gamma \to 1$ is given by*

$$\lim_{\gamma \to 1} (1-\gamma) k(\gamma)^{1+\frac{1}{m}} = C = C(m, N) \neq 0. \tag{5.9}$$

In agreement with (5.5), $C(m, N) \to \rho_\nu^2/2$ as $m \to 1$. He also showed that, with the appropiate scalings as $\gamma \to 1$, on its supporting ball the asymptotic similarity profile is a radial solution of the Dirichlet problem with zero boundary data for

$$\Delta U^m + U = 0. \tag{5.10}$$

The similarity solutions in Theorem 5.2 describe the intermediate asymptotics for compactly supported nonnegative solutions of (5.1), see [21]. Note however that (5.1) is not of the form $u_t = \Phi(\Delta u)$, which makes existence and uniqueness results considerably harder to obtain.

We conclude this section with another important example in which anomalous exponents appear, namely the Aronson-Graveleau solution [4] of the porous medium equation. This is a selfsimilar solution of the form

$$u(x, t) = (T-t)^{-\alpha} U(r), \quad r = |x|(T-t)^{-\beta}, \quad (m-1)\alpha + 2\beta = 1. \tag{5.11}$$

Solutions of the corresponding ordinary equation for $U(r)$ are always unbounded. Thus we cannot expect to find any compactly supported solutions of the form (5.11). However there is an equally interesting class of special solutions to be found here, namely the solutions whose pointwise limit as $t \uparrow T$ exists, and is not identically equal to zero. There is only one nonnegative (family of) solution(s) with this property. They are supported on the outside of a focussing ball, and their limit profile at the focussing time is given by $v(x) = u(x)^{m-1} = Cx^{2-1/\beta}$. In dimension $N = 1$ these are explicit colliding travelling wave solutions (whence $\beta = 1$), but in higher dimensions these solutions are of the second kind and have $\beta < 1$. The existence of this solution has important consequences for the regularity theory for solutions of (1.1) in dimension $N \geq 2$, because it shows that in general the pressure ($v = u^{m-1}$) is not Lipschitz continuous. They were announced in [1] after numerical experiments by Graveleau, but only recently their existence and uniqueness was established in [4]. It has also been shown by Angenent and Aronson [2], that these solutions describe the behaviour of radial solutions supported on an annulus as the inner free boundary collapses, see also [3]. Finally, considering β as a function $\beta(N)$ of N, and not restricting to integer values of N, the value of $\beta'(1)$ can be computed explicitly [5].

6 THE $k - \epsilon$ MODEL.

The $k - \epsilon$ model for turbulence reads

$$(KE) \quad \begin{cases} k_t = \alpha\Big(\dfrac{k^2}{\varepsilon}k_x\Big)_x - \varepsilon; \\[3mm] \varepsilon_t = \beta\Big(\dfrac{k^2}{\varepsilon}\varepsilon_x\Big)_x - \gamma\dfrac{\varepsilon^2}{k}. \end{cases}$$

Here α, β, and γ are positive parameters, and k and ε are unknown nonnegative functions of x and t. This system describes the evolution of turbulent bursts [6], k stands for turbulent energy density, and ε is the dissipation rate of turbulent energy. In the applications α and β are usually different [28], [15]. A related scalar problem is discussed in [16] and [24].

We note that (KE) is a coupled system of two quasilinear diffusion-absorption equations. The diffusion coefficients may, depending on k and ε, become degenerate (very small) or singular (very large), and the second absorption term is also singular. In the case of equal diffusion coefficients, (KE) allows a family of explicit selfsimilar compactly supported "source type" solutions found by Barenblatt et al. [9]. To be precise, put

$$k = \frac{1}{t^{2\mu}}f(\eta), \quad \varepsilon = \frac{1}{t^{2\mu+1}}g(\eta), \quad \eta = \frac{x}{t^{1-\mu}}, \tag{6.1}$$

where we restrict our attention to the case $0 < \mu < 1$. If

$$\alpha = \beta = 1, \quad \mu = \frac{\gamma}{3(\gamma-1)}, \quad \gamma > \frac{3}{2}, \tag{6.2}$$

then

$$f(\eta) = \frac{1}{6}\kappa(2-\kappa)(1-\eta^2)_+, \quad g(\eta) = \kappa f(\eta), \quad \kappa = \frac{1}{\gamma-1} \tag{6.3}$$

is an explicit solution of the equations for f and g, which read

$$\begin{cases} \alpha\Big(\dfrac{f^2}{g}f'\Big)' + (1-\mu)\eta f' + 2\mu f - g = 0; \\[3mm] \beta\Big(\dfrac{f^2}{g}g'\Big)' + (1-\mu)\eta g' + (1+2\mu)g - \gamma\dfrac{g^2}{f} = 0. \end{cases} \tag{6.4} \tag{6.5}$$

In [18] this solution is used to obtain a compactly supported solution for $\alpha \neq \beta$ with $|\alpha - \beta|$ small. The result is

THEOREM 6.1. [18] *There exists an open neighbourhood $\mathcal{O}$ of the set*

$$\{(\alpha,\beta,\gamma): \ \alpha = \beta > 0, \ \gamma > \frac{3}{2}\},$$

such that for every $(\alpha,\beta,\gamma) \in \mathcal{O}$ there is precisely one $0 < \mu < 1$ for which equations (6.4) and (6.5) have a solution pair (f,g), with f and g symmetric and positive on $(-1,1)$, and

$$f(\eta) \to 0, \quad g(\eta) \to 0, \quad \frac{f(\eta)}{g(\eta)}f'(\eta) \to -\alpha(1-\mu), \quad \frac{f(\eta)^2}{g(\eta)^2}g'(\eta) \to -\beta(1-\mu), \tag{6.6}$$

as $\zeta \uparrow 1$. Moreover, if we write

$$\kappa = \frac{g(0)}{f(0)}, \quad \lambda = \frac{\alpha}{\beta}, \tag{6.7}$$

then in $\lambda = 1$

$$\kappa = \frac{1}{\gamma - 1}, \quad \frac{d\mu}{d\lambda} = 0, \quad \frac{d\kappa}{d\lambda} = \frac{\kappa(2 - \kappa)}{\kappa + 1}\left(\kappa - 1 + \frac{2}{B_\kappa}\right). \tag{6.8}$$

Here B_κ is defined by

$$B_\kappa = \frac{\Gamma(\frac{1}{2})}{\Gamma(a)\Gamma(b)}, \quad a + b = \frac{3}{2}, \quad ab = \frac{3}{2(2 - \kappa)}. \tag{6.9}$$

The conditions in (6.6) are the usual free boundary conditions for porous medium type equations. In particular the fluxes at the free boundary are zero.

Theorem 6.1 was the first result for this system with $\alpha \neq \beta$. For $\alpha = \beta$, Bertsch, Dal Passo and Kersner [12] proved an existence result for the Cauchy problem in the class of compactly supported solutions. They also show in their paper that these solutions have intermediate asymptotics described by (6.3), provided $\gamma > 3/2$.

7   QUADRATIC SYSTEMS

Our results on similarity solutions depend for a large part on a transformation of the (radial) similarity equation into an autonomous quadratic system. In this section we shall discuss this method. As an illustration we sketch the proof of Proposition 5.3 about the behaviour of the selfsimilar compactly supported solutions of the modified porous medium equation as $\gamma \to 1$.

Let $U(\eta)$ be a radial solution of (2.10). Writing $U(r) = U(|\eta|)$ we let

$$\tau = \log r, \quad x(\tau) = \frac{rU'(r)}{U(r)}, \quad y(\tau) = r^2|U(r)|^{1-m}. \tag{7.1}$$

This transforms (2.10) into

$$(Q_k) \begin{cases} \dot{x} = (2 - N - mx)x - (x + k)y; \\ \dot{y} = y(2 + (1 - m)x). \end{cases}$$

Note that the scaling (2.3) amounts to a shift in τ. The instantaneous point source solutions correspond to $k = N$ and the straight line $x + y = 0$. It was shown in [17] that for any k the radial solutions are all mapped into a unique orbit $\Gamma_{r,k}$ coming out of the origin into the upper half plane along the eigenvector corresponding to the eigenvalue 2 of the linearisation of (Q_k) around the origin. Also, as long as $k > 0$, there is a unique orbit $\Gamma_{i,k}$ escaping in finite time to infinity in the second quadrant along the straight line $x + y = 0$. This orbit contains the solutions with a free boundary, on which the flux is easily seen to be zero. Using either polar coordinates

$$x = \frac{\rho}{1 - \rho}\cos\phi, \quad y = \frac{\rho}{1 - \rho}\sin\phi, \tag{7.2}$$

or homogeneous coordinates

$$x = \frac{X}{Z}, \quad y = \frac{Y}{Z}, \quad X^2 + Y^2 + Z^2 = 1, \tag{7.3}$$

we find $\Gamma_{i,k}$ is really the unique orbit going into a saddle point at infinity. In terms of (7.2) it is given by $\rho = 1$ and $\phi = 3\pi/4$. We note that the behaviour near infinity is completely determined by the quadratic terms in (Q_k).

Now let $y_r(k)$ and $y_i(k)$ be the intersections of respectively $\Gamma_{r,k}$ and $\Gamma_{i,k}$ with the line $x + k = 0$. Theorem 5.2 is based on the simple observation that for the compactly supported similarity profile of the modified porous medium equation, k and γ are related by

$$\frac{y_i(k)}{y_r(k)} = \frac{1 + \gamma}{1 - \gamma}. \tag{7.4}$$

The left hand side is a real analytic function of k, strictly increasing, defined on the interval $(\max(0, N - 2), \infty)$ with range $(0, \infty)$. Formula (5.8) follows from a computation of the first derivative of this function in $k = N$. In fact one can compute higher order derivatives just as well, in [20] this is done in the case that $m = N = 2$. All these computations are based on solving linear first order inhomogeneous ordinary differential equations with rational coefficients.

Proof of Proposition 5.3. This is based on on a careful inspection of $\Gamma_{i,k}$ and $\Gamma_{r,k}$ as $k \to \infty$. First we consider $\Gamma_{i,k}$. Scaling x, y and τ by

$$A = \frac{x}{k}; \quad B = \frac{y}{k}; \quad t = k\tau, \tag{7.5}$$

we obtain

$$(\tilde{Q}_k) \begin{cases} \dot{A} = (\frac{2-N}{k} - mA)A - (A + 1)B; \\ \dot{B} = B(\frac{2}{k} + (1 - m)A), \end{cases}$$

in which we can take the limit $k \to \infty$ to obtain

$$(\tilde{Q}_\infty) \begin{cases} \dot{A} = -mA^2 - (A + 1)B; \\ \dot{B} = (1 - m)AB. \end{cases}$$

The scaling has been chosen to preserve the behaviour at infinity. The orbit $\Gamma_{i,k}$ corresponds to an orbit $\tilde{\Gamma}_{i,k}$ going into a saddle point at infinity of $(\tilde{Q}_k)$ with the same coordinates as the original saddle point. Taking the limit as $k \to \infty$, we have $\tilde{\Gamma}_{i,k} \to \tilde{\Gamma}_{i,\infty}$, which intersects the line $A = -1$ at some $B = B_\infty > 0$. Here we use the Stable Manifold Theorem with parameter dependence, combined with the observation above that the quadratic terms determine the behaviour at infinity. For the original system this implies that

$$\lim_{k \to \infty} \frac{y_i(k)}{k} = B_\infty. \tag{7.6}$$

Next we consider $\Gamma_{r,k}$. Here we have to use a different scaling preserving the behaviour near the origin, which is determined by the linear terms. We leave x and τ unchanged and only set

$$b = ky, \tag{7.7}$$

which transforms (Q_k) into

$$(Q_k^*) \begin{cases} \dot{x} = (2 - N - mx)x - (\frac{x}{k} + 1)b; \\ \dot{b} = b(2 + (1 - m)x), \end{cases}$$

in which we can take again the limit $k \to \infty$ to obtain

$$(Q_\infty^*) \begin{cases} \dot{x} = (2 - N - mx)x - b; \\ \dot{b} = b(2 + (1 - m)x), \end{cases}$$

Corresponding to $\Gamma_{r,k}$ we have orbits $\Gamma_{r,k}^*$ and in the limit $\Gamma_{r,\infty}^*$. It was shown in [17] that these orbits escape to infinity in finite time, and that $b(-x)^{(1-m)/m}$ converges to a positive constant C_k. A continuity argument [17] then implies that the intersection height b_k of $x + k = 0$ and $\Gamma_{r,k}^*$ satisfies $b_k(-k)^{(1-m)/m} \to C_\infty$ as $k \to \infty$. Consequently we have

$$\lim_{k \to \infty} y_r(k)k^{\frac{1}{m}} = C_\infty. \tag{7.8}$$

Combining (7.4), (7.6) and (7.8) completes the proof of Proposition 5.3. We note in addition that a change of variables similar to (7.1) relates (Q_∞^*) to (5.10), which is why this semilinear elliptic equation describes the asymptotic similarity profile as $k \to \infty$.

We have seen that selfsimilar solutions of scalar diffusion equations with a two-parameter group of scalings are related to two-dimensional quadratic autonomous dynamical systems. For systems however, such scaling groups lead to quadratic systems with dimension equal to twice the number of equations in the system. The $k - \epsilon$ system is an example of such a system. If f and g are solutions of (6.4) and (6.5) we may set

$$t = \log \eta, \ x = \frac{\eta f'}{f}, \ y = \frac{\eta g'}{g}, \ z = \eta^2 \frac{g}{\alpha f^2}, \ u = \frac{g}{f}, \tag{7.9}$$

which transforms the two coupled nonautonomous second order equations (6.4) and (6.5) into the four-dimensional quadratic autonomous system

$$(Q_{\lambda,\mu,\gamma}) \begin{cases} \dfrac{dx}{dt} = x(1 - 3x + y) - z(x(1 - \mu) + 2\mu - u); \\ \dfrac{dy}{dt} = y(1 - 2x) - \lambda z(y(1 - \mu) + 2\mu + 1 - \gamma u); \\ \dfrac{dz}{dt} = z(2 + y - 2x); \\ \dfrac{du}{dt} = u(y - x). \end{cases}$$

In analogy to what we saw above for scalar equations, compactly supported solutions of (6.4) and (6.5) correspond to intersections of the two-dimensional fast unstable manifold of the u-axis, and the two-dimensional stable manifold of a saddle point at infinity. A global analysis of these intersections is still open, all the results in [18] are based on perturbation arguments in the spirit of [5] and [20]. The classification of the behaviour at infinity is done by using homogeneous coordinates rather than polar coordinates. The transformations are carried out with the help of the symbolic manipulation package Maple. Finally the perturbation argument boils down to solving linear inhomogeneous systems of three coupled first order differential equations with rational coefficients. These systems can be completely solved in terms of hypergeometric functions.

We conclude this section on quadratic systems from a completely different perspective. The basic idea here it not to put the emphasis on the similarity structure of (1.11) but rather on the fact that in (1.11) the pressure, u^{m-1}, is a polynomial. This seems to go back to Kersner who constructed explicit nonselfsimilar solutions of certain porous media equations with sink terms [26]. In [27] this idea is used to construct new solutions for the porous medium equation itself. Since for the normalized pressure,

$$v = \frac{m}{m-1}u^{m-1}, \tag{7.10}$$

the equation reads

$$v_t = (m-1)v\Delta v + |\nabla v|^2, \tag{7.11}$$

this method works for polynomials with degree at most equal to two. As an example consider in two dimensions solutions of the form

$$v(x, y, t) = C(t) - A(t)x^2 - B(t)y^2, \tag{7.12}$$

with $A, B, C > 0$. Substitution leads to the following two dimensional quadratic system for 1 and B

$$\begin{cases} \dfrac{dA}{dt} = -2A((m+1)A + (m-1)B); \\ \dfrac{dB}{dt} = -2B((m-1)A + (m+1)B), \end{cases}$$

coupled with

$$\frac{dC}{dt} = -2(m-1)(A+B)C.$$

Every solution curve of this system with $A, B, C > 0$ defines a compactly supported solution of the porous medium equation with an elliptical free boundary (the instantaneous point source solution corresponds to $A \equiv B$). This example shows that nonlinear equations may have rather nontrivial linear subspaces, see also [14].

References

[1] D.G. Aronson (1986), *The porous medium equation*, in: A. Fasano and M. Primicerio (eds.), Some Problems in Nonlinear Diffusion, Lecture Notes in Math. 1224, Springer.

[2] S.B. Angenent & D.G. Aronson (1996), *The focusing problem for the radially symmetric porous medium equation*, Comm. P.D.E. **20**, 1217-1240.

[3] S.B. Angenent & D.G. Aronson (1996), *Self-similarity in the post-focussing regime in porous medium flows*, Eur. J. Appl. Math. **7**, 277-285.

[4] D.G. Aronson & J. Graveleau (1993), *A selfsimilar solution for the focussing problem for the porous medium equation*, Eur. J. Appl. Math. **4**, 65-81.

[5] D.G. Aronson & J.L. Vazquez (1994), *Calculation of Anomalous Exponents in Nonlinear Diffusion*, Physical Review Letters **72(3)**, 348-351.

[6] G.I. Barenblatt (1952), *On selfsimilar motions of compressible fluids in porous media*, Prikl. Mat. Mekh. **16**, 679-698 (in russian).

[7] G.I. Barenblatt (1987), *Dimensional Analysis*, Gordon and Breach, New York.

[8] G.I. Barenblatt & Y.B. Zel'dovich (1957), *On dipole solutions in problems of nonstationary filtration of gas under polytropic regime*, Prikl. Mat. Mekh. **21**, 718-720.

[9] G.I. Barenblatt, N.L. Galerkina & M.V. Luneva (1987), *Evolution of a turbulent burst*, Inzherno-Fizicheskii Zh. **53**, 773-740 (in Russian).

[10] L.-Y. Chen, N. Goldenfeld & Y. Oono (1991), *Renormalization-group theory for the modified porous-medium equation*, Physical Review A **44**, 6544-6550.

[11] F. Bernis, J. Hulshof & J. L. Vazquez (1993), *A very singular solution for the dual porous medium equation and the asymptotic behaviour of general solutions*, J. reine und angewandte Math. **435**, 1-31.

[12] M. Bertsch, R. Dal Passo & R. Kersner (1994), *The evolution of turbulent bursts: the $b - \varepsilon$ model*, Eur. J. Appl. Math. **5**, 537-557.

[13] N. Goldenfeld, O. Martin, Y. Oono & F. Liu (1990), *Anomalous dimensions and the renormalization group in a nonlinear diffusion equation*, Physical Review Letters **64**, 1361-1364.

[14] V. Galaktionov (1991), *Invariant subspaces and new explicit solutions to evolution equations with quadratic nonlinearities*, Report AM-91-11, Univ. of Bristol.

[15] K. Hanjalic & B.E. Launder (1974), *A Reynolds stress model of turbulence and its applications to thin shear flows*, J. Fluid. Mech. **52**, 609-638.

[16] S.P. Hastings & L.A. Peletier (1992), *On the decay of turbulent bursts*, Eur. J. Appl. Math. **3**, 319-341.

[17] J. Hulshof (1991), *Similarity solutions of the porous medium equation with sign changes*, J. Math. Anal. Appl. **157**, 75-111.

[18] J. Hulshof (1997), *Selfsimilar solutions of Barenblatt's model for turbulence*, SIAM J. Math. Anal. **28(1)**, 33-48.

[19] J. Hulshof & J.L. Vazquez (1993), *The dipole solution for the porous medium equation in several space dimensions*, Ann. d. Scu. Norm. Sup. d. Pisa Serie IV **20**, 193-217.

[20] J. Hulshof & J.L. Vazquez (1994), *Selfsimilar solutions of the second kind for the modified porous medium equation*, Eur. J. Appl. Math. **5**, 391-403.

[21] J. Hulshof & J.L. Vazquez (1996), *Viscosity solutions of the modified porous medium equation*, Eur. J. Appl. Math. **7**, 453-471.

[22] S. Kamin & J.L. Vazquez (1988), *Fundamental solutions and asymptotic behaviour for the p-Laplacian equation*, Rev. Mat. Iberoamericana **4**, 339-354.

[23] S. Kamin & J.L. Vazquez (1991), *Asymptotic behaviour of solutions of the porous medium equation with changing sign*, SIAM J. Math. Anal. **22**, 34-45.

[24] S. Kamin & J.L. Vazquez (1992), *The propagation of turbulent bursts*, Eur. J. Appl. Math.**3**, 263-272.

[25] S. Kamin, L.A. Peletier& J.L. Vazquez (1991), *On the Barenblatt equation of elasto-plastic filtration* Indiana Univ. Math. Journal **40**, 1333-1362

[26] R. Kersner (1978),*The behaviour of temperature fronts in media with nonlinear thermal conductivity under absorption*, Moscow Univ. Math. Bull. **33**, 35-41.

[27] J.R. King (1993), *Exact polynomial solutions for some nonlinear diffusion equations*, Physica D **64**, 35-65.

[28] B.E. Launder, A.P. Morse, W. Rodi & D.B. Spalding (1972), *Prediction of free shear flows - a comparison of six turbulence models*, NASA SP, **321**.

[29] R.E. Pattle (1959), *Diffusion from an instanteneous point source with a concentration dependent coefficient* Quart. J. Appl. Math. **12** 407-409.

[30] L.A. Peletier (1994) *The anomalous exponent of the Barenblatt equation* Eur. J. Appl. Math. **5**, 165-175.

[31] Zel'dovich & Kompanyeets (1950), *On the theory of heat conduction depending on temperature*, Lectures dedicated on the 70th birthday of A.F. Joffe, Akad. Nauk SSSR, 61-71 (in russian).

Optimal L^∞ and Schauder Estimates for Elliptic and Parabolic Operators with Unbounded Coefficients

ALESSANDRA LUNARDI   Dipartimento di Matematica, Università di Parma, Via D'Azeglio 85/A, 43100 Parma, Italy

VINCENZO VESPRI   Dipartimento di Matematica Pura e Applicata, Università dell'Aquila, Via Vetoio, 67010 Coppito (AQ), Italy

1 INTRODUCTION

This paper deals with L^∞ and Hölder regularity results for a class of elliptic and parabolic equations of the type

$$\lambda u - \mathcal{A}u = f, \ x \in R^n, \tag{1.1}$$

$$\begin{cases} u_t = \mathcal{A}u + g, \ 0 \le t \le T, \ x \in R^n, \\ \\ u(0,x) = u_0(x), \ x \in R^n, \end{cases} \tag{1.2}$$

where $\lambda > 0$ and

$$\begin{aligned} \mathcal{A}u(x) &= \sum_{i,j=1}^{n} q_{ij}(x)D_{ij}u(x) + \sum_{i=1}^{n} f_i(x)D_i u(x) \\ \\ &= \mathrm{Tr}\,(Q(x)D^2 u(x)) + \langle F(x), Du(x)\rangle. \end{aligned} \tag{1.3}$$

The coefficients q_{ij} are bounded and differentiable with bounded first order derivatives. They satisfy the ellipticity condition

$$\sum_{i,j=1}^{n} q_{ij}(x)\xi_i\xi_j \ge \nu|\xi|^2, \ \forall x, \ \xi \in R^n, \tag{1.4}$$

with $\nu > 0$. The coefficients f_i are Lipschitz continuous in R^n. If in addition they are bounded, optimal Schauder estimates for problems (1.1), (1.2) are well known. We are interested here in the case where some of the f_i's are unbounded. Then we add a further assumption on the coefficients q_{ij}: we assume that

$$\sup_{x \in R^n} |\langle Dq_{ij}(x), F(x)\rangle| < \infty. \tag{1.5}$$

Such condition is obviously satisfied in the case where Q is independent of x.

Problems of the type (1.1) and (1.2) arise for instance in stochastic perturbations of finite dimensional dynamical systems. See e.g. [10, Ch. 5].

There is not much literature about equations with unbounded coefficients. The papers [4], [5] and [6] deal with operators with an unbounded potential term balancing in a certain sense the unbounded coefficients of the derivatives, so that they cannot be used in our situation. Problems (1.1) and (1.2) have been studied in the case of constant Q and linear F in [9], [12], [13].

Our main results are the following.

THEOREM 1 *Let $0 < \theta < 1$. If λ is sufficiently large, for every $f \in C^\theta(R^n)$ problem (1.1) has a unique solution $u \in C^{2+\theta}(R^n)$, and there exists $C > 0$, independent of f, such that*

$$\|u\|_{C^{2+\theta}(R^n)} \leq C\|f\|_{C^\theta(R^n)}.$$

THEOREM 2 *Let $T > 0$, $g \in C([0, T] \times R^n)$ be such that $g(t, \cdot) \in C^\theta(R^n)$ for every t and $\sup_{0 \leq t \leq T} \|g(t, \cdot)\|_{C^\theta(R^n)} < \infty$. Let moreover $u_0 \in C^{2+\theta}(R^n)$. Then problem (1.2) has a unique bounded solution u belonging to $C^{1,2}([0, T] \times R^n)$, and there is $C > 0$, independent of g, u_0, such that*

$$\sup_{0 \leq t \leq T} \|u(t, \cdot)\|_{C^{2+\theta}(R^n)} \leq C(\|u_0\|_{C^{2+\theta}(R^n)} + \sup_{0 \leq t \leq T} \|g(t, \cdot)\|_{C^\theta(R^n)}).$$

A result similar to the one of Theorem 1 holds also for data $f \in C^0(R^n)$ (the space of the uniformly continuous and bounded functions in R^n). However in this case the solution is not necessarily twice differentiable, but it belongs to $W^{2,p}_{loc}(R^n)$ for every $p < \infty$ and to the Zygmund space $\mathcal{C}^2(R^n)$. See Section 4 for the definition of such spaces.

In the proofs we use functional analytic tools. Precisely, we show that the realization of $\mathcal{A}$ in $X - C^0(R^n)$ generates (in a suitable sense) a semigroup $T(t)$ in X, which is not analytic nor strongly continuous, but satisfies estimates similar to the ones which hold in the case of constant coefficients: there are $\omega \in R$, $C > 0$ such that for $0 \leq \theta \leq \alpha < 3$, $\theta \neq 1, 2$,

$$\|T(t)\|_{L(C^\theta(R^n), C^\alpha(R^n))} \leq Ce^{\omega t}(1 + t^{-(\alpha-\theta)/2}), \quad t > 0. \tag{1.6}$$

Such estimates, together with the well known characterization

$$(C^\alpha(R^n), C^{2+\alpha}(R^n))_{\gamma,\infty} = C^{\alpha+2\gamma}(R^n), \quad \alpha \geq 0, \ 0 < \gamma < 1, \ \alpha + 2\gamma \notin N,$$

are sufficient to prove Theorems 1 and 2 by a general interpolation procedure, already used in [12] and in [13]. See [14]. So, the main point is the proof of estimates (1.6). They are obtained taking $g \equiv 0$ in (1.2) and setting $v(t, x) = u(t, \xi(-t, x))$, where $\xi(t, x)$ is the characteristic curve associated to the operator $u \mapsto u_t - \langle F, Du \rangle$, i.e. the solution of the ODE

$$\begin{cases} \xi_t(t, x) = F(\xi(t, x)), & t \in R, \\ \\ \xi(0, x) = x. \end{cases} \tag{1.7}$$

If F is thrice continuously differentiable with bounded derivatives, then v satisfies a parabolic PDE with coefficients depending on (t, x), which are bounded together with their first order derivatives. The general theory of parabolic initial value problems with regular bounded data may be applied to get estimates for v and its derivatives. Since $u(t, x) = v(t, \xi(t, x))$, (1.6) follows easily.

If F is Lipschitz continuous we construct the semigroup $T(t)$ by a perturbation method and a fixed point theorem in a suitable functional space, in such a way that $T(t)$ satisfies (1.6).

Estimates (1.6) allow also to study the realization A of $\mathcal{A}$ in X. As mentioned above, we prove that $D(A)$ is continuously embedded in $C^2(R^n)$. Moreover we characterize the interpolation space $(X, D(A))_{\theta,\infty}$ for $0 < \theta < 1$. It turns out to be the subspace of $C^{2\theta}(R^n)$ if $\theta \neq 1/2$, (of $C^1(R^n)$ if $\theta = 1/2$) consisting of the functions f such that

$$\sup_{0 < t < 1, \, x \in R^n} \frac{|f(t, \xi(t, x)) - f(x)|}{t^\theta} < \infty.$$

In the case where F is linear and Q is independent of x, our results have been proved in [9, 12] for elliptic equations and in [12] for parabolic equations.

2 ESTIMATES FOR $T(t)$ IN THE CASE WHERE $F \in C^3$

2.1 Estimates for the characteristics

Let $F : R^n \mapsto R^n$ be thrice differentiable with bounded first, second, and third order derivatives. For every $x \in R^n$ let $\xi = \xi(t, x)$ be the solution of (1.7).

In this subsection we collect some properties of ξ and of its derivatives which will be used later.

First, from the identity $\xi(t + s, x) = \xi(t, \xi(s, x))$, differentiating with respect to s and taking $s = 0$ we get

$$F(\xi(t, x)) = \xi_x(t, x) F(x), \quad t \in R, \, x \in R^n. \tag{2.8}$$

Differentiating (1.7) with respect to x we get

$$\xi_{tx}(t, x) = F'(\xi(t, x)) \xi_x(t, x), \quad t \in R, \, \xi(0, x) = I.$$

Therefore, ξ_x and ξ_{tx} are bounded on every bounded time interval, and for every (t, x) it holds

$$\|\xi_x(t, x)\|_{L(R^n)} \leq e^{|t| \, \|F'\|_\infty}, \quad \|\xi_{tx}(t, x)\|_{L(R^n)} \leq \|F'\|_\infty e^{|t| \, \|F'\|_\infty}, \tag{2.9}$$

where $\|F'\|_\infty = \sup_{x \in R^n} \|F'(x)\|_{L(R^n)}$.

From the identity $\xi(t, \xi(-t, x)) = x$, differentiating with respect to x we get

$$\xi_x(t, \xi(-t, x)) = \xi_x(-t, x)^{-1}. \tag{2.10}$$

Therefore,

$$\partial/\partial t\, \xi_x(t, \xi(-t, x)) = \xi(-t, x)^{-1}\xi_{tx}(-t, x)\xi(-t, x)^{-1}$$

$$= \xi_x(t, \xi(-t, x))\xi_{tx}(-t, x)\xi_x(t, \xi(-t, x))$$

is bounded on every bounded time interval, and using (2.9) we get

$$\|\partial/\partial t\, \xi_x(t, \xi(-t, x))\|_{L(R^n)} \le \|F'\|_\infty e^{3|t|\, \|F'\|_\infty}, \quad t \in R,\ x \in R^n. \tag{2.11}$$

Differentiating twice (1.7) with respect to x we get

$$\xi_{txx} = F''(\xi(t, x))(\xi_x(t, x), \xi_x(t, x)) + F'(\xi(t, x))\xi_{xx}(t, x); \quad \xi_{xx}(0, x) = 0, \tag{2.12}$$

so that for $t \in R$, $x \in R^n$ we have

$$\|\xi_{xx}(t, x)\|_{L(R^n, L(R^n))} \le e^{|t|\, \|F'\|_\infty}(e^{|t|\, \|F'\|_\infty} - 1)\|F''\|_\infty/\|F'\|_\infty,$$

$$\|\xi_{txx}(t, x)\|_{L(R^n, L(R^n))} \le (e^{|t|\, \|F'\|_\infty}(e^{|t|\, \|F'\|_\infty} - 1) + e^{2|t|\, \|F'\|_\infty})\|F''\|_\infty. \tag{2.13}$$

Differentiating (2.10) with respect to x we find

$$\xi_{xx}(t, \xi(-t, x)) = -\xi_x(-t, x)^{-1}\xi_{xx}(t, x)\xi_x(-t, x)^{-1}\xi_x(-t, x)^{-1} \tag{2.14}$$

so that from (2.11), (2.13), and again (2.10), we see that $\partial/\partial t\, \xi_{xx}(t, \xi(-t, x))$ is bounded on every bounded time interval. Differentiating again (2.12) with respect to x and recalling that F''' is bounded we get that also ξ_{xxx} is bounded in $[-T, T] \times R^n$ for every $T > 0$. Differentiating (2.14) with respect to x_i, $i = 1, \ldots, n$, we get that $\partial/\partial x_i \xi_{xx}(t, \xi(-t, x))$ is bounded on every bounded time interval. That is all what we need for the sequel.

2.2 Reduction to the case of bounded coefficients

Let us consider problem (1.2) with $g \equiv 0$ and $u_0 \in C^0(R^n)$. For $0 \le t \le 1$, $x \in R^n$ set $v(t, x) = u(t, \xi(-t, x))$, where ξ is the solution of (1.7). Then

$$Du(t, x) = \xi_x^*(t, x)Dv(t, \xi(t, x)),$$

$$D^2u(t, x)(h, k) =$$

$$= \langle \xi_x^*(t, x)D^2v(t, \xi(t, x))\xi_x(t, \xi(t, x))h, k\rangle + \langle Dv(t, \xi(t, x)), \xi_{xx}(t, x)(h, k)\rangle$$

$$u_t(t, x) = v_t(t, \xi(t, x)) + \langle Dv(t, \xi(t, x)), F(\xi(t, x))\rangle$$

$$= v_t(t, \xi(t, x)) + \langle Dv(t, \xi(t, x)), \xi_x(t, x)F(x)\rangle$$

(in the last equality we have used (2.8)). Setting $y = \xi(t, x)$ we get

$$\begin{cases} v_t(t, y) = \displaystyle\sum_{i,j=1}^n a_{ij}(t, y)D_{ij}v(t, y) + \sum_{i=1}^n b_i(t, y)D_iv(t, y), \quad t > 0,\ y \in R^n, \\[2mm] v(0, y) = u_0(y), \quad y \in R^n, \end{cases} \tag{2.15}$$

where the matrix $A(t, y) = [a_{ij}(t, y)]_{i,j=1,\ldots,n}$ is given by

$$A(t, y) = \xi_x^*(t, \xi(-t, y))Q(\xi(-t, y))\xi_x(t, \xi(-t, y)),$$

and the vector $B(t, y) = (b_1(t, y), \ldots, b_n(t, y))$ is given by

$$B(t, y) = \sum_{i=1}^{n} \xi_{xx}(t, \xi(-t, y))(e_i, Q(\xi(-t, y))e_i).$$

Since ξ_x is invertible with bounded inverse in $[0, T] \times R^n$ for every $T > 0$ by (2.9) and (2.10), and recalling that (1.4) holds, we get

$$\sum_{i,j=1}^{n} a_{ij}(t, y)\eta_i\eta_j \geq k|\eta|, \ 0 \leq t \leq 1, \ y, \ \eta \in R^n,$$

with $k > 0$, i.e. problem (2.15) is uniformly parabolic in $[0, T] \times R^n$. The coefficients a_{ij} and b_i are continuously differentiable in $[0, T] \times R^n$. Moreover, we have shown in Subsection 2.1 that the functions mapping (t, x) into $\xi_x(t, \xi(-t, x))$, $\xi_{xx}(t, \xi(-t, x))$, $\partial/\partial x\ \xi_x(t, \xi(-t, x))$, $\partial/\partial x\ \xi_{xx}(t, \xi(-t, x))$, $\partial/\partial t\ \xi_x(t, \xi(-t, x))$ and $\partial/\partial t\ \xi_{xx}(t, \xi(-t, x))$ are bounded in $[0, T] \times R^n$. Since the coefficients q_{ij} belong to $C^1(R^n)$ and satisfy (1.5), then $(t, y) \mapsto q_{ij}(\xi(-t, y))$ is bounded and differentiable with bounded derivatives. Therefore, a_{ij} and b_i are bounded and differentiable with bounded derivatives in $[0, T] \times R^n$, for every $T > 0$.

By the general theory of evolution equations in Banach spaces there exists a parabolic evolution operator $G(t, s)$ acting in the Banach space $X = C^0(R^n)$ for problem (2.15). Then $v(t, x) = (G(t, 0)u_0)(x)$. Coming back to u, we get $u(t, x) = v(t, \xi(t, x)) = (G(t, 0)u_0)(\xi(t, x))$. Therefore, the semigroup $T(t)$ associated to problem (1.2) is

$$(T(t)u_0)(x) = (G(t, 0)u_0)(\xi(t, x)), \ t \geq 0, \ x \in R^n. \tag{2.16}$$

2.3 Proof of estimates (1.6)

Our aim now is to prove that the semigroup $T(t)$ defined in (2.16) satisfies estimates (1.6). We refer to [2, 3, 1], who gave precise estimates for $G(t, 0)$ in several norms. To apply such abstract results we recall that for every $t \geq 0$ the realization of the elliptic operator

$$u \mapsto \Lambda(t)u = \text{Tr}\, A(t, \cdot)D^2u + \langle Du, B(\cdot)\rangle$$

in X generates an analytic semigroup, thanks to [15]. Moreover (see e.g. [11, Ch. 3]) $D(\Lambda(t))$ is dense in X, and for $0 < \theta < 3$, θ not integer, we have

$$C^\theta(R^n) = D_{\Lambda(t)}(\theta/2, \infty), \tag{2.17}$$

with equivalence of the respective norms. The computations made in [2, §6] in the case where R^n is replaced by a bounded open set with regular boundary may be done also in the present situation; things are even easier due to the lack of boundary conditions. In conclusion, one gets

$$\|(\lambda I - \Lambda(t))^{-1} - (\lambda I - \Lambda(s))^{-1}\|_{L(X, D_{\Lambda(0)}(\rho, \infty))} \leq C(t - s)^\beta, \ 0 \leq s \leq t,$$

for every ρ, β in $(0,1)$. By Theorem 7.9 of [3], the assumptions of [3, 1] are satisfied. Therefore there exists the parabolic evolution operator $G(t,s)$ for the abstract problem

$$v'(t) = \Lambda(t)v(t)$$

in the space X. By Theorems 2.3(i), 4.2(iii) of [1], and Theorem 6.2(iii) of [3], $\|G(t,s)\|_{L(X)}$, $\|G(t,s)\|_{L(D_{\Lambda(0)}(\rho,\infty))}$, and $\|G(t,s)\|_{L(D_{\Lambda(s)}(\rho+1,\infty),D_{\Lambda(t)}(\rho+1,\infty))}$ are bounded in every bounded time interval. Using (2.17) with $\rho = \theta/2$ we get, for every noninteger $\theta \in (0,3)$, and $0 \le s \le t \le 1$,

$$\|G(t,s)\|_{L(C(R^n))} + \|G(t,s)\|_{L(C^\theta(R^n))} \le C. \tag{2.18}$$

From Theorem 2.3(i)(v) of [1] one deduces by interpolation that

$$\|G(t,s)\|_{L(C^0(R^n),C^\theta(R^n))} \le C(t-s)^{-\theta/2}, \quad 0 \le s \le t \le 1, \, 0 < \theta < 1. \tag{2.19}$$

By Theorem 6.4 of [3], $(t-s)\|G(t,s)\|_{L(D_{\Lambda(0)}(\rho,\infty),D_{\Lambda(t)}(\rho+1,\infty))}$ is bounded for every $\rho \in (0,1)$. Taking $\rho = \theta/2$ and recalling (2.17) we get for every $\theta \in (0,1)$

$$\|G(t,s)\|_{L(C^\theta(R^n),C^{2+\theta}(R^n))} \le C(t-s)^{-1}, \quad 0 \le s < t \le 1. \tag{2.20}$$

We use now the interpolatory estimates $(0 \le \theta < \gamma \le \beta)$

$$\|\varphi\|_{C^\gamma} \le C(\|\varphi\|_{C^\theta})^{1-\frac{\gamma-\theta}{\beta-\theta}}(\|\varphi\|_{C^\beta})^{\frac{\gamma-\theta}{\beta-\theta}} \quad \forall \varphi \in C^\beta(R^n). \tag{2.21}$$

From (2.18), (2.20), and (2.21) we get

$$\|G(t,s)\|_{L(C^\theta,C^\alpha)}$$

$$\le C(\|G(t,s)\|_{L(C^\theta)})^{1-(\alpha-\theta)/2}(\|G(t,s)\|_{L(C^\theta,C^{2+\theta})})^{(\alpha-\theta)/2} \tag{2.22}$$

$$\le C(t-s)^{-(\alpha-\theta)/2}, \quad 0 \le s < t \le 1, \, 0 < \theta \le \alpha \le 2+\theta.$$

For $\theta = 0$, using (2.19) and (2.22) we get

$$\|G(t,s)\|_{L(C^0,C^\alpha)}$$

$$\le \|G(t,s+(t-s)/2)\|_{L(C^0,C^{\alpha/2})}\|G(s+(t-s)/2,0)\|_{L(C^{\alpha/2},C^\alpha)}$$

$$\le C(t-s)^{-\alpha/2}, \quad 0 < s < t \le 1, \, 0 \le \alpha < 3.$$

Therefore,

$$\|G(t,0)\|_{L(C^\theta,C^\alpha)} \le t^{-(\alpha-\theta)/2}, \quad 0 < t \le 1, \, 0 \le \theta \le \alpha < 3, \, \theta \ne 1, 2. \tag{2.23}$$

From the above estimates we get easily the estimates we need for $T(t)$. Indeed, since $T(t)u_0 = (G(t,0)u_0)(\xi(t,\cdot))$ then

$$\|T(t)u_0\|_{C^\alpha(R^n)} \le \|G(t,0)u_0\|_{C^\alpha(R^n)}\|\xi_x(t,\cdot)\|_\infty,$$

so that (2.22) and (2.9) imply

$$\|T(t)\|_{L(C^\theta,C^\alpha)} \le Ct^{-(\alpha-\theta)/2}, \quad 0 < t \le 1, \, 0 < \theta \le \alpha < 1, \tag{2.24}$$

whereas (2.19) implies

$$\|T(t)\|_{L(C^0, C^\alpha)} \leq Ct^{-\alpha/2}, \quad 0 < t \leq 1, \, 0 < \alpha < 1, \tag{2.25}$$

From the equalities

$$(DT(t)u_0)(x) = \xi_x^*(t, x)(DG(t, 0)u_0)(\xi(t, x)), \tag{2.26}$$

$$(D^2 T(t)u_0)(x) = \tag{2.27}$$

$$= \xi_x^*(t, x)(D^2 G(t, 0)u_0)(\xi(t, x))\xi_x(t, x) + \xi_{xx}^*(t, x)(DG(t, 0)u_0)(\xi(t, x)),$$

arguing similarly, we get

$$\|T(t)\|_{L(C^\theta, C^{2+\alpha})} \leq Ct^{-(\alpha-\theta)/2}, \quad 0 < t \leq 1, \, 0 \leq \theta \leq \alpha < 3, \, \theta \neq 1, 2. \tag{2.28}$$

By the semigroup property of $T(t)$, (2.28) implies that (1.6) holds for some $\omega > 0$.

2.4 Further results

We prove below some results which will be used and generalized in the next section.

Proposition 2.1 *Let $\varphi \in C^0(R^n)$. Then*

(i) *For every $T > 0$ the family $\{T(t)\varphi : 0 \leq t \leq T\}$ is equi-uniformly continuous in R^n.*

(ii) *For every compact set $K \subset R^n$ we have*

$$\lim_{t \to 0} \|(T(t)\varphi)_{|K} - \varphi_{|K}\|_{L^\infty(K)} = 0. \tag{2.29}$$

(iii) *The following equivalence holds true.*

$$\lim_{t \to 0} \|T(t)\varphi - \varphi\|_{L^\infty(R^n)} = 0 \iff \lim_{t \to 0} \|\varphi(\xi(t, \cdot)) - \varphi\|_{L^\infty(R^n)} = 0.$$

(iv) *If $\{\varphi_n\}_{n \in N}$ is a bounded sequence in $C^0(R^n)$ converging to φ uniformly on every compact subset of R^n, then for every $T > 0$ and for every compact set $K \subset R^n$ we have*

$$\lim_{n \to \infty} \sup_{0 \leq t \leq T} \|T(t)(\varphi_n - \varphi)_{|K}\|_{L^\infty(K)} = 0.$$

(v) *The derivatives $\partial/\partial t\,(T(t)\varphi)(x)$, $D_i(T(t)\varphi)(x)$, $D_{ij}(T(t)\varphi)(x)$ exist continuous in $(0, T] \times R^n$, and*

$$\partial/\partial t\,(T(t)\varphi)(x) = (AT(t)\varphi)(x), \quad t > 0, \, x \in R^n.$$

If in addition $\varphi \in C^{2+\theta}(R^n)$ for some $\theta \in (0, 1)$ then $\partial/\partial t\,(T(t)\varphi)(x)$, $D_i(T(t)\varphi)(x)$, $D_{ij}(T(t)\varphi)(x)$ exists continuous up to $t = 0$.

Proof — Statement (i) is a consequence of the continuity of $G(\cdot,0)\varphi$ and of the boundedness of ξ_x (see estimate (2.9)). For every $x \in R^n$ we have

$$(T(t)\varphi)(x) - \varphi(x) = (G(t,0)\varphi)(\xi(t,x)) - \varphi(\xi(t,x)) + \varphi(\xi(t,x)) - \varphi(x)$$

Since $G(t,0)\varphi$ goes to φ uniformly as $t \to 0$, statement (iii) follows. Since $\xi(t,x) - x$ goes to 0 uniformly on every compact set K as $t \to 0$, statement (ii) follows too.

Let us prove statement (iv). Set $f_n = \varphi_n - \varphi$, and fix any $R > 0$. Fixed any $\varepsilon > 0$, let θ be a smooth cutoff function, such that

$$0 \le \theta(x) \le 1, \quad \theta \equiv 1 \text{ on } B(0,R), \quad \theta \equiv 0 \text{ outside } B(0,R+1/\varepsilon),$$

$$\|D_i\theta\|_\infty \le \varepsilon, \quad \|D_{ij}\theta\|_\infty \le \varepsilon, \quad i,j = 1,\ldots,n. \tag{2.30}$$

The function $v_n(t,x) = (G(t,0)f_n)(x)\theta(x)$ is uniformly continuous and bounded, and it satisfies

$$\frac{\partial}{\partial t}v_n = \Lambda(t)v + \psi_n, \quad v_n(0,\cdot) = f_n\theta,$$

where

$$\psi_n = -2\sum_{i,j=1}^n a_{ij}D_i\theta D_j(G(t,0)f_n) - (G(t,0)f_n)\left(\sum_{i,j=1}^n a_{ij}D_{ij}\theta + \sum_{i=1}^n b_iD_i\theta\right).$$

$t \mapsto \psi_n(t,\cdot)$ belongs to $C((0,T];X) \cap L^1(0,T;X)$ for every n. Therefore,

$$v_n(t,\cdot) = G(t,0)(f_n\theta) + \int_0^t G(t,s)\psi_n(s,\cdot)ds, \quad 0 < t \le T.$$

Since $\|\psi_n(t,\cdot)\|_\infty \le C\varepsilon t^{-1/2}$ for every $t \in (0,T]$, with constant C independent of n, then

$$\|v_n(t,\cdot)\|_\infty \le C_1(\|f_n\theta\|_\infty + \varepsilon).$$

Since θ vanishes outside $B(0,R+1/\varepsilon)$ and f_n goes to 0 uniformly on every compact set, for n large enough we have $\|f_n\theta\|_\infty \le \varepsilon$. Therefore, v_n goes to 0 in X as n goes to ∞, which implies that $G(t,0)f_n$ goes to 0 uniformly on $B(0,R)$ as n goes to ∞. Statement (iv) follows from (2.16).

Concerning (v), the function $v(t,x) = (G(t,0)\varphi)(x)$ is continuous in $[0,T] \times R^n$ thanks to statements (i) and (ii). Moreover, from [1] we know that $t \mapsto G(t,0)$ is continuously differentiable in $(0,T]$ with values in $L(X)$. It follows that v_t is continuous in $(0,T] \times R^n$. Since for every $\varepsilon \in (0,T)$ the function $t \mapsto G(t,0)\varphi$ is continuous with values in X and bounded with values in $C^{2+\theta}(R^n)$ in $[\varepsilon,T]$, then by estimates (2.21) applied to the function $G(t,0)\varphi - G(s,0)\varphi$ we get that it is continuous with values in $C^2(R^n)$. It follows that the first and second order space derivatives of v are continuous in $(0,T] \times R^n$.

If $\varphi \in C^{2+\theta}(R^n)$ then $\varphi \in D(\Lambda(0))$ and $\Lambda(0)\varphi \in D_{\Lambda(0)}(\theta/2,\infty) \subset \overline{D(\Lambda(0))}$. By [3, Theorem 6.1(i)], $t \mapsto G(t,0)\varphi$ is continuously differentiable in $[0,T]$ with values in X. By (2.18), it is bounded with values in $C^{2+\theta}(R^n)$. Consequently, it is continuous with values in $C^2(R^n)$. Therefore, v_t and the first and second order space derivatives of v are continuous in $[0,T] \times R^n$.

Statement (v) follows. ∎

We denote by $\widetilde{C}([a,b];X)$ the space of all the bounded functions $f : [a,b] \mapsto X$ such that $(t,x) \mapsto f(t)(x)$ is continuous, and $f(t)$ is uniformly continuous in R^n, uniformly with respect to t. If Y is any Banach space and $\alpha \geq 0$, we denote by $B_\alpha((a,b];Y)$ the space of all functions $f : (a,b] \mapsto Y$ such that $t \mapsto \|(t-a)^\alpha f(t)\|_Y$ is bounded in (a,b). Moreover, if $I = [a,b]$ or $I = (a,b]$, $C^{1,2}(I \times R^n)$ denotes the vector space of all continuous functions $v : I \times R^n \mapsto R$ which are once differentiable with respect to t, twice differentiable with respect to x, with continuous (not necessarily bounded) derivatives.

Statements (i) and (ii) of Proposition 2.1 imply that for every $\varphi \in X$ the function $t \mapsto T(t)\varphi$ is in $\widetilde{C}([0,T];X)$ for every $T > 0$. Estimates (1.6) imply that it belongs to $B_{\theta/2}((0,T];C^\theta(R^n))$ for every $\theta \in [0,3)$ and $T > 0$. Statement (v) implies that $(t,x) \mapsto (T(t)\varphi)(x)$ belongs to $C^{1,2}((0,T] \times R^n)$.

We consider now problem (1.2) with initial datum $u_0 \in X$. We need a lemma about integrals of functions with values in Hölder spaces.

Lemma 2.2 *Let $\theta \in (0,3)$ be not integer, let I be a real interval, and let $\varphi : I \mapsto C^\theta(R^n)$ be such that for every $x \in R^n$ the real function $t \mapsto \varphi(t)(x)$ is continuous in I, and $\|\varphi(t)\|_{C^\theta} \leq c(t)$ with $c \in L^1(I)$. Then the function*

$$f(x) = \int_I \varphi(t)(x)dt, \ x \in R^n,$$

belongs to $C^\theta(R^n)$, and

$$\|f\|_{C^\theta} \leq \|c\|_{L^1(I)}.$$

Proof — We recall that $C^\theta(R^n)$ is the space of the functions $f \in C^0(R^n)$ such that

$$[f]_\theta = \sup_{x,h \in R^n, h \neq 0} |h|^{-\theta} \left| \sum_{l=0}^{3} (-1)^l f(x + lh) \right| < \infty,$$

and the norm

$$f \mapsto \|f\|_\infty + [f]_\theta$$

is equivalent to the C^θ norm. See e.g. [16, Sect. 2.7.2]. If $\varphi : I \mapsto C^\theta(R^n)$ is such that $t \mapsto \varphi(t)(x)$ is continuous for every $x \in R^n$ and $\|\varphi(t)\|_{C^\theta} \leq c(t)$ with $c \in L^1(I)$, then for every $x, h \in R^n$ we have

$$\left| \sum_{l=0}^{3} (-1)^l \int_I \varphi(t)(x+lh)dt \right| \leq \int_I \left| \sum_{l=0}^{3} (-1)^l \varphi(t)(x+lh) \right| dt \leq \int_I Kc(t)dt \, |h|^\theta,$$

so that $f(x) = \int_I \varphi(t)(x)$ belongs to $C^\theta(R^n)$, and the statement follows. ∎

Proposition 2.3 *Let $\alpha, \theta \in (0,1)$, $\theta < \beta \leq 2 + \theta$ and let $u_0 \in X$, $g \in B_\alpha((0,T]; C^\beta(R^n)) \cap \widetilde{C}([\varepsilon,T];X)$ for every $\varepsilon \in (0,T)$. Then problem (1.2) has a solution $u \in C^{1,2}((0,T] \times R^n)$ such that $t \mapsto u(t,\cdot) \in \widetilde{C}([0,T];X) \cap B_{1+\theta/2}((0,T];C^{2+\theta}(R^n))$. u is given by the variation of constants formula*

$$u(t)(x) = (T(t)u_0)(x) + \int_0^t (T(t-s)g(s))(x)ds, \ 0 \leq t \leq T, x \in R^n.$$

Moreover there is $C > 0$ such that for $0 < t \leq T$

$$\|u(t, \cdot)\|_\infty + t^{1+\theta/2}\|u(t, \cdot)\|_{C^{2+\theta}} \leq C(\|u_0\|_{C^{2+\theta}} + \|g\|_{B_\alpha((0,T];C^\beta(R^n))}).$$

If in addition $u_0 \in C^{2+\theta}(R^n)$ and $g \in B_0([0,T]; C^\beta(R^n))$, then $u \in C^{1,2}([0,T] \times R^n) \cap B_0([0,T]; C^{2+\theta}(R^n))$, and there is $C > 0$ such that

$$\|u(t, \cdot)\|_{C^{2+\theta}} \leq C(\|u_0\|_{C^{2+\theta}} + \|g\|_{B_0([0,T];C^\beta(R^n))}), \quad 0 \leq t \leq T.$$

Proof — Set

$$v(t)(x) = \int_0^t (T(t-s)g(s))(x)ds, \quad 0 \leq t \leq T, \ x \in R^n.$$

Let us prove that $v(t) \in C^{2+\theta}(R^n)$ for every $t \in (0, T]$. For every $x \in R^n$ the function $s \mapsto (T(t-s)g(s, \cdot))(x)$ is continuous in $(0, T)$ thanks to Proposition 2.1(ii)(iv). By estimates (1.6), $\|T(t-s)\|_{L(C^\beta(R^n),C^{2+\theta}(R^n))} \leq C(t-s)^{1-(\beta-\theta)/2}$ for $0 \leq s < t \leq T$. Therefore,

$$\|T(t-s)g(s)\|_{C^{2+\theta}(R^n)} \leq \frac{C}{s^\alpha(t-s)^{1-(\beta-\theta)/2}}\|g\|_{B_\alpha((0,T];C^\beta(R^n))}.$$

Since $s \mapsto 1/s^\alpha(t-s)^{1-(\beta-\theta)/2}$ is in $L^1(0,t)$, by Lemma 2.2 $v(t) \in C^{2+\theta}(R^n)$ for every $t \in (0, T]$, and

$$\|v(t)\|_{C^{2+\theta}} \leq C \int_0^t \frac{ds}{s^\alpha(t-s)^{1-(\beta-\theta)/2}} = \frac{C'}{t^{\alpha+(\beta-\theta)/2}}\|g\|_{B_\alpha((0,T];C^\beta(R^n))}.$$

Moreover, v belongs to $\widetilde{C}([0, T]; X)$. Indeed, since

$$|(T(t-s)g(s))(x_1) - (T(t-s)g(s))(x_2)|$$

$$\leq \|T(t-s)\|_{L(C^\beta(R^n))}s^{-\alpha}|x_1 - x_2|^\beta\|g\|_{B_\alpha((0,T];C^\beta(R^n))},$$

then u is bounded with values in $C^\beta(R^n)$. In particular, it is uniformly continuous in x, uniformly with respect to t. Moreover, for every $x \in R^n$ and $s \in (0, T)$ the function $t \mapsto (T(t-s)g(s))(x)$ is continuous in $[s, T]$, so that $t \mapsto v(t)(x)$ is continuos in $[\varepsilon, T]$ for every $\varepsilon \in (0, T)$. Since it is obviously continuous at $t = 0$, then it is continuous in $[0, T]$.

Since $(t, x) \mapsto v(t)(x)$ is continuos, then for every compact set $K \subset R^n$ v belongs to $C([0, T]; C(K))$. We know already that it belongs to $B([\varepsilon, T]; C^{2+\theta}(K))$ for every $\varepsilon \in (0, T)$. By [11, Prop. 1.1.3(iii), Prop. 1.1.4(iii)], it belongs to $C([\varepsilon, T]; C^2(K))$. Therefore, v and its first and second order space derivatives are continuous and bounded in $[\varepsilon, T] \times R^n$. Concerning the regularity with respect to t, we know from Proposition 2.1(v) that for every $\varphi \in X$ and $x \in R^n$ the function $t \mapsto (T(t)\varphi)(x)$ is continuously differentiable in $(0, +\infty)$, with $\partial/\partial t\,(T(t)\varphi)(x) = (AT(t)\varphi)(x)$. Therefore for every $s \in [0, T]$ the function $t \mapsto (T(t-s)g(s, \cdot))(x)$ is continuously differentiable in $(s, T]$, and $\partial/\partial t\,(T(t-s)g(s, \cdot))(x) = (AT(t-s)g(s, \cdot))(x)$. By (1.6), if $|x| \leq R$ we have

$$|(AT(t-s)g(s, \cdot))(x)| \leq \frac{C}{s^\alpha}\left(\frac{1}{(t-s)^{1-\beta/2}} + \frac{R}{(t-s)^{1/2-\beta/2}}\right)\|g\|_{B_\alpha((0,T];C^\beta(R^n))}$$

so that $\partial/\partial t \, (T(t-s)g(s,\cdot))(x)$ is in $L^1(0,t)$. Then v is continuously differentiable with respect to time and

$$v_t = \mathcal{A}v + g \quad \text{in } (0,T] \times R^n.$$

By (1.6), the function $t \mapsto t^{1+\theta/2}T(t)u_0$ is bounded with values in $C^{2+\theta}(R^n)$. By Proposition 2.1(v), $(t,x) \mapsto (T(t)u_0)(x)$ belongs to $C^{1,2}((0,T] \times R^n)$. Therefore, the function

$$u(t,\cdot) = T(t)u_0 + v(t), \quad 0 \le t \le T,$$

belongs to $C^{1,2}((0,T] \times R^n) \cap \tilde{C}([0,T];X)$, satisfies (1.2), and

$$\sup_{0<t\le T} t^{1+\theta/2}\|u(t,\cdot)\|_{C^{2+\theta}(R^n)} \le K(\|u_0\|_X + \sup_{0<t\le T} t^\alpha\|g(t,\cdot)\|_{C^\beta(R^n)}).$$

If in addition g is bounded with values in $C^\beta(R^n)$ then v is bounded with values in $C^{2+\theta}(R^n)$, since by (1.6)

$$\|T(t)\|_{L(C^\beta(R^n),C^{2+\theta}(R^n))} \le C(t-s)^{-1+(\beta-\theta)/2} \in L^1(s,t).$$

Due again to (1.6), if $u_0 \in C^{2+\theta}(R^n)$ then also $t \mapsto T(t)u_0$ is bounded with values in $C^{2+\theta}(R^n)$. Then, arguing as above one sees that $u \in C^{1,2}([0,T] \times R^n)$. ■

Note that the last statement of the proposition is not an optimal regularity result, since $\beta > \theta$. To prove optimal results we will need a more refined technique. See Section 4.

Uniqueness of the solution is a consequence of the next lemma, which we prove under weaker assumptions than the ones considered up to now.

Lemma 2.4 *Let q_{ij} be bounded coefficients satisfying the ellipticity condition (1.4), and let f_i be continuous functions having not more than linear growth at infinity. Define the operator $\mathcal{A}$ by (1.3).*
Let $u \in C([0,T] \times R^n) \cap C^{1,2}((0,T] \times R^n)$ be a bounded solution of

$$\begin{cases} u_t = \mathcal{A}u + \varphi, \quad 0 < t \le T, \; x \in R^n, \\ u(0,\cdot) = u_0, \end{cases}$$

with φ, u_0 bounded.
Then

$$\|u\|_\infty \le \max\{e^{\lambda T}\|u_0\|_\infty, \lambda^{-1}e^{\lambda T}\|\varphi\|_\infty\}, \quad \forall \lambda > 0.$$

In particular, if $\varphi \equiv 0$ then

$$\|u\|_\infty \le \|u_0\|_\infty.$$

Proof — For every $\lambda > 0$ the function $v(t,x) = e^{-\lambda t}u(t,x)$ satisfies

$$\begin{cases} v_t = \mathcal{A}v - \lambda v + e^{-\lambda t}\varphi, \quad 0 < t \le T, \; x \in R^n; \\ v(0,\cdot) = u_0. \end{cases}$$

If $\|v\|_\infty = \pm v(t_0, x_0)$ for some $(t_0, x_0) \in [0, T] \times R^n$ a standard application of the maximum principle gives

$$\|v\|_\infty \le \max\{\|u_0\|_\infty, \lambda^{-1}\|\varphi\|_\infty\}. \tag{2.31}$$

If $\|v\|_\infty \ne \pm v(t_0, x_0)$ for every (t_0, x_0), let (t_n, x_n) be a sequence such that $v(t_n, x_n) \to \|v\|_\infty$ as $n \to \infty$. We may assume without loss of generality that $t_n \to t_0 \in [0, T]$ and $v(t_n, x_n) \ge \|v\|_\infty - 1/n$. Let θ be a smooth cutoff function, such that $0 \le \theta(x) \le 1$ for every x, $\theta \equiv 1$ on $B(0, 1)$, $\theta \equiv 0$ outside $B(0, 2)$. Set $\theta_n(x) = \theta((x - x_n)/|x_n|)$, so that

$$\sup_{n \in N} \|\mathcal{A}\theta_n\|_\infty < \infty,$$

and set

$$v_n(t, x) = v(t, x) + \frac{2}{n}\theta_n(x), \quad 0 \le t \le T, \ x \in R^n.$$

Then v_n converges to v uniformly, and for every n, $\sup v_n = \max v_n$. Let (s_n, y_n) be any maximum point for v_n. Then $v_n(s_n, y_n)$ goes to $\|v\|_\infty$ as $n \to \infty$. Moreover,

$$\partial/\partial t\, v_n(s_n, y_n) - \mathcal{A}v_n(s_n, y_n) + \lambda v_n(s_n, y_n)$$

$$= \frac{2}{n}\left(\lambda \theta_n(s_n, y_n) - \mathcal{A}\theta_n(s_n, y_n)\right) + e^{-\lambda s_n}\varphi(s_n, y_n),$$

so that

$$v_n(s_n, y_n) \le \max\left\{\|u_0\|_\infty, \frac{2}{n\lambda}\|\lambda\theta_n - \mathcal{A}\theta_n\|_\infty + \frac{1}{\lambda}\|\varphi\|_\infty\right\}.$$

Letting $n \to \infty$ we get (2.31), and the statement follows. ∎

3 ESTIMATES FOR $T(t)$ IN THE CASE WHERE F IS LIPSCHITZ CONTINUOUS

In this section we show that if F is Lipschitz continuous and Q satisfies (1.4), (1.5), then for every $u_0 \in C^0(R^n)$ and $T > 0$ the problem

$$\begin{cases} u_t = \mathcal{A}u, \quad 0 \le t \le T, \ x \in R^n, \\ u(0, x) = u_0(x), \quad x \in R^n, \end{cases} \tag{3.32}$$

has a unique solution u in a suitable functional space. Setting then

$$(T(t)u_0)(x) = u(t, x), \quad t > 0, \ x \in R^n, \tag{3.33}$$

we prove that $T(t)$ maps continuously $C^0(R^n)$ into $C^{2+\theta}(R^n)$, for every $\theta \in (0, 1)$, and that estimates (1.6) hold.

The first step consists in solving problem (3.32) by a perturbation method which employs the results of Section 2.

Let ρ be any mollifier, and set

$$\widetilde{F}(x) = \int_{R^n} F(x - y)\rho(y)dy, \;\; x \in R^n. \tag{3.34}$$

Then $\widetilde{F}$ is smooth, with bounded any order derivatives. Moreover,

$$|\widetilde{F}(x) - F(x)| \leq \int_{R^n} |F(x - y) - F(x)|\rho(y)dy \leq [F]_{Lip} \int_{R^n} |y|\rho(y)dy,$$

so that $\widetilde{F} - F$ is bounded.

Set

$$\widetilde{\mathcal{A}}u(x) = \mathrm{Tr}\,(Q(x)D^2u(x)) + \langle \widetilde{F}(x), Du(x)\rangle, \tag{3.35}$$

and define $\widetilde{T}(t)$ by (2.16) with F replaced by $\widetilde{F}$.

Theorem 3.5 *Let $u_0 \in X$, $\theta \in (0,1)$. Then problem (3.32) has a unique solution $u \in C^{1,2}((0,T] \times R^n)$ such that $t \mapsto u(t,\cdot) \in B_{1+\theta/2}((0,T]; C^{2+\theta}(R^n)) \cap \widetilde{C}([0,T]; X)$, and*

$$\|u(t,\cdot)\|_X + t^{1+\theta/2}\|u(t,\cdot)\|_{C^{2+\theta}(R^n)} \leq C\|u_0\|_X, \;\; 0 < t \leq T.$$

If in addition $u_0 \in C^{2+\theta}(R^n)$, then $u \in C^{1,2}([0,T] \times R^n)$, $t \mapsto u(t,\cdot) \in B_0([0,T]; C^{2+\theta}(R^n))$, and

$$\|u(t,\cdot)\|_{C^{2+\theta}(R^n)} \leq C\|u_0\|_{C^{2+\theta}(R^n)}, \;\; 0 < t \leq T.$$

Proof — By the change of unknown $v(t) = e^{-\lambda t}u(t)$, with $\lambda > 0$, problem (3.32) is transformed into

$$\begin{cases} v_t = \mathcal{A}v - \lambda v, \;\; 0 < t \leq T, \; x \in R^n, \\ v(0, x) = u_0(x), \;\; x \in R^n, \end{cases} \tag{3.36}$$

which is equivalent to

$$\begin{cases} v_t - \widetilde{\mathcal{A}}v = -\lambda v + \langle \widetilde{F} - F, Dv\rangle, \;\; 0 < t \leq T, \; x \in R^n, \\ v(0, x) = u_0(x), \;\; x \in R^n, \end{cases} \tag{3.37}$$

where $\widetilde{F}$ and $\widetilde{\mathcal{A}}$ are defined in (3.34) and (3.35), respectively. Setting $v(t) = v(t, \cdot)$ and

$$(\Phi g)(t) = \int_0^t e^{-\lambda(t-s)}\widetilde{T}(t - s)g(s)ds,$$

for every $g \in \widetilde{C}([0,T]; X)$, problem (3.37) is equivalent to

$$v(t) = (\Gamma v)(t) = e^{-\lambda t}\widetilde{T}(t)u_0 + [\Phi(\langle \widetilde{F} - F, Dv\rangle)](t), \;\; 0 \leq t \leq T. \tag{3.38}$$

We are going to prove that Γ maps the space

$$Y = \widetilde{C}([0,T]; X) \cap B_{1+\theta/2}((0,T]; C^{2+\theta}(R^n))$$

into itself, and it is a contraction if λ is large enough. Y is endowed with the norm

$$\|u\|_Y = \sup_{0 \le t \le T} \|u(t)\|_{C^0(R^n)} + \sup_{0 < t \le T} t^{1+\theta/2} \|u(t)\|_{C^{2+\theta}(R^n)}.$$

For every $u \in Y$, each derivative $D_i u$ belongs to $B_{(1+\beta)/2}((0,T]; C^\beta(R^n)) \cap C((0,T);$ $C^\beta(K))$ for every $\beta \in (0,1)$ and for every compact set K. Indeed, using estimate (2.21) we get

$$\|u(t)\|_{C^{1+\beta}(R^n)} \le C(\|u(t)\|_{C^{2+\theta}(R^n)})^{(1+\beta)/(2+\theta)} (\|u(t)\|_X)^{1-(1+\beta)/(2+\theta)}$$

$$\le \frac{C}{t^{(1+\beta)/2}} \|u\|_Y, \ 0 < t \le T,$$

so that $D_i u \in B_{(1+\beta)/2}((0,T]; C^\beta(R^n))$. Moreover, since $(t,x) \mapsto u(t)(x)$ belongs to $B_0([\varepsilon,T]; C^{2+\theta}(K)) \cap C([\varepsilon,T]; C(K))$ for every compact set K, then by [11, Prop. 1.1.3(iii), 1.1.4(iii)] it belongs to $C([\varepsilon,T]; C^\alpha(K))$ for every $\alpha < 2+\theta$. Consequently, $D_i u \in C((0,T); C^\beta(K))$.

Since $F - \widetilde{F}$ is Lipschitz continuous and bounded, the function

$$g(s) = \langle F - \widetilde{F}, Du(s) \rangle, \ 0 < s \le T,$$

belongs to $B_{(1+\beta)/2}((0,T]; C^\beta(R^n)) \cap C((0,T]; C^\beta(K))$ for every $\beta \in (0,1)$.

Fix once and for all any $\beta \in (\theta,1)$. By Proposition 2.3 applied to the semigroup $e^{-\lambda t}\widetilde{T}(t)$, the operator Γ maps Y into itself. We revisit the estimates of Proposition 2.3 to prove that Γ is a contraction for λ large enough. Arguing as in the proof of Proposition 2.3 we see that for u, v in Y we have

$$\|(\Gamma u)(t) - (\Gamma v)(t)\|_{C^{2+\theta}(R^n)}$$

$$\le \int_0^t \frac{Ce^{-\lambda(t-s)}}{s^{(1+\beta)/2}(t-s)^{1-(\beta-\theta)/2}} \, ds \, \|g\|_{B_{(1+\beta)/2}((0,T]; C^\beta(R^n))}, \ 0 < t \le T,$$

with

$$g(s) = \langle F - \widetilde{F}, (Du(s) - Dv(s)) \rangle, \ 0 < s \le T.$$

It is not hard to see that

$$C_1(\lambda) = \sup_{0 < t \le T} t^{1+\theta/2} \int_0^t \frac{e^{-\lambda(t-s)}}{s^{(1+\beta)/2}(t-s)^{1-(\beta-\theta)/2}} \, ds$$

goes to 0 as λ goes to ∞. Moreover, for every $u \in Y$, each derivative $D_i u$ belongs to $B_{1/2}((0,T]; C^0(R^n))$. Indeed, from estimate (2.21) we get for $0 < t \le T$

$$\|u(t)\|_{C^1(R^n)} \le C(\|u(t)\|_{C^{2+\theta}(R^n)})^{1/(2+\theta)} (\|u(t)\|_X)^{1-1/(2+\theta)} \le \frac{C}{t^{1/2}} \|u\|_Y,$$

so that $D_i u \in B_{1/2}((0,T]; C^0(R^n))$. It follows that for every $x \in R^n$ we have

$$|(\Gamma u)(t)(x) - (\Gamma v)(t)(x)| \le C \int_0^t \frac{e^{-\lambda(t-s)}}{s^{1/2}} \, ds \, \|g\|_Y, \ 0 < t \le T.$$

Again, it is easy to see that

$$C_2(\lambda) = \sup_{0 < t \leq T} t^{1+\theta/2} \int_0^t \frac{e^{-\lambda(t-s)}}{s^{1/2}} \, ds$$

goes to 0 as λ goes to ∞. Therefore, for λ large enough Γ is a contraction in Y, so that it has a unique fixed point $v \in Y$.

From Proposition 2.3 we know that $(t, x) \mapsto v(t)(x)$ belongs also to $C^{1,2}((0, T] \times R^n)$ and it satisfies (3.37) pointwise. Therefore, $u(t, x) = e^{\lambda t} v(t, x)$ belongs to $C^{1,2}((0, T] \times R^n)$, it satisfies (1.2), and

$$\|u\|_\infty + \sup_{0 < t \leq T} t^{1+\theta/2} \|u(t, \cdot)\|_{C^{2+\theta}(R^n)} \leq e^{\lambda T} \|v\|_Y \leq e^{\lambda T} C \|u_0\|_{C^0(R^n)}.$$

In the case where $u_0 \in C^{2+\theta}(R^n)$ the operator Γ is defined in the space

$$Y_1 = \widetilde{C}([0, T]; X) \cap B_0([0, T]; C^{2+\theta}(R^n)).$$

Arguing as above it is not difficult to show that it maps Y_1 into itself and it is a contraction if λ is large enough.

The statement follows. ∎

By Theorem 3.5, for every $u_0 \in X$ problem (3.32) has a solution u such that $t \mapsto u(t, \cdot) \in \widetilde{C}([0, T]; X) \cap B_{1+\theta/2}((0, T]; C^{2+\theta}(R^n))$ for every $T > 0$, $\theta \in (0, 1)$. We define

$$T(t)u_0 = u(t, \cdot), \quad t > 0. \tag{3.39}$$

The estimates of Theorem 3.5 give

$$\|T(t)\|_{L(X, C^{2+\theta}(R^n))} \leq \frac{C}{t^{1+\theta/2}}, \quad 0 < t \leq 1,$$

$$\|T(t)\|_{L(C^{2+\theta}(R^n))} \leq C, \quad 0 < t \leq 1.$$

Since $T(t)$ is a semigroup we get

$$\|T(t)\|_{L(X, C^{2+\theta}(R^n))} \leq \frac{C}{t^{1+\theta/2}} e^{\omega t}, \quad t > 0,$$

$$\|T(t)\|_{L(C^{2+\theta}(R^n))} \leq C e^{\omega t}, \quad t > 0,$$

for some $\omega > 0$. Moreover by Lemma 2.4 we get

$$\|T(t)\|_{L(X)} \leq 1, \quad t > 0.$$

By interpolation we get

$$\|T(t)\|_{L(C^\alpha(R^n))} \leq C e^{\omega t}, \quad t > 0, \ \alpha \in (0, 3), \ \alpha \neq 1, 2.$$

(1.6) follows from the interpolatory estimates (2.21).

4 REMARKS ON THE GENERATOR OF $T(t)$ IN $C^0(R^n)$

From now on we assume that F is Lipschitz continuous and that (1.4), (1.5) hold.

Even in the case of constant Q and linear nonzero F the semigroup $T(t)$ is not strongly continuous and it is not analytic in $X = C^0(R^n)$, as shown in [9]. However, from Theorem 3.5 we get that $(t, x) \mapsto (T(t)u_0)(x)$ is continuously differentiable with respect to t and twice continuously differentiable with respect to the space variables x for $t > 0$, and

$$\frac{\partial}{\partial t}(T(t)u_0)(x) = \mathcal{A}T(t)u_0(x), \quad t > 0, \ x \in R^n. \tag{4.40}$$

Concerning the behavior of $T(t)$ near $t = 0$, and the restriction of $T(t)\varphi$ to the compact sets in R^n, we shall see that all the statements of Proposition 2.1 hold true. Statements (i) and (ii) follow easily from Theorem 3.5, which states that $(t, x) \mapsto (T(t)\varphi)(x)$ belongs to $\widetilde{C}([0, T]; X)$. This implies that the family of functions $\{T(t)\varphi : 0 \leq t \leq T\}$ are equi-uniformly continuos, and that the function $(t, x) \mapsto (T(t)\varphi)(x)$ belongs to $C([0, T] \times K)$ for every compact set K. Therefore, $\lim_{t \to 0} \|(T(t)\varphi)_{|K} - \varphi_{|K}\|_{L^\infty(K)} = 0$. Statements (iii) and (iv) will be shown later. Statement (v) is an immediate consequence of Theorem 3.5.

The realization of $\mathcal{A}$ in X generates $T(t)$ in a weak sense, which we explain below (see [7] for more details).

By Theorem 3.5 and estimate (1.6) with $\theta = 0$, for every $\varphi \in X$ and $x \in R^n$ the function $t \mapsto e^{-\omega t}(T(t)\varphi)(x)$ is continuous and bounded in $(0, +\infty)$. Then for $\lambda > \omega$ the integral

$$(R(\lambda)\varphi)(x) = \int_0^{+\infty} e^{-\lambda t}(T(t)\varphi)(x)dt, \quad x \in R^n, \tag{4.41}$$

makes sense, and defines a function belonging to $C^\alpha(R^n)$ for $0 \leq \alpha < 2$ thanks to lemma 2.2 and to estimates (1.6) with $\theta = 0$. So, $R(\lambda) \in L(X)$, and

$$\|R(\lambda)\|_{L(X)} \leq \frac{1}{\lambda}.$$

Moreover, $R(\lambda)$ satisfies the resolvent identity because $T(t)$ is a semigroup, and it is one to one because for every $x \in R^n$ $(R(\lambda)\varphi)(x)$ is the anti-Laplace transform of the real continuous function $t \mapsto (T(t)\varphi)(x)$, which takes the value $\varphi(x)$ at $t = 0$. Therefore there exists a closed operator

$$A : D(A) \mapsto X, \quad D(A) = \text{Range } R(\lambda) \text{ for } \lambda > \omega,$$

such that $R(\lambda) = R(\lambda, A)$ for $\lambda > \omega$. We are going to show that A is the realization of $\mathcal{A}$ in X. To this aim we need a uniqueness lemma for the solution of the elliptic equation $\lambda u - \mathcal{A}u = f$.

Lemma 4.6 *Let* $u \in \bigcap_{p \geq 1} W^{2,p}_{loc}(R^n) \cap X$ *satisfy*

$$\lambda u(x) - \mathcal{A}u(x) = f, \quad x \in R^n,$$

with $\lambda > 0$ and $f \in X$. Then

$$\|u\|_\infty \leq \frac{1}{\lambda}\|f\|_\infty.$$

Proof — In [11, §3.1] it has been proved that if $u \in \cap_{p \geq 1} W_{loc}^{2,p}(R^n) \cap X$ is such that $\mathcal{A}u$ is continuous, then at any relative maximum (respectively, minimum) point x_0 we have $\mathcal{A}u(x_0) \leq 0$ (respectively, $\mathcal{A}u(x_0) \geq 0$). The rest of the proof is similar to the one of Lemma 2.4 (and even simpler), and it is omitted. ∎

Proposition 4.7 *It holds*

$$D(A) = \left\{ f \in \bigcap_{p \geq 1} W_{loc}^{2,p}(R^n) \cap X : \mathcal{A}f \in X \right\}, \quad Af = \mathcal{A}f \ \forall f \in D(A).$$

Proof — First we prove the inclusion $\subset$. If $\phi \in C^1(R^n)$ and $\lambda > \omega$, estimates (1.6) imply that $R(\lambda)\phi \in C^2(R^n)$. Using the equality $(\mathcal{A}T(t)\phi)(x) = \partial/\partial t\, (T(t)\phi)(x)\ \forall x$ and estimate (1.6) with $\theta = 1$, $\alpha = 2$ we get

$$\lambda R(\lambda)\phi - \mathcal{A}R(\lambda)\phi = \phi.$$

From the general theory of elliptic differential equations with regular coefficients, for every $p > 1$ and $R > 0$ we get

$$\|R(\lambda)\phi\|_{W^{2,p}(B(0,R))} \leq C\|\phi\|_{L^p(B(0,R+1))}. \tag{4.42}$$

Let $f \in D(A)$, $\lambda > \omega$, and set $\phi = \lambda f - Af$. Let $\{\phi_n\} \subset C^1(R^n)$ be a sequence converging to ϕ in X. Set $f_n = R(\lambda)\phi_n$. Then $f_n \in C^2(R^n)$ and $f_n \to f$, $Af_n \to Af$ in X as $n \to \infty$. Applying estimate (4.42) to $\phi_n - \phi_m$ we see that $\{f_n\}_{n \in N}$ is a Cauchy sequence in $W^{2,p}(B(0,R))$ for every $p > 1$ and $R > 0$, so that $f \in W_{loc}^{2,p}(R^n)$ and the equality $Af_n = \mathcal{A}f_n\ \forall n$ implies $Af = \mathcal{A}f$. The inclusion $\subset$ is so proved.

Let now $f \in \cap_{p \geq 1} W_{loc}^{2,p}(R^n) \cap X$ be such that $\mathcal{A}f \in X$. Fix $\lambda > \omega$ and set $\phi = \lambda f - \mathcal{A}f$, $g = R(\lambda)\phi$. Our aim is to show that $f = g$. From the first part of the proof we know that $g \in \cap_{p \geq 1} W_{loc}^{2,p}(R^n) \cap X$ and that $\lambda g - \mathcal{A}g = \phi$. Therefore, $\lambda(f - g) - \mathcal{A}(f - g) = 0$. By Lemma 4.6, $f = g$, so that $f \in D(A)$. ∎

We are able now to prove that statements (iii) and (iv) of Proposition 2.1 hold true also in the case of Lipschitz continuous F.

Proposition 4.8 *Statements (iii) and (iv) of Proposition 2.1 hold. More precisely, for every $\varphi \in X$ we have*

$$\lim_{t \to 0} \|T(t)\varphi - \varphi\|_{L^\infty(R^n)} = 0 \iff \varphi \in \overline{D(A)} \iff \lim_{t \to 0} \|\varphi(\xi(t, \cdot)) - \varphi\|_{L^\infty(R^n)} = 0.$$

Moreover, if $\{\varphi_k\}_{k \in N}$ is a bounded sequence in $C^0(R^n)$ converging to a function φ uniformly on every compact subset of R^n, then for every $T > 0$ and for every compact $K \subset R^n$ we have

$$\lim_{k \to \infty} \sup_{0 \leq t \leq T} \|T(t)(\varphi_k - \varphi)_{|K}\|_{L^\infty(K)} = 0.$$

Proof — Let us prove that $\lim_{t\to 0} \|T(t)\varphi - \varphi\|_{L^\infty(R^n)} = 0 \iff \varphi \in \overline{D(A)}$. The arguments used in Proposition 3.3 of [9] show that

$$\|T(t)\varphi - \varphi\|_X \leq Ct\|\varphi\|_{D(A)}, \quad \forall \varphi \in D(A), \ 0 < t \leq 1.$$

Consequently, for every $\varphi \in \overline{D(A)}$, $T(t)\varphi$ converges to φ in X as t goes to 0. Conversely, if $\lim_{t\to 0} \|T(t)\varphi - \varphi\|_{L^\infty(R^n)} = 0$ then

$$\lambda R(\lambda, A)\varphi - \varphi = \int_0^\infty \lambda e^{-\lambda t}(T(t)\varphi - \varphi)dt$$

goes to 0 as λ goes to $+\infty$. Therefore, $\varphi = \lim_{\lambda\to\infty} \lambda R(\lambda, A)\varphi$ so that $\varphi \in \overline{D(A)}$.

We show now that there is $M > 0$ such that

$$|\xi(t, x) - \widetilde{\xi}(t, x)| \leq Mt, \quad 0 \leq t \leq 1, \ x \in R^n, \tag{4.43}$$

where $\widetilde{\xi}$ is the solution of

$$\begin{cases} \widetilde{\xi}_t(t, x) = \widetilde{F}(\widetilde{\xi}(t, x)), \quad t \in R, \\[2mm] \widetilde{\xi}(0, x) = x, \end{cases} \tag{4.44}$$

and $\widetilde{F}$ is defined in (3.34).

We have $\xi(0, x) - \widetilde{\xi}(0, x) = 0$ and

$$\xi_t - \widetilde{\xi}_t = \int_0^1 \widetilde{F}_x(\theta\xi + (1 - \theta)\widetilde{\xi})d\theta \ (\xi - \widetilde{\xi}) + F(\xi) - \widetilde{F}(\xi), \quad t > 0,$$

so that

$$|\xi_t - \widetilde{\xi}_t| \leq [F]_{Lip}|\xi - \widetilde{\xi}| + \|F - \widetilde{F}\|_\infty,$$

and (4.43) follows from the Gronwall Lemma.

Since $F - \widetilde{F} \in X$, and the functions in $D(A)$ and $D(\widetilde{A})$ have first order derivatives belonging to X, then $D(A) = D(\widetilde{A})$, so that $\overline{D(A)} = \overline{D(\widetilde{A})}$. By Proposition 2.1(iii), $\overline{D(\widetilde{A})} = \{\varphi \in X : \lim_{t\to 0} \|\varphi(\widetilde{\xi}(t, \cdot)) - \varphi\|_\infty = 0\}$. By (4.43), for every $\varphi \in X$ we have

$$\lim_{t\to 0} \|\varphi(\xi(t, \cdot)) - \varphi(\widetilde{\xi}(t, \cdot))\|_\infty = 0,$$

and statement (iii) follows.

To prove (iv), fix $\varepsilon > 0$ and define

$$\widetilde{F} = \frac{1}{\delta^n} \int_{R^n} F(y)\rho\left(\frac{x - y}{\delta}\right)dy, \quad x \in R^n,$$

with δ so small that

$$\|F - \widetilde{F}\|_\infty \leq \varepsilon.$$

Define then $\widetilde{A}$ and $\widetilde{T}(t)$ as in the case $\delta = 1$. The function

$$u_k(t, x) = (T(t) - \widetilde{T}(t))(\varphi_k - \varphi)(x), \quad 0 \leq t \leq T, \ x \in R^n,$$

is a solution of

$$
\begin{cases}
D_t u_k = \widetilde{A} u_k + \langle F - \widetilde{F}, DT(t)(\varphi_k - \varphi) \rangle, \ 0 \le t \le T, \ x \in R^n, \\[2mm]
u_k(0, x) = 0, \ x \in R^n.
\end{cases}
$$

The function in the right hand side $t \mapsto \langle F - \widetilde{F}, DT(t)(\varphi_k - \varphi) \rangle$ belongs to $\widetilde{C}([\sigma, T]; X) \cap B_{(1+\alpha)/2}((0,T]; C^\alpha(R^n))$ for every $\sigma \in (0, T)$, $\alpha \in (0,1)$. By Proposition 2.3 and Lemma 2.4 we have

$$
u_k(t, \cdot) = \int_0^t \widetilde{T}(t - s) \langle F - \widetilde{F}, DT(s)(\varphi_k - \varphi) \rangle ds, \ 0 \le t \le T,
$$

and since $\|\widetilde{T}(t - s)\|_{L(X)} \le 1$, $\|T(s)\|_{L(X, C^1(R^n))} \le Cs^{-1/2}$, then for $0 \le t \le T$

$$
\|(T(t) - \widetilde{T}(t))(\varphi_k - \varphi)\|_\infty = \|u_k(t, \cdot)\|_\infty \le \int_0^t \frac{\varepsilon C}{s^{1/2}} \|\varphi_k - \varphi\|_\infty ds \le K\varepsilon, \ k \in N.
$$

By Proposition 2.1(iv), $\sup_{0 \le t \le T} \|\widetilde{T}(t)(\varphi_k - \varphi)_{|K}\|_{L^\infty(K)}$ goes to 0 as k goes to ∞, for every compact set K. The statement follows. $\blacksquare$

Statements (i), (iii) and (iv), together with estimate (1.6) with $\theta = \alpha = 0$, imply that $T(t)$ is a weakly continuous semigroup in X, in the sense of [7, 8].

We are able to characterize the interpolation spaces $(D(A), X)_{\theta, \infty}$ between X and $D(A)$. To this aim we define the spaces X_θ for $0 < \theta < 1$ by

$$
X_\theta = \left\{ f \in X : [f]_{X_\theta} = \sup_{x \in R^n, 0 < t < 1} t^{-\theta} |f(\xi(t, x)) - f(x)| < \infty \right\}, \tag{4.45}
$$

and the spaces $\mathcal{C}^1(R^n)$, $\mathcal{C}^2(R^n)$ by

$$
\begin{cases}
\mathcal{C}^1(R^n) = \left\{ f \in C^0(R^n) : [f]_{\mathcal{C}^1} = \sup_{x \ne y} \frac{|f(x) + f(y) - 2f((x+y)/2)|}{|x - y|} < \infty \right\}, \\[3mm]
\|f\|_{\mathcal{C}^1} = \|f\|_\infty + [f]_{\mathcal{C}^1},
\end{cases}
$$

$$
\begin{cases}
\mathcal{C}^2(R^n) = \{ f \in \mathcal{C}^1(R^n) : D_i f \in \mathcal{C}^1(R^n), \ i = 1, \dots, n \}, \\[3mm]
\|f\|_{\mathcal{C}^2} = \|f\|_\infty + \sum_{i=1}^n \|D_i f\|_{\mathcal{C}^1}.
\end{cases}
$$

Proposition 4.9 *For $0 < \theta < 1$, $\theta \ne 1/2$ we have*

$$
(X, D(A))_{\theta, \infty} = \mathcal{C}^{2\theta}(R^n) \cap X_\theta
$$

and the norm $f \mapsto \|f\|_{\mathcal{C}^{2\theta}(R^n)} + [f]_{X_\theta}$ is equivalent to the $(D(A), X)_{\theta, \infty}$ norm. For $\theta = 1/2$,

$$
(X, D(A))_{1/2, \infty} = \mathcal{C}^1(R^n) \cap X_{1/2},
$$

and the norm $f \mapsto \|f\|_{\mathcal{C}^1(R^n)} + [f]_{X_{1/2}}$ is equivalent to the $(D(A), X)_{1/2, \infty}$ norm.

Proof — Arguing as in [9] we see that, as in the case of the strongly continuous semigroups,

$$(X, D(A))_{\theta,\infty} = D_A(\theta, \infty)$$

$$= \left\{ f \in X : [f]_{D_A(\theta,\infty)} = \sup_{0 < t \leq 1} \frac{\|T(t)f - f\|_X}{t^\alpha} < \infty \right\},$$

with equivalence of the respective norms.

Estimate (1.6) with $\theta = 0$ and α replaced by 2α implies (arguing again as in [9]) that for $1/2 < \alpha < 1$, $C^{2\alpha}(R^n)$ belongs to the class J_α between X and $D(A)$, i.e. there exists $C > 0$ such that

$$\|\varphi\|_{C^{2\alpha}(R^n)} \leq C \|\varphi\|_X^{1-\alpha} \|\varphi\|_{D(A)}^\alpha, \quad \forall \varphi \in D(A).$$

The Reiteration Theorem (see e.g. [16, §10.1]) yields

$$(X, D(A))_{\theta,\infty} \subset (X, C^{2\alpha}(R^n))_{\theta/\alpha,\infty} = \begin{cases} C^{2\theta}(R^n), & \text{if } \theta \neq 1/2, \\[2mm] \mathcal{C}^1(R^n), & \text{if } \theta = 1/2, \end{cases} \tag{4.46}$$

with continuous embeddings. The last equalities are well known, their proofs may be found in [16, §2.7].

From now on we may assume that F has bounded derivatives up to the third order. Indeed, if F is Lipschitz continuous let $\widetilde{F}$ be defined by (3.34) and let $\widetilde{A}$ be defined as A, with F replaced by $\widetilde{F}$. Since $D(A) = D(\widetilde{A})$ then $(X, D(A))_{\theta,\infty} = (X, D(\widetilde{A}))_{\theta,\infty}$ for every $\theta \in (0,1)$. On the other hand, by (4.43) if $\varphi \in C^\theta(R^n)$ then

$$|\varphi(\xi(t,x)) - \varphi(\widetilde{\xi}(t,x))| \leq [\varphi]_{C^\theta(R^n)}|\xi(t,x) - \widetilde{\xi}(t,x)|^\theta \leq C[\varphi]_{C^\theta(R^n)}t^\theta$$

Therefore, a function $\varphi \in C^{2\theta}(R^n)$ (respectively, $\varphi \in \mathcal{C}^1(R^n)$) belongs to X_θ if and only if it belongs to $\widetilde{X}_\theta$.

So, we assume that F has bounded derivatives up to the third order. We prove now that $(X, D(A))_{\theta,\infty}$ is continuously embedded in X_θ. By Theorem 4.1(iii) of [1] we have

$$\sup_{0 < t < 1} \|t^{-\theta}(G(t,0)\varphi - \varphi)\|_X < \infty \iff \varphi \in D_{\Lambda(0)}(\theta, \infty), \tag{4.47}$$

where the operators $\Lambda(t)$ have been defined in Subsection 2.1. By (2.17), $D_{\Lambda(0)}(\theta, \infty) = C^{2\theta}(R^n)$ if $\theta \neq 1/2$, $D_{\Lambda(0)}(1/2, \infty) = \mathcal{C}^1(R^n)$. Therefore, $(X, D(A))_{\theta,\infty}$ is continuously embedded in $D_{\Lambda(0)}(\theta, \infty)$. Then for every $x \in R^n$ and $\varphi \in (X, D(A))_{\theta,\infty}$,

$$|\varphi(\xi(t,x)) - \varphi(x)| \leq$$

$$\leq |\varphi(\xi(t,x)) - (G(t,0)\varphi)(\xi(t,x))| + |(G(t,0)\varphi)(\xi(t,x)) - \varphi(x)|$$

$$\leq \|\varphi - G(t,0)\varphi\|_\infty + \|T(t)\varphi - \varphi\|_\infty$$

$$\leq C_1 t^\theta \|\varphi\|_{(X,D(\Lambda(0))_{\theta,\infty}} + C_2 t^\theta \|\varphi\|_{(X,D(A))_{\theta,\infty}} \leq C t^\theta \|\varphi\|_{(X,D(A))_{\theta,\infty}}.$$

Let us prove the other inclusion. If $\varphi \in C^{2\theta}(R^n) \cap X_\theta$ for $\theta \neq 1/2$, or if $\varphi \in C^1(R^n) \cap X_{1/2}$ for $\theta = 1/2$, then for every $x \in R^n$ we have

$$|(T(t)\varphi)(x) - \varphi(x)| \leq$$

$$\leq |(G(t,0)\varphi)(\xi(t,x)) - \varphi(\xi(t,x))| + |\varphi(\xi(t,x)) - \varphi(x)|$$

$$\leq \|G(t,0)\varphi - \varphi\|_\infty + t^\theta[\varphi]_{X_\theta}.$$

So we get

$$\|T(t)\varphi - \varphi\|_\infty \leq \begin{cases} Ct^\theta(\|\varphi\|_{C^{2\theta}(R^n)} + [\varphi]_{X_\theta}), & \text{if } \theta \neq 1/2, \\ \\ Ct^{1/2}(\|\varphi\|_{C^1(R^n)} + \|\varphi\|_{X_{1/2}}), & \text{if } \theta = 1/2, \end{cases}$$

and the statement follows. ■

5 PROOF OF THEOREMS 1 AND 2

We shall use estimates (1.6) and some abstract interpolation results proved in [14], which we recall below.

Theorem 5.10 *Let Y_0, Y_1, Y_2 be Banach spaces, with $Y_2 \subset Y_1 \subset Y_0 \subset X = C^0(R^n)$, enjoying the following property: for $i = 1, 2$, if $\varphi : I \mapsto Y_0$ is such that for every $x \in R^n$ the real function $t \mapsto \varphi(t)(x)$ is continuous in I, and $\|\varphi(t)\|_{Y_i} \leq c(t)$ with $c \in L^1(I)$, then the function*

$$f(x) = \int_I \varphi(t)(x)dt, \quad x \in R^n,$$

belongs to Y_i, and

$$\|f\|_{Y_i} \leq \|c\|_{L^1(I)}.$$

Let moreover $T(t)$, $t \geq 0$, be a weakly continuous semigroup of linear bounded operators in X, such that for some $\omega \geq 0$, $0 \leq \gamma_1 < 1 < \gamma_2$ we have

$$\sup_{t>0} t^{\gamma_i} e^{-\omega t} \|T(t)\|_{L(Y_0, Y_i)} < \infty, \quad i = 1, 2. \tag{5.48}$$

Set $\theta = (1 - \gamma_1)/(\gamma_2 - \gamma_1)$. Then the domain of the part of A in Y_0 is contained in $(Y_1, Y_2)_{\theta, \infty}$, with continuous embedding.

Theorem 5.11 *Under the assumptions of Theorem 5.10, let $f \in \widetilde{C}([0,T]; X) \cap B([0,T]; Y_0)$. Then the function*

$$u(t)(x) = \int_0^t (T(t-s)f(s))(x)ds, \quad 0 \leq t \leq T, \ x \in R^n,$$

belongs to $B([0,T]; (Y_1, Y_2)_{\theta, \infty})$, with $\theta = (1 - \gamma_1)/(\gamma_2 - \gamma_1)$, and there is $C > 0$, independent of f, such that

$$\sup_{0 \leq t \leq T} \|u(t)\|_{(Y_1, Y_2)_{\theta, \infty}} \leq C \sup_{0 \leq t \leq T} \|f(t)\|_{Y_0}.$$

Proof of Theorem 1 — We have already shown that the spaces $Y_0 = C^\theta(R^n)$, $Y_1 = C^\alpha(R^n)$, $Y_2 = C^{2+\alpha}(R^n)$, with $0 \le \theta < \alpha < 1$, satisfy the assumptions of Theorem 5.10, with $\gamma_1 = (\alpha - \theta)/2$, $\gamma_2 = 1 + (\alpha - \theta)/2$.

By Theorem 5.10, the domain $D(A_\theta) = \{f \in D(A) : Af \in C^\theta(R^n)\}$ is continuously embedded in

$$(C^\alpha(R^n), C^{2+\alpha}(R^n))_{1-(\alpha-\theta)/2,\infty} = \begin{cases} C^{2+\theta}(R^n), & \text{if } \theta > 0, \\ \\ C^2(R^n), & \text{if } \theta = 0, \end{cases} \tag{5.49}$$

with continuous embeddings. The above equality is well known, see e.g. [16, §2.7].
∎

Proof of Theorem 2 — Uniqueness of the bounded solution of (1.2) in $C^{1,2}([0,T] \times R^n)$ follows from Lemma 2.4.

As far as existence and regularity are concerned, we apply Theorem 5.11. Its assumptions are satisfied by the spaces $Y_0 = C^\theta(R^n)$, $Y_1 = C^\alpha(R^n)$, $Y_2 = C^{2+\alpha}(R^n)$, with $0 \le \theta < \alpha < 1$, and $\gamma_1 = (\alpha - \theta)/2$, $\gamma_2 = 1 + (\alpha - \theta)/2$. By Theorem 5.11, the function

$$v(t)(x) = \int_0^t (T(t-s)g(s,\cdot))(x)ds, \ 0 \le t \le T, \ x \in R^n,$$

is bounded with values in $C^{2+\theta}(R^n)$, and

$$\sup_{0 \le t \le T} \|v(t)\|_{C^{2+\theta}(R^n)} \le C \sup_{0 \le t \le T} \|g(t,\cdot)\|_{C^\theta(R^n)}. \tag{5.50}$$

By (1.6), the function $T(\cdot)u_0$ is bounded with values in $C^{2+\theta}(R^n)$. Setting $u(t,x) = (T(t)u_0)(x) + v(t)(x)$, the arguments used in Proposition 2.3 to prove that $t \mapsto u(t,\cdot)$ belongs to $\widetilde{C}([0,T];X)$ and that $u \in C^{1,2}([0,T] \times R^n)$ work also in this case. The statement follows. ∎

Remark 5.12 The results of Theorems 1 and 2 still hold if the Lipschitz continuity assumption on F is replaced by

$$|F(x+y) - F(y)| \le C|y|^\theta, \ x, y \in R^n, \ |y| \le 1.$$

Indeed, defining $\widetilde{F}$ by (3.34), $\widetilde{F}$ is smooth with bounded derivatives of any order, and $F - \widetilde{F}$ belongs to $C^\theta(R^n)$, so that Theorems 1 and 2 hold if F is replaced by $\widetilde{F}$. The general case may be recovered by the procedure of Theorem 3.5, which works also in the elliptic case (of course, in the elliptic case the operator $D_t - \mathcal{A} + \lambda I$ has to be replaced by $-\mathcal{A} + \lambda I$).

References

[1] P. Acquistapace, Evolution operators and strong solutions of abstract linear parabolic equations, *Diff. Int. Eqns. 1*: 433-457 (1988).

[2] P. Acquistapace and B. Terreni, Linear Parabolic Equations in Banach Spaces with Variable Domains but Constant Interpolation Spaces, *Ann. Sc. Norm. Sup. Pisa, Serie IV, 13*: 75-107 (1986).

[3] P. Acquistapace and B. Terreni, A unified approach to abstract linear non-autonomous parabolic equations, *Rend. Sem. Mat. Univ. Padova, 78*: 47-107 (1987).

[4] D.G. Aronson and P. Besala, Parabolic equations with unbounded coefficients, *J. Diff. Eqns., 3*: 1-14 (1967).

[5] P. Besala, On the existence of a fundamental solution for a parabolic differential equation with unbounded coefficients, *Ann. Polon. Math., 29*: 403-409 (1975).

[6] P. Cannarsa and V. Vespri, Generation of analytic semigroups by elliptic operators with unbounded coefficients, *SIAM J. Math. Anal., 18*: 857-872 (1987).

[7] S. Cerrai, A Hille-Yosida theorem for weakly continuous semigroups, *Semigroup Forum, 49*: 349-367 (1994).

[8] S. Cerrai and F. Gozzi, Strong solutions of Cauchy problems associated to weakly continuous semigroups, *Diff. Int. Eqns., 8*: 465-486 (1995).

[9] G. Da Prato and A. Lunardi, On the Ornstein-Uhlenbeck operator in spaces of continuous functions, *J. Funct. Anal., 131*: 94-114 (1995).

[10] M.I. Freidlin and A.D. Wentzell, "Random perturbations of dynamical systems", Springer-Verlag, Berlin (1983).

[11] A. Lunardi, "Analytic semigroups and optimal regularity in parabolic problems", Birkhäuser Verlag, Basel (1995).

[12] A. Lunardi, Schauder estimates for a class of elliptic and parabolic operators with unbounded coefficients in R^n, *Ann. Sc. Norm. Sup. Pisa* (to appear).

[13] A. Lunardi, On the Ornstein-Uhlenbeck operator in L^2 spaces with respect to invariant measures, *Trans. Amer. Math. Soc.* (to appear).

[14] A. Lunardi, An interpolation method to characterize domains of generators of semigroups, *Semigroup Forum, 53*: 321-329 (1996).

[15] H.B. Stewart, Generation of analytic semigroups by strongly elliptic operators, *Trans. Amer. Math. Soc., 199*: 141-162 (1974).

[16] H. Triebel, "Interpolation Theory, Function Spaces, Differential Operators", North-Holland, Amsterdam (1978).

The Quasi-Isothermal Limit in Porous Catalysts

FRANCISCO J. MANCEBO   Departamento de Fundamentos Matemáticos, E.T.S.I. Aeronáuticos, Universidad Politécnica de Madrid, Spain

JOSÉ M. VEGA   Departamento de Fundamentos Matemáticos, E.T.S.I. Aeronáuticos, Universidad Politécnica de Madrid, Spain

1   INTRODUCTION

This paper deals with a well-known model of porous catalyst that after suitable nondimensionalization, see Aris 1975, may be written as

$$\partial u/\partial t = \Delta u - \phi^2 f(u,v) \quad \text{in } \Omega, \quad \partial u/\partial n = \sigma(1-u) \quad \text{at } \partial\Omega, \quad (1.1)$$
$$L^{-1}\partial v/\partial t = \Delta v + \beta\phi^2 f(u,v) \quad \text{in } \Omega, \quad \partial v/\partial n = \nu(1-v) \quad \text{at } \partial\Omega, \quad (1.2)$$

for $t > 0$, with appropriate initial conditions

$$u = u_0 > 0, \qquad v = v_0 > 0 \quad \text{in } \Omega, \quad \text{at } t = 0. \tag{1.3}$$

The functions $u > 0$ and $v > 0$ are the reactant concentration and the temperature respectively, Δ is the Laplacian operator, n is the outward unit normal to the smooth boundary of the bounded domain $\Omega \subset \mathbf{R}^m$ ($m \geq 1$) and the parameters ϕ^2 (Damköhler number), L (Lewis number), β (Prater number), σ and ν (material and thermal Biot numbers) are strictly positive. The function f is positive whenever $u > 0$ and $v > 0$.

The system (1.1)-(1.3) models also several physical problems ranging from

Combustion theory to Biology. In porous catalyst theory the most usual nonlinearities are the so called Arrhenius and Langmuir-Hinshelwood kinetic laws that lead to the following nonlinearities

$$f(u, v) = u^p \exp(\gamma - \gamma/v) \tag{1.4}$$

$$f(u, v) = u^p [u + k \exp(\gamma_a - \gamma_a/v)]^q \exp(\gamma - \gamma/v) \tag{1.5}$$

where the constants p, q, k are strictly positive and γ_a, γ are non-negative.

Porous catalysts usually exhibit a large thermal conductivity and thus β is usually small and the ratio σ/ν is large; ν is either small or of order unity, depending on the size of the catalyst, and L and ϕ^2 vary in a wide range, from small to large values. Therefore the limit

$$\beta \to 0, \qquad \sigma/\nu \to \infty \tag{1.6}$$

is realistic. In this paper we consider the sublimit of (1.6) $\nu \to 0$, which leads to the (so-called) quasi isothermal models. If in addition $\nu \to 0$ then two sub-models are obtained, that consist of a PDE (for the evolution of the reactant concentration) coupled with a nonlocal ODE (for the temperature v, that becomes spatially constant after some time). In the first sub-model the PDE, applies in the $m - D$ domain and corresponds to the limit $\phi \sim 1$. In the second sub-model, which corresponds to the limit $\phi \to \infty$, with $\beta\phi$ appropriately small, the chemical reaction is confined in a thin reaction layer near the boundary of the domain and, the PDE applies in a $1 - D$ semi-infinite interval. This sub-model is posed by

$$\partial \tilde{U}/\partial \tau = \partial^2 \tilde{U}/\partial \xi^2 - f(\tilde{U}, V) \qquad \text{in} \quad -\infty < \xi < 0, \tag{1.7}$$

$$u \to 0 \quad \text{as} \quad \xi \to -\infty, \qquad \partial u/\partial \xi = (\sigma/\phi)(1 - u) \quad \text{at} \quad \xi = 0, \tag{1.8}$$

$$(V_\Omega \phi^2/(S_\Omega \nu L))dV/d\tau = 1 - V + (\beta\phi/\nu) \int_{-\infty}^{0} f(\tilde{U}, V)d\xi, \tag{1.9}$$

with appropriate initial conditions, where

$$V_\Omega = \int_\Omega dx, \qquad S_\Omega = \int_{\partial\Omega} ds, \tag{1.10}$$

and $\tilde{U}$ is appropriately close to u. The new (re-scaled) variables τ and ξ are

$$\tau = \phi^2 t, \qquad \xi = \phi\eta, \tag{1.11}$$

where η is a coordinate along the outward unit normal to $\partial\Omega$. Notice that this sub-model is independent of the domain Ω. The first above-mentioned sub-model seems to have been first considered by Amundson and Raymond 1965, and the second by Parra and Vega 1988. The rigorous derivation was given in Mancebo and Vega 1993 and Mancebo and Vega 1996 ; see also Parra and Vega 1988, Vega 1988 and di Liddo and Maddalena 1992 for some results concerning the steady states, local bifurcations and global stability properties. For the rigorous justification of related sub-models of general reaction-diffusion equations see Conway et. al. 1978, Hale and Rocha 1987a-b and Hale and Sakamoto 1989.

Let us now briefly explain where the model (1.7)-(1.9) comes from. Since ϕ^2 is large, the chemical reaction is very strong and, after some time, the reactant is consumed and u becomes very small in Ω except in a thin boundary layer near the boundary of Ω. Moreover, if $\beta\phi$ and ν are small, the temperature v becomes spatially constant (in first approximation) after some time. Finally, if $f'_u > 0$, after some time, the reactant concentration in the boundary layer depends only on time and on the distance to the boundary of Ω in first approximation.

In this paper we state the results needed to provide a rigorous derivation of (1.7)-(1.9) and the sketch of the proofs of these results (see Mancebo and Vega 1996, for more details). More precisely, we shall prove that, after some time T, (i) u is quite small except in a thin boundary layer near $\partial\Omega$ and (ii) the temperature in the porous catalyst (1.9) is, in first approximation, spatially uniform if $t \geq T$, (iii) the reactant concentration in the boundary layer depends only on time and the distance to $\partial\Omega$ and (iv) the sub-model (1.7)-(1.9) applies in first approximation.

Let us now state precisely the assumptions to be made below. We shall consider the limit

$$\phi \to \infty, \qquad \sigma^{-1} = O(1), \qquad \beta\phi\sigma/(\phi + \sigma) \to 0. \tag{1.12}$$

The domain Ω and the nonlinearity f will be assumed such that

(H.1) $\Omega \subset \mathbf{R}^m$ $(m \geq 1)$ is a bounded domain, with a $C^{4+\alpha}$, conected boundary (for some $\alpha > 0$). Then Ω satisfies uniformly the interior and exterior sphere conditions: there are two constants, $\rho_1 > 0$ and $\rho_2 > 0$, such that, for each $x \in \partial\Omega$, two hyperspheres, of radii ρ_1 and ρ_2, S_1 and S_2, are tangent to $\partial\Omega$ at x and satisfy $S_1 \subset \Omega$ and $\bar{S}_2 \cap \bar{\Omega} = \{x\}$ (where overbars hereafter stands for closure).

(H.2) The C^1- function $f \colon [0, \infty[\times[0, \infty[\to \mathbf{R}$ is such that $f(0, v) = 0$ for all $v \geq 0$ and $f(u, v) > 0$ whenever $u > 0$ and $v \geq 0$.

(**H.3**) There is a continuous, increasing function, $g_1 : [0, \infty[\to \mathrm{R}$ such that

$$f(u, v) \le g_1(u) \qquad \text{if} \quad u \ge 0 \quad \text{and} \quad v \ge 0.$$

(**H.4**) There are two strictly positive constants, k_1 and k_2, and a positive, continuous, decreasing function, $g_2 : [0, \infty[\to \mathrm{R}$, such that

$$k_2 u \le f(u, v) \le k_1 u \qquad \text{if } 0 \le u \le 2 \quad \text{and } v \ge 1/2,$$
$$0 < u g_2(u) \le f(u, v) \qquad \text{if} \qquad u \ge 0 \quad \text{and } v \ge 1/2,$$

(**H.5**) There are three constants, $k_3 > 0$, $k_4 > 0$ and $k_5 > 0$, such that

$$k_3 \le f_u(u, v) \le k_4, \qquad |f_v(u, v)| \le k_5 u$$
$$\text{if } 0 \le u \le \sigma / \left(\sigma + \phi \sqrt{k_2 / 2m} \right) \quad \text{and } v \ge 1.$$

In addition, the initial conditions (1.3) will be assumed to be such that

(**H.6**) $\|u_0\|_{C(\bar{\Omega})} = O(1)$ and $\|v_0\|_{C(\bar{\Omega})} = O(1)$ in the limit (1.12).

The paper is organized as follows. In Section 2 we give bounds on the functions u and v that will be needed in the justification of the model. In Section 3 we give the results that lead to the justification of the isothermal model. In Section 4 we consider some sub-models of (1.7)-(1.9).

2   SOME PRELIMINARY ESTIMATES

Let us give some results concerning two linear singularly perturbed problems that will be used in the sequel.

LEMMA 2.1 Let the domain $\Omega \subset \mathrm{R}^m$ be such that assumption (H.1) holds, and let u and v be the unique solutions of

$$\Delta u = \Lambda^2 u \qquad \text{in } \Omega, \qquad \partial u / \partial n = \sigma(1 - u) \quad \text{at } \partial\Omega,$$
$$\Delta v + \varepsilon \Lambda^2 u = 0 \quad \text{in } \Omega, \qquad \partial v / \partial n = \nu(1 - v) \quad \text{at } \partial\Omega,$$

where Λ, ε, σ and ν are positive and $\sigma > \nu$. As $\Lambda \to \infty$, the following estimates hold

$$\frac{\sigma}{\sigma + \delta_1} \exp[-\delta_1 d(x)] \leq u(x) \leq \frac{\sigma}{\sigma + \delta_2} \left[\frac{\cosh\big(\delta_2(\rho_1 - d_1(x))\big)}{\cosh(\delta_2\rho_1)} \right],$$

$$\sigma S_\Omega \delta_2/(\sigma + \delta_2) \leq \Lambda^2 \int_\Omega u(x)dx \leq \sigma S_\Omega \delta_1/(\sigma + \delta_1),$$

$$1 < v(x) \leq 1 + \delta_3,$$

for all $x \in \bar{\Omega}$, where ρ_1 and S_Ω are as defined in (H.1) and (1.10) respectively, $d(x)$ is the distance from x to $\partial\Omega$, $d_1(x) = \min\{d(x), \rho_1\}$ and the positive constants δ_1, δ_2 and δ_3 satisfy

$$\delta_2 = \Lambda/\sqrt{m}, \qquad \delta_3 = \varepsilon\sigma\delta_1/(\sigma+\delta_1)\nu \quad \text{and} \quad |\delta_1 - \Lambda| = O(\Lambda^{-1}) \quad \text{as } \Lambda \to \infty$$

uniformly in $\varepsilon > 0$, $\sigma > 0$ and $\nu > 0$.

Sketch of the proof: These estimates are obtained by constructing supersolutions in an interior sphere and sub-solutions outside a exterior sphere.

LEMMA 2.2 Under the assumptions (H.1)-(H.4) and (H.6) at the end of Section 1, there is a constant T, depending only on

$$\|u_0\|_{C(\bar{\Omega})}, \quad \|v_0\|_{C(\bar{\Omega})}, \quad \phi, \ \sigma, \ L, \ \beta, \ \text{and} \ \nu,$$

and satisfying

$$T = O\left(\phi^{-2}\log(2 + \phi/\sigma + \sigma/\phi) + \phi^{-1} + (\nu L)^{-1}\log(2 + \beta\phi^2/\nu)\right)$$

in the limit (1.12), such that every solution of (1.1)-(1.3) satisfies

$$u_1 < u(\cdot, t) < u_2, \qquad 1/2 < v(\cdot, t) < 1 + v_1 \quad \text{in } \Omega \text{ if } t \geq T,$$

where u_1, u_2 and v_1 are the unique solutions of

$$\Delta u_1 = 2k_1\phi^2 u_1 \quad \text{in } \Omega, \qquad \partial u_1/\partial n = \sigma(1 - u_1) \quad \text{at } \partial\Omega,$$

$$\Delta u_2 = k_2\phi^2 u_2/2 \quad \text{in } \Omega, \qquad \partial u_2/\partial n = \sigma(1 - u_2) \quad \text{at } \partial\Omega,$$

$$\Delta v_1 + \beta k_1\phi^2 u_2 = 0 \quad \text{in } \Omega, \qquad \partial v_1/\partial n = \nu(1 - v_1) \quad \text{at } \partial\Omega,$$

with the constants $k_1 > 0$ and $k_2 > 0$ as defined in assumption (H.4).

Sketch of the proof: These bounds on u and v follow straighfordwardly by means of maximum principles. In proving these estimates, the assumptions (H.1)-(H.4) are essential whereas the assumption (H.5) is purely technical.

The following lemma is needed for obtaining bounds on the derivatives of u along lines on hyper-surfaces parallel to $\partial\Omega$.

LEMMA 2.3 Let $\tilde{t}_0$ be a unit vector that is tangent to a hypersurface, H, parallel to $\partial\Omega$ at $p \in N \cap \bar{\Omega}_1$, where

$$\Omega_1 = \{x \in \Omega : d(x) = \text{dist}\,(x, \partial\Omega) < \rho_1/2\}, \qquad (2.1)$$

and ρ_1 is as defined in (H.1). Then there are a neighborhood N of p in $\mathbf{R}^m$, a C^3 vector field $\tilde{t} : N \to \mathbf{R}^3$, two vectors a_1 and a_2 and two scalars, b_1 and b_2, such that the following properties holds:
   (i) a_1, a_2, b_1 and b_2 depend continuously on p and $\tilde{t}_0$
   (ii) $\tilde{t} = \tilde{t}_0$ at p, $\tilde{t} \cdot \tilde{t} = 1$ in N and, for each $q \in N \cap \Omega_1$, $\tilde{t}(q)$ is tangent to the hypersurface parallel to $\partial\Omega$ passing throught q.
   (iii) If $I \subset \mathbf{R}$ is an open interval and $U : (N \cap \bar{\Omega}_1) \times I \to \mathbf{R}$ is a $C^{3,1}$ function satisfying

$$\partial U/\partial t = \Delta U + \varphi \quad \text{in} \quad (N \cap \bar{\Omega}_1) \times I \qquad (2.2)$$

then the $C^{2,1}$ function $w = \nabla U \cdot \tilde{t}$ satisfies

$$\frac{\partial w}{\partial t} = \Delta w + a_1 \cdot \nabla w + a_2 \cdot \nabla U + b_1 w + \nabla\varphi \cdot \tilde{t} \quad \text{at } p \text{ for all } t \in I, \quad (2.3)$$

$$\frac{\partial w}{\partial n} = \nabla(\partial U/\partial n) \cdot \tilde{t} + b_2 w \qquad\qquad\qquad \text{at } p \text{ for all } t \in I, \quad (2.4)$$

where n is the outward unit normal to H at p. Sketch of the proof: The equations (2.3)-(2.4) are obtained upon writing (2.2) in Fermi geodesic coordinates, (see Guggenheimer 1977), in a neighborhood of p, and differentiating along an appropriate coordinate line on the hypersurface H.

3   MATHEMATICAL DERIVATION OF ISOTHERMAL MODELS

In this section we state the results, that lead to the justification of the isothermal model.

LEMMA 3.1 Under the assumptions of Lemma 2.2 there are two constants,

$\mu > 0$, and $T' \geq T$ such that (i) μ and $T' - T$ depend only on the domain Ω,

$$\phi, \ \sigma, \ L, \ \beta, \ \text{and} \ \nu \tag{3.1}$$

(ii) $\mu = (\nu + \beta\phi\sigma/(\sigma+\phi))$ and $T' - T = O(L^{-1})\log[(1 + (\beta\phi\sigma/(\sigma+\phi)\nu))/\mu]$ in the limit (1.12) and (iii) if $t > T'$ then

$$|v - V| \leq \mu \qquad \text{in } \bar{\Omega}, \tag{3.2}$$

where $V(t)$ is the spatial average of v, i.e.,

$$V(t) = V_\Omega^{-1} \int_\Omega v(x,t)dx \tag{3.3}$$

where V_Ω is as defined in (1.10).

Sketch of the proof: The function $z = v - V$ satisfies

$$\frac{\partial z}{\partial t} - \Delta z = \beta\phi^2 f(u,v) - V_\Omega^{-1}\left(\nu\int_{\partial\Omega}(1-v)ds + \beta\phi^2\int_\Omega f(u,v)dx\right) \quad \text{in } \Omega,$$

$$\frac{\partial z}{\partial n} = \nu(1-v) \qquad \text{at } \partial\Omega.$$

If we decompose z as

$$z = -\nu V\varphi + w_1 + w_2$$

where φ and w_1 are defined by

$$\Delta\varphi = \varphi \quad \text{in } \Omega, \qquad \partial\varphi/\partial n + \nu\varphi = 1 \quad \text{at } \partial\Omega, \tag{3.4}$$

$$\partial w_1/\partial\tau = \Delta w_1 - w_1 + \beta\phi^2 f(u,v) \quad \text{in } \Omega, \quad \text{if } \tau > 0, \tag{3.5}$$

$$\partial w_1/\partial n + \nu w_1 = 0 \quad \text{at } \partial\Omega, \text{ if } \tau \geq 0, \qquad w_1 = 0 \quad \text{in } \bar{\Omega}, \quad \text{if } \tau = 0, \tag{3.6}$$

and w_2 is such that

$$\partial w_2/\partial\tau - \Delta w_2 = (\nu\varphi - 1)V_\Omega^{-1}\left[\beta\phi^2\int_\Omega f(u,v)dx + \nu\int_{\partial\Omega}(1-v)ds\right] \tag{3.7}$$
$$+ w_1 - \nu V\varphi, \quad \text{in } \Omega,$$

$$\partial w_2/\partial n + \nu w_2 = \nu \quad \text{at } \partial\Omega. \tag{3.8}$$

Then (3.2) is obtained upon application of (a) local L_p estimates and imbedding theorems to (3.4) and (3.7)-(3.8), and maximum principles to (3.5)-(3.6) and (b) by means of an induction argument.

LEMMA 3.2 Under the assumptions (H.1)-(H.4) and (H.6) there is a constant $T_1'' \geq 2T'$ such that $T_1'' - 2T'$ depends only on the quantities (3.1) and satisfies $T_1'' - 2T' = O(\log(2 + \nu(\phi + \sigma)/\beta\phi\sigma)/\nu L)$ in the limit (1.12), and

$$V > 1, \quad v > 1 \text{ and } u_1 \leq U \leq u_2 \text{ in } \Omega, \quad \text{for all } t \geq T_1'', \qquad (3.9)$$

where u_1, u_2 and T' are as defined in Lemmata 2.2 and 3.1 and U is defined by

$$\frac{\partial U}{\partial t} = \Delta U - \phi^2 f(U, V) \quad \text{in } \Omega, \qquad \frac{\partial U}{\partial n} = \sigma(1 - U) \quad \text{at } \partial\Omega \quad (3.10)$$
$$U(x, 0) = u(x, 0) \qquad \text{for all } x \in \Omega.$$

Sketch of the proof: The result follows straighfordwardly when maximum principles are applied and the estimates in Lemmata 2.1 and 2.2 are taken into account.

LEMMA 3.3 Under the assumptions (H.1)-(H.6) let μ and T_1'' be as defined in Lemmata 3.1 and 3.2. Then there is a constant $T_2'' \geq T_1''$ such that $T_2'' - T_1''$ depends only on the quantities (3.1) and satisfies $T_2'' - T_1'' = O(\phi^{-1})\log(1 + 1/\mu)$ in the limit (1.12), and

$$|U - u| \leq 4k_5\mu u_2/k_3 \quad \text{in } \bar{\Omega}, \qquad \text{for all } t \geq T_2'', \qquad (3.11)$$

where k_3 and k_5 are as defined in (H.5), u_2 as defined in Lemma 2.2 and v is given by (3.10).

Sketch of the proof: The bound (3.11) is obtained by means of maximum principles.

LEMMA 3.4 Under the assumptions of Lemma 3.2, there is a constant $\mu_1 > 0$, depending only on the quantities (3.1), such that $\mu_1 = O(\sigma\phi/(\phi + \sigma))$ in the limit (1.12) and

$$|\nabla U(x, t)| \leq \mu_1 \exp\left[-\phi d_1(x)\sqrt{k_2/2m}\right] \quad \text{if } x \in \bar{\Omega} \text{ and } t \geq T_1'' + 2/\phi^2$$

where $d_1(x) = \min\{\rho_1, d(x)\}$, $d(x)$ is the distance from x to $\partial\Omega$ and ρ_1, k_1 and T_1'' are as defined in assumptions (H.1) and (H.4) and in Lemma 3.2.

Sketch of the proof: This lemma is proved by stretching the domain, and applying apriori L_p estimates and imbedding theorems.

LEMMA 3.5 Under the assumptions (H.1)-(H.6) let ρ_1, k_1, k_3, T_1'' and μ_1 be as defined in assumptions (H.1), (H.4) and (H.5) and in Lemmata 2.4 and 2.6. Then there are two constants, $T'' \geq T_1'' + 1/\phi^2$ and $\mu_2 > 0$, depending only on the quantities (3.1) such that $T'' - T_1'' = O(\phi^{-2}) \log \phi$ and $\mu_2 = O(\sigma/(\phi + \sigma)\phi)$ in the limit (1.12), and

$$|\tilde{\nabla} U| \leq \mu_2 \exp\left[-\sqrt{k}\phi d(x)\right] \qquad \text{if} \quad x \in \bar{\Omega}_1, \quad \text{and } t \geq T'', \qquad (3.12)$$

where U is a solution of (3.10), $\tilde{\nabla} U$ is the gradient of U along the hyper-surfaces parallel to $\partial\Omega$,

$$k = \min\{k_2/4m, \ k_3/3\} > 0,$$

Ω_1 is as defined in (2.1), k_2 and k_3 are as defined in (H.4) and (H.5) and $d(x)$ is the distance from x to $\partial\Omega$.

Sketch of the proof: The estimate (3.12) is obtained upon application of maximum principles and the results in Lemma 2.3 to the equation that is obtained when (3.10) is differentiated with respect the first Fermi coordinate (see sketch of the proof of Lemma 2.3).

THEOREM 3.6 Under the assumptions (H.1)-(H.6) there are two constants, $\lambda > 0$ and $\varepsilon > 0$, and for each solution of

$$\partial\tilde{U}/\partial t = \partial^2\tilde{U}/\partial\eta^2 - \phi^2\tilde{U}\exp(-1/V_1) \quad \text{in} \ -\infty < \eta < 0, \qquad (3.13)$$

$$\tilde{U} = 0 \quad \text{at } \eta = -\infty, \qquad \partial\tilde{U}/\partial\eta = \sigma(1 - \tilde{U}) \quad \text{at } \eta = 0, \qquad (3.14)$$

$$\frac{V_\Omega}{S_\Omega L}\frac{dV_1}{dt} = -\nu V_1 + \beta\phi^2 \exp(\frac{-1}{V_1})\int_{-\infty}^0 \tilde{U}d\xi + \nu/\gamma + \psi(t) \qquad (3.15)$$

and a constant $\tilde{T} > 0$ such that

(i) λ depends only on the domain Ω, ε depends only on Ω and on the quantities (3.1), and $\tilde{T}$ depends only on Ω, on the quantities (3.1) and on $\|u_0\|_{C(\bar{\Omega})}$, $\|v_0\|_{C(\bar{\Omega})}$

(ii) ε and $\tilde{T}$ are such that

$$\varepsilon = O(\beta\phi\sigma/(\phi + \sigma)),$$

$$\tilde{T} = O(\phi^{-1} + L^{-1})\log\left(\frac{1}{\varepsilon + \nu}\right) + O(L^{-1})\log\left(2 + \frac{\phi}{\sigma} + \frac{\varepsilon}{\nu(\varepsilon + \nu)}\right)$$

$$+ O((\nu L)^{-1})\log\left(2 + \frac{\nu}{\varepsilon} + \frac{\beta\phi^2}{\nu}\right),$$

in the limit (1.12).

(iii)   For all $t \geq \tilde{T}$ we have

$$|\tilde{U}(-d(x),t) - u(x,t)| \leq [\sigma(\varepsilon + \nu)/(\phi + \sigma)] \exp[-\lambda\phi d(x)] \quad \text{if } d(x) < \rho_1/2,$$

$$|V(t) - v(x,t)| \leq \varepsilon + \nu \quad \text{if } x \in \Omega, \qquad |\psi(t)| \leq (\varepsilon + \nu)^2 + \varepsilon/\phi^2,$$

where $d(x)$ is the distance from x to $\partial\Omega$ and ρ_1 is as defined in (H.1). Sketch of the proof: The result in this theorem follows when re-writting eq. (3.10) in terms of the averaged variable $\tilde{U} = S^{-1}(\eta) \int_{H(\eta)} U(s,t)ds$, in the domain Ω_1, where $S(\eta) = \int_{H(\eta)} ds$, Ω_1 is a s defined in (2.1), $H(\eta)$ is the hyper-surface parallel to $\partial\Omega$ at a distance $-\eta$ from $\partial\Omega$, and the results in Lemmata 2.1, 2.2, 3.1-3.5 are taken into account.

4   SUBMODELS OF THE ISOTHERMAL MODEL

In this section we will consider sub-models of the distinguished limit of (3.13)-(3.15). The distinguished limit of (3.13)-(3.15) is

$$\sigma/\phi \to s, \quad \nu L \to V_\Omega \ell/S_\Omega, \quad \beta\phi/\nu \to \lambda,$$

where s, ℓ and λ are strictly positive constants, and $\phi \to \infty$, $L \to \infty$ and $\nu \to 0$. In this case, the approximate sub-model is

$$\partial\tilde{U}/\partial\tau = \partial^2\tilde{U}/\partial\xi^2 - f(\tilde{U},V) \quad \text{in} \quad -\infty < \xi < 0, \tag{4.1}$$

$$\tilde{U} = 0 \quad \text{at} \ \xi = -\infty, \qquad \partial\tilde{U}/\partial\xi = s(1 - \tilde{U}) \quad \text{at} \ \xi = 0, \tag{4.2}$$

$$\ell^{-1}dV/d\tau = 1 - V + \lambda \int_{-\infty}^{0} f(\tilde{U},V)d\xi + \psi_1(\tau), \tag{4.3}$$

where

$$\xi = \phi\eta, \qquad \tau = \phi^2 t, \qquad \psi_1 = \psi/\nu.$$

Notice that, according the estimates in Section 3, $\tilde{U}$ and ψ are such that

$$0 < \tilde{U} < \left[\frac{2s}{(\sqrt{k_2/2m} + s)}\right] \exp(\sqrt{k_2/2m}\,\xi) \text{ if } -\infty < \xi < 0 \text{ and } \tau \geq 0, \tag{4.4}$$

$$|\psi_1(\tau)| \leq \varepsilon_1 = O(\nu) \quad \text{uniformly in } 0 \leq \tau < \infty.$$

The equation (4.3) and the estimate (4.4) define an invariant region of (4.1)-(4.3).

Let us now consider the model (4.1)-(4.3) in the sub-limits $s \to 0$, $s \to \infty$ and $\ell \to 0$.

a) In the limit $s \to 0$, $\tilde{U}$ is small, see the estimate (4.4), and according the assumption (H.5), the reaction term in (H.1) may be written as

$$f(\tilde{U}, V) = f_u(0, V)\tilde{U} + O(|\tilde{U}|^2). \tag{4.5}$$

For λ fixed the right hand side of (3.5) equals $1 - V$ in first approximation and the dynamics of the resulting model is trivial. If $s \to 0$, λ is large and such that

$$s\lambda \to \lambda_1 \neq 0, \infty, \quad \text{with } 0 \neq \ell = \text{fixed}, \tag{4.6}$$

then the model (4.1)-(4.3) may be rewritten as

$$\partial\tilde{U}_1/\partial\tau = \partial^2\tilde{U}_1/\partial\xi^2 - \tilde{U}_1 f_1(V) \tag{4.7}$$
$$+ \psi_2(\xi, \tau)\exp\left(\sqrt{k_2/2m}\,\xi\right) \text{ in } -\infty < \xi < 0,$$
$$\tilde{U}_1 = 0 \quad \text{at } \xi = -\infty, \qquad \partial\tilde{U}_1/\partial\xi = 1 + \psi_3(\tau) \quad \text{at } \xi = 0, \tag{4.8}$$
$$\ell^{-1}dV/d\tau = 1 - V + \lambda_1 f_1(V)\int_{-\infty}^{0} \tilde{U}_1 d\xi + \psi_1(\tau) + \psi_4(\tau), \tag{4.9}$$

where

$$\tilde{U}_1 = \tilde{U}/s, \qquad f_1(V) = f_u(0, V)$$

and, according to (4.4) and (4.5), the remainders ψ_2, ψ_2, ψ_3 and ψ_4 are such that

$$|\psi_2(\xi, \tau)| + |\psi_3(\tau)| + |\psi_4(\tau)| = O(s) \quad \text{uniformly in } -\infty < \xi < 0, \quad \tau \geq 0$$

in the limit $s \to 0$ and (4.6). Then, if the remainders are ignored we obtain an asymptotic model for (4.1)-(4.3).

b) In the limit

$$s \to \infty, \quad \text{with } l \neq 0 \quad \text{and} \quad \lambda \neq 0 \quad \text{fixed},$$

we have $|\partial\tilde{U}(0, \tau)/\partial\xi|$ is uniformly bounded in $\tau \geq 0$, this can be seen by an argument similar to that in the proof of Lemma 3.4. Then the boundary conditions may be written as

$$\tilde{U} = 0 \quad \text{at } \xi = -\infty, \qquad \tilde{U} = 1 + \psi_5(\tau) \quad \text{at } \xi = 0,$$

where ψ_5 is given by

$$\psi_5(\tau) = s^{-1}\partial \tilde{U}(0,\tau)/\partial \xi$$

that is of order s^{-1} uniformly in $\tau \geq 0$. If the remainders are ignored we obtain another asymptotic model of (1.1)-(1.2).

c) Finally we will consider the limit $\ell \to 0$, with s and λ fixed. Let us first replace eq. (4.1) by

$$0 = \partial^2 \tilde{U}/\partial \xi^2 - f(\tilde{U}, V) \quad \text{in} \quad -\infty < \xi < 0. \tag{4.10}$$

For each function $V = V(\tau)$ the equations (4.10) and (4.2) uniquely define

$$\tilde{U} = H(\xi, V). \tag{4.11}$$

By means of maximum principles, and the results in Section 3 we obtain

$$dV/d\tau = 1 - V + \lambda \int_{-\infty}^{0} f(H(\xi, V), V)d\xi + \psi_1(\tau) + \psi_6(\tau),$$

where H is as given in (4.11), and the remainder ψ_6 is small, i.e.

$$|\psi_6(\tau)| = O(\ell) \quad \text{uniformly in} \quad -\infty < \xi < 0, \ \tau > \tau_0,$$

in the limit $\ell \to 0$, s, λ fixed. This sub-model (if the remainders are ignored) cosists of an autonomous ODE that yields trivial dynamics, namely $V(\tau)$ converges to a steady state as $\tau \to \infty$.

5   CONCLUDING REMARKS

We have considered the model (1.1)-(1.2) in the limit (1.12), under the assumptions (H.1)-(H.6). The assumption (H.4) is purely technical; if the inequalities in (H.4) are replaced by $k_2 u^p < f(u,v) < k_1 u^p$ and $u^p g_2(u) \leq f(u,v)$ with $p > 1$, then the results in this paper still apply. The assumption $f'_u > k_3 > 0$ in (H.5) is necessary for some of the results in this paper to apply.

In Section 2 we have first obtained some estimates on the solutions of (1.1)-(1.2) implying that, after some time, the reactant concentration u becomes quite small except in a boundary layer, near the boundary of the domain $\partial\Omega$. In Section 3 we have proved (i) the temperature v becomes approximately spatially constant and (ii) the gradient of u along the hypersurfaces parallel to $\partial\Omega$ becomes small. Then the asymptotic model (1.7)-(1.9) was obtained.

This model consist in $1 - D$ parabolic semilinar equation (1.7) that gives the reactant concentration in the boundary layer, and the ODE (1.9) gives the spatial average of the temperature.

The asymptotic model was analyzed in Section 4 in the distinguished limit and in some representative sublimits. In the limits $s \to 0$ and $s \to \infty$, the mixed boundary condition in (1.8) can be replaced by Neumann and Dirichlet boundary conditions respectively, as intuition suggests. Finally, as $\ell \to 0$ the reactant concentration becomes quasi-steady and the asymptotic model is reduced to an ODE.

REFERENCES

1. Amundson, N.R. and Raymond, L.R. (1965). Stability in distributed parameter systems, *Am.Int. Chem. Engr.* **11**: 339-350.

2. Aris, R. (1975). *The Mathematical Theory of Diffusion and Reaction in Permeable Catalysts, vol. I and II*, Clarendon Press, Oxford.

3. Conway, E., Hoff, D. and Smoller, J. (1978). Large time behavior of solutions of nonlinear reaction-diffusion equations, *SIAM J. Appl. Math.* **35**: 1-16.

4. Guggenheimer, H.W. (1977). *Differential Geometry*, Dover, New York, 1977.

5. Hale, J.K. and Rocha, C. (1987-a). Varying boundary conditions with large diffusivity, *J. Math. Pures Appl.* **66**: 139-158.

6. Hale, J.K. and Rocha, C. (1987-b). Interaction of diffusion and boundary conditions, *Nonlinear Analysis TMA* **11**: 633-649.

7. Hale, J.K. and Sakamoto, K. (1989). Shadow systems and attractors in reaction-diffusion equations, *Appl. Anal.* **32**: 287-303.

8. di Liddo, A. and Maddalena, L. (1992). Mathematical analysis of a chemical reaction with lumped temperature and strong absorption, *J. Math. Anal. Appl.* **163**: 86-103.

9. Mancebo, F.J. and Vega, J.M. (1993). An asymptotic justification of a model of isothermal catalysts, *J. Math. Anal. Appl.* **175**: 523-536.

10. Mancebo, F.J. and Vega, J.M. An asymptotic justification of a non-local 1-D model arising in porous catalyst theory, *to appear in J.D.E.*.

11. Parra, I.E. and Vega, J.M. (1988). Local non-linear stability of the steady state in an isothermal catalysts, *SIAM J. Appl. Math.* **48**: 854-881.

12. Vega, J.M. (1988). Invariant regions and global asymptotic stability in an isothermal catalyst, *SIAM J. Math. Anal.* **19**: 774-796.

On the Reaction-Diffusion Electrolysis Nonlinear Elliptic Equations

S.I.POHOZAEV   Steklov Mathematical Institute, Moscow, Russia

1 INTRODUCTION

H.Amann (1992) proposed a sufficiently general mathematical model for electro-chemical reactions.

The model is described by a system of N nonlinear parabolic equations with non-linear boundary conditions, along with an additional relation—the condition that the solution be electrically neutral. This gives a system of $N + 1$ nonlinear equations with nonlinear boundary conditions for finding the nonnegative concentrations $u_1, u_2, \ldots, u_N$ of the N substances and the electric potential φ.

In Amann (1992) and Amann & Renardy (1994) local solvability theorems were obtained for the corresponding problems both in the time-dependent case and in the steady-state case. In the time-dependent case local solvability was established with respect to the time, while in the steady-state solvability was established for the elliptic equation of the potential φ with nonlinear boundary condition for prescribed nonnegative concentrations $u_1, u_2, \ldots, u_N$ when the total diffusion coefficient is positive.

In this lecture we consider the Amann model in the steady-state case.

In the first part we restrict ourselves to the simplest case $N = 2$ which does not require the general theory of a priori estimates for elliptic equations. Later we shall consider the general situation.

2 STATEMENT OF THE PROBLEM FOR THE SPECIAL CASE N = 2

In the case under consideration the problem is reduced to finding functions $u_1(x)$,

$u_2(x)$, and $\varphi(x)$ satisfying the equations

$$-a_r \Delta u_r - \nabla(\mu_r u_r \nabla \varphi) = 0, \quad u_r(x) \geq 0 \text{ in } \Omega \quad (r = 1, 2) \tag{2.1}$$

$$a_r \partial_\nu u_r + \mu_r u_r \partial_\nu \varphi = g_r(x, u_1, u_2, \varphi) \text{ on } \partial\Omega \quad (r = 1, 2) \tag{2.2}$$

$$z_1 u_1(x) + z_2 u_2(x) = 0 \text{ in } \Omega. \tag{2.3}$$

Here $a_r > 0$, and z_r $(r = 1, 2)$ are constant coefficients, with

$$\text{sign } \mu_r = \text{sign } z_r \quad \text{if } z_r \neq 0, \quad \text{and } \mu_r = 0 \quad \text{if } z_r = 0. \tag{2.4}$$

These coefficients have the following physical meaning: a_r is the diffusion coefficient, μ_r is the electrochemical mobility coefficient, and z_r is the charge of the rth component. Thus, $\mu_r = 0$ if the rth component is uncharged. The vector $\vec{u} = (u_1, u_2)$ is the concentration vector, so that $u_1 \geq 0$, and $u_2 \geq 0$.

The domain Ω is bounded in $\mathbf{R}^n$ with boundary $\partial\Omega$ consisting of several connected components corresponding to the outer wall and to the boundaries of electrodes. It is assumed that each of the components of the boundary is of class C^2. The vector ν is the outer unit normal vector to $\partial\Omega$, and ∂_ν is the derivative in the direction of this vector at a point $x \in \partial\Omega$.

The functions g_1 and g_2 satisfy the conditions

$$g_1(x, 0, u_2, \varphi) = 0, \qquad g_2(x, 0, u_2, \varphi) = 0$$

for $x \in \partial\Omega$ and for various values of $u_1 \geq 0$, $u_2 \geq 0$, and φ under consideration.

Before formulating the following conditions, we consider some possible cases of this problem.

The case $z_1 z_2 > 0$. In this case the problem under consideration has only the trivial solution $u_1(x) \equiv u_2(x) \equiv 0$ in the classes of nonnegative functions $u_1(x)$ and $u_2(x)$. This follows immediately from (2.3) and (2.5).

The case $z_1 = z_2 = 0$. Then $\mu_1 = \mu_2 = 0$ in view of (2.4), and the problem takes the form

$$\begin{cases} -a_r \Delta u_r = 0, \quad u_r(x) \geq 0 & \text{in } \Omega \ (r = 1, 2), \\ a_r \partial_\nu u_2 = g_r(x, u_1, u_2, \varphi) & \text{on } \partial\Omega \ (r = 1, 2). \end{cases} \tag{2.6}$$

Thus, the problem simplifies in the case of uncharged components, but the potential φ remains undetermined.

The case $z_1 = 0$, $z_2 \neq 0$. Then $u_2 \equiv 0$ in view of (2.3). It follows from (2.4) that $m_1 = 0$. Thus, in this case the original problem reduces to the problem

$$\begin{cases} -a_1 \Delta u_1 = 0, \quad u_1(x) \geq 0 & \text{in } \Omega, \\ a_1 \partial_\nu u_1 = g_1(x, u_1, u_2, \varphi) & \text{on } \partial\Omega. \end{cases} \tag{2.7}$$

Consequently, the original problem simplifies in this case too, but the potential φ also remain undetermined. Thus, the most interesting case is the one when

$$z_1 z_2 < 0. \tag{2.8}$$

It is this case of the problem (2.1)–(2.3) that will be treated below.
Setting

$$z = -\frac{z_1}{z_2} > 0, \tag{2.9}$$

we find from (2.3) that

$$u_2(x) = z\, u_1(x). \tag{2.10}$$

Then the problem (2.1)–(2.3) reduces to the determination of functions $u(x)$ and $\varphi(x)$ in $\Omega \in \mathbf{R}^n$ satisfying the relations

$$-\Delta u = 0 \quad \text{in } \Omega, \tag{2.11}$$
$$-\nabla(u\nabla\varphi) = 0 \quad \text{in } \Omega, \tag{2.12}$$
$$\partial_\nu u = h_1(x, u, \varphi) \quad \text{on } \partial\Omega, \tag{2.13}$$
$$u\,\partial_n u\varphi = h_2(x, u, \varphi) \quad \text{on } \partial\Omega. \tag{2.14}$$

Here

$$h_1(x, u, \varphi) = a_{11}g_1(x, u, zu, \varphi) + a_{12}g_2(x, u, zu, \varphi), \tag{2.15}$$
$$h_2(x, u, \varphi) = a_{21}g_1(x, u, zu, \varphi) + a_{22}g_2(x, u, zu, \varphi), \tag{2.16}$$

where

$$a_{11} = -\frac{z_1\mu_2}{\lambda}, \qquad a_{12} = -\frac{z_2\mu_1}{\lambda},$$
$$a_{21} = \frac{z_1 a_2}{\lambda}, \qquad a_{22} = \frac{z_2 a_1}{\lambda},$$
$$\lambda = z_1(\mu_1 a_2 - \mu_2 a_1).$$

The following condition is assumed about the boundary functions g_1 and g_2.

We remark that by virtue of our assumptions

$$a_{11} > 0, \qquad a_{12} > 0, \qquad \text{and } \lambda > 0.$$

The following condition is assumed about the boundary functions g_1 and g_2.

CONDITION (H). *Suppose that in the plane* $\mathbf{R}^2$ *of the independent variables* u *and* φ *there exists a rectangle*

$$D = \{(u, \varphi) \in \mathbf{R}^2 \,|\, 0 < m \leq u \leq M,\ \varphi_1 \leq \varphi \leq \varphi_2\}$$

such that the functions h_1 *and* h_2 *defined by (2.15) and (2.16) satisfy the relations* $h_1, h_2 \in C^1(\partial\Omega \times D)$ *and*

$$(\vec{h}, \vec{N}) < 0 \tag{2.17}$$

for all $x \in \partial\Omega$ *and* $(u, p) \in \partial\Omega$ *away from the corner points of* D.

Here $(\,.\,,\,.\,)$ is the inner product in $\mathbf{R}^2$, $\vec{h} = (h_1, h_2)$, and $\vec{N}$ is the outer unit normal vector to the boundary $\partial\Omega$ of the rectangle D except at the corner points, where the normal $\vec{N}$ is not defined.

3 MAIN THEOREM

THEOREM 3.1. *Let $z_1, z_2 < 0$. Assume the condition (H). Then the problem (2.1)–(2.3) with (2.4) has a solution*

$$(u_1, u_2, \varphi) \in W_p^2(\Omega) \times W_p^2(\Omega) \times W_p^2(\Omega) \quad \text{with } p > n,$$

and

$$0 < m \le u_1 \le M, \qquad 0 < z\,m \le u_2 \le z\,M, \qquad \varphi_1 \le \varphi \le \varphi_2$$

for all $x \in \bar{\Omega}$.

The proof of the theorem is based on a priori estimates and uses the Leray-Schauder method.

This proof is given in the sections to follow.

4 LEMMAS ON EXTENSION

We consider special extensions of the functions h_1 and h_2 satisfying the condition (H).

The following notation is used for the sides of the rectangle $D \subset \mathbf{R}^2$:

$$\partial D_1^- = \{(u, \varphi) \in \mathbf{R}^2 \,|\, u = m, \varphi_1 \le \varphi \le \varphi_2\},$$
$$\partial D_1^+ = \{(u, \varphi) \in \mathbf{R}^2 \,|\, u = M, \varphi_1 \le \varphi \le \varphi_2\},$$
$$\partial D_2^- = \{(u, \varphi) \in \mathbf{R}^2 \,|\, m \le u \le M, \varphi = \varphi_1\},$$
$$\partial D_2^+ = \{(u, \varphi) \in \mathbf{R}^2 \,|\, m \le u \le M, \varphi = \varphi_2\},$$

Let

$$D_{1,m}^- = \{(u, \varphi) \in \mathbf{R}^2 \,|\, u \le m, \}, \qquad D_{1,M}^+ = \{(u, \varphi) \in \mathbf{R}^2 \,|\, u \ge M, \},$$
$$D_{2,\varphi_1}^- = \{(u, \varphi) \in \mathbf{R}^2 \,|\, \varphi \le \varphi_1, \}, \qquad D_{2,\varphi_2}^+ = \{(u, \varphi) \in \mathbf{R}^2 \,|\, \varphi \ge \varphi_2, \}.$$

LEMMA 4.1. *There exists a function $\tilde{h}_1(x, u, \varphi)$ defined on $\bar{\Omega} \times \mathbf{R}^2$ such that*

$$\begin{aligned} &(i) && \tilde{h}_1 \in C^1(\bar{\Omega} \times \mathbf{R}^2), \\ &(ii) && \tilde{h}_1 > 0 \quad \text{on } \partial\Omega \times D_{1,m}^-, \\ &(iii) && \tilde{h}_1 < 0 \quad \text{on } \partial\Omega \times D_{1,M}^+, \\ &(iv) && \tilde{h}_1 = h_1 \quad \text{on } \partial\Omega \times D. \end{aligned}$$

LEMMA 4.2. *There exists a function $\tilde{h}_2(x, u, \varphi)$ defined on $\bar{\Omega} \times \mathbf{R}^2$ such that*

$$\begin{aligned} &(i) && \tilde{h}_2 \in C^1(\bar{\Omega} \times \mathbf{R}^2), \\ &(ii) && \tilde{h}_2 > 0 \quad \text{on } \partial\Omega \times D_{2,\varphi_1}^-, \\ &(iii) && \tilde{h}_2 < 0 \quad \text{on } \partial\Omega \times D_{2,\varphi_2}^+, \\ &(iv) && \tilde{h}_1 = h_2 \quad \text{on } \partial\Omega \times D. \end{aligned}$$

Proof of Lemma 4.1. By assumption, the function $h_1(x, u, \varphi)$, which is defined on $\partial\Omega \times D$, belongs to the class $C^1(\partial\Omega \times D)$, and

$$h_1 > 0 \quad \text{on } \partial\Omega \times D_1^-, \tag{4.1}$$

$$h_1 < 0 \quad \text{on } \partial\Omega \times D_1^+. \tag{4.2}$$

In view of the inequality (4.1) for the C^1-function h_1, there exist constants $\varepsilon_1 > 0$ and $\delta_1 > 0$ such that

$$h_1(x, u, \varphi) \geq \varepsilon_1 \quad \text{on } \partial\Omega \times [m, m + \delta_1] \times [\varphi_1, \varphi_2].$$

On this set we let

$$f_1(x, u, \varphi) = \ln h_1(x, u, \varphi).$$

Then on this set f_1 belongs to the class C^1, and there exists (Fikhtengol'ts) a C^1-extension F_1 of this function to

$$\partial\Omega \times D_{1,m+\delta_1}^-, \qquad D_{1,m+\delta_1}^- = \{(u, \varphi) \in \mathbf{R}^2 \mid u \leq m + \delta_1\}.$$

Let $H_1(x, u, \varphi) = \exp F_1(x, u, \varphi)$. Then $H_1 > 0$ is a C^1-extension of h_1 to $\partial\Omega \times D_{1,m+\delta_1}^-$. In view of the inequality (4.2) there exists for the C^1-function h_1 constants $\varepsilon_2 > 0$ and $\delta_2 > 0$ such that

$$-h_1(x, u, \varphi) \geq \varepsilon_2 \quad \text{on } \partial\Omega \times [M - \delta_2, M] \times [\varphi_1, \varphi_2].$$

On this set we let

$$f_2(x, u, \varphi) = \ln(-h_1(x, u, \varphi)).$$

Then on this set f_2 belongs to the class C^1, and there exists a C^1-extension F_2 of this function to

$$\partial\Omega \times D_{1,M-\delta_2}^+, \qquad D_{1,M-\delta_2}^- = \{(u, \varphi) \in \mathbf{R}^2 \mid u \geq M - \delta_2\}.$$

Let $H_2(x, u, \varphi) = -\exp F_2(x, u, \varphi)$. Then $H_2 < 0$ is a C^1-extension of h_1 to $\partial\Omega \times D_{1,M-\delta_2}^+$.

Let us now consider the function

$$H(x, u, \varphi) = \begin{cases} H_1(x, u, \varphi) & \text{on } \partial\Omega \times D_{1,m}^-, \\ h_1(x, u, \varphi) & \text{on } \partial\Omega \times D, \\ H_2(x, u, \varphi) & \text{on } \partial\Omega \times D_{1,M}^+. \end{cases}$$

On the closed set $(\partial\Omega \times D_{1,m}^-) \cup (\partial\Omega \times D) \cap (\partial\Omega \times D_{1,M}^+)$ this function belongs to the class C^1 and admits (Fikhtengol'ts) a C^1-extension $\tilde{H}(x, u, \varphi)$ to $\partial\Omega \times \mathbf{R}^2$.

Next, since the boundary $\partial\Omega$ is of class C^2, $\tilde{H}$ automatically admits a C^1-extension $\tilde{h}_1$ to $\bar{\Omega} \times \mathbf{R}^2$. By construction, $\tilde{h}_1$ satisfies all the conditions of Lemma 4.1. Lemma 4.1 is proved.

The proof of Lemma 4.2 is similar.

5 THE FIRST PROBLEM

Let us consider the following boundary value problem with parameter $t \in [0, 1]$:

$$\begin{cases} -\Delta u = 0 & \text{in } \Omega, \\ \partial_\nu u + (1 - t)u = (1 - t)u_0 + t\tilde{h}_1(x, u, \varphi) & \text{on } \partial\Omega \end{cases} \qquad (5.1)$$

for $\varphi \in W_p^1(\Omega)$ with $p > n$. here u_0 is a constant such that

$$m < u_0 < M \qquad (5.2)$$

and $\tilde{h}_1$ is the function in Lemma 4.1.

LEMMA 5.1. *Suppose that u is a $W_p^2(\Omega)$-solution of the problem (5.1) for $\varphi \in W_p^1(\Omega)$ with $p > n$. Then for any $t \in [0, 1]$*

$$m \le u(x) \le M \quad \text{in } \Omega \qquad (5.3)$$

Proof. We prove the lower estimate. Let $v(x) = u(x) - m$. Then the function $v \in W_p^2(\Omega)$ is a solution of the problem

$$\begin{cases} -\Delta v = 0 & \text{in } \Omega, \\ \partial_\nu v + (1 - t)v = (1 - t)(u_0 - m) + t\tilde{h}_1(x, m + v, \varphi) & \text{on } \partial\Omega. \end{cases} \qquad (5.4)$$

From this,

$$\int_\Omega \nabla v \nabla \zeta \, dx - (1 - t)(u_0 - m) \int_{\partial\Omega} \zeta \, ds + (1 - t) \int_{\partial\Omega} v\zeta \, ds$$
$$- t \int_{\partial\Omega} \tilde{h}_1(x, m + v, \varphi)\zeta \, ds = 0 \quad (5.5)$$

for any function $\zeta \in W_p^1(\Omega)$. For $v \in W_p^2(\Omega)$, and hence $v \in W_p^1(\Omega)$, we set

$$\zeta = v^- = \frac{|v| - v}{2} \ge 0.$$

Then (Gilbarg & Trudinger, Lemma 7.6, and Bourdaud & Meyer) we have $v^- \in W_p^1(\Omega)$, and $\nabla v^- = -\nabla v \chi[v < 0]$. Consequently, for $\zeta = v^-$ the relation (5.5) takes the form

$$- \int_\Omega |\nabla v^-|^2 \, dx - (1 - t)(u_0 - m) \int_{\partial\Omega} v^- \, ds - (1 - t) \int_{\partial\Omega} |v^-|^2 \, ds$$
$$- t \int_{\partial\Omega} \tilde{h}_1(x, m - v^-, \varphi)v^- \, ds = 0. \quad (5.6)$$

From this we get for $t = 0$ that

$$- \int_\Omega |\nabla v^-|^2 \, dx - (u_0 - m) \int_{\partial\Omega} v^- \, ds - \int_{\partial\Omega} |v^-|^2 \, ds = 0.$$

Then by (5.2),

$$v^- \equiv 0 \quad \text{for } t = 0. \tag{5.7}$$

From (5.6) we get for $t = 1$ that

$$-\int_\Omega |\nabla v^-|^2 \, dx - \int_{\partial\Omega} \tilde{h}_1(x, m - v^-, \varphi) v^- \, ds = 0.$$

By Lemma 4.1 (the property (ii)), this gives us that

$$v^- \equiv 0 \quad \text{for } t = 1.$$

By virtue of the same property (ii) of the function $\tilde{h}_1$ in Lemma 4.1 we conclude on the basis of (5.6) that

$$v^- \equiv 0 \quad \text{for } 0 < t \leq 1. \tag{5.8}$$

Thus, it follows from (5.7) and (5.8) that $v(x) \geq 0$ in the domain Ω for all $t \in [0, 1]$, and hence $u(x) \geq m$ in Ω all $t \in [0, 1]$. Similarly, we use the property (iii) of the function $\tilde{h}_1$ in Lemma 4.1, to establish the upper estimate

$$u(x) \leq M \quad \text{in } \Omega \quad \text{for all } t \in [0, 1].$$

Lemma 5.1 is proved.

6 THE SECOND PROBLEM

Let us consider the following boundary value problem with parameter $t \in [0, 1]$:

$$\begin{cases} -\nabla[(1 - t + tu)\nabla\varphi] = 0 & \text{in } \Omega, \\ (1 - t + tu)\partial_\nu\varphi + (1 - t)\varphi = (1 - t)\varphi_0 + t\tilde{h}_2(x, u, \varphi) & \text{on } \partial\Omega \end{cases} \tag{6.1}$$

for $u \in W_p^1(\Omega)$ with $p > n$ and

$$0 < m \leq u(x) \leq M \quad \text{in } \Omega. \tag{6.2}$$

Here φ_0 is a constant such that

$$\varphi_1 < \varphi_0 < \varphi_2 \tag{6.3}$$

and the function $\tilde{h}_2$ is from Lemma 5.2.

We remark that here the constants m and M and the constants φ_1 and φ_2 are from the condition (H) of the main theorem.

LEMMA 6.1. *Suppose that φ is a $W_p^2(\Omega)$-solution of the problem (6.1) for $u \in W_p^1(\Omega)$, $p > n$, under the conditions (6.2) and (6.3). Then for any $t \in [0, 1]$*

$$\varphi_1 < \varphi_0 < \varphi_2 \quad \text{in } \Omega. \tag{6.4}$$

Proof. We prove the lower estimate.

Let $\psi = \varphi(x) - \varphi_1$. Then the function $\psi \in W_p^2(\Omega)$ is a solution of the problem

$$
\begin{cases}
-\nabla[(1 - t + tu)\nabla\psi] = 0 \quad \text{in } \Omega, \\
(1 - t + tu)\partial_\nu\psi + (1 - t)\psi = (1 - t)(\varphi_0 - \varphi_1) + t\tilde{h}_2(x, u, \varphi_1 + \psi) \text{ on } \partial\Omega
\end{cases}
$$

for $u \in W_p^1(\Omega)$ with $p > n$ under conditions (6.2) and (6.3).

From this,

$$
\int_\Omega (1 - t + tu)\nabla\psi\nabla\zeta \, dx + (1 - t)\int_{\partial\Omega} \psi\zeta \, ds
$$

$$
- (1 - t)(\varphi_0 - \varphi_1)\int_{\partial\Omega} \zeta \, ds - t\int_{\partial\Omega} \tilde{h}_1(x, u, \varphi_1 + \psi)\zeta \, ds = 0
$$

for any function $\zeta \in W_p^1(\Omega)$. For $\psi \in W_p^2(\Omega)$, and hence for $\psi \in W_p^1(\Omega)$, we let

$$
\zeta = \psi^- = \frac{|\psi| - \psi}{2} \geq 0.
$$

Then $\psi^- \in W_p^1(\Omega)$ and $\nabla\psi^- = -\nabla\psi\chi[\psi < 0]$. Consequently, for $\zeta = \psi^-$ the relation (6.5) takes the form

$$
-\int_\Omega (1 - t + tu)|\nabla\psi^-|^2 \, dx - (1 - t)\int_{\partial\Omega} |\psi^-|^2 \, ds
$$

$$
- (1 - t)(\varphi_0 - \varphi_1)\int_{\partial\Omega} \psi^- \, ds - t\int_{\partial\Omega} \tilde{h}_2(x, u, \varphi_1 - \psi^-)\psi^- \, ds = 0. \quad (6.6)
$$

From this we get for $t = 0$ that

$$
-\int_\Omega |\nabla\psi^-|^2 \, dx - \int_{\partial\Omega} |\psi^-|^2 \, ds - (\varphi_0 - \varphi_1)\int_{\partial\Omega} \psi^- \, ds = 0.
$$

Then by (6.3) we have that $\psi^-(x) \equiv 0$ in Ω for $t = 0$. From (6.6) we get for $t = 1$ that

$$
-\int_\Omega u|\nabla\psi^-|^2 \, dx - \int_{\partial\Omega} \tilde{h}_2(x, u, \varphi_1 - \psi^-)\psi^- \, ds = 0.
$$

By (6.2) and the property (ii) of the function $\tilde{h}_2$ in Lemma 4.2 , this gives us that $\psi^- \equiv 0$ in Ω for $t = 1$. Further, by (6.2), (6.3), and the cited property (ii) of the function $\tilde{h}_2$ in Lemma 4.2, it follows from (6.6) that $\psi^- \equiv 0$ in Ω for any $t \in [0, 1]$. Thus, $\varphi(x) \geq \varphi_1$ in Ω for any $t \in [0, 1]$.

In the same way we establish the upper estimate $\varphi(x) \leq \varphi_2$ in Ω for any $t \in [0, 1]$ on the basis of the property (iii) of the function $\tilde{h}_2$ in Lemma 4.2.

Lemma 6.1 is proved.

7 THE GENERAL PROBLEM

Let us now consider the following problem with parameter $t \in [0, 1]$:

$$
-\Delta u = 0 \quad \text{in } \Omega, \tag{7.1}
$$

$$
\partial_\nu u + (1 - t)u = (1 - t)u_0 + t\tilde{h}_1(x, u, \varphi) \quad \text{on } \partial\Omega \tag{7.2}
$$

$$
-\nabla[(1 - t + tu)\nabla\varphi] = 0 \quad \text{in } \Omega, \tag{7.3}
$$

$$
(1 - t + tu)\partial_\nu\varphi + (1 - t)\varphi = (1 - t)\varphi_0 + t\tilde{h}_2(x, u, \varphi) \quad \text{on } \partial\Omega. \tag{7.4}
$$

Here $\tilde{h}_1$ and $\tilde{h}_2$ are the functions in Lemma 4.1 and 4.2, respectively, and u_0 and φ_0 are constants satisfying the inequalities (5.2) and (6.3), respectively.

We consider a solution (u, φ) of the system (7.1)–(7.4) in the space $W_p^2(\Omega) \times W_p^2(\Omega)$ with $p > n$.

LEMMA 7.1. *Let* $(u, \varphi) \in W_p^2(\Omega) \times W_p^2(\Omega)$ $(p > n)$ *be a solution of the system* (7.1)–(7.4). *Then*

$$0 < m \le u(x) \le M, \quad \varphi_1 \le \varphi(x) \le \varphi_2 \quad \text{in } \Omega \quad \text{for any } t \in [0, 1], \qquad (7.5)$$

and there exists a constant A_1 *such that*

$$\|u\|_{2,p} \le A_1(\|\varphi\|_{2,p}^{1/2} + 1) \quad \text{for any } t \in [0, 1]. \qquad (7.6)$$

Here the positive constant A_1 *does not depend on the solution* $(u, \varphi) \in W_p^2(\Omega) \times W_p^2(\Omega)$, $A_1 = A_1(m, M, \varphi_1, \varphi_2, p, \Omega)$, *and* $\| \cdot \|_{k,p} = \| \cdot \|_{W_p^k(\Omega)}$.

Proof. Let (u, φ) be a solution of the system under consideration in the space $W_p^2(\Omega) \times W_p^2(\Omega)$ $(p > n)$.

Then the function $u \in W_p^2(\Omega)$ is a solution of the boundary value problem (7.1)–(7.2) for some function $\varphi \in W_p^2(\Omega)$, and it satisfies the inequality (5.3) in view of Lemma 5.1, independently of this function $\varphi \in W_p^2(\Omega)$ $(p > n)$.

Similarly, we get the inequality (6.4) for $\varphi(x)$ by virtue of Lemma 6.1.

Further, by the familiar linear theory (Agmon, Douglis & Nirenberg and Amann (1993)),

$$\|u\|_{2,p} \le c_1 \left[(1 - t) \|u_0\|_{W_p^{1-1/p}(\partial\Omega)} + t \|\tilde{h}_1\|_{W_p^{1-1/p}(\partial\Omega)} + \|u\|_{0,p} \right]$$
$$\le c_1 (\|\tilde{h}_1\|_{W_p^{1-1/p}(\partial\Omega)} + M), \quad (7.7)$$

where the constant c_1 is independent of the solution $u(x)$ and the function $\tilde{h}_1 = \tilde{h}_1(x, u(x), \varphi(x))$.

The well-known theory of Sobolev-Besov function spaces (Besov, Il'in & Nikol'skii; see also Agmon, Douglis & Nirenberg, §14, (14.2)) tells us that for this function

$$\|\tilde{h}_1\|_{W_p^{1-1/p}(\partial\Omega)} \le c_2 \|\tilde{h}_1\|_{W_p^1(\Omega)}, \qquad (7.8)$$

where the constant $c_2 > 0$ does not depend on $\tilde{h}_1$.

Further, we have

$$\|\tilde{h}_1\|_{W_p^1(\Omega)} \le c_3 \left(\|\nabla_x \tilde{h}_1\|_{L_p(\Omega)} + \left\| \frac{\partial \tilde{h}_1}{\partial u} \nabla u \right\|_{L_p(\Omega)} + \left\| \frac{\partial \tilde{h}_1}{\partial \varphi} \nabla \varphi \right\|_{L_p(\Omega)} + \|\tilde{h}_1\|_{L_p(\Omega)} \right)$$
$$\le c_4 (\|\nabla u\|_{L_p(\Omega)} + \|\nabla \varphi\|_{L_p(\Omega)} + 1), \quad (7.9)$$

where $c_4 = c_4(m, M, \varphi_1, \varphi_2, p, \Omega)$ is a positive constant.

By the Gagliardo-Nirenberg interpolation inequalities (Nirenberg),

$$\|\nabla u\|_{L_p(\Omega)} \leq \text{const}(\|u\|^{1/2}_{W^2_p(\Omega)} \|u\|^{1/2}_{L_p(\Omega))} + \|u\|_{L_p(\Omega)}),$$
$$\|\nabla \varphi\|_{L_p(\Omega)} \leq \text{const}(\|\varphi\|^{1/2}_{W^2_p(\Omega)} \|\varphi\|^{1/2}_{L_p(\Omega))} + \|\varphi\|_{L_p(\Omega)}), \tag{7.10}$$

where const=const(p, Ω).

Then by virtue of the inequalities (7.7), (7.8), and (7.9) we get that for any $t \in [0, 1]$

$$\|u\|_{2,p} \leq c_5(\|u\|^{1/2}_{2,p} + \|\varphi\|^{1/2}_{2,p} + 1),$$

where $c_5 = c_5(m, M, \varphi_1, \varphi_2, p, \Omega)$ is a positive constant. The inequality (7.6) now follows immediately from this.

Lemma 7.1 is proved.

LEMMA 7.2. *Suppose that* $(u, \varphi) \in W^2_p(\Omega) \times W^2_p(\Omega)$ $(p > n)$ *is a solution of the system (7.1)–(7.4). Then there exists a constant* A_2 *such that*

$$\|\varphi\|_{2,p} \leq A_2(\|u\|^{1/2}_{2,p}\|\varphi\|^{1/2}_{2,p} + \|u\|^{1/2}_{2,p} + 1) \tag{7.11}$$

for any $t \in [0, 1]$. *Here the positive constant* A_2 *does not depend on the solution*

$$(u, \varphi) \in W^2_p(\Omega) \times W^2_p(\Omega), \qquad A_2 = A_2(m, M, \varphi_1, \varphi_2, p, \Omega).$$

Proof. Let $k(t, u) = 1 - t + tu(x)$. By the inequality (7.5), we have for u that

$$0 < m_0 \leq k(t, u) \leq M_0 \quad \text{in } \Omega \quad \text{for any } t \in [0, 1], \tag{7.12}$$

where $m_0 = \min\{1, m\}$ and $M_0 = \max\{1, M\}$.

We rewrite the boundary value problem (7.3)–(7.4) in the form

$$-\Delta\varphi = \frac{t}{k(t, u)}\nabla u \nabla \varphi \quad \text{in } \Omega, \tag{7.13}$$

$$\partial_\nu \varphi = \frac{t}{k(t, u)}\tilde{h}_2(x, u, \varphi) - \frac{1 - t}{k(t, u)}(\varphi - \varphi_0) \quad \text{on } \partial\Omega. \tag{7.14}$$

In view of the familiar theory of elliptic problems,

$$\|\varphi\|_{W^2_p(\Omega)} \leq \text{const}\left[t \left\|\frac{\nabla u \nabla \varphi}{k(t, u)}\right\|_{L_p(\Omega)} + \|\varphi\|_{L_p(\Omega)} + \right.$$
$$\left. t \left\|\frac{\tilde{h}_2}{k(t, u)}\right\|_{W^{1-1/p}_p(\partial\Omega)} + (1 - t) \left\|\frac{\varphi - \varphi_0}{k(t, u)}\right\|_{W^{1-1/p}_p(\partial\Omega)}\right],$$

where the positive constant does not depend on the functions under consideration in the indicated classes.

By the well-known theory of Sobolev-Besov function spaces, this gives us

$$\|\varphi\|_{W_p^2(\Omega)} \leq \mathrm{const}_1 \left[\left\|\frac{\nabla u \nabla \varphi}{k(t,u)}\right\|_{L_p(\Omega)} + \|\varphi\|_{L_p(\Omega)} + \right.$$
$$\left. \left\|\frac{\tilde{h}_2}{k(t,u)}\right\|_{W_p^1(\Omega)} + \left\|\frac{\varphi - \varphi_0}{k(t,u)}\right\|_{W_p^1(\Omega)} \right], \quad (7.15)$$

where the positive constant does not depend on the solution $(u,\varphi) \in W_p^2(\Omega) \times W_p^2(\Omega)$ under consideration $(p > n)$.

We now estimate the terms on the right-hand side of the inequality (7.15).

Estimation of $\left\|\frac{\nabla u \nabla \varphi}{k(t,u)}\right\|_{L_p(\Omega)}$. By the inequality (7.12),

$$\left\|\frac{\nabla u \nabla \varphi}{k(t,u)}\right\|_{L_p(\Omega)} \leq \frac{1}{m_0} \|\nabla u \nabla \varphi\|_{L_p(\Omega)}.$$

Then use of the Gagliardo-Nirenberg inequality (Nirenberg) gives us that

$$\left\|\frac{\nabla u \nabla \varphi}{k(t,u)}\right\|_{L_p(\Omega)} \leq \frac{1}{m_0} \|\nabla u\|_{L_{2p}(\Omega)} \|\nabla \varphi\|_{L_{2p}(\Omega)}$$
$$\leq \frac{\mathrm{const}_2}{m_0} \left(\|u\|_{W_p^2(\Omega)}^{1/2} \|u\|_{L_\infty(\Omega)}^{1/2} + \|u\|_{L_\infty(\Omega)} \right)$$
$$\left(\|\varphi\|_{W_p^2(\Omega)}^{1/2} \|\varphi\|_{L_\infty(\Omega)}^{1/2} + \|\varphi\|_{L_\infty(\Omega)} \right).$$

From this, in view of the inequalities (7.5),

$$\left\|\frac{\nabla u \nabla \varphi}{k(t,u)}\right\|_{L_p(\Omega)} \leq c_6 (\|u\|_{W_p^2(\Omega)}^{1/2} + 1)(\|\varphi\|_{W_p^2(\Omega)}^{1/2} + 1), \quad (7.16)$$

where $c_6 = c_6(m, M, \varphi_1, \varphi_2, p, \Omega)$.

Estimation of $\left\|\frac{\tilde{h}_2}{k(t,u)}\right\|_{W_p^1(\Omega)}$. On the basis of (7.5) and the fact that $\tilde{h}_2 \in C^1(\bar{\Omega} \times \mathbf{R}^2)$,

$$\left\|\frac{\tilde{h}_2}{k(t,u)}\right\|_{W_p^1(\Omega)} \leq c_7 (\|\nabla u\|_{L_p(\Omega)} + \|\nabla \varphi\|_{L_p(\Omega)} + 1),$$

where $c_7 = c_7(m, M, \varphi_1, \varphi_2, p, \Omega)$. By the Gagliardo-Nirenberg inequalities (7.10), we get that

$$\left\|\frac{\tilde{h}_2}{k(t,u)}\right\|_{W_p^1(\Omega)} \leq c_8 (\|u\|_{W_p^2(\Omega)}^{1/2} + \|\varphi\|_{W_p^2(\Omega)}^{1/2} + 1), \quad (7.17)$$

where $c_8 = c_8(m, M, \varphi_1, \varphi_2, p, \Omega)$.

Estimation of $\left\|\frac{\varphi - \varphi_0}{k(t,u)}\right\|_{W_p^1(\Omega)}$. As in the preceding estimation, we get

$$\left\|\frac{\varphi - \varphi_0}{k(t,u)}\right\|_{W_p^1(\Omega)} \leq c_9 (\|u\|_{W_p^2(\Omega)}^{1/2} + \|\varphi\|_{W_p^2(\Omega)}^{1/2} + 1), \quad (7.18)$$

where $c_9 = c_9(m, M, \varphi_1, \varphi_2, p, \Omega)$.

We continue the proof of Lemma 7.2. Let us substitute the estimates (7.16), (7.17) and (7.18) obtained into the inequality (7.15). then we get that

$$\|\varphi\|_{W_p^2(\Omega)} \leq c_{10}(\|u\|_{W_p^2(\Omega)}^{1/2}\|\varphi\|_{W_p^2(\Omega)}^{1/2} + \|u\|_{W_p^2(\Omega)}^{1/2} + \|\varphi\|_{W_p^2(\Omega)}^{1/2} + 1),$$

where $c_{10} = c_{10}(m, M, \varphi_1, \varphi_2, p, \Omega)$. The inequality (7.11) follows immediately from this.

Lemma 7.2 is proved.

THEOREM 7.1. *Let* $(u, \varphi) \in W_p^2(\Omega) \times W_p^2(\Omega)$ $(p > n)$ *be a solution of the system (7.1)–(7.4). Then there exists a constant* R *such that*

$$\|u\|_{W_p^2(\Omega)} + \|\varphi\|_{W_p^2(\Omega)} \leq R \tag{7.19}$$

for any $t \in [0, 1]$. *Here* $R = R(m, M, \varphi_1, \varphi_2, p, \Omega)$.

The proof follows immediately from Lemmas 7.1 and 7.2, since the inequality (7.19) is a consequence of the inequalities (7.6) and (7.11).

8 THE OPERATOR T_t

Let $X = W_p^1(\Omega) \times W_p^1(\Omega)$ with $p > n$, and let

$$X_1 = \{(v, \psi) \in X^p \mid 0 < \frac{m}{2} \leq v(x) \leq 2M \quad \text{in } \Omega\}.$$

With the nonlinear problem (7.1)–(7.4) parameterized by $t \in [0, 1]$ we associate the operator $T_t : X_1 \to X$ defined by the relations

$$u = T_{1,t}(v, \psi), \qquad \varphi = T_{2,t}(v, \psi),$$

where u is a solution of the problem

$$\begin{cases} -\Delta u = 0 & \text{in } \Omega, \\ \partial_\nu u + u = (1 - t)u_0 + tv + t\tilde{h}_1(x, v, \psi) & \text{on } \partial\Omega, \end{cases} \tag{8.1}$$

and φ is a solution of the problem

$$\begin{cases} -\nabla[k(t, v)\nabla\varphi] = 0 & \text{in } \Omega, \\ k(t, v)\partial_\nu\varphi + k(t, v)\varphi = (1 - t)(\varphi_0 - \psi) + k(t, v)\psi + t\tilde{h}_2(x, u, \psi) & \text{on } \partial\Omega \end{cases} \tag{8.2}$$

Here $k(t, v) = 1 - t + tv$, where $v \in W_p^1(\Omega)$ $(p > n)$, and

$$0 < \frac{m}{2} \leq v(x) \leq 2m \quad \text{in } \Omega. \tag{8.3}$$

Obviously, the operator $T_{1,t}$ acts from X to $W_p^2(\Omega)$ and is continuous. Consequently, since $W_p^2(\Omega)$ is compactly imbedded in $W_p^1(\Omega)$, the operator $T_{1,t} : X \to W_p^1(\Omega)$ is compact.

We now prove that the operator $T_{2,t}$ is well defined under the condition (8.3). To do this we rewrite the problem (8.2) in the form

$$\begin{cases} -\Delta\varphi = \dfrac{t}{k(t,v)}\nabla v \nabla\varphi & \text{in } \Omega, \\ \partial_\nu\varphi + \varphi = \Phi_*(t,x,v,\psi) & \text{on } \partial\Omega, \end{cases} \tag{8.4}$$

where

$$\Phi_*(t,x,v,\psi) = \frac{1}{k(t,v)}\Phi(t,x,v,\psi)$$

and $0 < m_1 \le k(t,v) \le M_1 = \max\{1, 2M\}$ under the condition (8.3). The familiar linear theory is applicable for this theorem when the functions v and ψ are sufficiently smooth, where v satisfies the condition (8.3). By virtue of this theory, the problem has a unique solution $\varphi \in W_p^2(\Omega)$ for such functions v and ψ.

Let us now pass to the closure of the smooth functions v and ψ in the space $W_p^1(\Omega)$ with $p > n$, where v satisfies the condition (8.3). Then we get the following assertion.

LEMMA 8.1. *For any pair* $(v,\psi) \in X$ *of functions, where* v *satisfies the inequality* (8.3), *there exists a unique solution* $\varphi \in W_p^2(\Omega)$ $(p > n)$ *of the problem* (8.2).

Thus, under the condition (8.3) the operator $T_{2,t}: X_1 \to W_p^2(\Omega)$ is defined:

$$\varphi = T_{2,t}(v,\psi).$$

On the basis of the linear theory, this operator is continuous with respect to (v,ψ) in $W_p^1(\Omega) \times W_p^1(\Omega)$ $(p > n)$ under the condition (8.3).

Under this condition the operator under consideration is also locally bounded and continuous with respect to $(v,\psi,t) \in W_p^1(\Omega) \times W_p^1(\Omega) \times [0,1]$ $(p > n)$.

By the Kondrashev-Sobolev theorem on compactness of the imbedding $W_p^2(\Omega) \to W_p^1(\Omega)$, we get that the operator $T_{2,t}: X_1 \to W_p^1(\Omega)$ is compact.

Accordingly, the operator $T_t: X_1 \to X$ introduced above is compact.

9 EXISTENCE OF A SOLUTION

We now consider the fixed point of the operator T_t. Let

$$B_{R_1} = \{(v,\psi) \in X \mid \|v\|_{W_p^1(\Omega)} + \|\psi\|_{W_p^1(\Omega)} \le R_1\} \quad \text{with } R_1 = c_0 R + 1.$$

Here the constants m and M are from Lemma 7.1, and c_0 is from the inequality $\|\cdot\|_{W_p^1(\Omega)} \le c_0\|\cdot\|_{W_p^2(\Omega)}$ from the imbedding $W_p^2(\Omega) \to W_p^1(\Omega)$ $(c_0 = 1$ under the natural norming of the space $W_p^2(\Omega)$).

Let $Q = B_{R_1} \cap X_1$. then Q is a closed convex set.

LEMMA 9.1. *Int* $Q \ne \emptyset$.

Proof. Any fixed point $(u,\varphi) \in X$ of the operator (with $t \in [0,1]$ belongs to the space $W_p^2(\Omega) \times W_p^2(\Omega)$ $(p > n)$ and satisfies the inequality (7.19) by Theorem 7.1. On the other hand, for $t = 0$ such a fixed point is a solution of the problem (7.1)–(7.4) for $t = 0$, that is,

$$\begin{cases} -\Delta u = 0 & \text{in } \Omega, \\ \partial_\nu u + u = u_0 & \text{on } \partial\Omega, \\ -\Delta\varphi = 0 & \text{in } \Omega \\ \partial_\nu\varphi + \varphi = \varphi_0 & \text{on } \partial\Omega. \end{cases} \tag{9.1}$$

Here u_0 and φ_0 are the constants in Section 7. There is obviously a solution of this problem: $(u, \varphi) = (u_0, \varphi_0)$. By Theorem 7.1, this solution (u_0, φ_0) satisfies the inequality (7.19), and hence

$$\|u_0\|_{W_p^1(\Omega)} + \|\varphi_0\|_{W_p^1(\Omega)} \le c_0 R < c_0 + 1 = R_1.$$

On the other hand, by the inequality (5.2) for u_0, the inequality (6.3) for φ_0, and the continuous imbedding $W_p^1(\Omega) \to C(\bar\Omega)$ for $p > n$, we get that $(u_0, \varphi_0) \in \text{Int } Q$.
  Lemma 9.1 is proved.

LEMMA 9.1. *For any $t \in [0,1]$ there exists a solution $(u, \varphi) \in W_p^2(\Omega) \times W_p^2(\Omega)$ of the problem (7.1)–(7.4) satisfying the inequalities*

$$0 < m \le u(x) \le M, \qquad \varphi_1 \le \varphi \le \varphi_2, \tag{9.2}$$

$$\|u\|_{W_p^2(\Omega)} + \|\varphi\|_{W_p^2(\Omega)} \le R \tag{9.3}$$

Here the constants M and m and the constants φ_1 and φ_2 are from Lemma 7.1 and R is from Theorem 7.1.

Proof. We consider the family of compact operators $T_t : Q \to X$ for $t \in [0,1]$. The boundary ∂Q of the closed convex bounded set Q with $\text{int } Q \ne \emptyset$ does not contain a fixed point $(u, \varphi) \in \partial Q$ of T_t for any $t \in [0,1]$.
  Since $\text{Im } T_t(Q) \subset W_p^2(\Omega) \times W_p^2(\Omega)$, this fixed point is in the space $W_p^2(\Omega) \times W_p^2(\Omega)$ and is a solution of the problem (7.1)–(7.4) for $t \in [0,1]$.
  Then in view of the inequality (7.19) in Theorem 7.1

$$\|u\|_{W_p^1(\Omega)} + \|\varphi\|_{W_p^1(\Omega)} \le c_0(\|u\|_{W_p^2(\Omega)} + \|\varphi\|_{W_p^2(\Omega)}) \le c_0 R \le R_1.$$

On the other hand, by Lemma 5.1,

$$0 < \frac{m}{2} \le u(x) \le M < 2m \quad \text{in } \Omega.$$

  Thus, the operator $T_t : Q \to X$ does not have a fixed point on ∂Q for any $t \in [0,1]$. On the other hand, for $t = 0$ the operator T_0 has a fixed point $(u_0, \varphi_0) \in \text{Int } Q$. This fixed point is a solution of the linear problem (9.1) and has a topological index 1.
  Then on the basis of the Leray-Schauder principle we conclude that there exists a fixed point (u, φ) of the compact operator $t_t : Q \to X$ for each $t \in [0,1]$.
  Further, since $\text{Im } T_t(Q) \subset W_p^2(\Omega) \times W_p^2(\Omega)$, we get that this fixed point (u, φ) is in the space $W_p^2(\Omega) \times W_p^2(\Omega)$ and is a solution of the problem (7.1)–(7.4).
  Then in view of Lemma 7.1 we see that inequality (9.2) is valid. The inequality (9.3) follows from Theorem 7.1.

Theorem 9.1 is proved.

We now prove the main theorem 3.1

In Theorem 3.1 let $t = 1$. Then we get a solution $(u, \varphi) \in W_p^2(\Omega) \times W_p^2(\Omega)$ of the problem (2.11)–(2.14) with the respective functions $\tilde{h}_1$ and $\tilde{h}_2$.

In view of Theorem 9.1 the inequalities (9.2) hold for this solution. It follows from these inequalities that the relation $\tilde{h}_1(x, u, \varphi) = h_1(x, u, \varphi)$ and $\tilde{h}_2(x, u, \varphi) = h_2(x, u, \varphi)$ hold for $x \in \partial\Omega$ on the solution (u, φ) obtained.

Consequently, $(u, \varphi) \in W_p^2(\Omega) \times W_p^2(\Omega)$ $(p > n)$ is a solution of the problem (2.11)–(2.14) satisfying the inequalities (9.2).

Let us now consider the functions

$$u_1(x) = u(x) \quad \text{and} \quad u_2(x) = z\, u(x) \quad \text{with } z = -\frac{z_1}{z_2} > 0.$$

We remark that the functions h_1 and h_2 are defined by the formulas (2.15) and (2.16) with $u_2 = z\, u$.

Then a direct check shows that $(u_1, u_2, \varphi) \in W_p^2(\Omega) \times W_p^2(\Omega) \times W_p^2(\Omega)$ $(p > n)$ is a solution of the problem (2.1)–(2.3) satisfying the inequality (3.1).

Theorem 3.1 is proved.

10 STATEMENT OF THE GENERAL PROBLEM N > 2

The problem is to find functions $u_1(x), u_2(x), \ldots, u_N(x)$, and $\varphi(x)$ that satisfy the equations

$$-a_r \Delta u_r - \nabla\left(\mu_r u_r \nabla\varphi\right) = f_r\left(x, u_1, u_2, \ldots, u_N, \varphi\right), \qquad u_r(x) \geq 0 \ \text{ in } \ \Omega,$$
$$\tag{10.1}$$

$$a_r \partial_\nu u_r + \mu_r u_r \partial_\nu \varphi = g_r\left(x, u_1, u_2, \ldots, u_N, \varphi\right) \ \text{ on } \ \partial\Omega, \tag{10.2}$$

$$\sum_{r-1}^N z_r u_r(x) = 0 \ \text{ in } \ \Omega. \tag{10.3}$$

Here $a_r > 0$, μ_r and z_r $(r = 1, 2, \ldots, N)$ are constant coefficients, and

$$\operatorname{sign}\mu_r = \operatorname{sign} z_r \quad \text{and} \quad \mu_r = 0 \text{ if } z_r = 0. \tag{10.4}$$

The physical meaning of these coefficients is as follows: a_r is the diffusion coefficient, μ_r is the electrochemical mobility coefficient, and z_r is the charge of the rth component. Thus, $\mu_r = 0$ if the rth component is neutral. The vector $u = (u_1, u_2, \ldots, u_N)$ is the vector of concentrations $u_1 \geq 0$, $u_2 \geq 0$, $\ldots$, $u_N \geq 0$.

Problem (10.1)–(10.3) is considered in bounded domain $\Omega \subset \mathbf{R}^n$ with boundary $\partial\Omega$ consisting of several connected components that correspond to the external wall and to the electrode boundaries. Here we assume that each component is of class C^2. The unit outward normal to $\partial\Omega$ is denoted by ν, and ∂_ν is the directional derivative along ν at a point $x \in \partial\Omega$.

We assume that

$$f_r(\,\cdot\,, u, \,\cdot\,) \geq 0, \qquad g_r(\,\cdot\,, u, \,\cdot\,) = 0$$

for $u_r = 0$ and for the considered values of the other arguments.

11 THE FIRST REDUCTION OF THE PROBLEM

We first consider several versions of problem (10.1)–(10.3).

I) $z_1 = z_2 = \cdots = z_N = 0$. In this case of neutral components the problem is reduced to the nonlinear elliptic system (10.1), (10.2) with a given arbitrary electric potential φ.

II) $z_1 \neq 0$, $z_2 = z_3 = \cdots = z_N = 0$. Equation (10.3) yields $u_1(x) \equiv 0$. This case, with the natural conditions $f_1 = 0$ and $g_1 = 0$ for $u_1 = 0$ is also reduced to system (10.1), (10.2) for the neutral components $u_2, u_3, \ldots, u_N$, with a given arbitrary electric potential φ.

III) $z_1 \neq 0, \ldots, z_k \neq 0$, $z_{k+1} = \cdots = z_N = 0$, and sign $z_1 =$ sign $z_2 = \cdots =$ sign z_k. Equation (10.3) yields $u_1(x) = u_2(x) = \cdots = u_k(x) = 0$. This case, with the natural conditions $f_1 = f_2 = \cdots = f_k = 0$ and $g_1 = g_2 = \cdots = g_k = 0$ for $u_1 = u_2 = \cdots = u_k = 0$, $k < N$, is also reduced to system (10.1), (10.2) with respect to the neutral components $u_{k+1}, \ldots, u_N$, with a given arbitrary electric potential φ.

Note that if the potential φ is specified, then for all these cases we obtain a system of elliptic equations and boundary conditions with identical principal parts (if each equation is divided by the corresponding positive coefficient a_r).

Thus, the case in which there are at least two components with opposite charges is of principal importance for electrochemical reactions.

Let u_i^+ denote the concentration of the ith component with a positive charge $z_i > 0$, let u_j^- denote the concentration of the jth component with a negative charge $z_j < 0$, and let u_k^0 denote the concentration of the kth component with the zero charge $z_k = 0$, if any. Furthermore, we denote

$$u_r' = |z_r|u_r \quad \text{and} \quad z_r' = \text{sign } z_r \quad \text{if} \quad z_r \neq 0,$$
$$u_r' = u_r \quad \text{if} \quad z_r = 0. \tag{11.1}$$

Then system (10.1)–(10.3) takes the form

$$-\Delta u_r' - \nabla \left(\mu_r' u_r' \nabla \varphi \right) = f_r' \left(x, u_1', \ldots, u_N', \varphi \right), \qquad u_r'(x) \geq 0 \quad \text{in} \quad \Omega, \tag{11.2}$$

$$\partial_\nu u_r' + \mu_r' u_r' \partial_\nu \varphi = g_r' \left(x, u_1', \ldots, u_N', \varphi \right) \quad \text{on} \quad \partial\Omega, \tag{11.3}$$

$$\sum u_i^+(x) - \sum u_j^-(x) = 0. \tag{11.4}$$

Here $\mu_r' = (|z_r|/a_r)\mu_r$ if $z_r \neq 0$ and $\mu_r' = 0$ if $z_r = 0$; $f_r^1(\cdots) = (|z_r|/a_r)f_r(\cdots)$, $g_r'(\cdots) = (|z_r|/a_r)g_r(\cdots)$ after the substitution (11.1) if $z_r \neq 0$ and $f_r'(\cdots) = f_r(\cdots)/a_r$, $g_r'(\cdots) = g_r(\cdots)/a_r$ after the substitution (11.1) if $z_r = 0$. Due to versions I–III we suppose that there exists at least one $z_r < 0$.

Let us now introduce a partial order for the indices. Let the first indices correspond to the concentration components with positive charges. We denote this set of indices by $I_+ = \{1, 2, \ldots, r_+\}$. The indices $r_+ + 1, \ldots, r_0$ correspond to the neutral components, if any. We denote the corresponding set of indices by I_0. Finally, the remaining indices $r_0 + 1, \ldots, N$, which correspond to negatively charged components, are particular ordered according to their relative mobility coefficients $\mu'_r < 0$, namely,

$$\mu'_{r_0+1} \le \mu'_{r_0+2} \le \cdots \le \mu'_N < 0. \tag{11.5}$$

Let I'_- denote the set of indices corresponding to negatively charged components except for the last one, i.e., $I'_- = \{r_0 + 1, \ldots, N - 1\}$. We do not exclude the case in which $I'_- = \emptyset$. We retain the old notation for this, probably different, numbering.

We now express the function $u'_N(x)$ from Eq. (11.4);

$$u'_N(x) = \sum_{r \in I_+} u'_r(x) - \sum_{r \in I'_-} u'_r(x), \tag{11.6}$$

and substitute it into system (11.2), (11.3). Then we take into account the remaining equations of the system and obtain

$$-\Delta u'_r - \nabla \left(\mu'_r u'_r \nabla \varphi\right) = f''_r\left(x, u', \varphi\right), \qquad u'_r(x) \ge 0 \quad \text{in} \quad \Omega, \tag{11.7}$$

$$\partial_\nu u'_r + \mu'_r u'_r \partial_\nu \varphi = g''_r\left(x, u', \varphi\right) \quad \text{on} \quad \partial\Omega \tag{11.8}$$

for $r = 1, 2, \ldots, N - 1$; $u' = \left(u'_1, u'_2, \ldots, u'_{N-1}\right)$,

$$-\nabla \left(k'(u') \nabla \varphi\right) = f''_N\left(x, u', \varphi\right) \quad \text{in} \quad \Omega, \tag{11.9}$$

$$k'\left(u'\right) \partial_\nu \varphi = g''_N\left(x, u', \varphi\right) \quad \text{on} \quad \partial\Omega, \tag{11.10}$$

where

$$k'\left(u'\right) = \sum_{r \in I_+} \left(\mu'_r - \mu'_N\right) u'_r + \sum_{r \in I'_-} \left(\mu'_N - \mu'_r\right) u'_r, \tag{11.11}$$

$$\begin{cases} \mu'_r > 0 & \text{for } r \in I_+, \\ \mu'_r \le \mu'_N < 0 & \text{for } r \in I'_- \end{cases}. \tag{11.12}$$

Here

$$f''_r\left(x, u', \varphi\right) = f'_r\left(x, u'_1, \ldots, u'_{N-1}, \sum_{r \in I_+} u'_r - \sum_{r \in I'_-} u'_r, \varphi\right),$$

$$g''_r\left(x, u', \varphi\right) = g'_r\left(x, u'_1, \ldots, u'_{N-1}, \sum_{r \in I_+} u'_r - \sum_{r \in I'_-} u'_r, \varphi\right)$$

for $r = 1, 2, \ldots, N - 1$ and

$$f''_N\left(x, u', \varphi\right) = -f'_N\left(x, u'_1, \ldots, u'_{N-1}, \sum_{r \in I_+} u'_r - \sum_{r \in I'_-} u'_r, \varphi\right) + \sum_{r \in I_+} f''_r\left(x, u', \varphi\right)$$

$$- \sum_{r \in I'_-} f''_r\left(x, u', \varphi\right),$$

$$g''_N\left(x, u', \varphi\right) = -g'_N\left(x, u'_1, \ldots, u'_{N-1}, \sum_{r \in I_+} u'_r - \sum_{r \in I'_-} u'_r, \varphi\right) + \sum_{r \in I_+} g''_r\left(x, u', \varphi\right)$$

$$- \sum_{r \in I'_-} g'_r\left(x, u', \varphi\right).$$

Let us indicate the main properties of the function $k'(u')$.

LEMMA 11.1. $k'(u') \geq 0$ for $u' \geq 0$.

Here the inequality $u' = (u'_1, \ldots, u'_{N-1}) \geq 0$ is understood in the sense that $u'_1 \geq 0, \ldots, u'_{N-1} \geq 0$.

LEMMA 11.2. $k'(u') > 0$ provided that $u' \geq 0$ and there exists at least one component u'_r that satisfies at least one of the inequalities

$$u'_r > 0 \quad \text{for} \quad r \in I_+,$$
$$(\mu'_N - \mu'_r) u'_r > 0 \quad \text{for} \quad r \in I'_-.$$

Proof of these Lemmas obviously follows from inequalities (11.12).

Thus, we have reduced the Amann problem (10.1)–(10.3) with condition (10.4), which describes an electrochemical reaction in a solution with at least two oppositely charged components (i.e., with $I_+ \neq \emptyset$) to problem (11.8)–(11.10) with the function $k'(u')$ specified by (11.11) under conditions (11.12).

12 THE SECOND REDUCTION OF THE PROBLEM

Consider the transformation

$$(u', \varphi) = (u'_1, \ldots, u'_{N-1}, \varphi) \to (v_1, \ldots, v_{N-1}, \varphi) = (v, \varphi), \qquad (12.1)$$

where $v_r = u'_r e^{\mu'_r \varphi}$ $(r = 1, 2, \ldots, N-1)$. Then $f''_l(x, u', \varphi) \to F_l(x, v, \varphi)$ and $g''_l(x, u', \varphi) \to G_l(x, v, \varphi)$ with

$$F_l(x, v, \varphi) = f''_l\left(x, v_1 e^{-\mu'_1 \varphi}, \ldots, v_{N-1} e^{-\mu'_{N-1} \varphi}\right),$$
$$G_l(x, v, \varphi) = g''_l\left(x, v_1 e^{-\mu'_1 \varphi}, \ldots, v_{N-1} e^{-\mu'_{N-1} \varphi}\right), \qquad (12.2)$$

$l = 1, 2, \ldots, N$. In this case, the boundary value problem (11.7)–(11.10) takes the form

$$-\nabla\left(e^{-\mu'_r \varphi} \nabla v_r\right) = F_r(x, v, \varphi), \qquad v_r(x) \geq 0 \quad \text{in} \quad \Omega, \qquad (12.3)$$

$$e^{-\mu'_r \varphi} \partial_\nu v_r = G_r(x, v, \varphi) \quad \text{on} \quad \partial\Omega \qquad (12.4)$$

with $r = 1, 2, \ldots, N-1$ and

$$-\nabla[k(v, \varphi)\nabla\varphi] = F_N(x, v, \varphi) \quad \text{in} \quad \Omega, \qquad (12.5)$$

$$k(v, \varphi)\partial_\nu \varphi = G_N(x, v, \varphi) \quad \text{on} \quad \partial\Omega, \qquad (12.6)$$

$$k(v, \varphi) = \sum_{r \in I_+} (\mu'_r - \mu'_N) v_r e^{-\mu'_r \varphi} + \sum_{r \in I'_-} (\mu'_N - \mu'_r) v_r e^{-\mu'_r \varphi}, \qquad (12.7)$$

where μ'_n satisfy (11.12).

13 MAIN THEOREM

Thus, the original problem (10.1)–(10.3) with condition (10.4) is reduced to (12.3)–(12.6) provided that there are at least two oppositely charged components (i.e., $z_j z_k < 0$ for some j, $k \in \{1, 2, \ldots, N\}$). We consider this problem under the following condition.

CONDITION (A). There exists a rectangular parallelepiped (a closed 'N-dimensional segment')

$$D = \{(v, \varphi) \in \mathbf{R}^N \mid 0 < m_1 \le v_1 \le M_1, \ldots, 0 < m_{N-1} \le v_{N-1} \le M_{N-1},$$
$$\varphi_1 \le \varphi \le \varphi_2\}$$

in the space $\mathbf{R}^N$ of the independent variables $v \in \mathbf{R}^{N-1}$ and $\varphi \in \mathbf{R}$ such that

i) $F_1, \ldots, F_N \in C^1(\overline{\Omega} \times D)$ and

$$(\vec{F}, \vec{\xi}) < 0 \tag{13.1}$$

for all $x \in \overline{\Omega}$ and $(v, \varphi) \in (\partial D)'$;
ii) $G_1, \ldots, G_N \in C^1(\partial\Omega \times D)$ and

$$(\vec{G}, \vec{\xi}) < 0 \tag{13.2}$$

for all $x \in \partial\Omega$ and $(v, \varphi) \in (\partial D)'$. Here $\vec{F} = (F_1, \ldots, F_N)$, $\vec{G} = (G_1, \ldots, G_N)$; $(\cdot, \cdot)$ is the inner product on $\mathbf{R}^N$; $\vec{\xi}$ is the unit outward normal vector to $(\partial D)'$; $(\partial D)'$ is the boundary ∂D with 'edges' of any dimension less than $N - 2$ or equal to $N - 2$ deleted, including the vertices.

REMARK 13.1. Condition i) and inequality (13.1) can be omitted if $F_1(x, v, \varphi) \equiv \cdots \equiv F_N(x, v, \varphi) \equiv 0$.

THEOREM 13.1. *Let Condition (A) be satisfied. Then problem (12.3)–(12.6) has a solution* $(v, \varphi) \in W_p^2(\Omega, \mathbf{R}^N)$ *with* $p > n$, *and moreover,*

$$(v(x), \varphi(x)) \in D \tag{13.3}$$

for all $x \in \overline{\Omega}$.

The proof of this theorem is given in the subsequent sections. It is based on *a priori* estimates and employs the Leray-Schauder method.

14 CONTINUATION LEMMAS

Let us consider special C^1-continuations of the vector-valued function $\vec{F}$ from $\overline{\Omega} \times D$ to $\overline{\Omega} \times \mathbf{R}^N$ and of the vector-valued function $\vec{G}$ from $\partial\Omega \times D$ to $\overline{\Omega} \times \mathbf{R}^N$. The essence of these continuations is as follows. By (13.1) and (13.2), each of the components of $\vec{F}$ and $\vec{G}$ has a constant sign on the corresponding face of the parallelepiped D. The special continuations preserve the sign of the continued component in the part of the half-space $\mathbf{R}^N$ which contains this face.

LEMMA 14.1. *There exists a vector-valued function* $\widetilde{F} = (\widetilde{F}_1, \ldots, \widetilde{F}_N)$ *on* $\overline{\Omega} \times \mathbf{R}^N$ *such that*

i) $\widetilde{F}_l \in C^1(\overline{\Omega} \times \mathbf{R}^N)$, $l = 1, 2, \ldots, N$;

ii) $\widetilde{F}_r > 0$ *on* $\overline{\Omega} \times \mathbf{R}^N \cap \{v_r \leq m_r\}$, $r = 1, 2, \ldots, N-1$, $\widetilde{F}_N > 0$ *on* $\overline{\Omega} \times \mathbf{R}^N \cap \{\varphi \leq \varphi_1\}$;

iii) $\widetilde{F}_r < 0$ *on* $\overline{\Omega} \times \mathbf{R}^N \cap \{v_r \geq M_r\}$, $r = 1, 2, \ldots, N-1$, $\widetilde{F}_N < 0$ *on* $\overline{\Omega} \times \mathbf{R}^N \cap \{\varphi \geq \varphi_2\}$;

iv) $\widetilde{F}_l = F_l$ *on* $\overline{\Omega} \times D$, $l = 1, 2, \ldots, N$.

The proof of this lemma is similar to that of Lemma 4.1. Along with the local Whitney continuation, techniques of continuation in neighborhoods of entering and outgoing right angles (Fikhtengol'ts) are essentially used. This procedure is performed by induction on the dimension.

LEMMA 14.2. *There exists a vector-valued function* $\widetilde{G} = (\widetilde{G}_1, \ldots, \widetilde{G}_N)$ *on* $\overline{\Omega} \times \mathbf{R}^N$ *such that*

i) $\widetilde{G}_l \in C^1(\overline{\Omega} \times \mathbf{R}^N)$, $l = 1, 2, \ldots, N$;

ii) $\widetilde{G}_r > 0$ *on* $\partial\Omega \times \mathbf{R}^N \cap \{v_r \leq m_r\}$, $r = 1, 2, \ldots, N-1$, $\widetilde{G}_N > 0$ *on* $\partial\Omega \times \mathbf{R}^N \cap \{\varphi \leq \varphi_1\}$;

iii) $\widetilde{G}_r < 0$ *on* $\partial\Omega \times \mathbf{R}^N \cap \{v_r \geq M_r\}$, $r = 1, 2, \ldots, N-1$, $\widetilde{G}_N < 0$ *on* $\partial\Omega \times \mathbf{R}^N \cap \{\varphi \geq \varphi_2\}$;

iv) $\widetilde{G}_l = G_l$ *on* $\partial\Omega \times \mathbf{R}^N$, $l = 1, 2, \ldots, N$.

The proof of this lemma first follows the proof of the preceding Lemma for $x \in \partial\Omega$. Then C^1-continuation from the boundary $\partial\Omega \subset C^2$ to the entire domain $\overline{\Omega}$ is performed (see Gilbarg & Trudinger, Sec. 6.9).

15 THE FIRST PROBLEM

Let us consider the following boundary value problem with a parameter $t \in [0, 1]$ in the space $W_p^2(\Omega, \mathbf{R}^{N-1})$ [i.e., $v_r \in W_p^2(\Omega)$, $r = 1, \ldots, N-1$] with $p > n$:

$$-t\nabla\left(e^{-\mu_r'\varphi}\nabla v_r\right) - (1-t)\Delta v_r = t\widetilde{F}_r(x, v, \varphi) \quad \text{in} \quad \Omega \tag{15.1}$$

$$te^{-\mu_r'\varphi}\partial_\nu v_r + (1-t)\partial_\nu v_r + (1-t)v_r = (1-t)v_r^0 + t\widetilde{G}_r(x, v, \varphi) \quad \text{on} \quad \partial\Omega, \tag{15.2}$$

where $r = 1, \ldots, N-1$ and $\varphi \in W_p^1(\Omega)$, $p > n$. Here v_r^0 are constants such that

$$0 < m_r < v_r^0 < M_r. \tag{15.3}$$

LEMMA 15.1. *Let* v *be a solution in* $W_p^2(\Omega, \mathbf{R}^{N-1})$ *to problem (15.1), (15.2) under condition (15.3). Then for any* $t \in [0, 1]$

$$m_r \leq v_r(x) \leq M_r \quad \text{in} \quad \Omega \quad (r = 1, 2, \ldots, N-1). \tag{15.4}$$

Proof. Let us prove the lower bound. Let $w_r(x) = v_r(x) - m_r$. Then the function $w_r \in W_p^2(\Omega)$ is the solution to the problem

$$-t\nabla\left(e^{-\mu_r'\varphi}\nabla w_r\right) - (1-t)\Delta w_r$$
$$= t\widetilde{F}_r\left(\ldots, w_r + m_r, \ldots\right) \quad \text{in} \quad \Omega,$$
$$te^{-\mu_r'\varphi}\partial_\nu w_r + (1-t)\partial_\nu w_r + (1-t)\left(w_r + m_r\right)$$
$$= (1-t)v_r^0 + t\widetilde{G}_r\left(\ldots, w_r + m_r, \ldots\right) \quad \text{on} \quad \partial\Omega.$$
$$(15.5)$$

Here $(\ldots, w_r + m_r, \ldots) = (x, v_1, \ldots, v_{r-1}, w_r + m_r, v_{r+1}, \ldots, v_{N-1}, \varphi)$. It follows that

$$(1-t)\int_{\partial\Omega}\left(w_r + m_r - v_r^0\right)\zeta\,ds - t\int_{\partial\Omega}\widetilde{G}_r\zeta\,ds$$
$$+ t\int_\Omega e^{-\mu_r'\varphi}\nabla w_r\nabla\zeta\,dx + (1-t)\int_\Omega\nabla w_r\nabla\zeta\,dx = t\int_\Omega\widetilde{F}_r\zeta\,dx$$
$$(15.6)$$

for any function $\zeta \in W_p^1(\Omega)$. For the function $w_r \in W_p^2(\Omega)$ and, hence, $w_r \in W_p^1(\Omega)$ we set $\zeta = w_r^- = 2^{-1}(|w_r| - w_r) \geq 0$. Then we have (Gilbarg & Trudinger, Lemma 7.6; Bourdaud & Meyer) $w_r^- \in W_p^1(\Omega)$ and $\nabla w_r^- = -\nabla w_r\chi[w_r < 0]$. Therefore, Eq. (15.6) for $\zeta = w_r^-$ takes the form

$$(1-t)\int_{\partial\Omega}\left(-w_r^- + m_r - v_r^0\right)w_r^-\,ds - t\int_{\partial\Omega}\widetilde{G}_r w_r^-\,ds$$
$$- t\int_\Omega e^{-\mu_r'\varphi}\left|\nabla w_r^-\right|^2\,dx - (1-t)\int_\Omega\left|\nabla w_r^-\right|^2\,dx = t\int_\Omega\widetilde{F}_r w_r^-\,dx.$$
$$(15.7)$$

From this we obtain, for $t = 0$,

$$-\int_{\partial\Omega}\left|w_r^-\right|^2\,ds + \int_{\partial\Omega}\left(m_r - v_r^0\right)w_r^-\,ds - \int_\Omega\left|\nabla w_r^-\right|^2\,dx = 0.$$

By (15.3), we have

$$w_r^-(x) = 0 \quad \text{in} \quad \Omega \quad \text{for} \quad t = 0. \qquad (15.8)$$

Equation (15.7) for $t = 1$ implies

$$-\int_{\partial\Omega}\widetilde{G}_r w_r^-\,ds - \int_\Omega e^{-\mu_r'\varphi}\left|\nabla w_r^-\right|^2\,dx = \int_\Omega\widetilde{F}_r w_r^-\,dx. \qquad (15.9)$$

Here

$$\widetilde{G}_r = \widetilde{G}_r\left(x, v_1(x), \ldots, v_{r-1}(x), m_r - w_r^-(x), v_{r+1}(x), \ldots, v_{N-1}(x), \varphi(x)\right),$$
$$\widetilde{F}_r = \widetilde{F}_r\left(x, v_1(x), \ldots, v_{r-1}(x), m_r - w_r^-(x), v_{r+1}(x), \ldots, v_{N-1}(x), \varphi(x)\right).$$

According to properties ii) in Lemmas 14.1 and 14.2, these integrands are strictly positive for $w_r^- \geq 0$, and consequently, $\widetilde{G}_r > 0$ and $\widetilde{F}_r > 0$ in (15.9). Therefore, $w_r^-(x) = 0$ in Ω for $t = 1$, $r = 1, 2, \dots, N-1$. By virtue of the same properties of the functions $\widetilde{F}_r$ and $\widetilde{G}_r$ and by (15.7) we conclude that $w_r^-(x) = 0$ in Ω for $0 < t \leq 1$, $r = 1, 2, \dots, N-1$. Thus, $w_r^-(x) = 0$ in Ω for all $t \in [0,1]$, i.e., $v_r(x) \geq m_r$ in Ω for all $t \in [0,1]$, $r = 1, 2, \dots, N-1$.

Similarly we can use properties iii) of the functions $\widetilde{F}_r$ and $\widetilde{G}_r$ in Lemmas 14.1 and 14.2 to derive the upper bound $v_r(x) \leq M_r$ in Ω for all $t \in [0,1]$, $r = 1, 2, \dots, N-1$. This completes the proof.

16 THE SECOND PROBLEM

Let us consider the following boundary value problem with a parameter $t \in [0,1]$:

$$-\nabla[(1 - t + tk(v, \varphi))\nabla\varphi] = t\widetilde{F}_N(x, v, \varphi) \quad \text{in} \quad \Omega,$$

$$(1 - t + tk(v, \varphi))\partial_\nu\varphi + (1 - t)\varphi = (1 - t)\varphi_0 + t\widetilde{G}_N(x, v, \varphi) \quad \text{on} \quad \partial\Omega \tag{16.1}$$

for $v \in W_p^1(\Omega, \mathbf{R}^{N-1})$ with $p > n$ and

$$0 < m_r \leq v_r(x) \quad \text{in} \quad \Omega \quad (r = 1, 2, \dots, N-1). \tag{16.2}$$

Here φ_0 is a constant such that

$$\varphi_1 < \varphi_0 < \varphi_2. \tag{16.3}$$

LEMMA 16.1. *Let* φ *be a* $W_p^2(\Omega)$*-solution to problem (16.1) for* $v \in W_p^1(\Omega, \mathbf{R}^{N-1})$ *$(p > n)$ under conditions (16.2) and (16.3). Then*

$$\varphi_1 \leq \varphi(x) \leq \varphi_2 \quad \text{in} \quad \overline{\Omega} \tag{16.4}$$

for any $t \in [0,1]$.

Proof. Let us prove the lower bound. Let $\psi(x) = \varphi(x) - \varphi_1$. Then the function $\psi \in W_p^2(\Omega)$ is a solution to the problem

$$-\nabla[(1 - t + tk(v, \psi + \varphi_1))\nabla\psi]$$
$$= t\widetilde{F}_N(\dots, \psi + \varphi_1) \quad \text{in} \quad \Omega,$$
$$(1 - t + tk(v, \psi + \varphi_1))\partial_\nu\psi + (1 - t)(\psi + \varphi_1)$$
$$= (1 - t)\varphi_0 + t\widetilde{G}_N(\dots, \psi + \varphi_1) \quad \text{on} \quad \partial\Omega.$$

Here $\widetilde{F}_N(\dots, \psi + \varphi_1) = \widetilde{F}_N(x, v(x), \psi(x) + \varphi_1)$ and $\widetilde{G}_N(\dots, \psi + \varphi_1) = \widetilde{G}_N(x, v(x), \psi(x) + \varphi_1)$. It follows that

$$(1 - t) \int_{\partial\Omega} (\psi + \varphi_1 - \varphi_0)\,\zeta\,ds - t \int_{\partial\Omega} \widetilde{G}_N\zeta\,ds + (1 - t) \int_{\Omega} \nabla\psi\nabla\zeta\,dx$$
$$+ t \int_{\Omega} k(v, \psi + \varphi_1)\,\nabla\psi\nabla\zeta\,dx = t \int_{\Omega} \widetilde{F}_N\zeta\,dx \tag{16.5}$$

for any function $\zeta \in W_p^1(\Omega)$. For a function $\psi \in W_p^2(\Omega)$ and hence $\psi \in W_p^1(\Omega)$ we put $\zeta = \psi^- = (|\psi| - \psi)/2 \geq 0$. Then $\psi^- \in W_p^1(\Omega)$ and $\nabla \psi^- = -\nabla \psi \chi[\psi < 0]$. Thus, Eq. (16.5) for $\zeta = \psi^-$ becomes

$$- (1-t) \int\limits_{\partial\Omega} \left| \psi^- \right|^2 ds + (1-t) \int\limits_{\partial\Omega} (\varphi_1 - \varphi_0) \, \psi^- \, ds - t \int\limits_{\partial\Omega} \widetilde{G}_N \psi^- \, ds$$

$$- (1-t) \int\limits_{\Omega} \left| \nabla \psi^- \right|^2 dx - t \int\limits_{\Omega} k \left(v, -\psi^- + \varphi_1 \right) \left| \nabla \psi^- \right|^2 dx = t \int\limits_{\Omega} \widetilde{F}_N \psi^- \, dx. \tag{16.6}$$

It follows that

$$- \int\limits_{\partial\Omega} \left| \psi^- \right|^2 ds + \int\limits_{\partial\Omega} (\varphi_1 - \varphi_0) \, \psi^- \, ds - \int\limits_{\Omega} \left| \nabla \psi^- \right|^2 dx = 0$$

for $t = 0$. Therefore, by (16.3) we have $\psi^-(x) = 0$ in Ω for $t = 0$. By (16.6) for $t = 1$ we find

$$- \int\limits_{\partial\Omega} \widetilde{G}_N \psi^- \, ds - \int\limits_{\Omega} k \left(v, \varphi_1 - \psi^- \right) \left| \nabla \psi^- \right|^2 dx = \int\limits_{\Omega} \widetilde{F}_N \psi^- \, dx. \tag{16.7}$$

Here

$$\widetilde{G}_N = \widetilde{G}_N \left(x, v(x), \varphi_1 - \psi^-(x) \right) > 0,$$

$$\widetilde{F}_N = \widetilde{F}_N \left(x, v(x), \varphi_1 - \psi^-(x) \right) > 0.$$

These inequalities are satisfied by virtue of property ii) of the functions $\widetilde{F}_N$ and $\widetilde{G}_N$ in Lemmas 14.1 and 14.2.

Equation (16.2) and Lemma 11.2 imply $k(v, \varphi_1 - \psi^-) > 0$. Then by (16.7) we obtain $\psi^-(x) = 0$ in Ω for $t = 1$. Further, by virtue of the above-mentioned properties of the functions k, $\widetilde{F}_N$, and $\widetilde{G}_N$ under conditions (16.2) and (16.3), we conclude from (16.6) that $\psi^-(x) = 0$ in Ω for all $t \in [0,1]$, i.e., $\varphi(x) \geq \varphi_1$ in Ω for all $t \in [0,1]$.

The same technique provides the upper bound $\varphi(x) \leq \varphi_2$ in Ω for all $t \in [0,1]$. This completes the proof.

17 THE GENERAL PROBLEM

We now consider the following problem with a parameter $t \in [0,1]$:

$$-\nabla \left(a_r(t, \varphi) \nabla v_r \right) = t \widetilde{F}_r(x, v, \varphi) \quad \text{in} \quad \Omega, \tag{17.1}$$

$$a_r(t, \varphi) \partial_\nu v_r + (1-t) v_r = (1-t) v_r^0 + t \widetilde{G}_r(x, v, \varphi) \quad \text{on} \quad \partial\Omega, \tag{17.2}$$

where $a_r(t, \varphi) = t e^{-\mu'_r \varphi} + 1 - t$, $r = 1, 2, \ldots, N - 1$;

$$-\nabla \left(a_N(t, v, \varphi) \nabla \varphi \right) = t \widetilde{F}_N(x, v, \varphi) \quad \text{in} \quad \Omega, \tag{17.3}$$

$$a_N(t, v, \varphi) \partial_\nu \varphi + (1-t) \varphi = (1-t) \varphi_0 + t \widetilde{G}_N(x, v, \varphi) \quad \text{on} \quad \partial\Omega. \tag{17.4}$$

where $a_N(t, v, \varphi) = tk(v, \varphi) + 1 - t$, and v_r^0 and φ_0 satisfy conditions (15.3) and (16.3).

LEMMA 17.1. *Let* $(v, \varphi) \in W_p^2(\Omega, \mathbf{R}^N)$ $(p > n)$ *be a solution to system* (17.1)–(17.4). *Then*

$$0 < m_r \leq v_r(x) \leq M_r \qquad (r = 1, 2, \ldots, N - 1), \tag{17.5}$$

$$\varphi_1 \leq \varphi(x) \leq \varphi_2 \tag{17.6}$$

in Ω *for any* $t \in [0, 1]$.

Proof. Under the conditions of the lemma, the vector-valued function $v \in W_p^2(\Omega, \mathbf{R}^{N-1})$ is a solution to (17.1), (17.2) for $\varphi \in W_p^2(\Omega)$. We obtain inequality (17.5) by Lemma 7.1.

Under the conditions of the lemma, $\varphi \in W_p^2(\Omega)$ is a solution to problem (17.3), (17.4) which satisfies inequalities (17.5). By Lemma 16.1 we obtain inequality (17.6).

LEMMA 17.2. *Let* $(v, \varphi) \in W_p^2(\Omega, \mathbf{R}^N)$ $(p > n)$ *be a solution to system* (17.1)–(17.4). *Then there exist constants* $\alpha \in (0, 1)$ *and* $A_r > 0$ *such that*

$$\|v_r\|_{C^\alpha(\overline{\Omega})} \leq A_r \qquad (r = 1, 2, \ldots, N - 1) \tag{17.7}$$

for all $t \in [0, 1]$.

Here the constants α and A_r do not depend on the solution $(v, \varphi) \in W_p^2(\Omega, \mathbf{R}^N)$ with $p > n$.

Proof. For the function $v_r \in W_p^2(\Omega)$ we have

$$-\nabla \left(a_r(t, \varphi) \nabla v_r \right) = t\widetilde{F}_r(x, v, \varphi) \quad \text{in} \quad \Omega, \tag{17.8}$$

$$a_r(t, \varphi) \partial_\nu v_r = (1 - t) \left(v_r^0 - v_r \right) + t\widetilde{G}_r(x, v, \varphi) \quad \text{on} \quad \partial\Omega, \tag{17.9}$$

where $a_r(t, \varphi) = te^{-\mu_r' \varphi} + (1 - t)$.

By Lemma 17.1, we obtain

$$0 < \delta_r \leq a_r(t, \varphi) \leq \beta_r < \infty \tag{17.10}$$

for $x \in \overline{\Omega}$ and $t \in [0, 1]$, where $\delta_r = \min\{a_r(t, \varphi) | t \in [0, 1], \; \varphi_1 \leq \varphi \leq \varphi_2\}$ and $\beta_r = \max\{a_r(t, \varphi) | \; t \in [0, 1], \; \varphi_1 \leq \varphi \leq \varphi_2\}$.

Furthermore, by virtue of the estimates (17.5) and (17.6), the right-hand sides of Eq. (17.8) for $r = 1, 2, \ldots, N - 1$ are bounded in $L_\infty(\Omega)$. The same estimates imply that the right-hand sides of the boundary conditions (17.9) for $r = 1, 2, \ldots, N - 1$ are bounded in $L_\infty(\partial\Omega)$ and their continuations from the boundary $\partial\Omega$ to the domain Ω are bounded in $L_\infty(\Omega)$. Then we derive the inner estimate for the norm $\|v_r\|_{C^\alpha(\Omega')}$, $\Omega' \subset\subset \Omega$ from Theorems 14.1 (Chap. 3), 1.1 (Chap. 4), and 2.1 (Chap. 9) in Ladyzhenskaya & Ural'tseva. The norm $\|v_r\|_{C^\alpha}$ can be estimated in the vicinity of the boundary $\partial\Omega$ by using the scheme in Ladyzhenskaya & Ural'tseva, Sec. 2, Chap. 10.

The boundary integral is estimated in a different way. This integral is first estimated via the maximum modulus of the right-hand side in (17.9) and via the corresponding boundary integral of the corresponding positive test function. Then the latter integral is represented as an integral over the corresponding domain and conventional techniques of estimating the norm $\|v_r\|_{C^\alpha}$ in the vicinity of the boundary are used.

REMARK. This Lemma essentially follows from Lemma 5.1 in Liebermann despite the fact that the latter was stated for $C^2(\overline{\Omega})$-solutions to nonlinear Neumann problems.

LEMMA 17.3. *Let* $(v, \varphi) \in W_p^2(\Omega, \mathbf{R}^N)$ $(p > n)$ *be a solution to system (17.1)–(17.4). Then there exist constants* $\beta \in (0, 1)$ *and* $B > 0$ *such that*

$$\|\varphi\|_{C^\beta(\overline{\Omega})} \le B \qquad (17.11)$$

for all $t \in [0, 1]$.

Here the constants β and B do not depend on the solution $(v, \varphi) \in W_p^2(\Omega, \mathbf{R}^N)$ with $p > n$.

Proof. For the function $\varphi \in W_p^2(\Omega)$ we obtain

$$-\nabla (a_N(t, v, \varphi)\nabla\varphi) = t\widetilde{F}_N(x, v, \varphi) \quad \text{in} \quad \Omega, \qquad (17.12)$$

$$a_N(t, v, \varphi)\partial_\nu\varphi = (1 - t)(\varphi_0 - \varphi) + t\widetilde{G}_N(x, v, \varphi) \quad \text{on} \quad \partial\Omega, \qquad (17.13)$$

where $a_N(t, v, \varphi) = tk(v, \varphi) + 1 - t$. By virtue of inequalities (17.5) and (17.6) we have

$$0 < \delta_N \le a_N(t, v, \varphi) \le \beta_N < \infty \qquad (17.14)$$

for $x \in \overline{\Omega}$, $t \in [0, 1]$, where

$$\delta_N = \min\{a_N(t, v, \varphi) \,|\, t \in [0, 1], \quad m_r \le v_r \le M_r, \quad \varphi_1 \le \varphi \le \varphi_2\},$$
$$\beta_N = \max\{a_N(t, v, \varphi) \,|\, t \in [0, 1], \quad m_r \le v_r \le M_r, \quad \varphi_1 \le \varphi \le \varphi_2\}.$$

Here $\delta_N > 0$, since $m_r > 0$ $(r = 1, 2, \ldots, N - 1)$.

The estimates (17.5) and (17.6) imply that the right-hand side in Eq. (17.12) is bounded in $L_\infty(\Omega)$, the right-hand side of the boundary condition in (17.13) is bounded in $L_\infty(\partial\Omega)$, and its continuation from the boundary to the entire domain Ω is bounded in $L_\infty(\Omega)$. The proof can be completed by analogy with that of the preceding lemma.

18 ESTIMATE FOR THE NORMS OF SOLUTIONS IN $C^{1,\alpha'}(\overline{\Omega})$

Let us consider the rth equation in (17.1) with the corresponding rth boundary condition of (17.2) for the function v_r $(r = 1, 2, \ldots, N - 1)$.

LEMMA 18.1. *Let*

$$(v, \varphi) \in W_p^2(\Omega, \mathbf{R}^N) \qquad (p > n)$$

be a solution to (17.1)–(17.4). Then there exist constants $\alpha' \in (0,1)$ and A_r' such that

$$\|v_r\|_{C^{1,\alpha'}(\Omega)} \leq A_r' \qquad (18.1)$$

for all $t \in [0,1]$. Here the constants α' and A_r' do not depend on the solution $(v,\varphi) \in W_p^2(\Omega, \mathbf{R}^N)$ with $p > n$.

Proof. The function $v_r \in W_p^2(\Omega)$ is a solution to the quasilinear Neumann problem (17.8), (17.9) with coefficient $a_r(t,\varphi)$ satisfying the uniform ellipticity condition (17.10). Thus, v_r can be uniformly estimated (with respect to $t \in [0,1]$) in $C^{\alpha'}(\overline{\Omega})$, $\alpha' = \min\{\alpha, \beta\}$, provided that the estimate (17.11) is known.

Furthermore, the right-hand sides in Eq. (17.8) and condition (17.9) are uniformly bounded in modulus, by virtue of (17.5) and (17.6) and can be uniformly estimated in $C^{\alpha'}(\overline{\Omega})$ and $C^{\alpha'}(\partial\overline{\Omega})$, respectively, by virtue of (17.7) and (17.11). Thus, by Theorem 8.32 in Gilbarg & Trudinger we find the inner estimates for $\|v_r\|_{C^{1,\alpha'}(\Omega')}$, $\Omega' \subset\subset \Omega$.

Estimates of the norm in $C^{1,\alpha'}$ in the vicinity of the boundary $\partial\Omega$ follow the scheme of the proof of Theorem 8.33 in Gilbarg & Trudinger and use Theorem 4.15 in Gilbarg & Trudinger.

LEMMA 18.2. Let $(v,\varphi) \in W_p^2(\Omega, \mathbf{R}^N)$ be a solution to (17.1)–(17.4). Then there exists a constant B' such that

$$\|\varphi\|_{C^{1,\alpha'}(\overline{\Omega})} \leq B^1 \qquad (18.2)$$

for all $t \in [0,1]$.

Here the constant B' does not depend on the solution $(v,\varphi) \in W_p^2(\Omega, \mathbf{R}^N)$ with $p > n$.

The proof of this lemma is similar to that of Lemma 18.1 and is based on the analysis of the solution $\varphi \in W_p^2(\Omega)$ to the Neumann problem (17.12), (17.13).

REMARK. Lunardi and Vespri, Theorem 2, obtained estimates of norms in $C^{1,\alpha}(\overline{\Omega})$ for solutions to a linear Neumann problem in terms of estimates in $C^{\alpha}(\overline{\Omega})$ of the coefficients of a linear operator in the divergent form for the case of a parabolic equation.

19 ESTIMATES OF THE NORMS OF SOLUTIONS IN $\mathbf{W}_p^2(\Omega, \mathbf{R}^N)$,   p > n

First, let us recall the conventional results (Agmon, Douglis & Nirenberg and Amann (1993)) on the solution to a linear Neumann problem in bounded domain $\Omega \subset \mathbf{R}^n$ with boundary $\partial\Omega \subset C^2$.

Consider the problem

$$\sum_{i,j=1}^{n} \frac{\partial}{\partial x_i}\left(a_{ij}(x)\frac{\partial u}{\partial x_j}\right) + a_0(x)u = f(x) \quad \text{in} \quad \Omega,$$

$$-\sum_{i,j=1}^{n} a_{ij}(x)\frac{\partial u}{\partial x_j}\cos(\nu, x_i) = g(x) \quad \text{on} \quad \partial\Omega. \qquad (19.1)$$

Let $a_{ij} = a_{ji}$, $\lambda_0|\xi|^2 \leq \sum_{i,j=1}^n a_{ij}(x)\xi_i\xi_j \leq \lambda_1|\xi|^2$ with constants $\lambda_1 \geq \lambda_0 > 0$ for all $x \in \overline{\Omega}$ and $\xi \in \mathbf{R}^n$. Let $a_{ij} \in C^1(\overline{\Omega})$, $a_0 \in L_\infty(\Omega)$, $f \in L_p(\Omega)$, and $g \in W_p^{1-1/p}(\partial\Omega)$ with $p > n$.

Then the inequality (Agmon, Douglis & Nirenberg and Amann (1993))

$$\|u\|_{W_p^2(\Omega)} \leq C \left(\|f\|_{L_p(\Omega)} + \|g\|_{W_p^{1-1/p}(\partial\Omega)} + \|u\|_{L_p(\Omega)} \right), \qquad (19.2)$$

holds for each $W_p^2(\Omega)$-solution to the Neumann problem (19.1), where the constant C depends only on λ_0, λ_1, on the norms $\|a_{ij}\|_{C^1(\overline{\Omega})}$ and $\|a_0\|_{L_\infty(\Omega)}$, on the continuity module of the first-order derivatives of the coefficients a_{ij}, on the index $p > n$, and on the domain Ω with $\partial\Omega \subset C^2$.

A priori estimates for the norms of possible $W_p^2(\Omega, \mathbf{R}^N)$-solutions to problem (17.1)–(17.4) in the spaces $L_\infty(\Omega, \mathbf{R}^N)$, $C^\alpha(\overline{\Omega}, \mathbf{R}^N)$, and $C^{1,\alpha'}(\overline{\Omega}, \mathbf{R}^N)$ are obtained in Sections 9 and 18. By virtue of these estimates, we have uniform (with respect to $t \in [0, 1]$) estimates for the coefficients a_r in $C^{1,\alpha'}(\overline{\Omega})$ and uniform estimates for the right-hand sides in $L_p(\Omega)$. From the same estimates we obtain uniform estimates for the functions $\widetilde{G}_r$ $(r = 1, 2, \ldots, N)$ in $W_p^1(\Omega)$.

Furthermore, we apply the theorem on traces (Besov, Il'in & Nikol'skii) for functions in $W_p^1(\Omega)$ and find that $\|\widetilde{G}_r\|_{W_p^{1-1/p}(\partial\Omega)} \leq \mathrm{const}\, \|\widetilde{G}_r\|_{W_p^1(\Omega)}$, where the constant does not depend on $\widetilde{G}_r \in W_p^1(\Omega)$.

We thus obtain uniform estimates for the right-hand sides of the Neumann boundary conditions in $W_p^{1-1/p}(\partial\Omega)$.

Finally, inequalities (17.10) and (17.14) provide uniform lower and upper bounds for the ellipticity constants of the equations considered. Then we apply inequality (19.2) to the solutions $u_r \in W_p^2(\Omega)$ to problem (17.1), (17.2) with $r = 1, 2, \ldots, N-1$ and to the solution $\varphi \in W_p^2(\Omega)$ to (17.3), (17.4) and obtain an estimate for the solution $(v, \varphi) \in W_p^2(\Omega, \mathbf{R}^N)$ to problem (17.1)–(17.4) in terms of the cited estimates in $L_\infty(\Omega, \mathbf{R}^N)$, $C^\alpha(\overline{\Omega}, \mathbf{R}^N)$ and $C^{1,\alpha'}(\overline{\Omega}, \mathbf{R}^N)$.

LEMMA 19.2. *Let* $(v, \varphi) \in W_p^2(\Omega, \mathbf{R}^N)$ $(p > n)$ *be a solution to system (17.1)–(17.4). Then there exists a constant R such that*

$$\|v_1\|_{2,p} + \|v_2\|_{2,p} + \cdots + \|v_{N-1}\|_{2,p} + \|\varphi\|_{2,p} \leq R \qquad (19.3)$$

for all $t \in [0, 1]$.

Here $R = R(m_1, \ldots, m_{N-1}, M_1, \ldots, M_{N-1}, \varphi_1, \varphi_2, p, \Omega)$ and $\|\cdot\|_{k,p} = \|\cdot\|_{W_p^k(\Omega)}$.

20 THE OPERATOR $\mathbf{T}_t$

Let $X = W_p^1(\Omega, \mathbf{R}^N)$ and $X_1 = \{(w, \psi) \in X \mid m_1/2 \leq w_1 \leq 2M_1, \ldots, m_{N-1}/2 \leq w_{N-1} \leq 2M_{N-1}, \varphi_1 - 1 \leq \varphi \leq \varphi_2 + 1\}$. To the nonlinear problem (17.1)–(17.4) with a parameter $t \in [0, 1]$ we assign the nonlinear operator $T_t : X_1 \to X$ specified by the relations $v_r = T_{r,t}(w, \psi)$, $r = 1, 2, \ldots, N-1$; $\varphi = T_{N,t}(w, \psi)$, where v_r is a solution to the problem

$$-\nabla\left(a_r(t, \psi)\nabla v_r\right) + v_r = t\widetilde{F}_r(x, w, \psi) + w_r \quad \text{in} \quad \Omega,$$

$$a_r(t, \psi)\partial_\nu v_r + a_r(t, \psi)v_r = t\widetilde{G}_r(x, w, \psi) + (1-t)\left(v_r^0 - w_r\right) + a_r(t, \psi)w_r \quad \text{on} \quad \partial\Omega$$
$$(20.1)$$

and φ is a solution to the problem

$$-\nabla\left(a_N(t,w,\psi)\nabla\varphi\right) + \varphi = t\widetilde{F}_N(x,w,\psi) + \psi \quad \text{in} \quad \Omega,$$

$$a_N(t,w,\psi)\partial_\nu\varphi + a_N(t,w,\psi)\varphi$$
$$= t\widetilde{G}_N(x,w,\psi) + a_N(t,w,\psi)\psi + (1-t)(\varphi_0 - \psi) \quad \text{on} \quad \partial\Omega.$$
$$(20.2)$$

Note that by the definition of X_1 we have

$$0 < \delta'_r \le a_r(t,\psi) \le \beta'_2 < \infty, \quad r = 1, 2, \dots, N-1,$$

for all $t \in [0,1]$ and $x \in \overline{\Omega}$ for $\varphi_1 - 1 \le \psi(x) \le \varphi_2 + 1$. Here

$$\delta'_r = \min\{a_r(t,\psi)\mid t \in [0,1], \ \varphi_1 - 1 \le \psi \le \varphi_2 + 1\}$$

and

$$\beta'_r = \max\{a_r(t,\psi)\mid t \in [0,1], \ \varphi_1 - 1 \le \psi \le \varphi_2 + 1\}.$$

By the definition of X_1, we also obtain

$$0 < \delta'_N \le a_N(t,w,\psi) \le \beta'_N < \infty$$

for all $t \in [0,1]$ and $x \in \overline{\Omega}$ for $(w,\psi) \in X_1$. Here

$$\delta'_N = \min\{a_N(t,w,\psi)\mid t \in [0,1], \ m_r/2 \le w_r \le 2M_r, \ \varphi_1 - 1 \le \psi \le \varphi_2 + 1\},$$

and

$$\beta'_N = \max\{a_N(t,w,\psi)\mid t \in [0,1], \ m_r/2 \le w_r \le 2M_r, \ \varphi_1 - 1 \le \psi \le \varphi_2 + 1\}.$$

Note that $\delta'_N > 0$ since $m_r > 0$ $(r = 1, \dots, N-1)$.

To establish the properties of the operator T_t, consider the linear Neumann problem

$$Lu \equiv -\nabla(a(x)\nabla u) + u = h(x) \quad \text{in} \quad \Omega,$$
$$Bu \equiv a(x)\partial_\nu u + a(x)u = g(x) \quad \text{on} \quad \partial\Omega.$$
$$(20.3)$$

Here $0 < a_0 \le a(x) \le a_0^{-1}$ in Ω and $a \in W_p^1(\Omega)$ with $p > n \ge 2$.

LEMMA 20.1. *Let* $h \in L_p(\Omega)$ *and* $g \in W_p^{1-1/p}(\partial\Omega)$. *Then problem (20.3) has a unique solution* $u \in W_p^2(\Omega)$.

Proof. The uniqueness of a solution to problem (20.3) in the space $W_p^2(\Omega)$ with $p > n \ge 2$ is obvious.

To prove the existence of a solution, we set $u(x) = v(x) + u_0(x)$, where $u_0 \in W_p^2(\Omega)$ is a solution to the problem

$$L_0 u_0 \equiv -\Delta u_0 + u_0 = h(x) \quad \text{in} \quad \Omega,$$
$$B_0 u_0 \equiv \partial_\nu u_0 + u_0 = g(x)/a(x) \quad \text{on} \quad \partial\Omega.$$

Then problem (20.3) is reduced to the problem

$$Lv \equiv -\nabla(a(x)\nabla v) + v = h_1(x) \quad \text{in} \quad \Omega,$$
$$B_0 v \equiv \partial_\nu v + v = 0 \quad \text{on} \quad \partial\Omega, \tag{20.4}$$

where $h_1(x) = h_0(x) - Lu_0(x) \in L_p(\Omega)$.

To prove that problem (20.4) is solvable in $W_p^2(\Omega)$ $(p > n)$ we use the method of continuation with respect to a parameter. To this end we introduce the space $W_{p,B_0}^2(\Omega) = \{v \in W_p^2(\Omega) | B_0 v|_{\partial\Omega} = 0\}$ and consider the family of operators $L_t = (1-t)L_0 + tL : W_{p,B_0}^2(\Omega) \to L_p(\Omega)$. Let us prove that there exists a constant C such that

$$\|v\|_{2,p} \leq C\|L_t v\|_p \tag{20.5}$$

for all $t \in [0,1]$ and $v \in W_{p,B_0}^2(\Omega)$. This inequality implies that the inequality

$$\|v\|_{2,p} \leq C\|h\|_p \tag{20.5'}$$

holds for all $t \in [0,1]$ for any $W_p^2(\Omega)$-solution v to the problem

$$L_t v \equiv -(1-t)\Delta v - t\nabla(a(x)\nabla v) + v = h(x) \quad \text{in} \quad \Omega,$$
$$B_0 v \equiv \partial_\nu v + v = 0 \quad \text{on} \quad \partial\Omega. \tag{20.6}$$

Here $C = C(a_0, \|a\|_{1,p}, n, p, \Omega)$.

Let us first obtain an *a priori* estimate in the space $W_2^1(\Omega)$. The energy identity

$$\int_\Omega [1 - t + ta(x)]v^2\, ds + \int_\Omega [1 - t + ta(x)]|\nabla v|^2\, dx + \int_\Omega v^2\, dx = \int_\Omega hv\, dx$$

yields

$$\|v\|_{1,2} \leq C_1 \|h\|_p \tag{20.7}$$

for all $t \in [0,1]$, where the constant $C_1 = C_1(a_0, p, \Omega)$, $p > n \geq 2$.

Next, we rewrite (20.6) as

$$-\Delta v = \frac{t\nabla a \nabla v - v + h(x)}{1 - t + ta(x)} \quad \text{in} \quad \Omega,$$
$$\partial_\nu v + v = 0 \quad \text{on} \quad \partial\Omega.$$

We apply the well-known *a priori* estimate (Agmon, Douglis & Nirenberg)

$$\|v\|_{2,p} \leq \text{const} \left(\|\Delta v\|_p + \|\partial_\nu v + v\|_{W_p^{1-1/p}(\partial\Omega)} + \|v\|_p \right)$$

to this problem. Here the constant does not depend on the function $v \in W_p^2(\Omega)$ $(p > n)$. We thus obtain

$$\|v\|_{2,p} \leq \text{const}_1 \left(\|a\|_{1,p} \|\nabla v\|_\infty + \|h\|_p + \|v\|_p \right). \tag{20.8}$$

Let us now use the Gagliardo-Nirenberg interpolation inequality (Nirenberg)

$$\|\nabla v\|_\infty \leq \text{const}_2 \left(\|v\|_{2,p}^\theta \|v\|_{g_*}^{1-\theta} + \|v\|_{g_*} \right). \tag{20.9}$$

Here $q_* = 2n/(n-2)$ and $\theta = np/((n+2)p - 2n)$ for $n > 2$, whereas $\theta = (q_* + 2)p/2(p + q_*(p-1))$ with sufficiently large q_* for $n = 2$, so that $1/2 < \theta < 1$.

Note that here, as usual, the coefficients in the inequalities do not depend on the functions explicitly written out.

Let us now use the well-known Sobolev inequality (Besov, Il'in & Nikol'skii) concerning the embedding $W_2^1(\Omega) \to L_q(\Omega)$ with $q = q_*$ for $n > 2$ and with any $q > 1$ for $n = 2$:

$$\|v\|_q \leq C(\Omega, n, q) \|v\|_{1,2}. \tag{20.10}$$

Then inequality (20.8), by virtue of (20.9) and (20.7), becomes

$$\|v\|_{2,p} \leq \text{const}_3 \left(\|a\|_{1,p} \|v\|_{2,p}^\theta \|h\|_p^{1-\theta} + \|a\|_{1,p} \|h\|_p + \|v\|_p + \|h\|_p \right). \tag{20.11}$$

Let us consider the first possible case $n < p \leq 2n/(n-2)$ if $n > 2$ (i.e., $n = 3, 4$) and an arbitrary $p > 2$ if $n = 2$. In this case inequalities (20.7), (20.10), and (20.11) yield inequality (20.5′).

Let us consider the second case, in which $p > 2n/(n-2)$ $(p > n > 2)$. We use the following Gagliardo-Nirenberg inequality (Nirenberg):

$$\|v\|_p \leq \text{const} \left(\|v\|_{2,p}^{\theta_1} \|v\|_{2n/(n-2)}^{1-\theta_1} + \|v\|_{2n/(n-2)} \right),$$

where $0 < \theta_1 = ((n-2)p - 2n)/((n+2)p - 2n) < 1$ for $p > 2n/(n-2)$ for $n > 2$.

According to this inequality, (20.7), (20.10), from (20.11) we obtain inequality (20.5′) for the second case, in which $p > 2n/(n-2)$ for $p > n > 2$.

Thus, inequality (20.5′) and hence the *a priori* estimate (20.5) hold for any $t \in [0, 1]$.

Consequently, we can use Theorem 5.2 in Gilbarg & Trudinger, since the operator L_0 maps the space $W_{p,B_0}^2(\Omega)$ onto $L_p(\Omega)$ bijectively $(p > n)$. By this theorem, we obtain the conclusion of Lemma 20.1.

LEMMA 20.2. *Equations (20.1) and (20.2) uniquely determine the operator* $T_t : X_1 \to W_p^2(\Omega, \mathbf{R}^N)$ *for all* $t \in [0, 1]$.

The proof immediately follows from Lemma 20.1.

21 PROPERTIES OF THE OPERATOR T_t

Let us prove that the operator $T_t : X_1 \to W_p^2(\Omega, \mathbf{R}^N)$ is continuous. To this end, we consider a 'perturbed' problem (20.3), namely,

$$\begin{aligned}
-\nabla \left(a_\delta(x) \nabla u_\delta \right) + u_\delta &= h_\delta(x) \quad \text{in} \quad \Omega, \\
a_\delta(x) \partial_\nu u_\delta + a_\delta(x) u_\delta &= g_\delta(x) \quad \text{on} \quad \partial\Omega.
\end{aligned} \tag{21.1}$$

Here

$$a_\delta \in W_p^1(\Omega), \quad p > n, \quad h_\delta \in L_p(\Omega),$$

$$g_\delta = g_\delta|_{\partial\Omega}, \quad \text{where} \quad g_\delta \in W_p^1(\Omega),$$

(21.2)

$0 < a_0 \le a_\delta(x) \le a_0^{-1}$ for all $x \in \overline{\Omega}$ and all δ from a neighborhood $U_0 \subset \mathbf{R}$ of the point $\delta_0 \in \mathbf{R}$. For the difference $w_\delta = u_\delta - u_{\delta_0}$ we have

$$-\nabla(a_{\delta_0}(x)\nabla w_\delta) + w_\delta = \nabla[(a_\delta(x) - a_{\delta_0}(x))\nabla u_\delta] + h_\delta(x) - h_{\delta_0}(x) \text{ in } \Omega,$$

$$a_{\delta_0}(x)\partial_\nu w_\delta + a_{\delta_0}(x)w_\delta = -(a_\delta(x) - a_{\delta_0}(x))(\partial_\nu u_\delta + u_\delta) + g_\delta(x) - g_{\delta_0}(x) \text{ on } \partial\Omega.$$

The *a priori* estimate $\|u_\delta\|_{2,p} \le c(a_0, \|a_\delta\|_{1,p})(\|h_\delta\|_p + \|g_\delta\|_{1,p})$ is valid for the solution $u_\delta \in W_p^2(\Omega)$ ($p > n \ge 2$). Then asymptotic relations $\|a_\delta - a_{\delta_0}\|_{1,p} \to 0$, $\|h_\delta - h_{\delta_0}\|_p \to 0$, and $\|g_\delta - g_{\delta_0}\|_{1,p} \to 0$ as $\delta \to \delta_0$ imply for w_δ $\|w_\delta - w_{\delta_0}\|_{2,p} \to 0$ as $\delta \to \delta_0$.

Thus, the operator $U = U(a_\delta; h_\delta, g_\delta)$ specified by problem (21.1) under condition (21.2) continuously acts from $W_p^1(\Omega) \times L_p(\Omega) \times W_p^{1-1/p}(\partial\Omega)$ into $W_p^2(\Omega)$ under condition (21.2). In fixing the operator of continuation of function from $W_p^{1-1/p}(\partial\Omega)$ to $W_p^1(\Omega)$ we use the equivalence of norms $\|\cdot\|_{W_p^{1-1/p}(\partial\Omega)} \approx \|\cdot\|_{W_p^1(\Omega)}$.

Since the operator U is continuous with respect to indicated arguments under condition (21.2), the operator $T_t : X_1 \to W_p^2(\Omega, \mathbf{R}^N)$ ($p > n$) is continuous as well. Similarly, we can state that the operator T_t is continuous with respect to the parameter $t \in [0, 1]$.

The continuity of the operator $T_t : X_1 \to W_p^2(\Omega, \mathbf{R}^N)$, its continuity with respect to $t \in [0, 1]$, and the compact embedding $W_p^2(\Omega, \mathbf{R}^N) \to W_p^1(\Omega, \mathbf{R}^N)$ imply that the operator $T_t : X_1 \to X = W_p^1(\Omega, \mathbf{R}^N)$ is completely continuous ($p > n$).

22 THE EXISTENCE OF SOLUTIONS

Let us consider fixed points of the operator T_t. We denote $B_{R_1} = \{(v, \varphi) \in X| \|v_1\|_{1,p} + \cdots + \|v_{N-1}\|_{1,p} + \|\varphi\|_{1,p} \le R_1\}$ with $R_1 = c_0 R + 1$.

Here R is the constant from Theorem 19.1 and c_0 is the constant in the inequality $\|\cdot\|_{1,p} \le c_0 \|\cdot\|_{2,p}$ for the embedding $W_p^2(\Omega) \to W_p^1(\Omega)$ [if the space $W_p^2(\Omega)$ is equipped with the usual norm, we have $c_0 = 1$].

Let $Q = B_{R_1} \cap X_1$. Then Q is a closed convex set.

LEMMA 22.1. *Int* $Q \ne \emptyset$.

Proof. Any fixed point $(v, \varphi) \in X$ of the operator T_t with $t \in [0, 1]$ belongs to the space $W_p^2(\Omega, \mathbf{R}^N)$ and by Theorem 19.1 satisfies inequality (19.3). On the other hand, by Lemma 17.1 inequalities (17.5) and (17.6) are valid for any fixed point. Therefore, any fixed point of the operator T_t with $t \in [0, 1]$ belongs to Int Q.

Let us now consider the operator T_t at $t = 0$. A fixed point of this operator T_0 is defined as a solution to the problem

$$-\Delta v_r = 0 \quad \text{in} \quad \Omega,$$

$$\partial_\nu v_r + v_r = v_r^0 \quad \text{on} \quad \partial\Omega \qquad (r = 1, 2, \ldots, N-1),$$

$$-\Delta\varphi = 0 \quad \text{in} \quad \Omega,$$

$$\partial_\nu\varphi + \varphi = \varphi_0 \quad \text{on} \quad \partial\Omega.$$

(22.1)

The unique solution to this linear problem in $W_p^2(\Omega, \mathbf{R}^N)$ is the constant vector with components $v_r(x) = v_r^0$ $(r = 1, 2, \ldots, N-1)$, $\varphi(x) = \varphi_0$, and this vector satisfies (7.3) and (16.3).

On the other hand, the *a priori* estimate (19.3) is satisfied for this solution, and consequently this solution belongs to $B_{R_1} \backslash \partial B_{R_1}$. Thus, this solution belongs to $\text{Int}\, Q$.

TEOREM 22.1. *For any* $t \in [0, 1]$ *there exists a solution* $(v, \varphi) \in W_p^2(\Omega, \mathbf{R}^N)$ $(p > n)$ *to problem (17.1)–(17.4). The inequalities*

$$0 < m_r \le v_r(x) \le M_r \qquad (r = 1, 2, \ldots, N-1),$$
$$\varphi_1 \le \varphi(x) \le \varphi_2 \quad \text{in} \quad \Omega \tag{22.2}$$

and

$$\|v_1\|_{2,p} + \|v_2\|_{2,p} + \cdots + \|v_{N-1}\|_{2,p} + \|\varphi\|_{2,p} \le R \tag{22.3}$$

hold for this solution. Here the constants m_r *and* M_r $(r = 1, 2, \ldots, N-1)$ *are as in Lemma 17.1, and* R *is as in Theorem 19.1.*

Proof. Consider the family of completely continuous operators $T_t : Q \to X$, $t \in [0, 1]$. The boundary ∂Q of the closed convex compact set Q with $\text{Int}\, Q \ne \oslash$ does not contain a fixed point of the operator T_t for any $t \in [0, 1]$.

Indeed, let us assume the contrary. Let there exist a fixed point $(v, \varphi) \in \partial Q$ of the operator T_t for some $t \in [0, 1]$. Since $\text{Im}\, T_t(\overline{Q}) \subset W_p^2(\Omega, \mathbf{R}^N)$, it follows that this fixed point belongs to the space $W_p^2(\Omega, \mathbf{R}^N)$ and is a solution to problem (17.1)–(17.4) for this $t \in [0, 1]$. Then by the inequality (19.3) and Theorem 19.1 we find that

$$\|v_1\|_{1,p} + \|v_2\|_{1,p} + \cdots + \|v_{N-1}\|_{1,p} + \|\varphi\|_{1,p}$$
$$\le c_0 \left(\|v_1\|_{2,p} + \|v_2\|_{2,p} + \cdots + \|v_{N-1}\|_{2,p} + \|\varphi\|_{2,p} \right) \le c_0 R < R_1.$$

On the other hand, by Lemma 17.1 we have

$$0 < m_2/2 < m_r \le v_r(x) \le M_r < 2M_r \qquad (r = 1, 2, \ldots, N-1),$$
$$\varphi_1 - 1 < \varphi_1 \le \varphi(x) \le \varphi_2 < \varphi_2 + 1, \qquad x \in \overline{\Omega}.$$

Thus, the fixed point of the operator $T_t : Q \to X$ does not belong to the boundary ∂Q for any $t \in [0, 1]$. On the other hand, for $t = 0$ the operator T_0 has a fixed point $(v_0, \varphi_0) \in \text{Int}\, Q$. This fixed point is a solution to the linear problem (22.1) and its index is equal to 1.

By the Leray-Schauder principle we conclude that for any $t \in [0, 1]$ there exists a fixed point (v, φ) of the completely continuous operator $T_t : Q \to X$.

Furthermore, since $\text{Im}\, T_t(Q) \subset W_p^2(\Omega, \mathbf{R}^N)$, this fixed point (v, φ) belongs to the space $W_p^2(\Omega, \mathbf{R}^N)$ and is a solution to (17.1)–(17.4). By Lemma 17.1, inequality (22.2) is satisfied.

Inequality (22.3) follows from Theorem 19.1. Theorem 22.1 is proved.

Now let us prove the main Theorem 13.1. Set $t = 1$ in Theorem 22.1. Then we obtain a solution $(v, \varphi) \in W_p^2(\Omega, \mathbf{R}^N)$ to problem (12.3)–(12.6) with functions $\widetilde{F}_r$ and $\widetilde{G}_r$ $(r = 1, 2, \ldots, N)$, respectively.

By Theorem 22.1, inequalities (22.2) are satisfied for this solution, whence it follows that the relations $\widetilde{F}_r(x, v, \varphi) = F_r(x, v, \varphi)$ and $\widetilde{G}_r(x, v, \varphi) = G_r(x, v, \varphi)$ $(r = 1, 2, \ldots, N)$ hold on the obtained solution (v, φ) for $x \in \overline{\Omega}$.

Therefore, the solution $(v, \varphi) \in W_p^2(\Omega, \mathbf{R}^N)$ is a solution to problem (12.3)–(12.6) and satisfies (13.3). This completes the proof of Theorem 13.1.

23 THE H. AMANN PROBLEM

Straightforward verification shows that the functions $u_k(x)$ $(k = 1, 2, \ldots, N)$ and $\varphi(x)$ expressed via the components $v_r(x)$ $(r = 1, 2, \ldots, N - 1)$ and $\varphi(x)$ of the solution (v, φ) to problem (12.3)–(12.6) according to the formulas given in Sections 11 and 12 satisfy the system of equations and boundary conditions (10.1)–(10.3). By virtue of (11.1) and (12.1) we have $u_k(x) > 0$ in Ω for $k = 1, 2, \ldots, N - 1$.

The sign of $u_N(x)$ is chosen as follows. If there exists only one component of opposite charge, say negative, so that $I'_- = \varnothing$, then, by (11.6), we find $u_N(x) > 0$ in Ω.

Thus, in this case the solution to problem (12.3)–(12.6) specifies, by means of the formulas introduced in Sections 11 and 12, the solution to the Amann problem (10.1)–(10.3).

If there exist more than one negatively charged component and more than one positively charged component, it suffices to impose, in addition to Condition (A), yet another condition, so as to specify the sign of $u_N(x)$.

CONDITION (A_N^+). Let at least one of the following conditions be satisfied:

$$\text{either} \quad f_N(x, u_1, \ldots, u_N, \varphi) > 0 \quad \text{and} \quad g_N(x, u_1, \ldots, u_N, \varphi) \geq 0,$$
$$\text{or} \quad f_N(x, u_1, \ldots, u_N, \varphi) \geq 0 \quad \text{and} \quad g_N(x, u_1, \ldots, u_N, \varphi) > 0$$
$$\text{for all} \quad x \in \overline{\Omega}, \quad u_1 > 0, \ldots, u_N > 0 \quad \text{and} \quad \varphi \in [\varphi_1, \varphi_2].$$

Then under Conditions (A) and (A_N^+) we have $u_N(x) \geq 0$ in Ω.

Indeed, by the preceding, the function $u_N(x)$ specified by the solution (v, φ) to problem (12.3)–(12.6), satisfies the system in (10.1)–(10.3) to within a sign. To determine the sign, we rewrite the Nth equation and the Nth boundary condition in (10.1), (10.2) in the form

$$-\nabla \left(a_N e^{(\mu_N / a_N)\varphi} \nabla v_N \right) = f_N(x, u, \varphi) \quad \text{in} \quad \Omega, \tag{23.1}$$

$$a_N e^{-(\mu_N / a_N)\varphi} \partial_\nu v_N = g_N(x, u, \varphi) \quad \text{on} \quad \partial\Omega. \tag{23.2}$$

Here $u = (u_1(x), \ldots, u_N(x))$ and $\varphi = \varphi(x)$ are specified by the solution (v, φ) to problem (12.3)–(12.6) according to the formulas of Sections 11 and 12. In problem (23.1), (23.2) the function $u_N(x)$ is replaced by the function $v_N(x) = e^{\mu_N \varphi(x)/a_N} u_N(x)$. Note that earlier we have specified only the functions $v_1(x), \ldots, v_{N-1}(x)$.

We multiply equation (23.1) by the function

$$\zeta(x) = v_N^-(x) = (|v_N(x)| - v_N(x))/2$$

and repeat the reasoning used in the proof of Lemma 15.1. Then we obtain $v_N(x) \geq 0$ in Ω, and therefore, $u_N(x) \geq 0$ in Ω under Conditions (A) and (A_N^+).

Thus, under the above conditions the solution to problem (12.3)–(12.6) specifies a solution to the Amann problem (10.1)–(10.3) according to the formula given in Sections 11 and 12.

This work was financially supported by the Russian Foundation for Fundamental Basic Research (project No. 96–01–00097).

REFERENCES

(1) S. Agmon, A. Douglis, and L. Nirenberg (1959) *Estimates near Boundary For Solutions of Elliptic Partial Differential Equations Satisfying General Boundary Conditions*, I, Comm. Pure Appl. Math. **12**, 632–727.

(2) H. Amann (1992) *Reaction-Diffusion Problems in Electrolysis*, Preprint, Math. Inst, Universitat Zürich.

(3) H. Amann (1993) *Nonhomogeneous linear and quasilinear elliptic and parabolic boundary value problems*, in: H.-J. Schmeisser, Function Spaces, Differential Operators, and Nonlinear Analysis, ed. by H. Treibel, Proceedings of the Conference in Germany, Teubner-Texte zur Mathematik, **133**, 9–126.

(4) H. Amann and M. Renardy (1994) *Reaction-Diffusion problems in Electrolysis,* Nonlinear Diff. Equations Appl. **94, 1–40.**

(5) O. V. Besov, V. P. Il'in, and S. M. Nikol'skii (1975) *Integral Representations and Embedding Theorems* [in Russian], Moscow; English transl., Vols. 1, 2, Wiley, New York, 1979.

(6) G. Bourdaud and Y. Meyer (1991) *Fonctions qui operent sur les espases de Sobolev*, J. Funct. Anal., **97**, No. 2, 351–360.

(7) G. M. Fikhtengol'ts, *Differential and Integral Calculus. Supplement,* [in Russian], Vol. 1, Moscow (1966).

(8) D. Gilbarg and N. Trudinger (1977) *Elliptic Partial Differential Equations of Second Order*, Springer Verlag, Berlin, New York.

(9) O. A. Ladyzhenskaya and N. N. Ural'tseva (1973) *Linear and Quasilinear Equations of Elliptic Type*, [in Russian], Moscow.

(10) G. M. Liebermann (1983) *The conformal derivative problem for elliptic equations of variational type*, J. Diff. Equat., **49**, 218–257.

(11) A. Lunardi and V. Vespri (1991) *Holder regularity in variational parabolic nonhomogeneous equations*, J. Diff. Equat., **94**, 1–40.

(12) L. Nirenberg (1966) *An extended interpolation inequality*, Ann. Scuola Normale Super. Pisa, **20**, No. 15, 733-737.

Existence of Nonnegative Solutions for Generalized p-Laplacians

HUMBERTO PRADO: Universidad de Santiago de Chile, Casilla 307, Correo 2, Santiago, Chile.[1]

PEDRO UBILLA: Universidad de Santiago de Chile, Casilla 307, Correo 2, Santiago, Chile.[2]

1 INTRODUCTION

Troughout the following sections we shall be concerned with the existence of solutions of the following boundary value problem

$$(P)_\lambda \quad \begin{cases} -div(a(|\nabla u|^p)|\nabla u|^{p-2}\nabla u) & = \lambda f(x, u) \text{ in } \Omega \\ u & = 0 \text{ on } \partial\Omega \end{cases}$$

where λ is a positive parameter, $p > 1$, Ω is a bounded smooth domain in $\mathbb{R}^n$, and ∇u denotes the gradient of u. The function f will always be assumed to be in

---

[1]This research was partailly supported by FONDECYT grant 1950605, and Dicyt 049633PC

[2]Partially supported by FONDECYT grant 1950605 and Dicyt 04-9533UL

$C(\Omega \times I\!R; I\!R)$ and $a \in C(I\!R^+; I\!R)$. Henceforth we denote

$$A(t) = \int_0^t a(s)ds \quad , \quad F(x,t) = \int_0^t f(x,s)ds \tag{1}$$

The problem P_λ has been studied by P. Ubilla [6], under the assumption that the function $t \to A(t)$ is convex. The results presented in [6] are an extension of the results obtained by K. Narukawa and T. Suzuki [3] for the modified capillary surface equation, notice that in their case the function $a(t)$ takes the form

$$a(|t|^p)|t|^{p-2}t = \frac{|t|^{2p-2}t}{(1+|t|^{2p})^{\frac{1}{2}}} \tag{2}$$

We show the existence of solutions for the problem $(P)_\lambda$ under the assumption that the function $t \to A(t^p)$ is convex and $t \to A(t)$ is not necessarily convex.

The fact that the function $A(t)$ is convex allows in [3,11] to use Jensen's inequality. Then the existence of weak solutions in [3] is obtained by an application of the mountain pass lemma without the Palais-Smale (PS) conditon (c.f. [3]). It is worth to mention that among the results obtained by [5,6] it is obtained as a particular case, that the functional of Euler-Lagrange associated to the boundary value problem does indeed verify the (PS) condition. We also remark that our results can be applied to a large class of operators, that is for example those operators which asymptotically behave as

$$a(|\nabla u|^p)|\nabla u|^{p-2}\nabla u \sim |\nabla u|^{p-2}\nabla u \tag{3}$$

when $|\nabla u| \to \infty$ and

$$a(|\nabla u|^p)|\nabla u|^{p-2}\nabla u \sim |\nabla u|^{q-2}\nabla u \tag{4}$$

when $|\nabla u| \to 0$, for $q < p$. The case when $q > p$ has been studied in [6]. In the context of this article we require $a(|t|^p)|t|^{p-2}t$ be asymptotic to a power of $p-1$ at infinity, furthermore, near the origin there is no asymptotic assumption. Instead we have assumed on the function $A(t)$ a weaker condition, which is given bellow by hypothesis II.

The scope of this article is to show the existence of nonnegative solutions for the superlinear and the sublinear problems respectively. The weak solutions of $(P)_\lambda$ are the critical points of the functional

$$I_\lambda(u) = \frac{1}{p} \int_\Omega A(|\nabla u|^p)dx - \lambda \int_\Omega F(x,u)dx \tag{5}$$

defined on the Sobolev space $W_0^{1,p}(\Omega)$ with the norm $\|u\|_{1,p} = (\int_\Omega |\nabla u|^p)^{\frac{1}{p}}$.
We also put the following hypothesis on A and F,

(I) The function $t \to A(|t|^p)$ is strictly convex.

(II) There are positive constants c_0, c_1, and c_2 such that

$$c_0 t \le A(t) \le c_1 t + c_2 \qquad t \ge 0$$

(III) There are positive constants b_0, b_1 such that for all $x \in \Omega$

$$|F(x,t)| \le b_0|t|^r + b_1 \qquad \text{for } r < p^*$$

where $p^* = \dfrac{pn}{n-p}$ is the critical exponent defined for $n > p$. We notice that if conditions (I) through (III) are satisfied then the functional I_λ is a C^1-functional on the space $W_0^{1,p}(\Omega)$. We shall also assume for the superlinear case the Ambrosetti-Rabinowitz type of condition, i.e., there exists

$0 < \theta < \dfrac{1}{p}, \quad t_0 > 0$ such that for all $x \in \Omega$

$(AR)_p \qquad\qquad \theta f(x,t)t > F(x,t) > 0 \qquad \text{for } 0 < t_0 < |t|$

Then, under hypothesis (I)~(III) and the $(AR)_p$ condition the (PS) condition is satisfied by the functional I_λ ([5,6]).

THEOREM 1 *(The superlinear case) Assume hypothesis (I)~(III) and the $(AR)_p$ condition. Then there exists a positive constant λ^* such that for any $0 < \lambda < \lambda^*$ there exists a non trivial solution u_λ in $W_0^{1,p}(\Omega)$ of $(P)_\lambda$. Moreover $\lim_{\lambda \to 0} \|u_\lambda\|_{1,p} = \infty$*

The proof of the above theorem is obtained by an aplication of the following two lemmas which assume hypothesis (II)~(III) and $(AR)_p$ condition.

LEMMA 1 *There exists positive constants $\alpha_\lambda, \rho_\lambda$ such that $\lim_{\lambda \to 0^+} \alpha_\lambda = \infty$, and $I_\lambda(u) > \alpha_\lambda > 0$ whenever $\|u\|_{1,p} = \rho_\lambda$.*

Proof. It follows from (II) and (III) that for each $u \in W_0^{1,p}(\Omega)$

$$I_\lambda(u) \geq c_0 \int_\Omega |\nabla u|^p dx - \lambda \int_\Omega (b_0|u|^r + b_1)dx \qquad (6)$$

Then the Poincare's inequality yields

$$I_\lambda(u) \geq \frac{c_0}{p}\|u\|_{1,p}^p - \lambda k_0\|u\|_{1,p}^r - \lambda k_1 \qquad (7)$$

for some positive constants k_0, k_1. Now, if we choose u so that

$$\|u\|_{1,p} = \lambda^{-\alpha} \qquad , 0 < \alpha < \frac{1}{r-p} \qquad (8)$$

then we define $\rho_\lambda := \lambda^{-\alpha}$. Thus we obtain

$$I_\lambda(u) \geq c_0\lambda^{-\alpha p} - k_0\lambda^{1-\alpha r} - \lambda k_1 \qquad (9)$$

Hence we define $\alpha_\lambda = c_0\lambda^{-\alpha p} - k_0\lambda^{1-\alpha r} - \lambda k_1$. Since $(AR)_p$ implies $p < r$. Thus α_λ verifies the assertion of the lemma.

LEMMA 2 *Let* $v \neq 0$ *in* $W_0^{1,p}(\Omega)$. *Then*

$$\lim_{t \to +\infty} I_\lambda(tv) = -\infty \qquad (10)$$

Proof. Notice that the $(AR)_p$ condition together with hypothesis (II) yields the estimate

$$I_\lambda(tv) \leq c_1 t^p \|v\|_{1,p}^p - k_0 t^{\frac{1}{\theta}} \|v\|_{1,p}^{\frac{1}{\theta}} - k_1 \qquad (11)$$

Where k_0, k_1 are positive constants. Now the desired limit follows from the fact that $p < \frac{1}{\theta}$.

We are now ready to proof theorem 1.
Proof of theorem 1. In addition to the Ambrosetti-Rabinowitz hypothesis $(AR)_p$, notice that assumptions (I) through (III) imply that the functional I_λ satisfy the Palais-Smale hypothesis [5]. Thus the foregoing lemmas allow us to apply the Mountain pass theorem [4]. Therefore there exists a non trivial critical point u_λ for I_λ such that

$$I_\lambda(u_\lambda) = c_\lambda \geq \alpha_\lambda \qquad (12)$$

Moreover from (II) and (III) we get

$$I(u_\lambda) \le c_1 \|u_\lambda\|_{1,p}^p + c_2 |\Omega| \tag{13}$$

Hence from equation (12) and lemma 1 we obtain $\lim_{\lambda \to 0^+} \|u_\lambda\|_{1,p} = +\infty$. This concludes with the proof of theorem 1.

The question of finding another weak solution v_λ of $(P)_\lambda$ such that norm $\|v_\lambda\|_{1,p}$ tends to zero when $\lambda \to 0$, can be answer by imposing certain conditions of regular variation at zero on the functions A and F respectively. This is the context of our Theorem 2 below. We recall that conditions of regular variation type have been previously introduced in the works of [1,2] in order to show the existence of positive radially symmetric solutions for quasilinear partial differential equations.

For our next result we need the following lemma which in turn is an immediate consequence of Ekeland's Variational Principle.

LEMMA 3 *Let X be a Banach space and $I : X \to \mathbb{R}$ a lower semicontinuous function which is Gateaux differentiable. In addition suppose that I is bounded below on the set $\overline{B(0,\delta)}$ and $\inf\{I(u) : u \in \overline{B(0,\delta)}\} < 0$. Moreover $I(u) \ge 0$ when $\|u\| = \delta$. Then for each $0 < \epsilon < -\inf\{I(u) : u \in \overline{B(0,\delta)}\}$ there exists u_ϵ such that $\|u_\epsilon\| < \delta$ and*
(i) $I(u_\epsilon) \le \inf\{I(u) : u \in \overline{B(0,\delta)}\} + \epsilon$
(ii) $\|I'(u_\epsilon)\| \le \epsilon$

Proof. We apply Ekeland's principle to the function I restricted to $\overline{B(0,\delta)}$. Hence for each $\epsilon > 0$ there exists a point $u_\epsilon \in \overline{B(0,\delta)}$ such that

$$I(u_\epsilon) - I(u) \le \epsilon \|u - u_\epsilon\| \tag{14}$$

for every $u \in \overline{B(0,\delta)}$, $u \ne u_\epsilon$. Now if $\|u_\epsilon\| = \delta$ then $I(u_\epsilon) \ge 0$. Moreover, (i) is automatically satisfied. Thus

$$0 \le I(u_\epsilon) \le \inf\{I(u) : u \in \overline{B(0,\delta)}\} + \epsilon \tag{15}$$

Since $c = \inf\{I(u) : u \in \overline{B(0,\delta)}\} < 0$ so whenever $0 < \epsilon < -c$ we arrive at a contradiction. Thus $\|u_\epsilon\| < \delta$. Moreover from inequality (14) we obtain (ii).

THEOREM 2 *Let us assume that $f(x,t) \ge 0$ for all $x \in \Omega$, and $t \ge 0$. Suppose the following holds for some $r_0 > 0$*

(a) $\lim_{t\to 0} \dfrac{F(x,t\sigma)}{F(x,t)} = \sigma^{r_0}$ *for all* $0 < \sigma < 1$, $x \in \Omega$

(b) $\lim_{t\to 0} \dfrac{F(x,t)}{A(t^p)} = +\infty$, $x \in \Omega$

Then under assumptions $(I) \sim (III)$ *there exists a positive constant* λ^* *such that for each* $0 < \lambda < \lambda^*$ *there exists a solution* u_λ. *Moreover*

$$\lim_{\lambda\to 0^+} \|u_\lambda\|_{1,p} = 0 \tag{16}$$

Proof. Let $\rho_\lambda = \lambda^\alpha$ for $\alpha > 0$. Then for a given $u \in B(0,\rho_\lambda)$ and recalling inequality (7) we obtain that

$$I(u) \geq c_0 \lambda^{\alpha p} - k_0 \lambda^{1+\alpha r} - k_1 \lambda \tag{17}$$

Now, if we let $0 < \alpha < \dfrac{1}{p}$ then choosing λ^* sufficiently small we obtain

$$I_\lambda(u) \geq 0 \tag{18}$$

for $0 < \lambda < \lambda^*$ whenever $\|u\|_{1,p} = \rho_\lambda = \lambda^\alpha$

Moreover, from inequality (7) it follows that I_λ is bounded below on the set $B[0,\rho_\lambda] = \{u \in W_0^{1,p}(\Omega) : \|u\|_{1,p} \leq \rho_\lambda\}$.

Now we define $\phi_t = tv$, $t \geq 0$, for $v \in C_0^\infty(\Omega)$, such that $0 < v \leq 1$ and $0 \leq |\nabla v| \leq 1$. Then

$$I_\lambda(\phi_t) = \frac{1}{p}\int_\Omega A(t^p|\nabla v|^p)dx - \lambda \int_\Omega F(x,tv)dx \tag{19}$$

so

$$I_\lambda(\phi_t) = A(t^p)\left[\frac{1}{p}\int_\Omega \frac{A(t^p|\nabla v|^p)}{A(t^p)}dx - \frac{\lambda F(x,t)}{A(t^p)}\int_\Omega \frac{F(x,tv)}{F(x,t)}dx\right] \tag{20}$$

Since $\dfrac{A(t^p|\nabla v|^p)}{A(t^p)} \leq 1$ for $t > 0$

and $\dfrac{F(x,tv)}{F(x,t)} \leq 1$ for $t > 0$, $x \in \Omega$.

Then (a) and (b), and the dominated convergence theorem yields the existence of $\delta > 0$ such that

$$I_\lambda(tv) < 0 \qquad\qquad \text{whenever } 0 < t < \delta$$

Thus $c_\lambda = \inf\{I^\lambda(u) : u \in \overline{B(0,\rho_\lambda)}\}$ is negative, and the assumptions of the preceding lemma are verified. Since I^λ satisfy the (PS) condition, there exists a non trivial minimizer u_λ in the interior of $B(0,\rho_\lambda)$, i.e., a non trivial weak solution of $(P)_\lambda$. Moreover $\|u_\lambda\|_{1,p} < \lambda^\alpha$ for $0 < \lambda < \lambda^*$. Hence, if $\alpha > 0$ then $u_\lambda \to 0$ when $\lambda \to 0^+$. This concludes with the proof.

COROLLARY 3 *Under the hypothesis of theorems 1, and 2, there exists at least two solutions u_λ and v_λ of P_λ such that $\lim_{\lambda \to 0} \|u_\lambda\|_{1,p} = +\infty$ and $\lim_{\lambda \to +0} \|v_\lambda\|_{1,p} = 0$*

THEOREM 4 *(The sublinear case) In addition to conditions $(I) \sim (III)$ assume that $r < p$ in (III), and suppose the following holds for all $x \in \Omega$ and all $0 < \sigma < 1$*

(i) $\lim_{t \to 0} \dfrac{F(x,t\sigma)}{F(x,t)} = \sigma^{r_0}$ *where $r_0 < p$*

(ii) $\lim_{t \to 0} \inf\left\{\dfrac{F(x,t)}{A(t^p)}\right\} > 0$

(iii) There are positive constants c_1, c_2, and $r_1 > 0$ such that

$$F(x,t) \geq c_1 t^{r_1} - c_2 \tag{21}$$

Then there is $\lambda^* > 0$ such that for $\lambda > \lambda^*$ there exists a solution u_λ such that $\|u_\lambda\|_{1,p} \to +\infty$ when $\lambda \to \infty$

Proof. Since

$$I_\lambda(u) \geq \frac{c_0}{p}\|u\|_{1,p}^p - \lambda k_0 \|u\|_{1,p}^r - \lambda k_1 \tag{22}$$

holds for some positive constants k_0, k_1 and $r < p$, then for each λ $c_\lambda = \inf\{I_\lambda(u) : u \in W_0^{1,p}(\Omega)\} > -\infty$. Thus there exist a minimizer u_λ since the functional I_λ satisfy the (PS) condition (for details see Prop. 3.2. in [5]). ¿From (ii) it follows the existence of μ, $\delta > 0$ such that

$$\frac{F(x,t)}{A(t^p)} \geq \mu, \qquad \text{whenever } 0 < t < \delta$$

Next we choose $\phi_t = tv$ then

$$I_\lambda(\phi_t) \leq A(t^p)\left(\frac{1}{p}\int_\Omega \frac{A(t^p|\nabla v|^p)}{A(t^p)}dx - \mu\lambda\int_\Omega \frac{F(x,tv)}{F(x,t)}dx\right) \tag{23}$$

for t sufficiently small. Thus there is $\lambda^* > 0$ such that for $\lambda > \lambda^*$ there exists t_λ so that

$$I_\lambda(t_\lambda v) < 0 \qquad (24)$$

Thus $-\infty < c_\lambda < 0$. Hence there exist u_λ a non trivial minimizer. From (iii) we obtain

$$I_\lambda(u) \leq \frac{c_0}{p} \int_\Omega |\nabla u|^p dx - \lambda c_1 \int_\Omega |u|^{r_0} \qquad (25)$$

Thus, fixing v and letting $u = tv$, we obtain

$$c_\lambda \leq \inf_{t \geq 0} \{ A t^p - \lambda B t^{r_0} \} \qquad (26)$$

where $A = \dfrac{c_0}{p} \displaystyle\int_\Omega |\nabla v|^p dx, \qquad B = c_1 \displaystyle\int_\Omega |v|^{r_0} dx$

Hence

$$c_\lambda \leq A^{-\frac{r_0}{p-r_0}} B^{\frac{p}{p-r_0}} \left[\left(\frac{r_0}{p} \right)^{\frac{p}{p-r_0}} - \left(\frac{r_0}{p} \right)^{\frac{r_0}{p-r_0}} \right] \lambda^{\frac{p}{p-r_0}} \qquad (27)$$

Since $r_0 < p$ and $\lambda > 0$ the right hand side of the inequality is less than zero. Thus

$$c_\lambda \leq -k \lambda^{\frac{p}{p-r_0}} \qquad \text{for } k > 0$$

On the other hand from hypothesis (II) and (III) we get

$$\begin{aligned} c_\lambda &= I_\lambda(u_\lambda) \qquad\qquad\qquad\qquad\qquad\qquad\qquad (28)\\ &= \frac{1}{p} \int_\Omega A(|\nabla u|^p) dx - \lambda \int_\Omega F(x, u_\lambda) dx \\ &\geq \frac{c_0}{p} \int_\Omega |\nabla u_\lambda|^p dx - \lambda \left(b_0 \int_\Omega |u_\lambda|^r dx + b_1 \right) \end{aligned}$$

Therefore

$$-k \lambda^{\frac{p}{p-r_0}} \geq c_\lambda \geq \frac{c_0}{p} \|u_\lambda\|_{1,p}^p - \lambda(C \|u_\lambda\|_{1,p}^r + b_1) \qquad (29)$$

where C is a positive constant. Now, if $\|u_\lambda\|_{1,p}$ is bounded for all $\lambda > 0$, then there exists a subsequence $\{\lambda_n\}$ such that $\lambda_n \to \infty$ and the sequence $\|u_{\lambda_n}\|_{1,p}$ converges when $\lambda_n \to \infty$. Then, from the above inequality and after dividing by λ_n we obtain

$$-k \lambda_n^{\frac{r_0}{p-r_0}} \geq \frac{c_0}{p \lambda_n} \|u_{\lambda_n}\|_{1,p}^p - (C \|u_{\lambda_n}\|_{1,p}^r + b_1) \qquad (30)$$

Now passing to the limit as $\lambda_n \to \infty$ we arrive to a contradiction.

Our next result gives the necessary conditions to ensure the existence of non-negative solutions for the problem (P_λ).

THEOREM 5 *Assume that f is such that for all* $x \in \Omega, f(x,t) = 0$ *if* $t \le 0$ *and* $f(x,t) > 0$, *for* $t > 0$. *Suppose that there exists s with* $s < p^*$ *such that for all* $x \in \Omega, t \ge 0$

$$f(x,t) \le d_0 t^{s-1} + d_1 \qquad (31)$$

where d_0, d_1 *are positive constants. If u is a weak solution of* (P_λ) *then u is non-negative.*

Proof. Let u be a weak solution of (P_λ). Then for every $\phi \in W_0^{1,p}(\Omega)$

$$\int (a(|\nabla u|^p)|\nabla u|^{p-2}\nabla u)\nabla \phi dx = \lambda \int f(x,u)\phi dx \qquad (32)$$

Let $u = u_+ - u_-$ and take $\phi = u_-$ then

$$-\int_\Omega a(|\nabla u_-|^p)|\nabla u_-|^p dx = \lambda \int f(x,-u_-)u_- dx \qquad (33)$$
$$= 0$$

Hence $|\nabla u_-| = 0$. Therefore $u \ge 0$ a.e. on Ω.

COROLLARY 6 *Under the hypothesis of theorems 1, and 2, there exists at least two nonnegative solutions* u_λ *and* v_λ *of* P_λ *such that* $\lim_{\lambda \to 0} \|u_\lambda\|_{1,p} = +\infty$ *and* $\lim_{\lambda \to +0} \|v_\lambda\|_{1,p} = 0$

COROLLARY 7 *Let f be such that* $f(x,t) \ge 0$ *for all* $t > 0$ *and all x. Then under the hypothesis of theorem 4 there is* $\lambda^* > 0$ *such that for* $\lambda > \lambda^*$ *there exists a nonnegative solution* u_λ *such that* $\|u_\lambda\|_{1,p} \to +\infty$ *when* $\lambda \to \infty$

1.1 CONCLUDING REMARKS

We conclude with some examples which are included among the class of non-linear differential operators that verify the conditions of the foregoing theorems. Thus the associated boundary value problem admits the existence of nonnegative solutions. For $1 < q < p$, we define

1. $L_p u = -\Delta_p u - c\Delta_q u$, where $c \geq 0$

2. $L_p u = -div((1 + |\nabla u|^q)^{\frac{p}{q}-1}|\nabla u|^{q-2}\nabla u)$

3. $L_p u = -div\left(\dfrac{(1 + |\nabla u|)^{\frac{-1}{2}}\nabla u}{ln(1 + |\nabla u|^{-p+1})} \right)$

We notice that examples (1) and (2) are of the type $-\Delta_q u$ at zero and $\Delta_p u$ at infinity. On the other hand example (3) is asymptotic to a p-Laplacian at infinity, while at zero is not asymptotic to a q-Laplacian for any choice of q.

REFERENCES

[1] GARCIA-HUIDOBRO M., MANASEVICH R. AND UBILLA P., *Existence of Positive Solutions for some Dirichlet Problems with an Asymptotically Homogeneous Operator.* Elec. J. of Diff. Equat., **10**, (1995), 1-22.

[2] GARCIA-HUIDOBBRO M., AND UBILLA P.,*Multiplicity of solutions for class of nonlinear second order equations.* To appear in Nonlinear Analysis T.M.A.

[3] NARUKAWA K. AND SUZUKI T., *Nonlinear Eigenvalue Problem for a Modified Capillary Surface Equation.* Funkcialaj Ekvacioj, **37**, (1994), 81-100.

[4] RABINOWITZ P.H.,*Minimax Methods in Critical Point Theory with Applications to Differential Equations,* C.B.M.S. Regional Confer. Ser. In Math, **65**, Am. Math. Soc., Providence, RI (1986).

[5] UBILLA P, *Multiplicity Results of Quasilinear Elliptic Equations,* Comm. on Appl. Nonlinear Analysis 3(1996); Number 2;35-49.

[6] UBILLA P, *Existence of Nonnegative solutions for Quasilinear Dirichlet Problem* ,(Submitted).

Stability and Blow-up for Dissipative Evolution Equations

PATRIZIA PUCCI Dipartimento di Matematica, Università degli Studi, Via Vanvitelli 1, 06123 Perugia, pucci@unipg.it

JAMES SERRIN Department of Mathematics, University of Minnesota, Minneapolis, MN 55455, serrin@math.mn.edu

1 INTRODUCTION

The problem of stability and blow–up for dissipative evolution equations will be treated by Lyapunov–type methods. The discussion will be carried out particularly in the context of evolution operators in a Banach space, with special care given to an appropriate definition of solution, and with the specific examples of degenerate damped wave equations and degenerate parabolic equations as principal applications.

Our treatment is expository in intent, based principally on the references [3, 7, 9, 10, 17, 18, 20]. Here we discuss mainly simplified versions of the results, in order to show the main ideas of the theory and to avoid technicalities.

2 GENERAL SETTING

The abstract evolution equation which we consider are of the type

$$[P(u_t)]_t + Q(t, u_t(t)) + A(u(t)) = F(u(t)), \quad t \in J = [0, \infty). \tag{2.1}$$

Simple examples, when $u : J \times \Omega \to$, with Ω an open bounded subset of ^{n}, are the following:

1. The wave equation: when $P = I$, $Q = 0$, $A = -\Delta$,

$$u_{tt} - \Delta u = f(x, u);$$

2. The wave equation with linear dissipation: as above, with $Q(t,v) = bv$ and $b > 0$,

$$u_{tt} + bu_t - \Delta u = f(x,u);$$

3. The parabolic equation: when $P = 0$, $Q(v) = v$, $A = -\Delta$,

$$u_t = \Delta u + f(x,u).$$

Of course the operators corresponding to these concrete cases must be understood in the sense of Nemitsky, as being defined by appropriate distribution relations. We shall discuss this in detail a little later.

First let us clarify the meaning of (2.1) as an abstract evolution equation. In particular we suppose that

$$P : V \to V', \quad A : W \to W', \quad F : X \to X'$$

for real Banach spaces V, W, X, and their dual spaces V', W', X'. Moreover, we suppose that P, A, F are the Fréchet derivatives of real C^1 potentials

$$\mathcal{P} : V \to , \quad \mathcal{A} : W \to , \quad \mathcal{F} : X \to ,$$

with $\mathcal{P}(0) = \mathcal{A}(0) = \mathcal{F}(0) = 0$ (normalization); thus from the definition of derivative we immediately get, in particular, that

$$\mathcal{F}(u) = \int_0^1 \langle F(\tau u), u \rangle_X d\tau,$$

where $\langle x', x \rangle_X = x'(x)$ for all $x \in X$, $x' \in X'$; and so on. Finally

$$Q : J \times Y \to X',$$

with $Y \hookrightarrow V$ continuously as Banach spaces. Some mild simplifications from the original papers have been made in order to clarify this exposition.

EXAMPLE 1. $A = -\Delta$, $W = H_0^1(\Omega)$, $\Omega \subset {}^n$. By definition, for $u \in H_0^1(\Omega)$, the expression $-\Delta u$ denotes the element w' of $[H_0^1(\Omega)]'$ such that

$$\langle w', \varphi \rangle_W \equiv w'(\varphi) = \int_\Omega (Du, D\varphi)dx \quad \text{for all } \varphi \in H_0^1(\Omega).$$

It is easy to check that A is the Fréchet derivative of the potential

$$\mathcal{A}(u) = \tfrac{1}{2} \int_\Omega |Du|^2 dx = \tfrac{1}{2}\|u\|_W^2.$$

Similarly, for the degenerate s–Laplace operator $A = -\Delta_s = -\text{div}(|Du|^{s-2}Du)$, we take $W = W_0^{1,s}(\Omega)$ and $-\Delta_s u$ as the element w' of W' such that

$$\langle w', \varphi \rangle_W = \int_\Omega |Du|^{s-2}(Du, D\varphi)dx \quad \text{for all } \varphi \in W_0^{1,s}(\Omega).$$

It follows at once that

$$A(u) = \frac{1}{s} \int_\Omega |Du|^s dx = \frac{1}{s} \|u\|_W^s.$$

EXAMPLE 2. Let P be a symmetric operator from a real Hilbert space V into V', that is

$$\langle Pv, \varphi \rangle_V = \langle P\varphi, v \rangle_V \quad \text{for all } v, \varphi \in V.$$

It is easy to prove that P must then be linear and also, by the uniform boundedness principle, continuous. Moreover, as is readily verified, by virtue of the symmetry of P we have the important formula

$$\mathcal{P}(v) = \tfrac{1}{2}\langle Pv, v \rangle_V.$$

As a special case, let $V = [L^2(\Omega)]^N$ and define the Riesz identity $I : V \to V'$ by

$$\langle Iv, \varphi \rangle_V = \int_\Omega (v, \varphi) dx.$$

Then $\mathcal{I}(v) = \tfrac{1}{2}\|v\|_{[L^2(\Omega)]^N}^2$. By abuse of notation one frequently writes $\langle v, \varphi \rangle_V$ instead of $\langle Iv, \varphi \rangle_V$.

EXAMPLE 3. Let $f : \overline{\Omega} \times \to$ be continuous. Define $\Phi(x, u) = \int_0^u f(x, z)dz$, so that $f(x, u) = \dfrac{\partial \Phi}{\partial u}(x, u)$ for all $x \in \overline{\Omega}$, $u \in$. For vector functions $f : \overline{\Omega} \times {}^N \to {}^N$ this definition is an assumption. Assume that

$$|f(x, u)| \le \text{Const.}\,(1 + |u|^{p-1}), \quad p > 1.$$

Then we can take $X = L^p(\Omega)$, and for $u \in X$ define $F(u)$ to be the element x' (Nemitsky operator) of $X' = L^{p'}(\Omega)$ such that

$$\langle x', \varphi \rangle_X \equiv w'(\varphi) = \int_\Omega f(x, u(x))\varphi(x)dx \quad \text{for all } \varphi \in L^p(\Omega).$$

(Here p' is the Hölder conjugate of p.) That this is well–defined follows from the calculation

$$|f(x, u(x))|^{p'} = |f(x, u(x))|^{p/(p-1)} \le \text{Const.}\,(1 + |u(x)|^p), \quad \text{in } \Omega,$$

so that $f(\cdot, u(\cdot)) \in L^{p'}(\Omega)$ for each $u \in L^p(\Omega)$. Furthermore we calculate easily that

$$\mathcal{F}(u) = \int_0^1 \int_\Omega f(x, \tau u(x))u(x)dx d\tau = \int_0^1 \int_\Omega \frac{d\Phi}{d\tau}(x, \tau u(x))dx d\tau$$

$$= \int_\Omega \int_0^1 \frac{d\Phi}{d\tau}(x, \tau u(x))d\tau dx = \int_\Omega \Phi(x, u(x))dx.$$

Note that $\mathcal{F}(u) = \|u\|_{L^p(\Omega)}^p/p$ when $f(u) = |u|^{p-2}u$. Moreover of course $F \in C(X \to X')$.

EXAMPLE 4. For the continuous function $\tilde{Q} : J \times \; \to \;$, defined by

$$\tilde{Q}(t, v) = b(t)|v|^{m-2}v, \quad m > 1,$$

the corresponding (Nemitsky) operator $Q : J \times Y \to Y'$ is given by

$$\langle Q(t, v), \varphi \rangle_Y = \int_\Omega \tilde{Q}(t, v(x))\varphi(x)dx$$

for $Y = L^m(\Omega)$, $Y' = L^{m'}(\Omega)$. Various more general functions $\tilde{Q}$ can also be allowed; see [9, 17].

Let us now return to the abstract problem (2.1), and give meaning to the idea of a solution. To begin with, we note that, formally,

$$\int_0^t \langle [P(u_t(\tau))]_t, \varphi(\tau) \rangle_V d\tau = \int_0^t \left\{ \frac{d}{d\tau} \langle P(u_t(\tau)), \varphi(\tau) \rangle_V - \langle P(u_t(\tau)), \varphi_t(\tau) \rangle_V \right\} d\tau$$

$$= \langle P(u_t(\tau)), \varphi(\tau) \rangle_V \Big|_0^t - \int_0^t \langle P(u_t(\tau)), \varphi_t(\tau) \rangle_V d\tau,$$

where the right hand side of this relation is well–defined for $u_t \in V$ and φ, $\varphi_t \in V$.

We assume that V, W, X have a common subspace G – not necessarily closed. If $W \hookrightarrow X \hookrightarrow V$, as is commonly the case in applications, one takes $G = W$. We now define the principal set

$$K = \{\varphi : J \to G \mid \varphi \in C(J \to W) \cap C(J \to X) \cap C^1(J \to V)\}$$

and say that $u \in K$ is a *(strong) solution* of (2.1) if,

(a) $u_t(t) \in Y$ for a.a. $t \in J$, and $\langle Q(\cdot, u_t(\cdot)), \varphi(\cdot) \rangle_X : J \to \; \cup \{-\infty, \infty\}$ is measurable for all $\varphi \in K$;

(b)   Distribution Identity:

$$\langle P(u_t(\tau)), \varphi(\tau) \rangle_V \Big|_0^t = \int_0^t \langle \{P(u_t(\tau)), \varphi_t(\tau) \rangle_V - \langle Q(\tau, u_t(\tau)), \varphi(\tau) \rangle_X$$

$$- \langle A(u(\tau)), \varphi(\tau) \rangle_W + \langle F(u(\tau)), \varphi(\tau) \rangle_X \} d\tau$$

for all $t \in J$ and $\varphi \in K$.

A third, and crucial, element in the definition of a solution of (2.1) is an appropriately formulated *conservation law* for the *energy* of a solution. To this end, the first task is to set up an appropriate energy functional. Proceeding formally, we write

$$\frac{d}{dt}\mathcal{A}(u(t)) = \langle A(u(t)), u_t(t) \rangle_W, \quad \frac{d}{dt}\mathcal{F}(u(t)) = \langle F(u(t)), u_t(t) \rangle_X$$

by virtue of the concept of Fréchet derivative. For $v \in V$ we introduce the *Hamiltonian* of the potential $\mathcal{P}$ by the formula

$$\mathcal{P}^*(v) = \langle P(v), v \rangle_V - \mathcal{P}(v) \tag{2.2}$$

(for Example 2 it is easy to see that $\mathcal{P}^*(v) = \mathcal{P}(v) = \frac{1}{2}\langle Pv, v\rangle_V$). Then, again formally,

$$\frac{d}{dt}\mathcal{P}^*(u_t(t)) = \langle\frac{d}{dt}P(u_t(t)), u_t(t)\rangle_V = \frac{d}{dt}\langle P(u_t(t)), u_t(t)\rangle_V - \langle P(u_t(t)), u_{tt}(t)\rangle_V;$$

this can be checked by assuming that $P(u_t)$ and u_t are both differentiable functions of t. Now put $\varphi = u_t$, formally, into the distribution identity (b); with the help of the above formal relations, we then get

$$\int_0^t \left\{-\frac{d}{dt}\mathcal{P}^*(u_t(\tau)) - \frac{d}{dt}\mathcal{A}(u(\tau)) + \frac{d}{dt}\mathcal{F}(u(\tau)) - \langle Q(\tau, u_t(\tau)), u_t(\tau)\rangle_X\right\} d\tau = 0,$$

that is,

$$\mathcal{P}^*(u_t(\tau)) + \mathcal{A}(u(\tau)) - \mathcal{F}(u(\tau))\Big|_0^t = -\int_0^t \langle Q(\tau, u_t(\tau)), u_t(\tau)\rangle_X d\tau.$$

Since $u_t \notin X$, in general, the right hand side of the last relation is of course only formal. We now take the major step of introducing the following *third part* to the definition of solution of (2.1):

(c)   Conservation Law: Let

$$\mathcal{E}u(t) = \mathcal{P}^*(u_t(t)) + \mathcal{A}(u(t)) - \mathcal{F}(u(t)) \tag{2.3}$$

be the *total energy of u*. There exists a function $\mathcal{D} : J \times Y \to [0, \infty]$, called the *dissipation rate*, such that $\mathcal{D}(\cdot, u_t(\cdot)) : J \to [0, \infty]$ is measurable and

$$\mathcal{E}u(t) - \mathcal{E}u(0) \le -\int_0^t \mathcal{D}(\tau, u_t(\tau))d\tau, \qquad t \in J. \tag{2.4}$$

Clearly $\mathcal{D}(\cdot, u_t(\cdot))$ is then locally integrable on J.

Condition (c) in this form first appears in [9, 17] – see particularly the discussion in [17, Section 2], and the related papers [12] and [23] concerning the existence of (strong) solutions of damped wave systems.

One must add to (c) a minimal relation between the dissipation rate $\mathcal{D}$ and the damping norm $\|Q(t, v)\|_{X'}$:

(C)   *There is an exponent $m > 1$ and a positive locally integrable function $\delta = \delta(t)$ such that*

$$\|Q(t, v)\|_{X'} \le [\delta(t)]^{1/m}[\mathcal{D}(t, v)]^{1/m'} \quad \text{for all } (t, v) \in J \times Y,$$

where m' denotes the Hölder conjugate of m.

Condition (C) allows us to prove (what has so far not been done) that the term $\langle Q(\cdot, u_t(\cdot)), \varphi(\cdot)\rangle_X$ in the distribution identity (b) is locally integrable on J – note that it is measurable on J by (a). Indeed, for a.a. $t \in J$ we have

$$|\langle Q(t, u_t(t)), \varphi(t)\rangle_X| \le \|Q(t, u_t(t))\|_{X'}\|\varphi(t)\|_X \le [\delta(t)]^{1/m}[\mathcal{D}(t, u_t(t))]^{1/m'}\|\varphi(t)\|_X;$$

so, using Hölder's inequality,

$$\int_0^t |\langle Q(\tau, u_t(\tau)), \varphi(\tau) \rangle_X| d\tau \le \left(\int_0^t \delta(\tau) d\tau \right)^{1/m} \left(\int_0^t \mathcal{D}(\tau, u_t(\tau)) d\tau \right)^{1/m'} \sup_{[0,t]} \|\varphi(\tau)\|_X.$$

But this is locally finite, as required, since $\varphi \in C(J \to X)$ and $\mathcal{D}(\cdot, u_t(\cdot))$ is locally integrable on J.

Since the damping term Q is of crucial importance, we show how condition (C) arises from the natural choice

$$\mathcal{D}(t, v) = \langle Q(t, v), v \rangle_Y$$

in the case when

$$Q : J \times Y \to Y', \qquad X \hookrightarrow Y \quad \text{continuously,}$$

and when, for all $t \in J$ and $v \in Y$,

$$\|Q(t, v)\|_{Y'} \le \hat{\delta}(t) \|v\|_Y^{m-1} \tag{i}$$

$$\|Q(t, v)\|_{Y'} \|v\|_Y \le \hat{\gamma}(t) \langle Q(t, v), v \rangle_Y \tag{ii}$$

(reverse pairing inequality).

The proof goes as follows, in the nontrivial case $v \ne 0$:

$$\begin{aligned} \|Q(t, v)\|_{Y'} &= \|Q(t, v)\|_{Y'}^{1/m} \|Q(t, v)\|_{Y'}^{1/m'} \\ &\le \{\hat{\delta}(t) \|v\|_Y^{m-1}\}^{1/m} \{\hat{\gamma}(t) \langle Q(t, v), v \rangle_Y / \|v\|_Y\}^{1/m'} \\ &= \{[\hat{\gamma}(t)]^{m-1} \hat{\delta}(t)\}^{1/m} [\mathcal{D}(t, v)]^{1/m'}, \end{aligned}$$

the second step following in view of (i) and (ii). Finally, since $X \hookrightarrow Y$ continuously, we have

$$\|Q(t, v)\|_{X'} \le d \|Q(t, v)\|_{Y'} \quad \text{for all } (t, v) \in J \times Y$$

where d is a positive constant. Hence (C) holds with $\delta(t) = d^m [\hat{\gamma}(t)]^{m-1} \hat{\delta}(t)$.

3 THE PROBLEM OF BLOW–UP

This is, more precisely, the problem of global non–continuation of solutions for all $t \in J = [0, \infty)$. In many cases, however, the two problems can be the same – see Levine & Serrin [9].

For simplicity we provide a detailed discussion of the blow–up problem in only two cases,

$$(Pu_t)_t = -A(u) + F(u), \qquad t \in J, \tag{3.1}$$

and

$$Q(t, u_t) = -A(u) + F(u), \qquad t \in J, \tag{3.2}$$

where P is a symmetric operator from a Hilbert space V into V' (see Example 2 in Section 2, and recall that P must be linear and continuous), which moreover is assumed to be non–negative definite, namely $\mathcal{P} \geq 0$ on V. Assume also that A, F and Q are abstract operators on appropriate Banach spaces, as in the previous section.

Case (3.1) has no dissipation present, namely $Q \equiv 0$, while case (3.2) is the *parabolic* analogue of the main evolution equation (2.1). For both (3.1) and (3.2) the relation $Y \hookrightarrow V$, which was introduced at the beginning of the previous section in connection with the definition of the operator Q, is no longer needed since these spaces do not appear together.

CASE (3.1). We adjoin the energy relation (c). Since $Q \equiv 0$, and so $\mathcal{D} \equiv 0$, this relation takes the form

$$\mathcal{E}u(t) = \tfrac{1}{2}\langle Pu_t, u_t \rangle + \mathcal{A}(u(t)) - \mathcal{F}(u(t)) \leq \mathcal{E}u(0); \tag{3.3}$$

for simplicity in printing we have dropped the space subscript V from $\langle \cdot, \cdot \rangle$.

We now proceed following an idea of Levine [7], showing non–continuation whenever the initial energy is *negative*, that is $\mathcal{E}u(0) < 0$.

Thus let u be a solution on J. Define

$$\mathcal{I}(t) = \tfrac{1}{2}\langle Pu(t), u(t) \rangle + \beta(t),$$

where $\beta = \beta(t)$ is a twice differentiable function which we shall fix later in an appropriate way. Then since P is symmetric,

$$\mathcal{I}'(t) = \langle Pu(t), u_t(t) \rangle + \beta'(t),$$
$$\mathcal{I}''(t) = \langle Pu_t(t), u_t(t) \rangle - \langle A(u(t)), u(t) \rangle_W + \langle u(t), F(u(t)) \rangle_X + \beta''(t),$$

by the main distribution identity with $\varphi = u \in K$.

Let us now introduce the following principal structure conditions on A and F:

$$q\mathcal{A}(u) \geq \langle A(u), u \rangle_W, \tag{A}$$

$$\langle F(u), u \rangle_X \geq q\mathcal{F}(u), \tag{D}$$

where $q > 0$. Hence

$$\mathcal{I}''(t) \geq \langle Pu_t(t), u_t(t) \rangle - q\mathcal{A}(u(t)) + q\mathcal{F}(u(t)) + \beta''(t)$$
$$\geq \left(1 + \tfrac{1}{2}q\right) \langle Pu_t(t), u_t(t) \rangle - q\mathcal{E}u(0) + \beta''(t), \tag{3.4}$$

by virtue of the energy relation (3.3).

We now suppose that $\mathcal{E}u(0) < 0$, and choose

$$\beta(t) = \beta_0(t + t_0)^2,$$

with $\beta_0 = |\mathcal{E}u(0)|$ and $t_0 > 0$. Then

$$-q\mathcal{E}u(0) + \beta''(t) = (q + 2)|\mathcal{E}u(0)| = 4\beta_0(1 + \alpha),$$

where $\alpha = \frac{1}{4}(q - 2)$ – note that $q + 2 = 4(1 + \alpha)$.

At $t = 0$ we have

$$\mathcal{I}(0) = \tfrac{1}{2}\langle Pu(0), u(0)\rangle + \beta_0 t_0^2, \quad \mathcal{I}'(0) = \langle Pu(0), u_t(0)\rangle + 2\beta_0 t_0.$$

If t_0 is chosen sufficiently large, then $\mathcal{I}(0), \mathcal{I}'(0) > 0$. Moreover, of course, $\mathcal{I}''(t) > 0$ so that $\mathcal{I}'(t) > 0$, $\mathcal{I}(t) > 0$ for all $t \in J$.

Now set

$$\mathcal{J}(t) = [\mathcal{I}(t)]^{-\alpha}, \quad t \in J.$$

Then

$$\mathcal{J}'(t) = -\alpha[\mathcal{I}(t)]^{-(\alpha+1)}\mathcal{I}'(t)$$

$$\mathcal{J}''(t) = \alpha[\mathcal{I}(t)]^{-(\alpha+2)}\{(\alpha + 1)[\mathcal{I}'(t)]^2 - \mathcal{I}(t)\mathcal{I}''(t)\}.$$

LEMMA. *We have*

$$\mathcal{K}(t) = \mathcal{I}(t)\mathcal{I}''(t) - (\alpha + 1)[\mathcal{I}'(t)]^2 \geq 0, \quad t \in J.$$

Proof: By the earlier calculations, see in particular (3.4), together with the symmetry of P,

$$\mathcal{K}(t) \geq \left\{\tfrac{1}{2}\langle Pu(t), u(t)\rangle + \beta(t)\right\}\left\{2(1 + \alpha)\langle Pu_t(t), u_t(t)\rangle + 4\beta_0(1 + \alpha)\right\}$$
$$- (\alpha + 1)\left\{\langle Pu(t), u_t(t)\rangle + \beta'(t)\right\}^2$$
$$= (\alpha + 1)\{4\mathcal{P}(u(t))\mathcal{P}(u_t(t)) - |\langle Pu(t), u_t(t)\rangle|^2$$
$$+ \beta_0\mathcal{P}(u(t) - (t + t_0)u_t(t))\} \geq 0$$

by the Cauchy–Schwarz inequality and the condition $\mathcal{P} \geq 0$ on V.

Assume $q > 2$. Then $\alpha > 0$. In turn

$$\mathcal{J}''(t) \leq 0, \quad \mathcal{J}(0) > 0, \quad \mathcal{J}'(0) < 0,$$

and therefore $\mathcal{J}$ reaches zero at a finite time t. But this is impossible, since $\mathcal{J}(t) > 0$ for all $t \in J$. This gives the following non–continuation result.

THEOREM 1. *Let (A), (D) be satisfied, with $q > 2$. Then no solution u of (3.1) can exist on J when $\mathcal{E}u(0) < 0$.*

EXAMPLE 1. If $A = -\Delta$ and $W = H_0^1(\Omega)$ in (3.1), then

$$\langle A(u(t)), u(t)\rangle_W = \int_\Omega |Du(t,x)|^2 dx = 2\mathcal{A}(u(t)),$$

so that the structure condition (A) is verified for any $q \geq 2$. When $Au = -\Delta_s u = -\text{div}(|\nabla u|^{s-2}\nabla u)$, $s > 1$, and $W = W_0^{1,s}(\Omega)$, we have

$$\langle A(u(t)), u(t)\rangle_W = \int_\Omega |Du(t,x)|^s dx = s\mathcal{A}(u(t)),$$

so that condition (A) is satisfied whenever $q \geq s$.

Thus to guarantee blow–up in these cases it is enough that F satisfies (D) with

$$q > 2 \quad \text{if } s \leq 2, \qquad q = s \quad \text{if } s > 2.$$

It should be noted particularly that in this example the domain Ω can be allowed to be unbounded.

CASE (3.2). For the equation (3.2) the solution set K should be slightly modified, by replacing the space $C^1(J \to V)$ by $AC(J \to Y)$, where AC denotes *absolutely continuous*, that is represented by the integral of an $L^1_{\text{loc}}(J)$ function.

Since $P = 0$, the energy conservation law (c) takes the form

$$\mathcal{E}u(t) \equiv \mathcal{A}(u(t)) - \mathcal{F}(u(t)) \leq \mathcal{E}u(0) - \int_0^t \mathcal{D}(\tau, u_t(\tau))d\tau. \tag{3.5}$$

We assume in addition to the earlier hypothesis (A) also the natural condition

$$\mathcal{A}(u) \geq 0 \quad \text{for any } u \in W \tag{A}'$$

and strengthen condition (D) to the form:

For every $\varepsilon > 0$ there exist two positive constant $c_1 = c_1(\varepsilon)$ and $c_2 = c_2(\varepsilon)$ and an exponent $p > 1$ such that

$$c_1\mathcal{F}(u) \leq c_2\|u\|_X^p \leq \langle F(u), u\rangle_X - q\mathcal{F}(u) \tag{D}'$$

for all $u \in G$ such that $\mathcal{F}(u) \geq \varepsilon$.

THEOREM 2 (see [10]). *Let (A), (A)$'$, (D)$'$ hold. Suppose that $1 < m < p$, where m is defined in (C) and*

$$\delta^{1/(1-m)} \notin L^1(J). \tag{3.6}$$

Then no solution u of (3.2) can exist on $J = [0, \infty)$ when $\mathcal{E}u(0) < 0$.

Proof: As in the previous demonstration we suppose the contrary, i.e., the existence of a solution on J. Consider the AC function

$$\mathcal{H}(t) = \int_0^t \mathcal{D}(\tau, u_t(\tau))d\tau - \mathcal{E}u(0).$$

Writing $-\mathcal{E}u(0) = \varepsilon > 0$, we get, since $\mathcal{A}, \mathcal{D} \geq 0$,

$$\mathcal{F}(u(t)) = \mathcal{A}(u(t)) - \mathcal{E}u(t) \geq -\mathcal{E}u(t) \geq \mathcal{H}(t) \geq \varepsilon > 0$$

on J. By the main distribution identity (b), with $P = 0$ and $\varphi = u \in K$, we obtain thanks to (A) and (D)$'$

$$\begin{aligned}
0 &= \langle F(u(t)), u(t)\rangle_X - \langle Q(t, u_t(t)), u(t)\rangle_X - \langle A(u(t)), u(t)\rangle_W \\
&\geq c_2\|u(t)\|_X^p + q\mathcal{F}(u(t)) - \langle Q(t, u_t(t)), u(t)\rangle_X - q\mathcal{A}(u(t)) \\
&= c_2\|u(t)\|_X^p - q\mathcal{E}u(t) - \langle Q(t, u_t(t)), u(t)\rangle_X.
\end{aligned}$$

Thus, recalling that $\mathcal{E}u(t) < 0$ on J, we see from the previous line that

$$\|Q(t, u_t(t))\|_{X'}\|u(t)\|_X \geq \langle Q(t, u_t(t)), u(t)\rangle_X > c_2\|u(t)\|_X^p.$$

This gives, since $p' = p/(p-1)$,

$$\|Q(t, u_t(t))\|_{X'} \geq c_2\|u(t)\|_X^{p-1} = c_2(\|u(t)\|_X^p)^{1/p'} \geq c_2\left(\frac{c_1}{c_2}\mathcal{F}(u(t))\right)^{1/p'}$$
$$\geq C[\mathcal{H}(t)]^{1/p'},$$

where $C = c_1^{1/p'}c_2^{1/p}$. Note particularly that C depends on ε and thus on $\mathcal{E}u(0)$.
On the other hand, by (C),

$$0 < \|Q(t, u_t(t))\|_{X'} \leq [\delta(t)]^{1/m}[\mathcal{D}(t, u_t(t))]^{1/m'} = [\delta(t)]^{1/m}[\mathcal{H}'(t)]^{1/m'}.$$

Combining the last two lines yields

$$\mathcal{H}'(t) \geq [\delta(t)]^{-m'/m}\|Q(t, u_t(t))\|_{X'}^{m'} \geq C^{m'}[\delta(t)]^{1/(1-m)}[\mathcal{H}(t)]^{m'/p'}.$$

Now, since $1 < m < p$ by assumption,

$$\frac{m'}{p'} - 1 = m'\left(\frac{1}{p'} - \frac{1}{m'}\right) = m'\left(\frac{1}{m} - \frac{1}{p}\right) > 0,$$

so we can write $m'/p' = 1 + \vartheta$, $\vartheta > 0$. Then

$$\frac{\mathcal{H}'}{\mathcal{H}^{1+\vartheta}} \geq C^{m'}\delta^{1/(1-m)}$$

and by integration, setting $\mathcal{H}_0 = \mathcal{H}(0) = \varepsilon$,

$$\frac{1}{\vartheta\mathcal{H}_0^\vartheta} \geq \frac{1}{\vartheta[\mathcal{H}(t)]^\vartheta} + C^{m'}\int_0^t [\delta(\tau)]^{1/(1-m)}\,d\tau.$$

This is impossible, since the left hand side is finite and the right hand side goes to ∞ as $t \to \infty$.

If we take $\delta(t) = (1+t)^\beta$, then

$$[\delta(t)]^{1/(1-m)} = (1+t)^{\beta/(1-m)}$$

and the divergence condition is exactly the request that $\beta \leq m - 1$. It should be noted that this range includes $\beta = 0$, i.e. $\delta \equiv 1$.

COROLLARY. *Suppose $\int_0^\infty [\delta(\tau)]^{1/(1-m)} d\tau = I$. Then no global solution of (3.2) can exist if $\mathcal{E}u(0) < 0$ and, even more,*

$$\mathcal{E}u(0) = -\mathcal{H}_0 < -(\vartheta C^{m'} I)^{-1/\vartheta}, \qquad \vartheta = \frac{p-m}{(m-1)p}. \qquad (3.7)$$

Note that (3.7) trivially holds whenever $\mathcal{E}u(0) < 0$ and $I = \infty$.

The operator Q can be allowed to depend on u provided that (C) is replaced by the following condition:

There are two exponents $m > 1$, $\kappa > 0$ and a positive locally integrable function $\delta = \delta(t)$ such that

$$\|Q(t,u,v)\|_{X'} \leq [\delta(t) \cdot \|u\|_X^\kappa]^{1/m} [\mathcal{D}(t,u,v)]^{1/m'} \quad \text{for all } (t,u,v) \in J \times X \times Y.$$

Then the same results hold if we suppose

$$p > m + \kappa.$$

In this case

$$\vartheta = \frac{p - m - \kappa}{(m-1)p}.$$

EXAMPLE 2. The previous theorem can be applied to show blow–up for the degenerate parabolic equation

$$\delta(t)|u_t|^{m-2} u_t = a \operatorname{div}(|Du|^{s-2} Du) + c|u|^{p-2} u, \quad (t,x) \in J \times \Omega, \quad \Omega \subset {}^n,$$

where δ is a positive locally integrable function satisfying (3.6), $a, c > 0$, $1 < m < p$, $s > 1$, and Ω bounded.

In particular, we take $W = W_0^{1,s}(\Omega)$ and $X = L^p(\Omega)$. Then conditions (A), (A)' hold with $q = s$, while (D)' becomes

$$\frac{cc_1}{p} \|u\|_X^p \leq c_2 \|u\|_X^p \leq c \left(\|u\|_X^p - \frac{s}{p} \|u\|_X^p \right).$$

Thus we must also have $1 < s < p$, and can take

$$c_2 = c \left(1 - \frac{s}{p} \right) > 0, \qquad c_1 = \frac{pc_2}{c} = p - s > 0.$$

Of course

$$Q(t,v) = \delta(t)|v|^{m-2} v, \qquad Q : J \times Y \to Y',$$

where $Y = L^m(\Omega)$. Then, taking $\mathcal{D}(t, y) = \delta(t)\|y\|^m_{L^m(\Omega)}$, it is easy to check that the main condition (C) is verified, for example by proceeding as in the discussion at the end of the previous section. Indeed (i) and (ii) hold with $\hat{\delta} = \delta$ and $\hat{\gamma} = 1$, while of course $X \hookrightarrow Y$ by the hypothesis $1 < m < p$ and the fact that Ω is bounded.

EXAMPLE 3. A second important case of (3.2) occurs when $Q(t, v) = Qv$, where Q is a linear continuous operator from Y to Y', and $X \hookrightarrow Y$. Then condition (C) holds with $\delta(t) = \text{Constant}$ and $\mathcal{D}(t, v) = \langle Qv, v \rangle_Y$ provided Q satisfies the additional requirement

$$\langle Qv, v \rangle_Y \geq \text{Pos. Const.} \|v\|^2_Y \quad \text{for all } v \in Y.$$

(See also [6] when $A : W \to W'$ is linear, $Q : Y \to Y'$ is symmetric, and W, Y are real Hilbert spaces.)

Finally, we state without proof the corresponding non–continuation theorem for the abstract equation (2.1); see [9, Theorem 1].

THEOREM 3. *Assume that the previous conditions* (A), (A)$'$, (C), (D)$'$ *hold, that* $\mathcal{P}^* \geq 0$ *in* V, *and that* $X \hookrightarrow V$ *continuously. Suppose furthermore that there are constants* $\ell > 1$, $c_3 > 0$ *such that for all* $v \in V$

$$c_3\|P(v)\|^{\ell'}_{V'} \leq (q+1)\langle P(v), v \rangle_V - q\mathcal{P}(v), \tag{E}$$

where q *is the exponent in* (A) *and* (D)$'$. *Assume finally that*

$$1 < \ell < p, \quad 1 < m < p \tag{3.8}$$

and

$$\int_0^\infty \frac{\min\{1, \delta^{(1+\vartheta)/(m-1)}\}}{\delta^{1/(m-1)}} dt = \infty \tag{3.9}$$

for some ϑ *such that*

$$0 < \vartheta < \min\left\{ \frac{p-\ell}{p\ell - p + \ell}, \frac{p-m}{pm - p + m} \right\}.$$

Then no solution u *on* J *of* (2.1) *can exist with* $\mathcal{E}u(0) < 0$.

In case $\delta(t) = (1+t)^\beta$ the divergence condition (3.9) is satisfied provided that $\beta \leq m - 1$.

EXAMPLE 4. Consider the degenerate wave equation with dissipation

$$u_{tt} + b|u_t|^{m-2}u_t = \Delta_s u + c|u|^{p-2}u, \quad (t, x) \in J \times \Omega, \quad \Omega \subset {}^n.$$

Here we suppose Ω bounded, m, p, $s > 1$, and take $P = I$, $V = L^2(\Omega)$, $W = W_0^{1,s}(\Omega)$, $X = L^p(\Omega)$; then $q = s$ for (A) and $\ell = 2$, $c_3 = 1 + q/2$ for (E), see Example 2 of Section 2.

As in Example 2 above, for (D)$'$ we need $p > s$. Recalling the hypothesis (3.8) then gives the principal exponent condition

$$p > \max\{2,\ m,\ s\}. \tag{3.10}$$

Of course $Q(t,v) = Q(v) = b|v|^{m-2}v$; the choice of an appropriate space Y is here complicated by the condition $Y \hookrightarrow V$ for the definition of Q and the requiremnt that $Q : Y \to X'$. A little reflection shows that the required space Y is given by

$$Y = \begin{cases} L^2(\Omega), & \text{if } 1 < m < 2 \\ L^m(\Omega), & \text{if } \quad m \geq 2. \end{cases}$$

Then it is easy to check that (C) holds with $\mathcal{D}(t,v) = \mathcal{D}(v) = b\|v\|_{L^m(\Omega)}^m$ and $\delta(t) = b|\Omega|^{(p-m)/m}$.

In [4] the special case $p > m > 2$ was obtained when $s = 2$, that is $\Delta_s = \Delta$.

Generalizations to time dependent potentials $\mathcal{A}$ and $\mathcal{F}$ will appear in work of Levine, Pucci & Serrin [11].

4   STABILITY

Here we are interested in the converse of the blow–up problem, that is, under which conditions does $\mathcal{E}u(t) \to 0$ as $t \to \infty$? We shall concentrate on perhaps the most interesting situation – in which stability holds when the initial data is small, and blow–up when the data is large, i.e. the case of the previous section. We shall treat the general equation (2.1), our discussion being based on work of Pucci & Serrin [18, 20] and Boccuto & Vitillaro [3].

Naturally, one cannot expect $\mathcal{E}u(t) \to 0$ as $t \to \infty$ unless the damping term Q is sufficiently strong. Thus for stability we shall require the following further condition on the dissipation function $\mathcal{D}$.

(C)$'$ *There is a non–negative function σ on J, and a wedge function ω, i.e. a non–decreasing function on $[0,\infty)$ with $\omega(0) = 0$, $\omega(\tau) > 0$ if $\tau \neq 0$, such that*

$$\omega(\|v\|_Y) \leq \sigma(t)\mathcal{D}(t,v) \quad on \ J \times Y.$$

We assume the *existence of numbers $a_0 > 0$, $s > 1$ such that*

$$\langle A(u), u \rangle_W \geq a_0 \|u\|_W^s \quad for \ u \in G, \tag{A}''$$

and we suppose that *there exist constants $p > s$, $\mu \in [0, a_0)$, $c > 0$ such that*

$$\langle F(u), u \rangle_X \leq \mu \|u\|_W^s + c\|u\|_X^p \quad for \ u \in G. \tag{D}''$$

It will be assumed that $\mathcal{P}^* \geq 0$ on V, while instead of (E) we now suppose for stability that

(E)′ *For some $d > 0$ the sets $P(E)$ and $\mathcal{P}(E)$ are bounded in V' and , respectively, where $E = E_d = \{v \in V : \mathcal{P}^*(v) \leq d\}$.*

Note that (A)″ implies $s\mathcal{A}(u) \geq a_0\|u\|_W^s$ on G, which is related to (A) and stronger than (A)′. Condition (E)′ can be replaced by the simpler but stronger requirement

$$\mathcal{P}^*(v) \to 0 \quad implies \quad \|v\|_V \to 0.$$

We shall finally require the continuous embeddings

$$W \hookrightarrow X \hookrightarrow V.$$

Thus we can take $G = W$; moreover there is $C > 0$ such that $\|u\|_X \leq C\|u\|_W$ for all $u \in G$.

We can now state the following main result of local asymptotic stability.

THEOREM ([18, 20, 3]). *Let (A)″, (C), (C)′, (D)″, (E)′ hold, and suppose also that*

$$\liminf_{t \to \infty} \frac{1}{t^m} \int_0^t \{\delta(\tau) + [\sigma(\tau)]^{m-1}\}d\tau < \infty.$$

Let u be a solution of (2.1) such that

$$\mathcal{E}u : J \to \quad \text{is non–increasing on } J. \tag{4.1}$$

Then

$$\mathcal{E}u(t) \to 0 \quad as \quad t \to \infty, \tag{4.2}$$

provided that the initial data $u(0)$, $u_t(0)$ has sufficiently small norms in W and V, respectively.

Note that (4.1) does *not* follow from (2.4), but *would* hold if in (2.4) the *equality* is enforced.

The particular case $\delta(t) = (1+t)^\beta$, $\sigma(t) = (1+t)^\gamma$ yields the conditions

$$\beta \leq m - 1, \qquad \gamma \leq 1.$$

Since generally $\sigma\delta \geq 1$ we expect $-1 \leq -\gamma \leq \beta \leq m - 1$. If δ, σ are constant (i.e. autonomous damping) these conditions are satisfied with $\beta = \gamma = 0$.

Proof of the theorem: Our first goal is to obtain an estimate for the required smallness of the initial data, after which the actual limit of the energy must be established. In view of the previous blow–up result this first part is crucial – since large data lead to blow–up. We divide the proof into several steps.

LEMMA 1. *Under the hypotheses above, for any solution u of (2.1) we have, for every $t \in J$,*

$$\mathcal{E}u(t) \geq \mathcal{P}^*(u_t(t)) + 2a\|u(t)\|_X^s - \frac{c}{p}\|u(t)\|_X^p, \qquad \text{where } a = \frac{a_0 - \mu}{2sC^s}.$$

Proof: From (2.3) we get by (A)″ and (D)″

$$\begin{aligned}
\mathcal{E}u(t) &= \mathcal{P}^*(u_t(t)) + \mathcal{A}(u(t)) - \mathcal{F}(u(t)) \\
&\geq \mathcal{P}^*(u_t(t)) + \frac{a_0}{s}\|u(t)\|_W^s - \frac{\mu}{s}\|u(t)\|_W^s - \frac{c}{p}\|u(t)\|_X^p \\
&= \mathcal{P}^*(u_t(t)) + \frac{a_0 - \mu}{s}\|u(t)\|_W^s - \frac{c}{p}\|u(t)\|_X^p,
\end{aligned} \qquad (4.3)$$

which gives the result since $\|u(t)\|_W \geq \|u(t)\|_X/C$.

Let

$$\Sigma_0 = \{(\lambda, \mathcal{E}) \in {}^2 : 0 \leq \lambda < \lambda_0, \ 2a\lambda^s - \frac{c}{p}\lambda^p \leq \mathcal{E} < \mathcal{E}_0\},$$

where

$$\lambda_0 = \left(\frac{2as}{c}\right)^{1/(p-s)}, \qquad \mathcal{E}_0 = 2a\lambda_0^s\left(1 - \frac{s}{p}\right).$$

By the lemma, if the initial data $u(0)$, $u_t(0)$ is such that the point $(\|u(0)\|_X, \mathcal{E}u(0))$ is in Σ_0, then, because $\mathcal{E}u$ is non–increasing on J by (4.1) and $\mathcal{P}^* \geq 0$ on V, it is easy to see that *for all $t \in J$* the point $(\|u(t)\|_X, \mathcal{E}u(t))$ remains in Σ_0. Thus Σ_0 is a *potential well* for the problem.

It is worth noting that one cannot have $\mathcal{E}u(0) < 0$ unless the data is fairly large, e.g., by Lemma 1,

$$\|u(0)\|_X \geq \left(\frac{2ap}{c}\right)^{1/(p-s)} = \lambda_2;$$

clearly, $\lambda_2 > \lambda_0$ since $p > s$ by (D)″.

The next problem is to ensure that $\mathcal{E}u(t) \to 0$ as $t \to \infty$. To this end, it is necessary that the initial point $(\|u(0)\|_X, \mathcal{E}u(0))$ lies in the *smaller set* Σ defined by

$$\Sigma = \{(\lambda, \mathcal{E}) \in {}^2 : 0 \leq \lambda < \lambda_1, \ 2a\lambda^s - \frac{c}{p}\lambda^p \leq \mathcal{E} < \mathcal{E}_1\},$$

where

$$\lambda_1 = \left(\frac{a}{c}\right)^{1/(p-s)}, \qquad \mathcal{E}_1 = \min\left\{d, \ a\lambda_1^s\left(2 - \frac{1}{p}\right)\right\}.$$

As above, once the *phase trajectory* $(\|u(t)\|_X, \mathcal{E}u(t))$ enters Σ it remains there for all larger $t \in J$. This remark is crucial in the following proofs.

LEMMA 2. *Let $(\|u(0)\|_X, \mathcal{E}u(0)) \in \Sigma$. Then, under the hypotheses of the theorem,*

$$\mathcal{E}u(t) \geq \mathcal{P}^*(u_t(t)) + \frac{a_0 - \mu}{2s}\|u\|_W^s \quad \text{in } J.$$

Proof: Clearly

$$a\lambda^s - \frac{c}{p}\lambda^p \geq a\lambda^s\left(1 - \frac{1}{p}\right) \geq 0 \quad \text{whenever } \lambda \leq \lambda_1. \tag{4.4}$$

Also

$$\frac{a_0 - \mu}{s}\|u(t)\|_W^s \geq \frac{a_0 - \mu}{2s}\|u(t)\|_W^s + a\|u(t)\|_X^s, \tag{4.5}$$

so the result follows from (4.3), (4.5) and (4.4).

LEMMA 3. *Let* $(\|u(0)\|_X, \mathcal{E}u(0)) \in \Sigma$. *Then, under the hypotheses of the theorem,* (4.2) *holds.*

Proof: We outline the ideas involved. Assume for contradiction that (4.2) fails. Then by (4.1) there exists $l > 0$ such that

$$\mathcal{E}u(t) \geq l \quad \text{for all } t \in J.$$

By the distribution identity (b) with $\varphi = u \in K$, together with the definition (2.2) of $\mathcal{P}^*$, we obtain

$$\frac{d}{dt}\langle P(u_t(t)), u(t)\rangle_V = \{\mathcal{P}(u_t(t)) + 2\mathcal{P}^*(u_t(t))\}$$
$$- \{\mathcal{P}^*(u_t(t)) + \langle A(u(t)), u(t)\rangle_W - \langle F(u(t)), u(t)\rangle_X\}$$
$$- \langle Q(t, u_t(t)), u(t)\rangle_X. \tag{4.6}$$

Next we assert that for any $\theta > 0$ there is $\gamma(\theta) > 0$ such that for all $t \geq T > 0$

$$\int_T^t \{\mathcal{P}(u_t(\tau)) + 2\mathcal{P}^*(u_t(\tau))\}d\tau \leq \theta t + \gamma(\theta)\varepsilon(T)\left(\int_0^t [\sigma(\tau)]^{m-1}d\tau\right)^{1/m} \tag{4.7}$$

$$\left|\int_T^t \langle Q(\tau, u_t(\tau)), u(\tau)\rangle_X d\tau\right| \leq \varepsilon(T)\left(\int_0^t \delta(\tau)d\tau\right)^{1/m}, \tag{4.8}$$

where $\varepsilon(T) \to 0$ as $t \to \infty$. To obtain (4.7), note by Lemma 2 and (4.1) that

$$0 \leq \mathcal{P}^*(u_t(t)) \leq \mathcal{E}u(t) \leq \mathcal{E}u(0) < \mathcal{E}_1 \leq d, \quad t \in J,$$

so by (E)$'$ also $\mathcal{P}(u_t)$ is bounded on J. Moreover $\mathcal{D}(\cdot, u_t(\cdot)) \in L^1(J)$ by the conservation law (c) and the fact that $\mathcal{E}u$ is bounded on J. The result then follows with the help of (C)$'$ and the fact that both $\mathcal{P}$ and $\mathcal{P}^*$ are continuous and vanish at $v = 0$; see [18, inequality (3.7)].

Next, (4.8) is an easy consequence of (C), and the facts that $\mathcal{D}(\cdot, u_t(\cdot)) \in L^1(J)$ and $\|u(t)\|_X < \lambda_1$ on J; see [18, inequality (3.10)].

Furthermore, from Lemma 2, the fact that $\mathcal{P}^* \geq 0$ on V, and the continuity of $\mathcal{A} : W \to$ and $\mathcal{F} : X \to$, it can be shown (see [3, Lemma 4.6; 20]) that for all $l > 0$ there is $\alpha(l) > 0$ such that

$$\mathcal{P}^*(u_t(t)) + \langle A(u(t)), u(t)\rangle_W - \langle F(u(t)), u(t)\rangle_X \geq \alpha(l) \tag{4.9}$$

for all $t \geq T$.

It now follows from (4.6)–(4.9) that for $t \geq T$

$$\langle P(u_t(\tau)), u(\tau) \rangle_V \Big|_T^t \leq [\theta - \alpha(l)]t$$
$$+ \, \varepsilon(T) \left\{ \gamma(\theta) \left(\int_0^t [\sigma(\tau)]^{m-1} d\tau \right)^{1/m} + \left(\int_0^t \delta(\tau) d\tau \right)^{1/m} \right\}.$$

Choose a sequence $(t_i) \nearrow \infty$ such that, for an appropriate constant M,

$$\frac{1}{t_i^m} \int_0^{t_i} \{ \delta(\tau) + [\sigma(\tau)]^{m-1} \} d\tau \leq M^m.$$

Consequently for all $t_i \geq T$

$$\langle P(u_t(\tau)), u(\tau) \rangle_V \Big|_T^{t_i} \leq \{ \theta - \alpha(l) + M\varepsilon(T)[1 + \gamma(\theta)] \} t_i.$$

Take $\theta = \alpha(l)/4$ and T so large that

$$M[1 + \gamma(\alpha(l)/4)]\varepsilon(T) \leq \alpha(l)/4.$$

Hence

$$\langle P(u_t(\tau)), u(\tau) \rangle_V \Big|_T^{t_i} \leq -\tfrac{1}{2}\alpha(l)t_i. \tag{4.10}$$

We claim that $\|u(\cdot)\|_V$ and $\|P(u_t(\cdot))\|_{V'}$ are bounded on J, when $(\|u(0)\|_X, \mathcal{E}u(0))$ is in Σ. Indeed, see above,

$$0 \leq \mathcal{P}^*(u_t(t)) < d \quad \text{on } J.$$

Hence by (E)′ it follows that $\|P(u_t(\cdot))\|_{V'}$ is bounded on J.

Moreover, by the embedding $X \hookrightarrow V$ we get

$$\|u(t)\|_V \leq \text{Const.}\, \|u(t)\|_X \leq \text{Const.}\, \lambda_1 \quad \text{on } J.$$

This completes the proof of the claim. In turn, we reach a contradiction with (4.10) when $t_i \to \infty$.

This contradiction concludes the proof of the theorem provided we show that $(\|u(0)\|_X, \mathcal{E}u(0))$ is in Σ whenever $\|u(0)\|_W$ and $\|u_t(0)\|_V$ are sufficiently small. This is, however, a consequence of the fact that $\|u(0)\|_X \leq C\|u(0)\|_W$, the definition (2.3) and the fact that the potentials $\mathcal{A}$ and $\mathcal{F}$, and the Hamiltonian $\mathcal{P}^*$ are continuous and normalized so that $\mathcal{A}(0) = 0$, $\mathcal{F}(0) = 0$, $\mathcal{P}^*(0) = 0$ in W, X and V, respectively.

EXAMPLE. Consider the degenerate wave equation

$$u_{tt} - \Delta_s u + \tilde{Q}(t, x, u_t) = f(x, u), \quad (t, x) \in J \times \Omega, \quad \Omega \subset {}^n.$$

Here Ω is bounded, $P = I$, $V = L^2(\Omega)$ and $s > 1$, $W = W_0^{1,s}(\Omega)$. For simplicity take $X = L^p(\Omega)$, $p \leq r$, where r is the Sobolev exponent for $W_0^{1,s}(\Omega)$. The embedding $W \hookrightarrow X$ is then the Sobolev theorem and C is the Sobolev constant. We suppose that the Nemitsky operator Q corresponding to the damping term $\tilde{Q}$ verifies (C), (C)$'$, see Example 4 in Section 2.

The conditions (A), (A)$'$, (A)$''$ are easily checked, in particular with $q = s$ and $a_0 = 1$. Moreover, for the specific function

$$f(x,u) = \hat{\mu}|u|^{s-2}u + c|u|^{p-2}u, \quad s < p \leq r, \quad c > 0,$$

the condition (D)$''$ is verified provided that $\hat{\mu} < \mu_0$. Here μ_0 is the Poincaré constant, that is the reciprocal of the *first eigenvalue* of $-\Delta_s$ in Ω with Dirichlet homogeneous boundary conditions, namely,

$$\|u\|_{L^s} \leq \mu_0^{-1/s}\|u\|_{W_0^{1,s}},$$

and we can take $\mu = \hat{\mu}/\mu_0 < 1 = a_0$. To verify (D)$'$ when $q = s$ one takes $c_2 = c(1 - s/p)$ as in Example 2 of Section 3, and $c_1 = c_1(\varepsilon) > 0$ sufficiently small, depending on p, s, $\hat{\mu}$, c, $|\Omega|$ – see [11, Section 4] for a complete discussion.

Finally, (E) holds with $\ell = 2$, $c_3 = 1 + q/2$, while (E)$'$ is clear since in the present case $\|P(v)\|_{V'} = \|v\|_{L^2(\Omega)}$ and $\mathcal{P}^*(v) = \mathcal{P}(v) = \frac{1}{2}\|v\|_{L^2(\Omega)}^2$. Thus if $\mathcal{P}^*(v) \leq d$ we get $\|P(v)\|_{V'} \leq \sqrt{2d}$ and $\mathcal{P}(v) \leq d$.

This example is of particular interest because *both* the blow–up Theorem 3 of Section 3 *as well as* the stability theorem above are applicable.

ACKNOWLEDGMENTS. P. Pucci is a member of *Gruppo Nazionale di Analisi Funzionale e sue Applicazioni* of the *Consiglio Nazionale delle Ricerche*. This research has been partly supported by the Italian *Ministero dell'Università e della Ricerca Scientifica e tecnologica*.

Various results contained in Sections 3 and 4 were presented in Trieste, as part of this conference, October 1995.

The paper in preliminary form was the basis of a lecture series given by J. Serrin at the *Universidad Internacional Menéndez Pelayo – International School of Mathematics: Recents Trends in Elliptic Equations and Related Topics*, Santander, July 1996.

REFERENCES

[1] J.M. Ball, *Stability theory for an extensible beam*, J. Differential Equations **14** (1973), 399–418.

[2] J. Ball, *Remarks on blow–up and nonexistence theorems for nonlinear evolution equations*, Quart. J. Math. Oxford (2) **28** (1977), 473–486.

[3] A. Boccuto & E. Vitillaro, *Asymptotic stability for abstract evolution equations and applications to partial differential systems*, to appear, Rendiconti Circolo Mat. Palermo.

[4] V. Georgiev & G. Todorova, *Existence of a solution of the wave equation with nonlinear damping and source terms*, J. Differential Equations **109** (1994), 295–308.

[5] J.K. Hale, *Asymptotic behavior of dissipative systems*, Mathematical Surveys and Monographs, **25**, American Mathematical Society, RI, 1988.

[6] H.A. Levine, *Some nonexistence and instability theorems for solutions of formally parabolic equations of the form* $Pu_t = Au + \mathcal{F}(u)$, Archive Rational Mech. Anal. **51** (1973), 371–386.

[7] H.A. Levine, *Instability and nonexistence of global solutions of nonlinear wave equations of the form* $Pu_{tt} = Au + \mathcal{F}(u)$, Trans. Amer. Math. Soc. **192** (1974), 1–21.

[8] H.A. Levine, *Nonexistence of global solutions of nonlinear wave equations*, in *Improperly posed boundary value problems*, Res. Notes Math., **1**, 94–104; Pitman, London, 1975.

[9] H.A. Levine & J. Serrin, *Global nonexistence theorems for quasilinear evolution equations with dissipation*, Archive Rational Mech. Anal. (in press).

[10] H.A. Levine, S.R. Park & J. Serrin, *Global nonexistence theorems for quasilinear evolution equations of formally parabolic type*, to appear.

[11] H.A. Levine, P. Pucci & J. Serrin, *Some remarks on the blow-up problem for nonautonomous abstract evolution equations*, to appear, Contemporary Math., 1997.

[12] J.L. Lions & W.A. Strauss, *On some nonlinear evolution equations*, Bull. Soc. Math. France **93** (1965), 43–96.

[13] P. Marcati, *Decay and stability for nonlinear hyperbolic equations*, J. Differential Equations **55** (1984), 30–58.

[14] P. Marcati, *Stability for second order abstract evolution equations*, Nonlinear Anal. **18** (1984), 237–252.

[15] M. Nakao, *Asymptotic stability for some nonlinear evolution equations of second order with unbounded dissipative terms*, J. Differential Equations **30** (1978), 54–63.

[16] L.E. Payne & D. Sattinger, *Saddle points and instability of nonlinear hyperbolic equations*, Israel Math. J. **22** (1981), 273–303.

[17] P. Pucci & J. Serrin, *Asymptotic stability for non-autonomous dissipative wave systems*, Comm. Pure Appl. Math. **XLIX** (1996), 177–216.

[18] P. Pucci & J. Serrin, *Stability for abstract evolution equations*, in *Partial Differential Equations and Applications*, edited by P. Marcellini, G. Talenti and E. Vesentini, pp. 279–288, M. Dekker, New York, 1996.

[19] P. Pucci & J. Serrin, *Asymptotic stability for nonlinear parabolic systems*, in: Proc. Conference on *Energy Methods in Continuum Mechanics*, Kluwer, Dordrecht, in press.

[20] P. Pucci & J. Serrin, *Local asymptotic stability for dissipative wave systems*, to appear.

[21] M.C. Salvatori & E. Vitillaro, *Decay for the solutions of nonlinear abstract damped evolution equations with applications to partial and ordinary differential systems*, to appear, Differential and Integral Equations.

[22] G. Webb, *Existence and asymptotic behavior for a strongly damped nonlinear wave equation*, Canadian J. Math. **32** (1980), 631–643.

[23] X. Zhu, *Existence of global solutions for wave systems*, Thesis, University of Minnesota, 1996.

A Nonlinear Schrödinger Equation Defined on the Whole Space

IAN SCHINDLER Ceremath, Université de Toulouse 1, Place Anatole France, 31042 Toulouse, France

INTRODUCTION

In this paper, we study the equation $\Omega \subset I\!\!R^n$, $n \geq 3$,

$$
\begin{aligned}
-\Delta u + q(x)u &= \lambda f(x, u) \\
u|_{\partial \Omega} &= 0 \\
\lim_{|x| \to \infty} u &= 0.
\end{aligned}
\tag{0.1}
$$

where $\lambda > 0$, $\Omega \in I\!\!R^n$ is a smooth unbounded domain, and the potential, $q(x)$ satisfies

$$
\lim_{|x| \to \infty} q(x) = \infty.
\tag{0.2}
$$

The problem with a bounded potential has been studied in [17] for bounded domains and has been studied by many people in unbounded domains among which we note [5, 24, 3, 4, 8, 13, 15, 12, 21, 23]. The problem is motivated by mathematical physics where certain stationary waves in nonlinear Klein-Gordon or Schrödinger equations can be reduced to this form (see [3] and [13]). Unbounded potentials occur in the theory of laser beams. The treatment of the problem in this case is more restricted. The linear case is discussed in [18] and [9]. The eigenvalues and vectors of the Schrödinger operator $-\Delta + q(x)$ and their asymptotic behavior when $q(x)$ satisfies (0.2) has been studied in [10, 11] and systems of Schrödinger equations involving unbounded potentials have been studied in [1, 2]. Hamiltonian systems with such potentials are studied in [16]. Among recent publications, F. Clarke [7, 6] and M. Schechter with K. Tintarev [20] have explored the relationship of critical value functions to eigenvalues. M. Schechter and K. Tintarev studied eigenvalues of nonlinear functionals on an infinite dimensional Hilbert space of the type

$$g'(u) = \rho u. \tag{0.3}$$

Assuming $g(u)$ and $(g'(u), u)$ to be weakly continuous, they showed that among the ρ's for which (0.3) has a solution are twice the directional derivatives of the critical value function

$$\gamma(t) = \sup_{\|u\|^2 = t} g(u).$$

The loss of compactness of the Sobolev embeddings on unbounded domains renders variational techniques more delicate. This paper uses a version of the concentration compactness lemma from [21] (similar in spirit to that introduced by P.L. Lions in [14]) to obtain the precompactness of critical sequences permitting the application of results of the type obtained by Schechter and Tintarev. The fact that the potential satisfies (0.2) permits a wider range of right hand sides in (0.1) than would otherwise be possible [22].

1 ABSTRACT THEORY

1.1 Main Results

The novelty of this section is more in the presentation than the content. We modify results in [20, 28, 23, 25] to suite our needs, reworking some of the proofs.

Let H be a Hilbert space. Let $g(u)$ be a real valued continuously Frechet differentiable function on H such that $g(0) = 0$. In this section we will be interested in the nonlinear eigenvalue problem

$$g'(u) = \rho u. \tag{1.4}$$

Specifically, we would like to know for which ρ equation (1.4) has a solution. This is equivalent to knowing for which ρ the functional

$$G_\rho(u) := \frac{\rho}{2} \|u\|^2 - g(u) \tag{1.5}$$

admits a critical point, i.e. a $u \in H \setminus \{0\}$ such that $G'_\rho(u) = 0$.

We define

$$S_t := \{u \in H | \, \|u\|^2 = t\}, \qquad B_t := \{u \in H | \, \|u\|^2 \leq t\}, \qquad (1.6)$$

$$\gamma(t) := \sup_{u \in S_t} g(u) \quad I := (2 \inf_{t \neq s} \frac{\gamma(t) - \gamma(s)}{t - s}, 2 \sup_{t \neq s} \frac{\gamma(t) - \gamma(s)}{t - s}) \quad (1.7)$$

Note that $g(0) = 0$ implies $\gamma(0) = 0$.

We will use a modification of the Palais-Smale condition. For $\rho \neq 0$, we say $G_\rho(u)$ satisfies property (P) if for $u_k \in H$ such that

$$\|u_k\|^2 \;\rightarrow\; t \neq 0 \qquad (1.8)$$

$$G'_\rho(u_k) \;\rightarrow\; 0 \qquad (1.9)$$

then there is a relabeled subsequence $u_k \to u$.

REMARK 1.1 If $\rho \neq 0$, u_k satisfies (1.9), is bounded and

$$G_\rho(u_k) \to b \neq 0 \qquad (1.10)$$

then modulo a subsequence, we have (1.8).

A convenient way to check property (P) is the following:

PROPOSITION 1.2 If $(g'(u), u)$ is weakly continuous and $g'(u)$ is continuous in the weak topology on sequences satisfying (1.9) and (1.8), then G_ρ satisfies property (P) for all $\rho \neq 0$.

Proof: Suppose $u_k \in H$ is a sequence satisfying equations (1.9) and (1.8). Then we can find a subsequence such that $u_k \rightharpoonup u \in H$. Note that (1.9) and the fact that the u_k are bounded imply $(G'_\rho(u_k), u_k) \to 0$, that is

$$\rho\|u_k\|_H^2 - (g'(u_k), u_k) \to 0. \qquad (1.11)$$

By hypothesis,

$$\rho u - g'(u) = \text{w-} \lim_{k \to \infty} G'_\rho(u_k) = 0. \qquad (1.12)$$

This implies $g'(u) = \rho u$. Again by hypothesis, $(g'(u_k), u_k) \to (g'(u), u)$ so equation (1.11) implies

$$(g'(u), u) = \rho t. \qquad (1.13)$$

From 1.12 and 1.13 we conclude that $\rho\|u\|_H^2 = \rho t$ which implies $\|u\|_H^2 = t$ and so $u_k \to u$. .

The main result of this section is the following:

THEOREM 1.3 Assume $G_\rho(u)$ satisfies property (P) for all $\rho \in \mathbb{R}$. Then for almost every $\rho \in I$ equation (1.4) admits a solution in $H \backslash \{0\}$. If every sequence satisfying (1.10) and (1.9) is bounded then equation (1.4) admits a solution for every nonzero $\rho \in I$.

1.2   Mountain Pass and Impass

We will now focus on the mountain pass and impass theorems. Very important in the use of the mountain pass theorem is the Palais-Smale condition. A functional $G_\rho(u)$ satisfies the Palais-Smale condition if every critical sequence (that is sequence satisfying (1.10) and (1.9)) is precompact. A simple and important way for a functional to fail to satisfy the Palais-Smale condition is for the functional to satisfy property (P) but to admit unbounded critical sequences. In this case one may not be able to obtain a critical point but one can obtain an almost" critical point. Among recent papers dealing with this subject are [19, 28].

When we say $G_\rho(u)$ has mountain pass geometry, for clarity, we will be assuming the classical has mountain pass geometry, though the results apply to more general minimax geometries. By classical mountain pass geometry we mean:

$$G_\rho(0) = 0 \tag{1.14}$$

there exists $e \in H \setminus \{0\}$ such that

$$G_\rho(e) \leq 0 \tag{1.15}$$

and

$$c = \inf_{\phi \in \Phi} \max_{s \in [0,1]} G_\rho(\phi(s)) > 0 \tag{1.16}$$

where

$$\Phi := \{\phi \in C([0,1], H) | \phi(0) = 0 \text{ and } \phi(1) = e\} \tag{1.17}$$

We have the following mountain impass lemma:

LEMMA 1.4 Assume G_{ρ_0} has mountain pass geometry and satisfies property (P). Then either there is a solution $u_0 \in H \setminus \{0\}$ such that $G'_\rho(u_0) = 0$ or there is a sequence $(u_k, h_k) \in H \times \mathbb{R}$ such that

$$0 < h_k \;\to\; 0, \tag{1.18}$$
$$\|u_k\|^2 = t_k \;\nearrow\; \infty \tag{1.19}$$
$$G_{\rho_0}(u_k) = c(t_k) \;\searrow\; c \tag{1.20}$$
$$G'_{\rho_0 + h_k}(u_k) \;=\; 0 \tag{1.21}$$

Where $c(t)$ is a monotone decreasing function of t.

Proof: Lemma (1.4) is essentially contained in [28], though it is not formulated in exactly the same way. We sketch an alternate proof motivated by Clarkes one line justification of the Langrange multiplier rule [6]. We begin with some definitions.

For $t \geq \|e\|^2$ let

$$\Phi_t := \{\phi \in C([0,1] \to B_t) | \phi(0) = 0 \text{ and } \phi(1) = e\}, \qquad (1.22)$$

and for $0 \leq t < \|e\|^2$ let

$$\Phi_t := \{\phi \in C([0,1] \to B_t) | \phi(0) = 0 \text{ and } \|\phi(1)\|^2 = t\}. \qquad (1.23)$$

For $t \geq 0$ let

$$c(t) := \inf_{\phi \in \Phi_t} \max_{s \in [0,1]} G_{\rho_0}(\phi(s)). \qquad (1.24)$$

Note that for $t \geq \|e\|^2$, $c(t)$ is a monotone decreasing function of t and hence differentiable almost everywhere. Let

$$\tilde{\Phi}_t := \{\phi \in \Phi_t | \max_{s \in [0,1]} G_{\rho_0}(\phi(s)) = G_{\rho_0}(\phi(\tilde{s})) \Rightarrow \|\phi(\tilde{s})\|^2 = t\} \qquad (1.25)$$

and let

$$\tilde{\Phi} := \bigcup_{t \geq \|e\|^2} \tilde{\Phi}_t \qquad (1.26)$$

Let

$$\tilde{c}(t) := \inf_{\phi \in \tilde{\Phi}_t} \max_{s \in [0,1]} G(\phi(s)). \qquad (1.27)$$

If G_{ρ_0} has no critical point, $\tilde{\Phi}_t$ is equivalent to Φ_t with respect to critical levels, that is $\tilde{c}(t)$ is well defined for all t and

$$c(t) = \tilde{c}(t) \text{ for all } t \geq \|e\|^2. \qquad (1.28)$$

For if for some t, we could find a sequence of paths $\phi_k \in \Phi_t$ such that

$$\max_{s \in [0,1]} \phi_k(s) = \phi_k(\tilde{s}) \to c(t)$$

and

$$\|\phi_k(\tilde{s})\| \to \tilde{t} \leq t,$$

by standard methods we could find a critical point of $G_{\rho_0}(u)$ (see [19]). Let

$$F(u) := G_{\rho_0}(u) - \tilde{c}(\|u\|^2). \qquad (1.29)$$

Mountain pass methods may be applied to $F(u)$ for $\phi(s) \in \tilde{\Phi}$ and for $\phi \in \tilde{\Phi}_t$ for $t \geq \|e\|^2$. To see this note first that

$$
\begin{aligned}
F(0) &= 0 & (1.30) \\
F(e) &= G_{\rho_0}(e) - c(\|e\|^2) < 0. & (1.31)
\end{aligned}
$$

Furthermore by the monotonicity of $c(t)$ for all $\phi \in \tilde{\Phi}_t$, $t \geq \|e\|^2$,

$$
\begin{aligned}
\max_{s \in [0,1]} F(\phi(s)) &\geq \max_s [G_{\rho_0}(\phi(s)) - c(t)] \\
&= G_{\rho_0}(\phi(\tilde{s})) - c(\phi(\tilde{s})) \\
&\geq 0
\end{aligned}
$$

For some $\phi(\tilde{s}) \in S_t$. The fact that $F(0)$ equals the critical level is not a problem because by construction of $\tilde{\Phi}_t$, the norms of critical sequences will not converge to 0. In particular, by construction we can find a critical sequence for F in each S_t, $t \geq \|e\|^2$. Now choose t_k such that $c(t)$ is differentiable at t_k. Let

$$
h_k := -c'(t_k)/2 \geq 0. \tag{1.32}
$$

As remarked above mountain pass methods give us a critical sequence u_j such that

$$
\begin{aligned}
\|u_j\|^2 &= t_k & (1.33) \\
F'(u_j) &= G'_{\rho_0}(u_j) + h_k u_j \to 0 & (1.34) \\
F(u_j) &= G_{\rho_0}(u_j) + c(t_k) \to 0. & (1.35)
\end{aligned}
$$

Because G satisfies property P, (1.33), (1.34), and (1.35) imply that u_j has a convergent subsequence, $u_j \to u_k$ satisfying

$$
\begin{aligned}
\|u_k\|^2 &= t_k & (1.36) \\
G_{\rho_0}(u_k) &= c(t_k) & (1.37)
\end{aligned}
$$

and (1.21). It follows from (1.21) that $h_k > 0$. It now suffices to take a sequence $t_k \to \infty$ such that $c'(t_k)$ exists and $c'(t_k) \to 0$ which is possible because $c(t)$ is bounded below by c. The sequence (u_k, h_k) will satisify (1.18) and (1.19), and Equation (1.37) implies (1.20) .

For a more detailed analysis of $c(t)$, see [27].

REMARK 1.5   1. *It is possible that $c(t) > c(\infty)$ for all $t < \infty$, but that G_{ρ_0} has a "local" min max critical point. We might have, for example, $c'(t_0) = 0$ for some t_0.*

2. *For $h > 0$ and $\phi \in \tilde{\Phi}$, if $\max_{s \in [0,1]} G_{\rho_0}(\phi(s)) = G_{\rho_0}(\phi(\tilde{s}))$, then $\max_{s \in [0,1]} G_{\rho_0+h}(\phi(s)) = G_{\rho_0+h}(\phi(\tilde{s}))$ since $\|\phi(\tilde{s})\|^2 = t$.*

The mountain impass lemma has the following corollary:

COROLLARY 1.6 *Assume $G_\rho(u)$ satisfies property (P) and that $G_{\rho_0}(u)$ has mountain pass geometry. Then there is a neighborhood I_{ρ_0} of ρ_0 in which the ρ for which (1.4) has a solution are dense.*

Proof: We first remark that mountain pass geometry is an open property of ρ and therefore there is a neighborhood I_{ρ_0} of ρ_0 for which $G_\rho(u)$ also has mountain pass geometry. Lemma 1.4 implies that the $\rho \in I_{\rho_0}$ for which G_ρ has critical point form a dense subset of I_{ρ_0}.

With the aid of mountain impass, it is now easy to establish the following result, which exploits an idea first used by Struwe [25]. (The result does not require mountain impasse).

LEMMA 1.7 Let $G_{\rho_0}(u)$ have mountain pass geometry and satisfy (H1). Then for almost every ρ in some neighborhood of ρ_0, there is a $u_\rho \in H \setminus \{0\}$ such that

$$G'_\rho(u_\rho) = 0. \qquad (1.38)$$

Proof: As noted above, mountain pass geometry as defined above is an open property of ρ and thus we can find an interval I_{ρ_0} containing ρ_0 such that for all $\rho \in I_{\rho_0}$, $G_\rho(u)$ also has mountain pass geometry. Consider $c(t, \rho)$ where $c(t, \rho)$ is defined as is $c(t)$ in (1.24) with $G_{\rho_0}(u)$ replaced by G_ρ.

Suppose that for $\rho \in I_{\rho_0}$, $G_\rho(u)$ does not have a critical point. By Lemma 1.4 we may assume that there exists a sequence $(\rho + h_k, u_k) \in \mathbb{R} \times H$ satisfying (1.18), (1.19), and (1.21), such that

$$G_{\rho_0}(u_k) = c(t_k, \rho_0) \to c(\infty, \rho_0). \qquad (1.39)$$

We claim that we may assume

$$c(t_k, \rho_0) = c(\infty, \rho_0 + h_k). \qquad (1.40)$$

By construction $G_{\rho_0+h_k}(u_k)$ (see Remark 1.5) represents a local min max critical level of $G_{rho_0+h_k}$ so clearly $c(t_k, \rho_0) \geq c(\infty, \rho_0 + h_k)$. Suppose that there was no sequence $(\rho + h_k, u_k)$ for which (1.40) held. Then by eliminating all the local critical points, one could obtain a geometry for which Lemma 1.4 failed to hold. Therefore it must be possible to find such a sequence and our assumption is valid.

We compute:

$$\frac{c(\infty, \rho_0 + h_k) - c(\infty, \rho_0)}{h_k} \geq \frac{G_{\rho_0+h_k}(u_k) - G_{\rho_0}(u_k)}{h_k}$$

$$= \frac{\frac{\rho_0+h_k}{2}\|u_k\|^2 - g(u_k) - [\frac{\rho_0}{2}\|u_k\|^2 - g(u_k)]}{h_k}$$

$$= \frac{h_k}{h_k}\|u_k\|^2 \to \infty.$$

Which implies $\frac{\partial}{\partial \rho} c(\infty, \rho_0)$ is not defined. But clearly $c(\infty, \rho)$ is a monotone increasing function in ρ for $\rho \in I_{\rho_0}$, and hence differentiable almost everywhere, therefore G_ρ admits a critical point for almost every $\rho \in I_{\rho_0}$.

The following lemma was excised from a proof in [20]. We feel that the result should be elevated to the status of a lemma. We use the fact that $\gamma(t)$ is continuous which is proved in [23].

LEMMA 1.8 Assume that $\rho \in I$. Then $G_\rho(u)$ admits a critical sequence u_k. The critical sequence tends either to a local minimum of $G_\rho(u)$ or satisfies equations (1.10) and (1.9) and is of mountain pass type. If the sequence tends towards a local minimum, then for some $t_0 \neq 0$, $\|u_k\| \to t_0$.

Proof: Let $\rho \in I$. Let $\Gamma(t) := \frac{1}{2}\rho t - \gamma(t)$. Because ρ satisfies

$$\inf_{t \neq s} \frac{\gamma(t) - \gamma(s)}{t - s} < \rho/2 < \sup_{t \neq s} \frac{\gamma(t) - \gamma(s)}{t - s}, \tag{1.41}$$

we know that $\Gamma(t)$ is not monotone. Indeed, suppose $\Gamma(t)$ were nondecreasing. Then for $t_1 < t_2$,

$$\frac{1}{2}\rho t_1 - \gamma(t_1) \leq \frac{1}{2}\rho t_2 - \gamma(t_2).$$

This would mean that

$$[\gamma(t_2) - \gamma(t_1)]/(t_2 - t_1) \leq \rho/2$$

for all t_1, t_2. This contradicts (1.41). Similar reasoning works for $\Gamma(t)$ nonincreasing. Thus $\Gamma(t)$ must have either a local minimum or a global maximum.

Suppose Γ has a local minimum at $t_0 > 0$. Then $G_\rho(u)$ is bounded below for $\|u\|^2 = t$ near t_0. In fact

$$\frac{1}{2}\rho t_0 - g(u_0) = \Gamma(t_0) \leq \Gamma(t) \leq \frac{1}{2}\rho\|u\|^2 - g(u).$$

By Ekeland's principle we can find a critical sequence for $G_\rho(u)$ such that $\|u_k\| \to t_0$.

If $\Gamma(t)$ does not have any local minima, it must have a global maximum at a unique point $t_0 > 0$. Thus $\Gamma(t) < \Gamma(t_0)$ for $t \neq t_0$. By the definition of $\gamma(t)$, we may choose $u_1 \in S_{t_1}$, with $t_1 > t_0$, such that

$$G_\rho(u_1) < \Gamma(t_0).$$

Clearly

$$0 = G_\rho(0) = \Gamma(0) < \Gamma(t_0)$$

and

$$G_\rho(u) \geq \Gamma(t_0), \ u \in S_{t_0}.$$

Thus $G_\rho(u)$ has mountain pass geometry and thus admits a critical sequence.

Proof of Theorem 1.3: The second statement in Theorem 1.3 follows from Lemma 1.8 and the assumption that $G_\rho(u)$ satisfies the Palais Smale condition. The first statement follows from Lemma 1.8, Remark 1.1, Lemma 1.7, and the fact that $G_\rho(u)$ satisfies property (P).

2 APPLICATION

We will study the following problem:

$$\begin{aligned}
-\Delta u + q(x)u &= \lambda f(x, u) \\
u|_{\partial\Omega} &= 0 \\
\lim_{|x|\to\infty} u &= 0
\end{aligned} \tag{2.42}$$

where $\Omega \subset I\!\!R^n$ is a smooth unbounded domain and $n \geq 3$. We make the following assumptions, denoting the critical Sobolev exponent by $2^* := 2n/(n-2)$.

(C1) For some $K < \infty$, $\limsup_{s\to 0} f(x,s)/s \leq K$.

(C2) We have $\lim_{|s|\to\infty} f(x,s)/|s|^{2^*-1} = 0$ uniformly in x.

(C3) The function $q(x) \in C^0(I\!\!R^n)$, $q(x) \geq 1$, and $\lim_{|x|\to\infty} q(x) = \infty$.

$$\|u\|_H^2 := \int_\Omega (q(x)|u|^2 + |\nabla u|^2)\, dx.$$

Let $f : \Omega \times I\!\!R \mapsto I\!\!R$ be continuous and let

$$F(x, s) := \int_0^s f(x, t)\, dt.$$

We define

$$S_t := \{u \in H : \|u\|^2 = t\}, \qquad g(u) := \int_\Omega F(x, u)\, dx$$
$$\gamma(t) := \sup_{u \in S_t} g(u) \quad I := (2\inf_{t\neq s} \tfrac{\gamma(t)-\gamma(s)}{t-s}, 2\sup_{t\neq s} \tfrac{\gamma(t)-\gamma(s)}{t-s}).$$

Our main result is the following:

THEOREM 2.1 For almost every λ such that $1/\lambda \in I$, equation (2.42) admits a nonzero solution.

2.1   Proofs of Main Results

The proof of Theorem 2.1 will consist in verifying the properties of Section 2. The following estimates are elementary and we omit their proofs.

LEMMA 2.2 Assume (C1) and (C2). Then there are positive constants C_1 and C_2, and for any $\epsilon > 0$ and $q \in (2, 2^*)$, there exists a $C_{\epsilon,q}$ such that:

$$|F(x,s)| \ \leq \ |s|^2/4 + C_1|s|^{2^*} \tag{2.43}$$
$$|F(x,s)| \ \leq \ C_2(1 + |s|^{2^*}) \tag{2.44}$$
$$|F(x,s)| \ \leq \ C_\epsilon|s|^2 + \epsilon|s|^{2^*} \tag{2.45}$$

REMARK 2.3 The estimate (2.44) along with the fact that $F \in C^1(\mathbb{R})$ imply that $g(u)$ is a C^1 functional on H with $(g'(u), v) = \int_\Omega f(u)v\,dx$ by Theorem C.1 in [26].

LEMMA 2.4 For $u \in S_t$ there is a positive constant C such that

$$|(g'(u), u)| \leq C(t + t^{2^*/2}). \tag{2.46}$$

We recall the following definitions which will be needed to state a version of the local case of the concentration compactness lemma.

Let $D_R := \Omega \cap B_R$ were B_R is the ball of radius R centered at 0. We will say that a sequence of nonnegative functions ϕ_k in $L^1(\Omega)$ is *tight* if given $\epsilon > 0$ there is an $R < \infty$ such that,

$$\int_{D_R} \phi_k\,dx \geq \|\phi_k\|_{L^1} - \epsilon \quad \text{for all } k.$$

The ϕ_k are tight up to translation if there exist $x_k \in E$ such that $\phi_k(\cdot + x_k)$ is tight.

LEMMA 2.5 Let ϕ_k be a sequence of nonnegative functions in $L^1(\Omega)$, with $\|\phi_k\|_{L^1} = t$. Then either
1^0) ϕ_k vanishes, i.e. for all $R < \infty$,

$$\lim_{k \to \infty} \sup_{x \in E} \int_{D_R + x} \phi_k\,dx = 0, \ \ or,$$

2^0) There is a renumbered subsequence ϕ_k, a $\lambda \in (0, t]$, and sequences of nonnegative functions r_k, ϕ_k^1, and ϕ_k^2 in $L^1(\Omega)$, such that the ϕ_k^1 are

tight up to translations and

$$\begin{cases} \phi_k = \phi_k^1 + \phi_k^2 + r_k \\ \|\phi_k - (\phi_k^1 + \phi_k^2)\|_{L^1} = \|r_k\|_{L^1} \to 0 \\ \|\phi_k^1\|_{L^1} \to \lambda \qquad\qquad\qquad \|\phi_k^2\|_{L^1} \to t - \lambda \\ dist\ (supp\ \phi_k^1,\ supp\ \phi_k^2) \to \infty \qquad \phi_k^i r_k = 0\ a.e.\ i = 1, 2, . \end{cases}$$

Furthermore, if for some $R < \infty$

$$\limsup_{k\to\infty} \int_{D_R} \phi_k\, dx = \delta > 0,$$

then we may assume the ϕ_k^1 *to be tight.*

Proof: Lemma 2.5 is proved with the exception of the last statement in [21] (see also [14]). The proof of the last statement is identical to that of the general case 2, using the density function

$$Q_t(R) := \limsup_{k\to\infty} \int_{D_R} \phi_k\, dx$$

rather than

$$Q_t(R) := \sup_{x\in\mathbb{R}^n} \limsup_{k\to\infty} \int_{D_{R+x}} \phi_k\, dx.$$

We have also the following smooth version of Case 2 of Lemma 2.5.

LEMMA 2.6 Let u_k *be a sequence in* S_t. *Let* $\phi_k = |q(x)u_k|^2 + |\nabla u_k|^2$. *Assume* ϕ_k *is as in Lemma 2.5 case 2. Then there exist sequences* $u_k^i \subset H$, $i = 1, 2$ *satisfying*

$$\|u_k - (u_k^1 + u_k^2)\|_{II} \to 0$$
$$dist\ (supp\ u_k^1,\ supp\ u_k^2) \to \infty$$
$$\|u_k^1\|_H^2 = \lambda, \quad \|u_k^2\|_H^2 = t - \lambda.$$

Moreover $\psi_k^1 = |q(x)u_k^1|^2 + |\nabla u_k^1|^2$, *will be tight up to translation.*

If $u_k \in S_t$ dichotomizes into u_k^1 and u_k^2 as in Lemma 2.6, we have

$$g'(u_k) = g'(u_k^1) + g'(u_k^2) + o(1). \tag{2.47}$$

Furthermore, because the supports of the u_k^i are disjoint by the definition of $(g'(u), v)$, $g'(u_k^1)$ is orthogonal to $g'(u_k^2)$.

The following lemma proved in [21] illustrates the importance of sequences in V_q the densities of whose norms are tight.

LEMMA 2.7 Assume (2). Then $g(u)$ *and* $(g'(u), u)$ *are weakly continuous on sequences the densities of whose norms are tight.*

We will also use the following from [23].

LEMMA 2.8 Assume (C2). If $u_k \in S_t$ dichotomizes into u_k^1 and u_k^2 as in Lemma 2.6, then

$$g(u_k) - (g(u_k^1) + g(u_k^2)) \to 0 \tag{2.48}$$

LEMMA 2.9 For $\rho \neq 0$, $G_\rho(u)$ satisfies property (P).

Proof: Let u_k be a sequence satisfying (1.8) and (1.9) for $G_\rho(u)$, $\rho \neq 0$. By multiplying u_k by appropriate constants $\theta_k \to 1$, we may assume that $\|u_k\|^2 = t$ for all k. Let ϕ_k be as in Lemma 2.6. We claim that Case 2 of Lemma 2.5 holds and that we can assume the densities of the norms of the u_k^1 to be tight, that is that for some $R < \infty$

$$\limsup_{k \to \infty} \int_{D_R} (|\nabla u_k|^2 + q u_k^2)\, dx = \delta > 0.$$

Suppose this were not the case. Then we could find a relabeled subsequence u_k such that for all $R < \infty$,

$$\int_{D_R} (|\nabla u_k| + q u_k^2) dx \to 0. \tag{2.49}$$

But this implies

$$\int_\Omega u_k^2 \, dx \to 0. \tag{2.50}$$

To see this, let $\epsilon > 0$ and let $M \geq t/\epsilon$. Choose R such that $q(x) > M$ for $|x| > R$. Then

$$t \geq \int_{\Omega \setminus B_R} (|\nabla u_k|^2 + q u_k^2) dx \geq \int_{\Omega \setminus B_R} M u_k^2 \, dx.$$

From which we conclude

$$\int_{D_R} u_k^2 \, dx \leq \epsilon. \tag{2.51}$$

Together with (2.49) this implies (2.50). Now (2.50) and (2.45) imply

$$(g'(u_k), u_k) \to 0. \tag{2.52}$$

Furthermore, (1.9) and the fact that that u_k is bounded imply

$$\rho \|u_k\|^2 - (g'(u_k), u_k) \to 0. \tag{2.53}$$

But (2.52) and (2.53) contradict the fact that $t \neq 0$ which proves the claim. This is sufficient for obtaining a solution (see [21, 23]. We can establish property (P) by simply noting that Lemmas 2.6 and (2.47) imply that the u_k^2 also form a critical sequence for $G_\rho(u)$. If one supposes $\liminf \|u_k^2\|^2 > \delta > 0$, the preceding analysis applied to the u_k^2 leads to the conclusion that part of the densities of the norms of the u_k^2 remain tight, contradicting the fact that $\mathrm{dist}(\mathrm{supp}\ u_k^1, \mathrm{supp}\ u_k^2) \to \infty$.

Proof of Theorem 2.1: The proof of Theorem 2.1 follows from Theorem 1.3 and Lemma 2.9.

References

[1] Abakhti-Mchachti, A. (1993). *Systèmes semilinéaires d'équations de Schrodinger*. Ph. D. thesis, University of Toulouse 3. These numero 1338.

[2] Alziary, B., L. Cardoulis, and J. Flekinger-Pellé (to appear). The maximum principle and existence of solutions for elliptic systems involving Schrodinger operators. *Real Acad. Cienc.*.

[3] Berestycki, H. and P.-L. Lions (1983a). Nonlinear scalar field equations 1. *Arch. Rat. Mech. Anal. 82*, 313–346.

[4] Berestycki, H. and P.-L. Lions (1983b). Nonlinear scalar field equations 2. *Arch. Rat. Mech. Anal. 82*, 347–376.

[5] Berger, M. and M. Schechter (1972). Embedding theorems and quasilinear elliptic boundary value problems for unbounded domains. *Trans. Amer. Math. Soc. 172*, 261–278.

[6] Clarke, F. H. (1989). *Methods of Dynamic and Nonsmooth Optimiztion*. CBMS-NSF Regional Conference Series in Applied Mathematics.

[7] Clarke, F. H. (1990). *Optimization and Nonsmooth Analysis*. Classics in Applied Mathematics SIAM.

[8] Ding, W.-Y. and W.-M. Ni (1986). On the existence of positive entire solutions of a semilinear elliptic equation. *Arch. Rat. Mech. Anal. 91*, 283–308.

[9] Edmunds, D. and W. Evans (1987). *Spectral Theory and Differential Operators*. Oxford University Press.

[10] Fleckinger-Pellé, J. (1984). Asymptotics of eigenvalues for some non-definite elliptic problems. In *Proc. Dundee conference on Diff Equations*, Volume 1151 of *Lecture Notes in Mathematics*, pp. 148–156.

[11] Fleckinger-Pellé, J. (1985). Valeurs propres de problèmes elliptiques indéfinis. *C. R. Acad. Sci. Paris 301.*

[12] Kavian, O. (1986). Minimum action solutions of nonlinear elliptic
equations in unbounded domains. *Proc. Roy. Soc. Edin. 102*, 327–
343.

[13] Lions, P. L. (1984). The concentration-compactness principle in
the calculus of variations. The locally compact case. part 2. *Ann.
Inst. H. Poincaré 1*, 223–283.

[14] Lions, P.-L. (1987). Solutions of Hartree-Fock equations for
Coulomb systems. *Comm. Math. Phys. 109*, 33–97.

[15] Lions, P.-L. (1988). On positive solutions of semilinear elliptic
equations in unbounded domains. In W.-M. Ni (Ed.), *Nonlinear
Diffusion Equations and Their Equilibrium States*, Volume 2, pp.
85–122. Springer-Verlag.

[16] Omana, W. and M. Willem (1992). Homoclinic orbits for a class
of hamiltonian systems. *Diff. and Int. Eqs. 5*, 1115–1120.

[17] Pohožaev, S. (1965). Eigenfunctions of the equation $\triangle u + \lambda f(u) = 0$. *Soviet Math. Dokl. 165*, 1408–1412.

[18] Reed, M. and B. Simon (1978). *Methods of Modern Mathematical
Physics IV: Analysis of Operators*. Academic Press.

[19] Schechter, M. (1991). The mountain pass alternative. *Advances in
Appl. Math. 12*, 91–105.

[20] Schechter, M. and K. Tintarev (1991). Eigenvalues for semilinear
boundary value problems. *Archives for Rational Mechanics Analysis 113*, 197–208.

[21] Schindler, I. (1992). Quasilinear elliptic boundary value problems
on unbounded cylinders and a related mountain pass lemma. *Arch.
Rat. Mec. and Anal. 120*, 363–374.

[22] Schindler, I. (1995). A critical value function and applications to
semilinear elliptic equations on unbounded domains. *Nonlinear
Analysis Theory, Methods, and Applications 24*, 947–959.

[23] Schindler, I. (95). A critical value function and applications to
translation invariant semilinear elliptic equations on unbounded
domains. *Differential and Integral Equations 8*, 813–828.

[24] Strauss, W. (1977). Existence of solitary waves in higher dimensions. *Comm. Math. Phys. 55*, 149–162.

[25] Struwe, M. (1988). The existence of surfaces of constant mean curvature with free boundaries. *Acta Mathematica 160*, 19–64.

[26] Struwe, M. (1990). *Variational Methods. Applications to Nonlinear Partial Differential Equations and Hamiltonian Systems.* Springer-Verlag.

[27] Tintarev, K. (1994). A relation between critical values and eigenvalues in nonlinear minimax problems. *Applicable Anal. 54*, 57–73.

[28] Tintarev, K. (1996). Mountain pass and impasse for non-smooth functionals with constraints. *Nonlinear Analysis, Theory, Methods, and Applications 26*, 798–803.

Asymptotic Behavior for Large Time of Solutions of Nonlinear Evolution Equations with Dissipation

I.A. SHISHMAREV   Moscow State University, Russia

The aim of this lecture is to present comparatively new results (not included in the book [1]) about large time behaviour of solutions for different kind of nonlinear evolution equations and system of equations.

1 NONLINEAR EQUATIONS

1.1 The equation $u_t + uu_x + \mathbf{K}(u) = 0$

We start with the Cauchy problem for the very general model nonlinear equation

$$u_t + uu_x + \mathbf{K}(u) = 0, \ u\mid_{t=0} = \overline{u}(x), \ x \in \mathbf{R}_1, \ t \geq 0. \tag{1}$$

Here $\mathbf{K}(u)$ is linear pseudodifferential operator, defined by the inverse Fourier transform

$$\mathbf{K}(u) = \frac{1}{2\pi} \int_{-\infty}^{\infty} \exp(ipx)K(p)\hat{u}(p,t)dp, \tag{2}$$

where

$$\hat{u}(p,t) = \int_{-\infty}^{\infty} \exp(-ipx)u(x,t)dx$$

is the Fourier transform, and $K(p)$ is the symbol of the linear operator $\mathbf{K}(u)$. Equation (1) is a remarkable model for the description of various wave motions in dispersive media. Due to the presence of the pseudodifferential operator $\mathbf{K}$, equation (1) includes many equations well-known in physics such as those of Burgers, Korteweg-de Vries, Korteweg-de Vries-Burgers, Benjamin-Ono, Kuramoto-Sivashinsky, Joseph, Smith, Klimontovich, Ott-Sudan-Ostrovsky, Leibovich and many others (see the survey in [1]).

In the case the real part of the symbol $K(p)$ grows fast enough at infinity (strong dissipation)

$$Re\, K(p) \geq c|p|, \ |p| \geq p_o > 0, \ c > 0 \tag{3}$$

the asymptotics as $t \to \infty$ of the solution $u(x,t)$ of the Cauchy problem for equation (1) was obtained in [1] by applying perturbation theory. The condition (3) of strong dissipation clearly holds, for example, for the Burgers equation, the Korteweg-de Vries-Burgers equation and the Kuramoto-Sivashinsky equation. If the condition (3) is not satisfied but the imaginary part of the symbol $K(u)$ grows sufficiently fast at infinity (strong dispersion), as it is in the case of the Korteweg-de Vries equation with linear dissipation, the asymptotics as $t \to \infty$ of the solution of the Cauchy problem can be found again by using perturbation theory [1]. Under conditions of strong dissipation or strong dispersion the linear operator $\mathbf{K}(u)$ in (1) prevails over the nonlinear operator, and therefore the nonlinearity can be considered a perturbation of the linear operator. However many physics effects do not have strong dissipation or strong dispersion. For example, in the exact potential theory of water waves with linear damping, wave motions are modeled by equation (1) with symbol

$$K(p) = \lambda - 2bip\sqrt{\frac{\tanh(p)}{p}}, \tag{4}$$

Klimontovich showed that linear hydrodynamics of a plasma is described by equation (1) with symbol

$$K(p) = \lambda - bip(1 + p^2)^{-\frac{1}{2}}, \tag{5}$$

in the Ott-Sudan-Ostrovsky equation the symbol has the form

$$K(p) = \lambda\sqrt{|p|} + ibp^3, \tag{6}$$

Leibovich derived equation (1) with symbol

$$K(p) = \lambda + ibp^3 \mathcal{K}_o(|p|) \tag{7}$$

to describe the evolution of a mixture of liquid and gas bubbles (here $\lambda > 0$ and b are constants, $\mathcal{K}_o$ is the Macdonald function of order zero). If the conditions of strong dissipation or strong dispersion are not fulfilled, the nonlinearity in equation (1) plays an essential role, so that sufficiently large initial data lead to blowing up the solution in finite time [1].

In the case of small initial data $(\overline{u}(x) \in H^\infty(\mathbf{R}_1), \| \overline{u} \|_{H^\infty(\mathbf{R}_1)}$ is small) for the global in time existence of a classical solution $u(x,t) \in C^\infty([0,\infty), H^\infty(\mathbf{R}_1))$ of the Cauchy problem (1) merely strict dissipation suffices (see [1])

$$Re\, K(p) \geq 0 \; for \; |p| \leq p_o, \; Re\, K(p) \geq a > 0 \; for \; |p| \geq p_0 > 0.$$

Now we consider the Cauchy problem (1) with weak (but strict) dissipation and arbitrary dispersion. In this case we cannot apply the perturbation theory and the method of obtaining the asymptotics as $t \to \infty$ is based on the possibility to get accurate estimate of time-dependence for some integral norms of solutions to the Cauchy problem (1).

THEOREM 1 Suppose that

(i) the symbol $K(p) \in C^0(\mathbf{R}_1)$ satisfies the conditions

$$K(p) = \lambda + a|p|^\delta + O(|p|^{\delta+\sigma}), \; |p| \leq 1 \tag{8}$$
$$K^1(p) = Re\, K(p) \geq \lambda + \theta m^\delta(p), \; p \in \mathbf{R}_1, \tag{9}$$

where $\lambda > 0$, $a \geq \theta > 0$, $\delta \in (0,2)$, $\sigma > 0$, $m(p) = \min(1,|p|)$

(ii) the initial data $\bar{u}(x) \in H^\infty(\mathbf{R}_1)$, $x\bar{u}(x) \in L^1(\mathbf{R}_1)$ is small in the following sense

$$\| \bar{u}(x) \|_{H^3(\mathbf{R}_1)} + \int_{-\infty}^{\infty} |\bar{u}(x)|dx \leq c_1, \qquad (10)$$

where $c_1 = c_1(\lambda, \theta, \delta) > 0$ is sufficiently small.

Then as $t \to \infty$ the solution $u(x,t)$ of the Cauchy problem (1) has the asymptotics

$$u(x,t) = \exp(-\lambda t)t^{-\frac{1}{\delta}} \left(2\hat{\bar{u}}(0) \int_0^\infty \exp(-ap^\delta)\cos(p\xi)dp + O(t^{-\gamma}) \right) \qquad (11)$$

uniformly with respect to $\xi = xt^{-\frac{1}{\delta}} \in \mathbf{R}_1$, where $\gamma > 0$ is some constant [2].

In the case of strong dissipation a smallness of initial data is no longer necessary for studying asymptotic behavior of solutions. The following statement is valid, [3].

THEOREM 2 Let

(i) the symbol $K(p) \in C^0(\mathbf{R}_1)$ and satisfy the conditions

$$\begin{aligned} K(p) &= \lambda + a|p|^\delta + O(|p|^{\delta+\sigma}), \quad |p| \leq 1 & (12)\\ Re\, K(p) &\geq \lambda + \theta m^\delta(p)M^\alpha(p), \quad p \in \mathbf{R}_1, & (13) \end{aligned}$$

where $\lambda > 0$, $a \geq \theta > 0$, $\delta \in (0,\frac{3}{2})$, $\sigma > 0$, $\alpha > \frac{3}{2}$, $M(p) = \max(1,|p|)$;

(ii) the initial data $\bar{u}(x) \in H^\infty(\mathbf{R}_1)$, $x\bar{u}(x) \in L^1(\mathbf{R}_1)$.

Then for solution $u(x,t)$ of the Cauchy problem (1) the asymptotic formula (11) holds.

EXAMPLE. The symbol $K(p)$ of the Ott-Sudan-Ostrovsky-Burgers equation

$$u_t + uu_x + a \int_{\mathbf{R}_1} \frac{sgn(s-x)}{\sqrt{|s-x|}} u_s(s,t)ds - u_{xx} - bu_{xxx} = 0$$

is equal to $K(p) = p^2 + a|p|^{\frac{1}{2}} + ip^3 b$. The asymptotics as $t \to \infty$ of solution $u(x,t)$ of the Cauchy problem for this equation has the form

$$u(x,t) = \frac{1}{t^2}(2\hat{\bar{u}}(0) \int_0^\infty \exp(-a\sqrt{p})\cos(p\xi)dp + O(t^{-\gamma})), \quad \gamma > 0.$$

1.2 The Korteweg-de Vries-Burgers equation

In this subsection we consider the Cauchy problem for the Korteweg-de Vries-Burgers (KdVB) equation, which is a very important particular case of equation (1)

$$u_t + 2uu_x - u_{xx} + \frac{a}{3}u_{xxx} = 0, \quad u\mid_{t=0} = \bar{u}(x), \quad x \in \mathbf{R}_1, \ t \geq 0. \qquad (14)$$

We produce here the first two terms of the asymptotic expansion as $t \to \infty$ of the solution to the Cauchy problem for the KdVB equation (14). As far as we know this is the first case in which the second term of asymptotics is found for a nonlinear equation that cannot be reduced to a linear equation by a change of variables. We present two theorems: in the first of them $t \to \infty$, $\frac{x}{\sqrt{t}} = O(1)$ and in the second one $t \to \infty$, $\frac{x}{\sqrt{t}} \to \infty$.

THEOREM 3 Let $\bar{u}(x) \in W_1^2(\mathbf{R}_1) \cap W_2^5(\mathbf{R}_1)$ and $x\bar{u}(x) \in L_1(\mathbf{R}_1)$, let the norms $\| \bar{u} \|_{W_1^2}$, $\| \bar{u} \|_{W_2^5}$ and $\| x\bar{u} \|_{L_1}$ be sufficiently small and $U = \int_{\mathbf{R}_1} \bar{u} dx \neq 0$.

Then the solution $u(x,t)$ of the Cauchy problem for the KdVB equation (14) has the following asymptotics as $t \to \infty$ uniformly with respect to $\xi = \frac{x}{\sqrt{t}} \in \mathbf{R}_1$

$$u(x,t) = A(\xi)\frac{1}{\sqrt{t}} + \tilde{A}(\xi)\frac{\log(t)}{t} + O(\frac{\sqrt{\log(t)}}{t}), \tag{15}$$

where

$$A(\xi) = \frac{1}{\sqrt{\pi}H(\xi)} \sinh(\frac{U}{2}) \exp(-\frac{\xi^2}{4}),$$

$$\tilde{A}(\xi) = -\frac{a(A(\xi) - \frac{\xi}{2})\exp(-\frac{\xi^2}{4})}{12\sqrt{\pi}H(\xi)} \int_{\mathbf{R}_1} A^3(y)H(y)dy,$$

$$H(y) = \exp(-\frac{U}{2})(\cosh(\frac{U}{2}) - \sinh(\frac{U}{2})Erf(\frac{y}{2})).$$

Note that we have explicit exprssions for the functions $A(\xi)$ and $\tilde{A}(\xi)$ and they depend only on total mass of the initial data $\int_{\mathbf{R}_1} \bar{u}(x)dx$.

Although the asymptotics (15) is formally valid uniformly with respect to ξ as $t \to \infty$, the coefficients $A(\xi)$ and $\tilde{A}(\xi)$ rapidly decay as $\xi \to \infty$, so the remainder in (15) becomes greater than the leading terms, thus the formula (15) makes sense only in the domain $\xi = O(1)$ as $t \to \infty$. In the next theorem we study the asymptotics of the solution of the Cauchy problem (14) as $t \to \infty$ and $\xi = \frac{x}{\sqrt{t}} \to \infty$ simultaneously. To be definite let $a > 0$ in the KdVB equation (if $a < 0$ we use the substitution $u(x,t) = -\tilde{u}(-x,t)$). In contrast to the case $\xi = O(1)$, in which the rate of decay of the linear and nonlinear terms in equation (14) are the same, the solution $u(x,t)$ decays faster as $\xi \to \infty$ and linear terms prevail over the nonlinear term in the KdVB. Therefore the asymptotics of $u(x,t)$ is quasilinear and is determined by equation $v_t - v_{xx} + \frac{a}{3}v_{xxx} = 0$, the nonlinearity influences the form of the coefficients in the asymptotic expansion.

THEOREM 4 Let the Fourier transform of the initial data $\bar{u}(x)$ be analytic in the domain $Im\, p \geq -\frac{1}{a}$, $a > 0$, and satisfy the estimate

$$\left| \frac{d^l \hat{\bar{u}}(p)}{dp^l} \right| \leq \epsilon M^{-8}(p) \exp(b\mu(p)), \quad l = 0, 1, 2, 3,$$

where $b > 0$, $0 < \epsilon < c$ are constants, $\mu(p) = \eta^2(1 + a\frac{\eta}{3})$, $\eta = Im\, p$. Then the following asymptotic formulas as $t \to \infty$, $\xi \to \infty$ are valid

(i) for $X \equiv \frac{x}{t} > 0$ and for $-\frac{1}{a} < X < 0$

$$u(x,t) = \frac{\exp(-\nu t)}{\sqrt{t}} \left(A(i\sigma) + \frac{1}{t}\tilde{A}(i\sigma) + O(\frac{\exp(b\mu(i\sigma))}{t^{\frac{3}{2}}m^8(\sigma)m^5(1 + a\sigma)}) \right)$$

where $\sigma = \frac{\sqrt{1+aX}-1}{a}$, $\nu = \frac{2}{3a^2}((1+aX)^{\frac{3}{2}} - 1 - \frac{3a}{2}X)$,

(ii) for $X = -\frac{1}{a}$

$$u(x,t) = \exp(-\frac{t}{3a^2})t^{-\frac{1}{3}}(A(-\frac{i}{a}) + t^{-\frac{1}{3}}\tilde{A}(-\frac{i}{a}) + O(t^{-\frac{2}{3}}\log(t)) \qquad (16)$$

(iii) for $X < -\frac{1}{a}$

$$\begin{aligned} u(x,t) &= \frac{\exp(-\omega t)}{\sqrt{t}}(Re\,(A(\beta - \frac{i}{a})\exp(i\frac{\pi}{4} - 2a\beta^3\frac{t}{3})) + \\ &+ \frac{1}{t}Re\,(\tilde{A}(\beta - \frac{i}{a})\exp(3i\frac{\pi}{4} - 2a\beta^3\frac{t}{3})) + O(\frac{\log(t)}{t^{\frac{3}{2}}m^2(\beta)}), \end{aligned}$$

where $\omega = -\frac{2}{3a^2} - \frac{X}{a}$ and $\beta = \frac{1}{a}\sqrt{-1-aX} > 0$. The coefficients A and $\tilde{A}$ are determined by the Fourier transform of the initial data $\overline{u}(x)$ and could be expressed explicitly, [4].

1.3 The generalized Kolmogorov-Petrovsky-Piskunov (KPP) equation

Consider the Cauchy problem for the generalized KPP equation

$$u_t + \sum_{k=1}^{n} a_k u^k + K(u) = 0, \; u\,|_{t=0} = \overline{u}(x), \; x \in \mathbf{R}_1, \; t \geq 0, \qquad (17)$$

here the linear pseudodifferential operator $K(u)$ is defined by (2).
We will study two cases: the case $a_1 > 0$, a_2, ..., a_n are arbitrary real numbers, i.e. the equation (17) has linear dissipation, and the case when $a_1 = a_2 = a_3 = 0$, $a_4, ..., a_n$ are arbitrary real numbers, so the minimal power of a polynomial $\sum_{k=1}^{n} a_k u^k$ is four. The following theorem describes the asymptotics of the solution $u(x,t)$ of the Cauchy problem (17) in both cases.

THEOREM 5 Suppose that

(i) the symbol $K(p) \in C^0(\mathbf{R}_1)$ satisfies the conditions

$$K(p) = \theta|p|^\delta + O(|p|^{\delta+\sigma}), \; |p| \leq 1, \; \sigma > 0, \; \theta > 0, \; \delta > 0$$

in the first case and $\delta \in (0,3)$ in the second one,

$$K^1(p) \equiv Re\,K(p) \geq 2\chi(p), \; p \in \mathbf{R}_1, \; \chi(p) \equiv c_1 m^\delta(p), \; c_1 > 0,$$

$$|K(p) - K(q)| \leq c_2|p - q|^{\sigma_1}M^\alpha(p), \; |p - q| \leq 1, \; p, q \in \mathbf{R}_1,$$

$\alpha \geq 0$, $\sigma_1 \in (0,1]$, $c_2 > 0$,

(ii) the Fourier transform $\hat{\overline{u}}(p)$ of the initial data $\overline{u}(x) \in L_1(\mathbf{R}_1)$ satisfies the requirements

$$|\hat{\overline{u}}(p)| \leq \epsilon M^{-n-\alpha}(p), \; p \in \mathbf{R}_1,$$

$$|\hat{\overline{u}}(p) - \hat{\overline{u}}(q)| \leq \epsilon|p - q|^{\sigma_2}M^{-n}(p), \; |p - q| \leq 1,$$

$$p, \; q \in \mathbf{R}_1, \; \sigma_2 \in (0,1],$$

where $\epsilon < c$ and $c > 0$ is determined by the symbol $K(p)$.

Then as $t \to \infty$ the asymptotics

$$u(x,t) = \exp(-\beta t)t^{-\frac{1}{\delta}}(A\int_0^\infty \exp(-\theta p^\delta)\cos(p\xi)dp + O(t^{-\gamma})) \qquad (18)$$

takes place uniformly with respect to $\xi = xt^{-\frac{1}{\delta}} \in \mathbf{R}_1$, $\gamma > 0$ is a constant. The constant A depends on the symbol $K(p)$ and the initial data $\overline{u}(x)$ and can be expressed explicitly by diagram technique (see [1]). The constant β is equal $a_1 > 0$ in the first case and is equal zero in the second one, [5].

REMARK The theorem 5 is true for the KPP equation itself and in this case we have $\delta = 2$, $\gamma = \frac{1}{2}$.

2 NONLINEAR SYSTEMS

2.1 Asymptotic representation of the surface waves

We study the system of equations, describing surface waves in an incompressible fluid

$$\psi_t + (\psi\phi)_x + \mathbf{K}_1(\psi) + \mathbf{K}_2(\phi) = 0 \qquad (19)$$

$$\phi_t + \frac{1}{2}(\phi^2)_x + \mathbf{K}_3(\psi) + \mathbf{K}_4(\phi) = 0 \qquad (20)$$

where ψ and ϕ are real functions, $x \in \mathbf{R}_1$, $t \geq t$. The symbols $K_j(p)$ of operators $\mathbf{K}_j$ are scalar complex-valued functions, defined by (2). The behavior of solutions of linearized system (19)-(20)

$$\psi_t + \mathbf{K}_1(\psi) + \mathbf{K}_2(\phi) = 0 \qquad (21)$$

$$\phi_t + \mathbf{K}_3(\psi) + \mathbf{K}_4(\phi) = 0 \qquad (22)$$

is determined by eigenvalues of the matrix

$$\begin{pmatrix} K_1(p) & K_2(p) \\ K_3(p) & K_4(p) \end{pmatrix}.$$

If we choose the symbols $K_j(p)$ as follows:

$$K_1(p) = K_4(p) = ap^2, \; K_3(p) = ip, \; K_2(p) = ib^2(1 + \delta p^2)\frac{\tanh(p\omega)}{p\omega},$$

a, b, ω 0, $\delta \geq 0$, then the system (19)-(20) represents the system of equations for surface water waves with viscosity and surface tension

$$\psi_t + (\psi\phi)_x - a\psi_{xx} + \frac{1}{2\pi}\int_{\mathbf{R}_1} \exp(ipx)K_2(p)\hat{\phi}(p,t)dt = 0 \qquad (23)$$

$$\phi_t + \frac{1}{2}(\phi^2)_x + \psi_x - a\phi_{xx} = 0. \qquad (24)$$

In this case eigenvalues $\lambda_j(p)$ are equal to

$$\lambda_j(p) = ap^2 + ibp(-1)^j\sqrt{(1 + \delta p^2)\frac{\tanh(p\omega)}{p\omega}}, \; j = 1,2 \qquad (25)$$

and correspond to dispersion well-known in the potential theory of water waves

$$c(p) = b\sqrt{(1 + \delta p^2)\frac{\tanh(p\omega)}{p\omega}}$$

and to viscosity described by the summand ap^2.

Another important example of system (19)-(20) is the well-known Boussinesq system of equations with viscosity

$$\psi_t + (\psi\phi)_x - a\psi_{xx} + b^2\phi_x + \frac{1}{3}b^2\omega^2\phi_{xxx} = 0 \tag{26}$$

$$\phi_t + \frac{1}{2}(\phi^2)_x + \psi_x - a\phi_{xx} = 0. \tag{27}$$

It is derived from (19)-(20) if $K_1(p)$, $K_3(p)$, $K_4(p)$ are the same as above and $K_2(p) = ib^2p(1 - \frac{1}{3}p^2\omega^2)$. The eigenvalues $\lambda_j(p)$ of the matrix $K(p)$ in this case have the form:

$$\lambda_j(p) = ap^2 + ibp(-1)^j\sqrt{1 - \frac{1}{3}\omega^2p^2}, \; j = 1, 2. \tag{28}$$

It is worth noting that the formula (25) passes into (28), if we take the expansion of $\tanh(\omega p)$ in the power series in the neighborhood of the origin and limit ourselves only by the first two terms of expansion and moreover put $\delta = 0$ (this means that effect of surface tension is sufficiently small and could be neglected). That correspond to so-called long waves Boussinesq approximation. It follows from (28) that if the viscosity is small in comparison with dispersion ($0 < a < \frac{\omega b}{\sqrt{3}}$), then $Re\,\lambda_j(p)$ becomes negative for large $|p|$. This presents an obstacle for solvability of the linearized Boussinesq system (26)-(27). To avoid this difficulty the mathematicians introduced in consideration so-called "good Boussinesq system"

$$\psi_t + (\psi\phi)_x - a\psi_{xx} + b^2\phi_x - \frac{1}{3}b^2\omega^2\phi_{xxx} = 0 \tag{29}$$

$$\phi_t + \frac{1}{2}(\phi^2)_x + \psi_x - a\phi_{xx} = 0. \tag{30}$$

which differs from (26)-(27) by the sign before the third derivative (in distinction from (29)-(30) mathematicians call the system (26)-(27) "bad" Boussinesq system).

Now the eigenvalues have the form

$$\lambda_j(p) = ap^2 + ibp(-1)^j\sqrt{1 + \frac{1}{3}\omega^2p^2}, \; j = 1, 2 \tag{31}$$

so $Re\,\lambda_j(p) > 0$ for all $|p| > 0$ and solvability of linearized system (29)-(30) is evident. From physical point of view the formula (31) can be derived from (25) in approximation of long waves ($|p| \leq 1$), when the forces of surface tension play essential role. Historically the systems (26)-(27) and (29)-(30) arose in connection with the intention of mathematicians to deal with simpler local equations instead of the exact nonlocal system of equations (23)-(24).

Our approach allows us to consider nonlocal version (19)-(20) and (24)-(24) of the system of equations for surface waves, which gives the adequate description of nature.

To study the asymptotic behavior of solutions to the Cauchy problem for system (19)-(20) as $t \to \infty$, we first pay attention to the system (29)(30). Heuristic arguments about contribution of each term in the system (29)-(30) as $t \to \infty$ show that the solution of system (29)-(30) should be asymptoticaly close to the solution of the following system

$$\psi_t + (\psi\phi)_x - a\psi_{xx} + b^2\phi_x = 0 \tag{32}$$

$$\phi_t + \frac{1}{2}(\phi^2)_x + \psi_x - a\phi_{xx} = 0 \tag{33}$$

(we omitted the last term in (29)-(30)). If we substitute here $\psi = b(u_1 + u_2)$, $\phi = u_2 - u_1$, we get

$$u_{1t} + (\frac{1}{4}u_2^2 - \frac{3}{4}u_1^2 + \frac{1}{2}u_1u_2))_x - au_{1xx} - bu_{1x} = 0 \tag{34}$$

$$u_{2t} + (\frac{1}{4}u_1^2 - \frac{3}{4}u_2^2 + \frac{1}{2}u_1u_2))_x - au_{2xx} - bu_{2x} = 0. \tag{35}$$

Further since the wave u_j has the largest amplitude ($\sim t^{-\frac{1}{2}}$) close to the wave front $x = (-1)^j bt$, then the main contributions in the system (34)-(35) are given by the summands $\pm\frac{3}{4}u_j^2$ (among other nonlinear terms). So the solutions $u_j(x,t)$ of (34)-(35) must be asymptotically close to the solution $v_j(x,t)$ of the Burgers equation with shift,

$$v_{jt} + (-1)^j\frac{3}{4}(v_j^2)_x - av_{jxx} + (-1)^jbv_{jx} = 0, \; j = 1,2 \tag{36}$$

and the solution of the system (29)-(30) as $t \to \infty$ can be represented as two travelling in opposite directions Burgers waves v_1 and v_2. This unexpected and surprising result turns out to be valid in general case of the system (19)-(20).

Namely under a rather broad conditions which by all means include systems (23)-(24) and (29)-(30), we proved that the solution (ψ, ϕ) to the Cauchy problem for the system (20) uniformly with respect to $x \in \mathbf{R}_1$ the following asymptotic representation is valid as $t \to \infty$

$$\psi(x,t) = bv_1(x,t) + bv_2(x,t) + O(t^{-\frac{1}{2}-\gamma}) \tag{37}$$

$$\phi(x,t) = v_2(x,t) - v_1(x,t) + O(t^{-\frac{1}{2}-\gamma}), \tag{38}$$

where $\gamma > 0$, $v_j(x,t)$, $j = 1,2$, are solutions of the Cauchy problem for the Burger equations (36). So the surface waves are represented asymptotically as two Burgers waves travelling in opposite directions.

With the help of the Hopf-Cole substitution the asymptotic behavior of functions $v_j(x,t)$ can be easily calculated

$$v_j(x,t) = (at)^{-\frac{1}{2}}A_j(\chi_j) + O(t^{-\frac{1}{2}-\gamma}) \tag{39}$$

where

$$\chi_j = \frac{x - (-1)^j bt}{\sqrt{at}}, \; A_j(z) = -\frac{4}{3}a(-1)^j\frac{\partial}{\partial z}\log(H_j(z)),$$

$$H_j(z) = \exp(-\frac{U}{2}j)(\cosh(\frac{U}{2}j) - \sinh(\frac{U}{2}j)Erf(\frac{z}{2})),$$

$$U_j = \frac{1}{2b} \int_{\mathbf{R}_1} \overline{\psi}(x)dx + \frac{(-1)^j}{2} \int_{\mathbf{R}_1} \overline{\phi}(x)dx.$$

The asymptotics (37) and (38) make sense (the principal term in (37) and (38) bigger than remainder) only in the region $\chi_j = O(1)$, $t \to \infty$, i.e. in the neighborhood of fronts $\chi = \pm bt$ of two traveling waves (under condition $U_j \neq 0$, if U_1 (or U_2) is equal zero, then corresponding component v_1 (or v_2) vanishes in (37), (38), so that in the vicinity of the first front (second front) we obtain only estimate of the rate of decay of the solution).

The question about asymptotics far from fronts, that is $t \to \infty$ and $\chi_j \to \infty$, is connected with applying of saddle-point method and technically complicated even in the linear case. Nevertheless if ratio $X = \frac{x}{t}$ and parameters ω, δ are small in comparison with the coefficient of viscosity $a > 0$ (see (25,28)), then the root $\rho = i\sigma_j$ of the equation $iX + \lambda_j'(p) = 0$, defining the saddle-point, will be close to the solution $\rho = (-1)^j i(2\sqrt{2}a)^{-1}$ of this equation with $X = 0$, $\omega = 0$, $\delta = 0$. In this case asymptotics can be calculated in explicit form, namely, if $t \to \infty$, $x \to \infty$, $X = \frac{x}{t}$ is fixed arbitrary, then

$$
\begin{aligned}
\psi(x,t) &= A_1 y_2(i\sigma_1)\frac{\exp(-\xi_1 t)}{\sqrt{t}} + A_2 y_1(i\sigma_2)\frac{\exp(-\xi_2 t)}{\sqrt{t}} + \\
&\quad O(\frac{\exp(-\xi_1 t) + \exp(-\xi_2 t)}{t})
\end{aligned}
\tag{40}
$$

$$
\begin{aligned}
\phi(x,t) &= A_2 \frac{\exp(-\xi_2 t)}{\sqrt{t}} - A_1 \frac{\exp(-\xi_1 t)}{\sqrt{t}} + \\
&\quad O(\frac{\exp(-\xi_1 t) + \exp(-\xi_2 t)}{t})
\end{aligned}
\tag{41}
$$

where $\xi_j = X\sigma_j + \operatorname{Re}\lambda_j(i\sigma_j) > 0$, $y_j(p) = \frac{\lambda_j(p) - K_1(p)}{K_3(p)}$, A_j are real numbers, which could be calculated through initial data. After this explanation we give the exact formulation (rather bulky) of two theorems, [6].

THEOREM 6 Let

(i) characteristic roots $\lambda_j(p)$ of the matrix $K(p)$ satisfy the conditions:

$$\lambda_j(p) = (-1)^j ibp + ap^2 + r_j(p), \ |r_j(p)| \leq c_1 m^{2+\nu}(p) M^\alpha(p), \ p \in \mathbf{R}_1;$$

$$\operatorname{Re}\lambda_j(p) \geq c_2 p^2, \ p \in \mathbf{R}_1;$$

$$y_j(p) = \frac{\lambda_j(p) - K_1(p)}{K_3(p)} = (-1)^j b + O(|p|^\nu), \ |p| \leq 1;$$

$$\Delta(p) \equiv y_2(p) - y_1(p), \ |\Delta(p)| \geq c_3 M^{-\alpha}(p), \ |y_j(p)| \leq c_4 M(p), \ p \in \mathbf{R}_1;$$

$$|\mathcal{G}_j^{k,l}(p,q)| \leq c_5(M(p) + M(q)),$$

$$|\mathcal{F}_j(p,q)| \leq c_6(M(p) + M(q))(m^\nu(p) + m^\nu(q)), \ p, q \in \mathbf{R}_1, \ j,k,l = 1,2;$$

where

$$\mathcal{G}_j^{1,1}(p,q) = -\frac{1}{2\Delta(p)}(2y_2(q) - y_j(p)),$$

$$\mathcal{G}_j^{2,2}(p,q) = \frac{1}{2\Delta(p)}(y_j(p) - 2y_1(q)),$$

$$\mathcal{G}_j^{1,2}(p,q) = \mathcal{G}_j^{2,1}(p,q) = -\frac{1}{2}(\mathcal{G}_j^{1,1}(p,q) + \mathcal{G}_j^{2,2}(p,q)),$$

$$\mathcal{F}_j(p,q) = \mathcal{G}_j^{j,j}(p,q) - \mathcal{G}_j^{j,j}(0,0), \ j = 1, 2, \ p, q \in \mathbf{R}_1,$$

$$a, b, c_1, ..., c_6 > 0, \ \nu \in (0,1), \ \alpha \geq 1.$$

(ii) the Fourier transform of the initial data satisfy the estimate

$$|\hat{\psi}(p)| + |\hat{\phi}(p)| \leq \epsilon M^{-5\alpha}(p),$$

where $0 < \epsilon < c$, $c > 0$ is some small constant. Then the classical solution (ψ, ϕ) to the Cauchy problem for system (19)-(20) uniformly with respect to $x \in \mathbf{R}_1$ has the asymptotcal representation as $t \to \infty$

$$\psi(x,t) = bv_1(x,t) + bv_2(x,t) + O(t^{-\frac{1}{2}-\gamma})$$
$$\phi(x,t) = v_2(x,t) - v_1(x,t) + O(t^{-\frac{1}{2}-\gamma}), \ \gamma > 0,$$

functions $v_j(x,t)$ are solutions to the Cauchy problem for the Burgers equation (36).

REMARK The systems of equations (23)-(24) and (29)-(30) satisfy all assumptions of Theorem 6.

Fix the ratio $X = \frac{x}{t}$ and let $p = i\sigma_j$ is the root of equation $iX + \lambda_j'(p) = 0$, $\lambda_j(p)$ are the characteristic roots of matrix $K(p)$, let also $\mu_j(p) = -Re\,\lambda_j(i\eta)$, $p = \rho + i\eta$, $\rho, \eta \in \mathbf{R}_1$, $h_1 = \min(0, \sigma_1, \sigma_2)$, $h_2 = \max(0, \sigma_1, \sigma_2)$, $P \equiv \{p \in \mathbf{C} : Im\,p \in [h_1, h_2]\}$.

THEOREM 7 Let

(i) there exists the solution $p = i\sigma_j$, $\sigma_j \in \mathbf{R}_1 \setminus 0$ of the equation $iX + \lambda_j'(p) = 0$ such that

$$\xi_j = \sigma_j X - \mu_j(i\sigma_j) > 0, \lambda_j''(i\sigma_j) > 0;$$

$$|\lambda_j'(p)| \leq c_1 M^2(p), 2\mu_j(\frac{p}{2}) \leq \mu_j(p) - \frac{\eta^2}{4}, \ p \in P;$$

$$Re\,\lambda_j(p) \geq c_2\rho^2 - \mu_j(p), \ p \in P;$$

$$|y_j(p)| \leq c_3 M(p), \ |\Delta(p)| \geq c_4 M^{-1}(p), \ p \in P;$$

$$\left|\frac{\partial^s}{\partial p^s}\mathcal{G}_j^{k,l}(p, \frac{p}{2} + q)\right| \leq c_5(M(p) + M(q))$$

$$p \in P, \ q \in \mathbf{R}, \ s = 0, 1, \ j, k, l = 1, 2, \ c_1, ...c_5 > 0;$$

(ii) the Fourier transform of initial data be analytical in the strip P and

$$|\frac{\partial^s}{\partial p^s}\hat{\psi}(p)| + |\frac{\partial^s}{\partial p^s}\hat{\phi}(p)| \leq \epsilon M^{-5}(p), \ s = 0, 1.$$

Then as $t \to \infty$ and $x \to \infty$ with $X = \frac{x}{t} = const$ for the solution to the Cauchy problem for the system (19)-(20) the following asymptotic representation holds

$$
\psi(x,t) = A_1 y_2(i\sigma_1)\frac{\exp(-\xi_1 t)}{\sqrt{t}} + A_2 y_1(i\sigma_2)\frac{\exp(-\xi_2 t)}{\sqrt{t}} +
$$
$$
O(\frac{\exp(-\xi_1 t) + \exp(-\xi_2 t)}{t})
$$
$$
\phi(x,t) = A_2\frac{\exp(-\xi_2 t)}{\sqrt{t}} - A_1\frac{\exp(-\xi_1 t)}{\sqrt{t}} +
$$
$$
O(\frac{\exp(-\xi_1 t) + \exp(-\xi_2 t)}{t}).
$$

The numbers A_1 and A_2 can be calculated in explicit form.

The same remark is valid, if parameters δ, $\omega > 0$ and X are chosen small enough in comparison with coefficient of viscosity $a > 0$ and $X \neq \pm b$, so that we find the asymptotics far from fronts $x = \pm bt$ of the traveling waves.

2.2 System of equations for nerve conduction in cells We consider now the periodic problem for the very general model system of nonlinear equations, describing processes of propagation of nerve impulses in cells. It is quite natural to consider sets of cells as a periodic structure

$$
u_t + \mathbf{\Gamma}(u) + \mathbf{K}(u) = 0, \ u \mid_{t=0} = \overline{u}(x), \ x \in \mathbf{R}_n, n \geq 1, \ t \geq 0. \tag{42}
$$

Here $u(x,t)$ is a real two-dimensional vector-function $u = (u_1, u_2)$, $\mathbf{K}$ is a 2×2 matrix, consisting of the pseudodifferential linear operators

$$
\left(\begin{array}{cc} \mathbf{K}^{(1)}(p) & \mathbf{K}^{(2)}(p) \\ \mathbf{K}^{(3)}(p) & \mathbf{K}^{(4)}(p) \end{array} \right),
$$

$\mathbf{\Gamma}(u)$ is the sum of the quadratic and cubic nonlinearities

$$
\mathbf{\Gamma}(u) = \sum_{k,j=1}^{2} (f^{(k,j)} u_k u_j + \sum_{l=1}^{2}(g^{(k,j,l)} u_k u_j u_l), \tag{43}
$$

$f^{(k,j)}$ and $g^{(k,j,l)}$ are the sets of constant vectors.
We will assume that the initial data $\overline{u}(x)$ and $u(x,t)$ are 2π-periodic functions with respect to spatial variable $x = (x^{(1)}, ..., x^{(n)}) \in \mathbf{R}_n$. Then the pseudodifferential operators $\mathcal{K}^{(j)}$ are defined by the Fourier series

$$
\mathcal{K}^{(j)}(\phi) = \frac{1}{(2\pi)^n} \sum_{p=-\infty}^{\infty} \exp(i(p,x))K_p^{(j)}\hat{\phi}_p, \tag{44}
$$

where $1 \leq j \leq 4$, $p = (p^{(1)}, ..., p^{(n)}) \in \mathbf{Z}^n$, $\hat{\phi}_p = \int_{\Omega} \exp(-i(p,x))\phi(x)dx$ are the Fourier coefficients of 2π-periodic function $\phi(x)$, $\Omega = [0, 2\pi]^n$ is the n-dimensional cube.
The well-known FitzHugh-Nagumo system of equations

$$
u_t^{(1)} - u^{(1)}(u^{(1)} - \gamma_1)(1 - u^{(1)}) + u^{(2)} - \Delta u^{(1)} = 0 \tag{45}
$$
$$
u_t^{(2)} - \gamma_2 u^{(1)} + \gamma_3 u^{(2)} - \gamma_4 \Delta u^{(2)} = 0, \tag{46}
$$

is a very particular case of system (42). This system can be obtained from (42) as

$$K_p^{(1)} = \gamma_1 + |p|^2, \ K_p^{(2)} = 1, \ K_p^{(3)} = -\gamma_2, \ K_p^{(4)} = \gamma_3 + \gamma_4|p|^2, \ p \in \mathbf{Z}^n,$$

$$f^{(1,1)} = (-1 - \gamma_1, 0), \ g^{(1,1,1)} = (1, 0)$$

and the other coefficients $f^{(k,l)}$ and $g^{(k,j,l)}$ are equal to zero.

To describe the asymptotic behavior of the solutions of the periodic problem (42) we introduce some notations. Consider the Cauchy problem for the system of ordinary differential equations depending on $p \in \mathbf{Z}^n$ as on a parameter

$$\frac{d\phi}{dt} + K_p\phi = 0, \ \phi \mid_{t=0} = \overline{\phi}_p, \tag{47}$$

where ϕ ia a two-dimensional vector $\phi = (\phi^{(1)}, \phi^{(2)})$, K_p is a 2×2 matrix, defining the symbol of the operator $\mathbf{K}$ of the system (42). Denote by $\mathbf{R}_p(t) = \exp(-K_p t)$ the fundamental (Cauchy) matrix-solution of the problem (47). Also denote by $|x| = (\sum_{l=1}^{n}(x^{(l)})^2)^{\frac{1}{2}}$ the norm of the vector $x = (x^{(1)}, ..., x^{(n)}) \in \mathbf{R}_n$ and by $\| \mathcal{L} \|$ the norm of the matrix $\mathcal{L}$ as the norm of the linear operator from $\mathbf{R}_2$ to $\mathbf{R}_2$. Let $\lambda_j(p)$ be the eigenvalues of the matrix K_p, $\lambda_p^{(1)} \neq \lambda_p^{(2)}$ and

$$Re \, \lambda_p^{(1)} \leq Re \, \lambda_p^{(2)}, \ p \in \mathbf{Z}^n.$$

We represent the fundamental matrix $\mathbf{R}_p(t)$ in the form

$$\mathbf{R}_p(t) = \sum_{j=1}^{2} \exp(-\lambda_p^{(j)}t)\mathcal{D}_p^{(j)}$$

where the matrix $\mathcal{D}_p^{(j)}$ is equal for $j = 1, 2$ to

$$\mathcal{D}_p^{(j)} = \frac{1}{\lambda_p^{(1)} - \lambda_p^{(2)}} \begin{pmatrix} (\lambda_p^{(j)} - K_p^{(4)})K_p^{(3)} & K_p^{(2)}K_p^{(3)} \\ K_p^{(2)}K_p^{(3)} & (\lambda_p^{(j)} - K_p^{(1)})K_p^{(2)} \end{pmatrix}.$$

The following statement is valid [7].

THEOREM 8 Let

(1) the matrix K_p be such that for $j = 1, 2$, $Re \, \lambda_j^{(j)} \geq a_j + \beta|p|^\alpha$ for all $p \in \mathbf{Z}^n \setminus 0$, where $a_j = Re \, \lambda_0^{(j)}$, $0 < a_1 < a_2$, $\alpha \geq 0, \beta > 0$ and $\| \mathcal{D}_j^{(j)} \| \leq c_1$ as $p \in \mathbf{Z}^n$, $c_1 > 0$.

(2) the Fourier coefficients $\hat{\overline{u}}_p$ of the initial data $\overline{u}(x)$ satisfy the estimate

$$|\hat{\overline{u}}_p| \leq \epsilon(1 + |p|)^{-2n}, \ p \in \mathbf{Z}^n,$$

where $\epsilon > 0$ is sufficiently small.

Then for the solution $u(x, t)$ of the periodic problem (42) the following asymptotics as $t \to \infty$ holds uniformly with respect to $x \in \Omega$

$$u(x, t) = \Psi \exp(-\lambda_0^{(1)}t) + O(\exp(-\lambda_0^{(1)}t - \gamma t)), \tag{48}$$

where $\gamma > 0$ is a constant, the vector Ψ can be expressed explicitly in terms of the symbol K_p and the initial data $\overline{u}(x)$ as an expansion on the parameter ϵ:

$$\Psi = \sum_{n=1}^{\infty} \epsilon^n \Psi_n.$$

Let us give the two first terms of this expansion

$$\Psi_1 = \mathcal{D}_0^{(1)} \frac{\hat{\overline{u}}_0}{\epsilon}, \quad \Psi_2 = \sum_{q=-\infty}^{\infty} \sum_{l,k=1}^{2} (\lambda_0^{(1)} - \lambda_{-q}^{(l)} - \lambda_q^{(k)})^{-1} \mathcal{G}(\mathcal{D}_{-q}^{(l)} \frac{\hat{\overline{u}}_{-q}}{\epsilon}, \mathcal{D}_q^{(k)} \frac{\hat{\overline{u}}_q}{\epsilon}),$$

where

$$\mathcal{G}(u,v) = \sum_{j,r=1}^{2} f^{(r,j)} u^{(r)} v^{(j)}.$$

REMARK The FitzHugh-Nagumo system of equations satisfy all the conditions of Theorem 8. For it we have $\alpha = 2$, $a_1 = \frac{1}{2} Re\,(\gamma_1 + \gamma_2 - \sqrt{(\gamma_1 - \gamma_3)^2 - 4\gamma_2}) > 0$.

References

[1] Naumkin, P.I. and Shishmarev, I.A. (1994), Nonlinear nonlocal equations in the theory of waves, Trans.of Math. monographs, 133 p. 289.

[2] Naumkin, P.I. and Shishmarev, I.A. (1994), Asymptotics as $t \to \infty$ of the solution of a nonlinear equation with weak dissipation and dispersion, Russian Acad. Sci. Izv. Math., 43, 3, p.443-453.

[3] Naumkin, P.I. and Shishmarev, I.A. (1996), Asymptotics as $t \to \infty$ of the solutions of nonlinear equations with considerable initial data, Mat. Zametki, 59, 6,p.855-864.

[4] Naumkin, P.I. and Shishmarev, I.A. (1994), Asymptotic relations between solutions to various nonlinear equations at large time, II, Diff. Eq., 30, 8, pp.1329-1340.

[5] Konotop V.I. and Shishmarev, I.A. (to appear) Asymptotic behavior of solutions of the generalized Kolmogorov-Petrovsky-Piskunov equation for large time.

[6] Naumkin, P.I. and Shishmarev, I.A. (1995), Asymptotic representation of surface waves in a form of two travelling Burgers waves, Functional Anal. i Prolozhen., 29, pp. 25-40.

[7] Naumkin, P.I. and Shishmarev, I.A. (to appear), Periodic problem for system of equations describing nerve conductivity.

Hardy's Inequality for the Stokes Problem

PAVEL E. SOBOLEVSKII, Institute of Mathematics, Hebrew University of Jerusalem, Givat Ram Campus, 91904 Jerusalem, Israel

0. INTRODUCTION

0.1. Uniqueness condition for Stationary boundary value problem, Stability condition for Non-stationary problem.

Let us consider stationary Navier-Stokes boundary value problem

$$-\nu\Delta\bar{u} + \sum_{k=1}^{n} u_k \cdot \partial\bar{u}/\partial x_k - \mathrm{grad}p = \bar{f}(x \in \bar{\Omega}); \bar{u} = \bar{0}(x \in \partial\Omega);$$

$$(p,1)_0 = (p,1)_{0,\Omega} \equiv \int_\Omega p(x)dx \tag{0.1}$$

in bounded domain $\Omega \in R^n$ with boundary $\partial\Omega \in C^2$. Here $\bar{u} = (u_1, \cdots, u_n)$ and p are unknown vector, and scalar-function respectively, $\bar{f} = (f_1, \cdots, f_n)$ is a given vector-function and ν is a given positive constant. It is well known (see [3] Ladyzenskaja, O. A. (1970)), that for any $\bar{f} \in \bar{L}_2(\Omega)$ problem (0.1) has (in the cases $n = 2, 3$) solution $(\bar{u}, p)$, such that

$$\bar{u} \in \overset{0}{\overline{W}}{}^2_2(\Omega), p \in W^1_2(\Omega). \tag{0.2}$$

It is established also, that problem (0.1) has unique solution in class (0.2) of strong solutions (or in more wide class of generalized solution) under assumption, that value

$$\|\bar{f}\|_{-1} = \|\bar{f}\|_{-1,\Omega} \equiv \sup_{\bar{z}\in\overset{0}{\overline{W}}{}^1_2(\Omega),\bar{z}(x)\neq\bar{0}} (\bar{f},\bar{z})_0 \cdot \|\bar{z}\|_1^{-1} \tag{0.3}$$

349

is sufficiently small. Here we set

$$(\bar{f}, \bar{z})_0 = (\bar{f}, \bar{z})_{0,\Omega} = \int_\Omega \sum_{i=1}^n f_i(x) \cdot z_i(x) dx, \tag{0.4}$$

$$\|\bar{z}\|_1^2 = \|\bar{z}\|_{1,\Omega}^2 = \int_\Omega |\nabla \bar{z}|^2 dx, |\nabla \bar{z}|^2 = \sum_{i,k=1}^n |\partial z_i|\partial x_k|^2. \tag{0.5}$$

Further let us consider nonstationary initial boundary value problem

$$\partial \bar{v}/\partial t - \nu \Delta \bar{v} + \sum_{k=1}^n v_k \cdot \partial \bar{v}/\partial x_k - \text{grad } q = \bar{F}, \text{div } \bar{v} = 0 (t \geq 0, x \in \Omega);$$

$$\bar{v} = \bar{0}(t \geq 0, x \in \partial\Omega); \bar{v}(0, x) = \bar{v}^0(x)(x \in \bar{\Omega} = \Omega \cup \partial\Omega);$$

$$(q, 1)_0 = 0(t \geq 0). \tag{0.6}$$

The asymptotic stability (when $t \to +\infty$) of solutions $(\bar{v}, q)$ of problem (0.6) is investigated under assumption, that values

$$\|\bar{v}^0(x) - \bar{u}(x)\|_0, \sup_{t \geq 0} \|\bar{F}(t, x) - \bar{f}(x)\|_0 \tag{0.7}$$

are sufficiently small. Here $\bar{u}$ is the first component of solution $(\bar{u}, p)$ of problem (0.1), $\bar{f}$ is a right part of equation (0.1) and

$$\|\bar{z}\|_0^2 = \|\bar{z}\|_{0,\Omega}^2 = \int_\Omega \sum_{i=1}^n |z_i(x)|^2 dx. \tag{0.8}$$

It is established (see [3] Ladyzenskaja, O. A. (1970), [5] Sobolevskii, P. E. (1996)) that solution $(\bar{v}, q)$ of problem (0.6) is asymptotic stable, if value (0.3) is sufficiently small also.

The proof of uniqueness for stationary problem (0.1) and the proof of asymptotic stability for nonstationary problem (0.6) are based (see [5] Sobolevskii, P. E. (1996)) on the positive definiteness of self-adjoint spectral problem

$$-\nu \Delta \bar{z} + \sum_{k=1}^n (z_k \cdot \partial u/\partial x_k + u_k \cdot \partial \bar{z}/\partial x_k) - \text{grad } r = \lambda \bar{z}, \text{div } \bar{z} = 0 \ (x \in \Omega);$$

$$\bar{z} = \bar{0} \ (x \in \partial\Omega); \ (r, 1)_0 = 0; \bar{z} \in \overset{0}{W_2^2}(\Omega), r \in W_2^1(\Omega). \tag{0.9}$$

¿From definition (0.3) and Fridrichs inequality (see [3] Ladyenskaja, O. A. (1970)) it follows, that

$$\|\bar{f}\|_{-1} \leq M \cdot \|\bar{f}\|_0 \tag{0.10}$$

for some $1 \leq M < +\infty$, does not depending on $\bar{f}$. Therefore, evidently, value (0.3) will be small, if value

$$\|\bar{f}\|_0 \tag{0.11}$$

is small, but it is rough condition.

Let $\rho_{\partial\Omega}(x)$ be the Euclidean distance from point $x \in \Omega$ to boundary $\partial\Omega$. From the classical Hardy's inequality (see [2] Hardy, G. M., Littlewood, J. E., Polya, G. (1934)) and from the partition of unity it follows, that value

$$H = H_\Omega \equiv \inf_{\bar{z}\in\overset{0}{W}{}_2^1(\Omega), \operatorname{div}\bar{z}=0, \bar{z}(x)\not\equiv\bar{0}} \|\bar{z}\|_1 \cdot \|\bar{z} \cdot \rho_{\partial\Omega}^{-1}\|_0^{-1} \tag{0.12}$$

is positive and finite. Further from definition (0.3), evidently, it follows, that

$$\|\bar{f}\|_{-1} \leq H^{-1} \cdot \|\rho_{\partial\Omega} \cdot \bar{f}\|_0. \tag{0.13}$$

This means, that value (0.3) will be small, if value

$$\|\rho_{\partial\Omega} \cdot \bar{f}\|_0 \tag{0.14}$$

is small. This condition is more useful for applications, since it permits to consider the right parts $\bar{f}$, having large values near boundary $\partial\Omega$.

0. 2. The content of the paper.

On the set of vector-functions

$$\bar{z} \in \overset{0}{W}{}_p^1(\Omega)(1 < p < +\infty); \operatorname{div}\bar{z} = 0 \ (x \in \Omega); \ \bar{z}(x) \not\equiv \bar{0} \tag{0.15}$$

let us define generalized Hardy's functional

$$\ell_{p,\Omega}(\bar{z}) = [\int_\Omega |\nabla\bar{z}|^p dx]^{1/p} \cdot [\int_\Omega |\bar{z} \cdot \rho_{\partial\Omega}^{-1}|^p dx]^{-1/p}. \tag{0.16}$$

Here we set

$$|\nabla\bar{z}|^2 = \sum_{i,k=1}^n |\partial z_i/\partial x_k|^2, \ |\bar{z}|^2 = \sum_{i=1}^n |z_i|^2. \tag{0.17}$$

From classical Hardy's inequality and from partitions of unity it follows, that value

$$H_p = H_p(\Omega) = \inf_{\bar{z} \in \overset{0}{W}{}^1_p(\Omega), \mathrm{div}\bar{z}=0, \ \bar{z}(x) \not\equiv \bar{0}} \ell_{p,\Omega}(\bar{z}) \tag{0.18}$$

is positive and finite. It turns out, that it is possible to compute the value $H_p(\Omega)$ in the case of convex domains $\Omega \in R^n$. We will remind [see p. 0.1], that the value $H_2(\Omega)$ defines the conditions of uniqueness and stability.

The obtained results are formulated in the following theorems:

Theorem 1 (§1). *Let Ω be any (bounded or unbounded) convex domain in $R^n (n \geq 2)$. Then inequality*

$$\ell_{p,\Omega}(\bar{z}) \geq \frac{p-1}{p} \tag{0.19}$$

is true for any vector-function $\bar{z} \in \overset{0}{W}{}^1_0(\Omega), \ \bar{z}(x) \not\equiv \bar{0}$.

The proof of inequality (0.19) is the vector generalization of the proof of the result, obtained in ([4] Matskweich, T., Sobolevskii, P. E. (1994)).

Theorem 2 (§2). *Let Ω be any (bounded or unbounded) convex domain in R^2 with boundary $\partial\Omega$, belonging to C^2-class of smoothness in the neighbourhood of at least one point $x^0 \in \partial\Omega$. Then*

$$H_p = H_p(\Omega) = \frac{p-1}{p}. \tag{0.20}$$

The estimate from below of functional $\ell_{p,\Omega}(\bar{z})$, evidently, follows from Theorem 1. For the estimate from above of functional $\ell_{p,\Omega}(\bar{z})$, defined on the set of solenoidal vector-functions, we can use the property, that in the case $n = 2$ this functional can be transform to (other) functional, defined on the corresponding set of scalar functions. Further we apply the method from ([4] Matskewich, T., Sobolevskii, P. E. (1994)).

Theorem 3 (§4). *Let Ω be any (bounded or unbounded) convex domain in $R^n (n \geq 2)$ with boundary $\partial\Omega$, belonging to C^1-class of smoothness in the neighborhood of at least one point $x^0 \in \partial\Omega$. Then*

$$H_2 = H_{2,\Omega} = \frac{1}{2}. \tag{0.21}$$

The estimate from below of functional $\ell_{2,\Omega}(z)$ also follows from Theorem 1. For the proof of estimate from above we consider the ε-approximation of Stokes problem, introduced by Roger

Teman (see [6] Teman, R. (1979)). The sharp estimates (when $\varepsilon \to +0$) of convergence rate are used (§3). The proof of these estimates is based on the estimate from below of value $\|\text{grad}p\|_{-1}$, obtained by Ivo Babuska (see [1] Babuska, I., Szabo, B. (1991)) under assumption, that $(p, 1)_0 = 0$.

1. THE PROOF OF THEOREM 1

The proof uses classical Hardy's method and following statement, established in ([4] Matskewich, T., Sobolevskii, P. E. (1994)):

Lemma 1.1. *Let Ω be convex polyhedron in R^n with faces $e_1, \cdots, e_m$. Then $\Omega = \bigcup\limits_{k=1}^{m} \Omega_k$, where sets Ω_k have following properties:*

I. *Ω_k is convex polyhedron in R^n, containing face e_k.*

II. *The part $\partial\Omega_k \setminus e_k$ of boundary $\partial\Omega_k$ of polyhderon Ω_k is graph of some (piecewise-linear) scalar positive function φ_k, defined on e_k. Namely, if the origin and axes $\vec{x}_1, \cdots, \vec{x}_{n-1}$ of Cartesian coordinate system in R^n lies on e_k, then $\partial\Omega_k \setminus e_k$ is the set of points $[x', \varphi_k(x')]$, $x' = (x_1, \cdots x_{n-1}), (x', 0) \in e_k$.*

III. *The n-dimensional Lebesgue's measure of set $\Omega_i \cap \Omega_k$ is equal to zero for $i \neq k$.*

IV. *(Main property). If $x \in \Omega_k$, then*

$$\rho_{\partial\Omega}(x) = \rho_{e_k}(x). \tag{1.1}$$

It is evident, that in the case, when $n = 2$ and $m = 3$, i.e. in the case, when Ω is a triangle on the plane, then the partition $\Omega = \Omega_1 \cup \Omega_2 \cup \Omega_3$ is defined by the intersection of bisectors of the inner angles of triangle.

Lemma 1.2. *Under the assumption of Lemma 1.1 the following inequality*

$$\left[\int\limits_{\Omega k} |\bar{z} \cdot \rho_{\partial\Omega}^{-1}|^p dx \right]^{1/p} \leq \frac{p}{p-1} \left[\int\limits_{\Omega_k} |\nabla \bar{z}|^p dx \right]^{1/p} \tag{1.2}$$

is true for any vector-function $\bar{z} \in \overset{0}{\overline{W}}{}_2^1(\Omega), \bar{z}(x) \not\equiv \bar{0}$.

Proof. Since values $|\bar{z}|, \rho_{\partial\Omega}$ and $|\nabla \bar{z}|$ do not depend on parallel transfer and rotation of coordinate system, then we can suppose, that origin and axes $\vec{x}_1, \cdots, \vec{x}_{n-1}$ lie on face e_k. Then property IV of set Ω_k means [see formula (1.1)], that

$$\rho_{\partial\Omega}(x) = x_n \quad \text{for } x \in \Omega_k. \tag{1.3}$$

Therefore the integration by parts (in virtue of boundary condition) permits to show, that

$$\int\limits_{\Omega_k} |\bar{z} \cdot \rho_{\partial\Omega}^{-1}|^p dx = \int\limits_{e_k} dx' \{ \int\limits_{0}^{\varphi_k(x')} \Big[\sum_{i=1}^{n} |z_i|^2 \Big]^{p/2} \cdot x_n^{-p} dx_n \} \le$$

$$\frac{p}{p-1} \int\limits_{e_k} dx' \{ \int\limits_{0}^{\varphi_k(x')} |\bar{z}|^{p-1} \cdot x_n^{1-p} \cdot |\partial\bar{z}/\partial x_n| dx_n \}. \tag{1.4}$$

Further the application of Hölder's inequality permits to establish inequality (1.2). Lemma 1.2 is proved.

Since polyhedrons Ω_k satisfy property III of Lemma 1.1, then from formula $\Omega = \bigcup\limits_{k=1}^{m} \Omega_k$ it follows, that inequality (0.19) is true in the case, when Ω is convex polyhedron. Finally from this statement, evidently, inequality (0.19) follows for any convex (bounded or unbounded) domain $\Omega \subset R^n$. Theorem 1 is proved.

2. THE PROOF OF THEOREM 2

2.1. Reduction to the functional on the set of scalar functions.

It is well known (see [3] Ladyzenskaja, O. A. (1970)), that in case $n = 2$ any solenoidal vector-function $\bar{z} = (z_1, z_2) \in \overset{0}{W}_p^1(\Omega)$ is defined by formulas

$$z_1 = \partial v/\partial x_2, z_2 = -\partial v/\partial x_1; \ v \in \overset{00}{W}_p^2(\Omega). \tag{2.1}$$

Therefore

$$\ell_{p,\Omega}(\bar{z}) = \mathcal{L}_{p,\Omega}(v) \equiv \Big[\int\limits_{\Omega} |\nabla^2 v|^p dx \Big]^{1/p} \cdot \Big[\int\limits_{\Omega} |\nabla v|^p \cdot \rho_{\partial\Omega}^{-p} dx \Big]^{-1/p}. \tag{2.2}$$

Here we set

$$|\nabla v|^2 = |\partial v/\partial x_1|^2 + |\partial v/\partial x_2|^2, \ |\nabla^2 v|^2 = |\partial^2 v/\partial x_1^2|^2 + 2|\partial^2 v/\partial x_1 \partial x_2|^2 + |\partial^2 v/\partial x_2^2|^2. \tag{2.3}$$

Since values $|\nabla^2 v|$, $|\nabla v|$ and $\rho_{\partial\Omega}$ do not depend on parallel transfer and rotation of coordinate system, then we can suppose, that origin is in point $x^0 \in \partial\Omega$, axis $\vec{x}_1$ is tangent to $\partial\Omega$ at point x^0 and axis $\vec{x}_2$ is the inner normal to $\partial\Omega$ in point x^0. According the definition of C^2-smoothness of $\partial\Omega$ at point $x_0 \in \partial\Omega$ there exists the number $h_0 > 0$, such that the part of $\partial\Omega$ in the some neighborhood of point x^0 is defined by equation

$$x_2 = \varphi(x_1) \ (-h_0 \le x_1 \le h_0) \tag{2.4}$$

for some function $\varphi \in C^2([-h_0, h_0])$, such that

$$\varphi(0) = \varphi'(0) = 0. \tag{2.5}$$

For any $h \in (0, h_0]$ let us define set $D(h)$ of points $x = (x_1, x_2)$, such that

$$x_1 \in [-h, h], \ x_2 \in [\varphi(x_1), \varphi(x_1) + h]. \tag{2.6}$$

For any sufficiently small number $\varepsilon > 0$ we will define number $h = h(\varepsilon) \in (0, h_0]$ and function $v \in \overset{oo}{W}{}_p^2(D[h(\varepsilon)])$, such that its continuation v_c by zero on the whole domain Ω satisfies inequality

$$\mathcal{L}_{p,\Omega}(v_c) \le \frac{p-1}{p} + M \cdot \varepsilon \tag{2.7}$$

for some $1 \le M < +\infty$, does not depending on ε. Then, evidently, Theorem 2 will follow from Theorem 1 and inequality (2.7).

Substitution

$$y_1 = x_1, y_2 = x_2 - \varphi(x_1); \ x_1 = y_1, \ x_2 = y_2 + \varphi(y_1) \tag{2.8}$$

transform domain $D(h)$ in coordinates $x = (x_1, x_2)$ on rectangle

$$\tilde{D}(h) = (-h \le y_1 \le h, \ 0 \le y_2 \le h) \tag{2.9}$$

in coordinates $y = (y_1, y_2)$, and functions $v(x) = v(x_1, x_2)$, defined on $D(h)$, to functions

$$\tilde{v}(y) = \tilde{v}(y_1, y_2) \equiv v[x_1, x_2 - \varphi(x_1)], \tag{2.10}$$

defined on $\tilde{D}(h)$. Further the formulas for the derivatives by substitution (2.8) and triangle inequality lead us to inequality

$$\left[\int_{D(h)} |\nabla^2 v|^2 dx \right]^{1/2} \le \left[\int_{\tilde{D}(h)} |\nabla^2 \tilde{v}|^2 dy \right]^{1/2} + \left\{ \int_{\tilde{D}(h)} [|\partial^2 \tilde{v}/\partial y_1 \partial y_2|^2 + \right.$$

$$\left. + |\partial^2 \tilde{v}/\partial y_2^2|^2]^{p/2} dy \right\}^{1/p} \cdot 2\varphi_h' + \left[\int_{\tilde{D}(h)} |\partial^2 \tilde{v}/\partial y_2^2|^p dy \right]^{1/p} \cdot (\varphi_h')^2 +$$

$$+ \left[\int_{\tilde{D}(h)} |\partial \tilde{v}/\partial y_2|^p dy \right]^{1/p} \cdot \varphi_{h_0}'', \tag{2.11}$$

since, evidently, Jacobian J of transform (2.8) equal to 1. Here we set

$$\varphi'_h = \max_{0 \le |x_1| \le h} |\varphi'(x_1)|, \quad \varphi''_h = \max_{0 \le |x_1| \le h} |\varphi''(x_1)|. \tag{2.12}$$

¿From the definition of C^2-smoothness of $\partial\Omega$ at point $x^0 \in \partial\Omega$ it follows [see (2.5)], that

$$\varphi'_h \to +0 \quad \text{for} \quad h \to +0. \tag{2.13}$$

Further for functions $v \in \overset{00}{W^2_p}[D(h)]$, evidently, inequality

$$\left[\int_{\tilde{D}(h)} |\partial\tilde{v}/\partial y_2|^p dy \right]^{1/p} \le 2h \cdot \left[\int_{\tilde{D}(h)} |\partial^2\tilde{v}/\partial y_1 \partial y_2|^p dy \right]^{1/p}$$

is true. Then from (2.11) it follows, that

$$\left[\int_{D(h)} |\nabla^2 v|^p dx \right]^{1/p} \le \left[(1 + \varphi'_h)^2 + 2\varphi''_{h_0} \cdot h \right] \cdot \left[\int_{\tilde{D}(h)} |\nabla^2\tilde{v}|^p dy \right]^{1/p}. \tag{2.14}$$

Since, in virtue of (2.8),

$$\rho_{\partial\Omega}(x) \le (1 + \varphi'_h) \cdot y_2 \tag{2.15}$$

for correspondent points x and y of domains $D(h)$ and $\tilde{D}(h)$ respectively, then, analogously to (2.11) inequality

$$\left[\int_{D(h)} |\nabla v|^p \cdot \rho_{\partial\Omega}^{-p} dx \right]^{1/p} \ge [1 - (\varphi'_h)^2] \cdot \left[\int_{\tilde{D}(h)} |\nabla\tilde{v}|^p \cdot y_2^{-p} dy \right]^{1/p} \tag{2.16}$$

can be established. From inequalities (2.14) and (2.16), evidently, it follows, that

$$\left[\int_{D(h)} |\nabla^2 v|^p dx \right]^{1/p} \cdot \left[\int_{D(h)} \left| \frac{\nabla v}{\rho_{\partial\Omega}} \right|^p dy \right]^{-1/p} \le [(1 + \varphi'_h)^2 +$$

$$+ 2 \cdot \varphi''_h \cdot h] \cdot [1 - (\varphi'_h)^2]^{-1} \cdot \left[\int_{\tilde{D}(h)} |\nabla^2\tilde{v}|^p dy \right]^{1/p} \cdot \left[\int_{\tilde{D}(h)} \left| \frac{\nabla\tilde{v}}{y_2} \right|^p dy \right]^{-1/p}. \tag{2.17}$$

So, we must construct the minimizing sequence for rectangle $\tilde{D}(h)$. Namely, we will set

$$\tilde{v}(y_1, y_2) = (h^2 - y_1^2) \cdot (h - y_2) \cdot y_2^{2 - \frac{1}{p} - \frac{\eta}{p}} \tag{2.18}$$

for some $\eta > 0$. Then it is easy to check, that

$$\left[\int\limits_{\tilde{D}(h)} |\nabla^2 \tilde{v}|^p dy\right]^{1/p} \cdot \left[\int\limits_{\tilde{D}(h)} \left|\frac{\nabla \tilde{v}}{y_2}\right|^p dy\right]^{-1/p} \leq (1 - \frac{1}{p} - \eta^{1/p} \cdot M_1) \cdot (1 - \eta^{1/p} \cdot M_1)^{-1} \qquad (2.19)$$

for some $1 \leq M_1 < +\infty$, does not depends on η (and $1 < p < +\infty$). Then from inequalities (2.17) and (2.19), in virtue of (2.13), evidently, it follows, that for given sufficiently small number $\varepsilon > 0$ there exists numbers $h = h(\varepsilon) > 0$ and $\eta = \eta(\varepsilon) > 0$, such that function $\tilde{v}(y)$, defined by formula (2.18), in virtue of (2.8), is transformed into function $v(x)$, such that for its continuation $v_c(x)$ inequality (2.7) holds. Theorem 2 is proved.

3. THE APPROXIMATION OF THE STOKES PROBLEM

3.1. The rate of the approximation.

It is well known (see [3] Ladyzenskaja, O. A. (1970)), that boundary value problem (Stokes problem)

$$-\Delta \bar{u}^0 - \text{grad } p^0 = \bar{f}, \text{div} \bar{u}^0 = 0 \ (x \in \Omega); \ \bar{u}^0 = 0 (x \in 0\Omega); \ (p^0, 1)_0 = 0 \qquad (3.1)$$

has unique solution $(\bar{u}^0, p^0), \bar{u}^0 \in \overset{0}{W}{}_2^2(\Omega), p \in W_2^1(\Omega)$ under assumptions, that $\bar{f} \in \bar{L}_2(\Omega)$ and $\partial\Omega \in C^2$. If $\bar{f} \in W_2^{-1}(\Omega)$ and $\partial\Omega \in C^1$, then problem (3.1) has unique generalized solution $(\bar{u}^0, p^0), \bar{u}^0 \in \overset{0}{W}{}_2^1(\Omega), p^0 \in L_2(\Omega)$. The same results hold for introduced by Roger Teman ([5] (1979) elliptic boundary value problem

$$-\Delta \bar{u}^\varepsilon - \text{grad } \frac{\text{div} \bar{u}^\varepsilon}{\varepsilon} = \bar{f}(x \in \Omega); \ \bar{u}^\varepsilon = \bar{0}(x \in \partial\Omega); \ \varepsilon > 0 \qquad (3.2)$$

for the approximation (for $\varepsilon \to +0$) of solution of problem (3.1).

Lemma 3.1. *Following estimates of approximation rate*

$$\|\bar{u}^0 - \bar{u}^\varepsilon\|_1 \leq 2\varepsilon \cdot \delta_B^{-2} \cdot \|\bar{f}\|_{-1}, \|p^0 - \frac{\text{div} \bar{u}^\varepsilon}{\varepsilon}\|_0 \leq 2\varepsilon \cdot \delta_B^{-3} \cdot \|\bar{f}\|_{-1} \qquad (3.3)$$

are true.

Here $\delta_B = \delta_B(\Omega)$ is a finite positive number (Ivo Babuska's constant), defined by formula

$$\delta_B = \inf_{p \in W_2^1(\Omega), (p,1)_0 = 0, p(x) \not\equiv 0} \|\text{grad } p\|_{-1} \cdot \|p\|_0^{-1} \qquad (3.4)$$

Proof. From (3.2), evidently, identity

$$\|\bar{u}^\varepsilon\|_1^2 + \frac{1}{\varepsilon}\|\operatorname{div}\bar{u}^\varepsilon\|_0^2 = (\bar{f}, \bar{u}^\varepsilon)_0 \tag{3.5}$$

follows, which, in virtue of definition (0.3), leads us to the first a priory estimate

$$\|\bar{u}^\varepsilon\|_1 \le \|\bar{f}\|_{-1}. \tag{3.6}$$

Further from evident identity $[\bar{z} \in \overset{0}{\overline{W}}{}_2^1(\Omega)]$

$$(-\Delta\bar{u}^\varepsilon, \bar{z})_0 = (\nabla\bar{u}^\varepsilon, \nabla\bar{z})_0 \equiv \sum_{k-1}^{n}(\partial\bar{u}^\varepsilon/\partial x_k, \partial\bar{z}/\partial x_k)_0, \tag{3.7}$$

definition (0.3) and estimate (3.6), it follows, that

$$\| -\Delta\bar{u}^\varepsilon\|_{-1} \le \|\bar{f}\|_{-1}. \tag{3.8}$$

Therefore, in virtue of equation (3.2), estimate

$$\|\operatorname{grad}\frac{\operatorname{div}\bar{u}^\varepsilon}{\varepsilon}\|_{-1} \le 2\|\bar{f}\|_{-1} \tag{3.9}$$

is true. Since, evidently,

$$(\frac{\operatorname{div}\bar{u}^\varepsilon}{\varepsilon}, 1)_0 = 0, \tag{3.10}$$

then, in virtue of definition (3.4), the second a priory estimate

$$\|\frac{\operatorname{div}\bar{u}^\varepsilon}{\varepsilon}\|_0 \le 2 \cdot \delta_B^{-1} \cdot \|\bar{f}\|_{-1} \tag{3.11}$$

holds. We will mark, that estimate (3.11) is essentially more exact (for $\varepsilon \to +0$), than estimate

$$\|\frac{\operatorname{div}\bar{u}^\varepsilon}{\sqrt{\varepsilon}}\|_0 \le \|\bar{f}\|_{-1}, \tag{3.12}$$

which immediately follows from identity (3.5) and estimate (3.6). Further from (3.1) and (3.2), evidently, identity

$$-\Delta(\bar{u}^0 - \bar{u}^\varepsilon) - \operatorname{grad}(p^0 - \frac{\operatorname{div}\bar{u}^\varepsilon}{\varepsilon}) = \bar{0} \tag{3.13}$$

follows, which, in virtue of identity (3.7) and estimate (3.11), leads us to inequality

$$\|\bar{u}^0 - \bar{u}^\varepsilon\|_1^2 \le \|p^0 - \frac{\operatorname{div}\bar{u}^\varepsilon}{\varepsilon}\|_0 \cdot \frac{2\varepsilon}{\delta_B} \cdot \|\bar{f}\|_{-1}. \tag{3.14}$$

From the other side identity (3.13), in virtue of definition (3.4), implies

$$\|p^0 - \frac{\operatorname{div}\bar{u}^\varepsilon}{\varepsilon}\|_0 \le \frac{1}{\delta_B} \cdot \|\bar{u}^0 - \bar{u}^\varepsilon\|_1. \tag{3.15}$$

From inequalities (3.14) and (3.15), evidently, estimates (3.3) follows. Lemma 3.1 is proved.

3.2. The functional $\ell_{2,\Omega}^\varepsilon(z)$.

For the vector-functions $\bar{z} \in \overset{0}{W}_2^1(\Omega), \bar{z}(x) \not\equiv \bar{0}$ formula

$$\ell_{2,\Omega}^\varepsilon(\bar{z}) = \{\int\limits_\Omega [|\nabla\bar{z}|^2 + \frac{1}{\varepsilon}|\mathrm{div}\bar{z}|^2]dx\}^{1/2} \cdot \{\int\limits_\Omega |\frac{\bar{z}}{\rho_{\partial\Omega}}|^2 dx\}^{-1/2} (\varepsilon > 0) \tag{3.16}$$

defines functional, corresponding to boundary value problem (3.2).

Lemma 3.2. *The following formula is true:*

$$\inf_{\bar{z}\in\overset{0}{W}_2^1(\Omega),\bar{z}(x)\not\equiv\bar{0}} \ell_{2,\Omega}^\varepsilon(\bar{z}) = \frac{1}{2} \tag{3.17}$$

Proof. From definition (3.16), evidently, it follows, that $\ell_{2,\Omega}^\varepsilon(\bar{z}) \geq \ell_{2,\Omega}(\bar{z})$, and therefore, in virtue of theorem 1, estimate from below

$$\ell_{2,\Omega}^\varepsilon(\bar{z}) \geq 1/2 \ [\bar{z} \in \overset{0}{W}_2^1(\Omega), \ \bar{z}(x) \not\equiv \bar{0}] \tag{3.18}$$

takes place. In order to establish the estimate from above we mark, that functional $\ell_{2,\Omega}^\varepsilon(\bar{z})$ does not depend on parallel transfer and rotation of coordinate system. Therefore we can suppose, that origin is in point $x^0 \in \partial\Omega$ of C^1- smoothness, axes $\vec{x}_1, \cdots, \vec{x}_{n-1}$ lie in tangent hyperplane to boundary $\partial\Omega$ at point x^0 and axis $\vec{x}_n$ is directed along inner normal to $\partial\Omega$ at point x^0. The estimate from above will be obtained for vector-function $\bar{z}$ of type

$$\bar{z} = (z_1, 0, \cdots, 0). \tag{3.19}$$

In this case, evidently,

$$\ell_{2,\Omega}^\varepsilon(\bar{z}) = \{\int\limits_\Omega [(1 + \frac{1}{\varepsilon})|\partial z_1/\partial x_1|^2 + \sum_{k=2}^n |\partial z_1/\partial x_k|^2]dx\}^{1/2} \cdot \{\int\limits_\Omega |\frac{z_1}{\rho_{\partial\Omega}}|^2 dx\}^{1/2}, \tag{3.20}$$

i.e. it is necessary to obtain the estimate from above for functional

$$\mathcal{L}_{2,\Omega}^\varepsilon(v) = \{\int\limits_\Omega [(1 + \frac{1}{\varepsilon})|\partial v/\partial x_1|^2 + \sum_{k=2}^n |\partial v/\partial x_k|^2]dx\}^{1/2} \cdot \{\int\limits_\Omega |\frac{v}{\rho_{\partial\Omega}}|^2 dx\}^{-1/2}, \tag{3.21}$$

defined on scalar functions $v \in \overset{0}{W}_2^1(\Omega), v(x) \not\equiv 0$. Here also it will be used the method from ([4] Matskewish, T., Sobolevskii, P. E. (1994)). For simplicity we will consider the case $n = 2$.

Let in domain $D(h)$ [see formula (2.6)] be defined function $v(x) = v(x_1, x_2) \in \overset{0}{W}{}^1_2(\Omega)$, $v(x) \not\equiv 0$, and let v_c be its continuation by zero on the whole domain Ω. Further let substitution (2.8) transform function $v(x)$ into function $\tilde{v}(y) = \tilde{v}(y_1, y_2)$, defined on rectangle $\tilde{D}(h)$. Then from formulas for derivatives under substitution (2.8), estimate (2.15) and triangle inequality, evidently, it follows, that

$$\Big\{ \int\limits_{D(h)} \Big[(1 + \frac{1}{\varepsilon})|\partial v/\partial x_1|^2 + |\partial v/\partial x_2|^2\Big] dx \Big\}^{1/2} \cdot \Big\{ \int\limits_{D(h)} |\frac{v}{\rho_{\partial\Omega}}|^2 dx \Big\}^{-1/2} \leq$$

$$\frac{1 + \varphi'_n}{1 - \varphi'_n} \cdot \Big\{ \int\limits_{\tilde{D}(h)} \Big[(1 + \frac{1}{\varepsilon})|\partial\tilde{v}/\partial y_1|^2 + |\partial\tilde{v}/\partial y_2|^2\Big] dy \Big\}^{1/2} \cdot \Big\{ \int\limits_{\tilde{D}(h)} |\frac{\tilde{v}}{y_2}|^2 dy \Big\}^{-1/2}. \tag{3.22}$$

Further for $\eta > 0$ we will set

$$\tilde{v}(y_1, y_2) = (h^2 - y_1^2) \cdot (h - y_2) \cdot y_2^{\frac{1}{2} + \frac{\eta}{2}}. \tag{3.23}$$

Then, evidently,

$$\int\limits_{\tilde{D}(h)} |\tilde{v}/y_2|^2 dy = \frac{16}{15} \cdot h^{7+\eta} \cdot (1 - \frac{2\eta}{1+\eta} + \frac{\eta}{2+\eta}) \cdot \eta^{-1}, \tag{3.24}$$

$$\int\limits_{\tilde{D}(h)} |\partial\tilde{v}/\partial y_1|^2 dy = \frac{8}{3} \cdot h^{7+\eta} \cdot (\frac{1}{2+\eta} - \frac{2}{3+\eta} + \frac{1}{4+\eta}) \tag{3.25}$$

$$\int\limits_{\tilde{D}(h)} |\partial\tilde{v}/\partial y_2|^2 dy = \frac{16}{15} \cdot h^{7+\eta} \cdot \Big[(\frac{1}{2} + \eta)^2 - \frac{2\eta(\frac{1}{2} + \eta) \cdot (\frac{3}{2} + \eta)}{1 + \eta} + \frac{(\frac{3}{2} + \eta)^2 \cdot \eta}{2 + \eta}\Big] \cdot \eta^{-1}. \tag{3.26}$$

Therefore for $\eta = \eta(\varepsilon) = \varepsilon^{1+2\delta}, \delta > 0$ estimate

$$\Big\{ \int\limits_{\tilde{D}(h)} \Big[(1 + \frac{1}{\varepsilon}) \cdot |\partial\tilde{v}/\partial y_1|^2 + |\partial\tilde{v}/\partial y_2|^2\Big] dy \Big\}^{1/2} \cdot \Big\{ \int\limits_{\tilde{D}(h)} |\frac{\tilde{v}}{y_2}|^2 dy \Big\}^{-1/2} \leq 1/2 + \tilde{M}_\delta \cdot \varepsilon^\delta \tag{3.27}$$

is true with some $1 \leq \tilde{M}_\delta < +\infty$, does not depending on ε. Therefore, in virtue of (2.5), for sufficiently small $\varepsilon > 0$ there exists $h = h(\varepsilon)$, such that from estimates (3.22) and (3.27) estimate

$$\Big\{ \int\limits_{D[h(\varepsilon)]} \Big[(1 + \frac{1}{\varepsilon})|\partial v/\partial x_1|^2 + |\partial v/\partial x_2|^2\Big] dx \Big\}^{1/2} \cdot \Big\{ \int\limits_{D[h(\varepsilon)]} |\frac{v}{\rho_{\partial\Omega}}|^2 dx \Big\}^{-1/2} \leq 1/2 + M_\delta \cdot \varepsilon^\delta \tag{3.28}$$

follows with some $1 \leq M_\delta < \infty$, does not depending on ε. From estimates (3.18) and (3.28), evidently, formula (3.17) follows. So, Lemma 3.2 proved in the case $n = 2$. The general case $n \geq 2$ can be considered by the same method.

4. THE PROOF OF THEOREM 3

4.1. The construction of solenoidal approximant.

Here we, also for simplicity, will consider the case $n = 2$. For defined by formulas (3.23) and (2.8) scalar function $v = v(x) = v(x_1, x_2)$ let us define vector - function

$$\bar{v} = (v, 0) \tag{4.1}$$

and let us set

$$\bar{f} = -\Delta v - \frac{1}{\varepsilon} \text{grad div} \bar{v}. \tag{4.2}$$

¿From formulas (3.25), (3.26) and (2.8), in virtue of definition (0.3), for $\eta = \varepsilon^2$ estimate

$$\|\bar{f}\|_{-1,D(h)} \leq M \cdot \varepsilon^{-1} \cdot h^{\frac{7+\varepsilon^2}{2}} \tag{4.3}$$

follows with some $1 \leq M < +\infty$, does not depending on ε and h. Further, let (u^0, p^0) be (generalized) solution [see (3.1)] of boundary value problem

$$-A\bar{u}^0 - \text{grad } p^0 = \bar{f}, \text{div } \bar{u}^0 = 0 [x \in D(h)]; \ \bar{u}^0 = \bar{0} [x \in \partial D(h)]; \ (p_0, 1)_{0,D(h)} = 0. \tag{4.4}$$

The comparison of vector - functions $\bar{v}$ and $\bar{u}^0$ is based, in virtue of Lemma 3.1, on the following statement.

Lemma 4.1. *For any scalar function p, such that*

$$p \in W_2^1[D(h)], \ (p,1)_{0,D(h)} = 0 \tag{4.5}$$

following inequality is true

$$\|\text{grad } p\|_{-1,D(h)} \geq \{\delta_B[\tilde{D}(1)] - \varphi_h'\} \cdot (1 + \varphi_h') \cdot \|p\|_{0,D(h)}. \tag{4.6}$$

Proof. For any vector-function $\bar{z} \in \overset{0}{W}{}^1_2[D(h)]$, in virtue of (2.8), following identity

$$\int\limits_{D(h)} \operatorname{grad} p \cdot \bar{z} dx = \int\limits_{\tilde{D}(h)} \operatorname{grad} \tilde{p} \cdot \tilde{\bar{z}} dy + \int\limits_{\tilde{D}(h)} \tilde{p} \cdot \varphi'(y_1) \cdot \partial \tilde{z}_2 / \partial y_2 dy \qquad (4.7)$$

takes place, since Jacobian $J = 1$. Further, in virtue of (2.8),

$$\|\bar{z}\|_{1,D(h)} \leq (1 + \varphi'_n) \cdot \|\tilde{\bar{z}}\|_{1,\tilde{D}(h)}. \qquad (4.8)$$

Then from Cauchy's inequality it follows, that

$$\left| \int\limits_{D(h)} \operatorname{grad} p \cdot \bar{z} dx \right| \cdot \|\bar{z}\|^{-1}_{1,D(h)} \geq \left| \int\limits_{\tilde{D}(h)} \operatorname{grad} \tilde{p} \cdot \tilde{\bar{z}} dy \right| \cdot \|\tilde{\bar{z}}\|^{-1}_{1,\tilde{D}(h)} \cdot (1 + \varphi'_h)^{-1}$$

$$- \|p\|_{0,\tilde{D}(h)} \cdot (1 + \varphi'_h)^{-1}, \qquad (4.9)$$

if $\bar{z}(x) \not\equiv \bar{0}$. Further, since $J = 1$,

$$(\tilde{p}, 1)_{0,\tilde{D}(h)} = 0. \qquad (4.10)$$

Therefore, in virtue of definition (0.3) and inequality (3.1), we will come to inequality

$$\|\operatorname{grad} p\|_{-1,D(h)} \geq \delta_B[\tilde{D}(h)] \cdot \|\tilde{p}\|_{0,\tilde{D}(h)} \cdot (1 + \varphi'_n)^{-1} -$$

$$- \varphi'_h \|\tilde{p}\|_{0,\tilde{D}(h)} \cdot (1 + \varphi'_n)^{-1}. \qquad (4.11)$$

¿From the structure of $\tilde{D}(h)$, evidently, it follows, that

$$\delta_B[\tilde{D}(h)] = \delta_B[\tilde{D}(1)]. \qquad (4.12)$$

Finally, since $J = 1$, then, evidently,

$$\|\tilde{p}\|_{0,\tilde{D}(h)} = \|p\|_{0,D(h)}. \qquad (4.13)$$

¿From (4.11)-(4.13) inequality (4.6) follows. Lemma 4.1 is proved.

Inequality (4.6) (for sufficiently small $h > 0$) permits to apply Lemma 3.1 to domain $D(h)$ and to obtain, in virtue of (4.3), estimate

$$\|\bar{u}^0 - \bar{v}\|_{1,D(h)} \leq M_1 \cdot h^{\frac{7+\varepsilon^2}{2}} \qquad (4.14)$$

with some $1 \leq M_1 < +\infty$, does not depending on ε and h.

4.2. The estimate from above of the value $\ell_{2,\Omega}(\bar{u}_C^0)$.

From definition of continuation $\bar{u}_C^0$ of vector - function $\bar{u}^0$ on domain Ω, evidently, it follows, that $\bar{u}_C^0 \in \overset{0}{W}{}_2^1(\Omega)$, div $\bar{u}_C^0 = 0$ and

$$\ell_{2,\Omega}(\bar{u}_C^0) = \ell_{2,D(h)}^\varepsilon(\bar{u}^0). \tag{4.15}$$

Further the application of triangle inequality gives

$$\ell_{2,\Omega}(\bar{u}_C^0) \le \langle \{ \int_{D(h)} [|\nabla \bar{v}|^2 + \frac{1}{\varepsilon}|\mathrm{div}\bar{v}|^2]dx \}^{1/2} +$$

$$+ [\int_{D(h)} |\nabla(\bar{u} - \bar{v})|^2 dx]^{1/2} \rangle \cdot \{ [\int_{D(h)} |\frac{\bar{v}}{\rho_{\partial\Omega}}|^2 dx]^{1/2} - [\int_{D(h)} |\frac{\bar{u}^0 - v}{\rho_{\partial\Omega}}|^2 dx]^{1/2} \}^{-1}. \tag{4.16}$$

¿From (2.8) it follows, that

$$\rho_{\partial\Omega}(x) \ge (1 - \varphi_h') \cdot [x_2 - \varphi(x_1)] \; [x = (x_1, x_2) \in D(h)]. \tag{4.17}$$

Therefore, in virtue of classical Hardy's inequality [3] and (2.8),

$$\left[\int_{D(h)} |\frac{\bar{u}^0 - \bar{v}}{\rho_{\partial\Omega}}|^2 dx \right]^{1/2} \le 2\frac{1 + \varphi_h'}{1 - \varphi_h'} \cdot \left[\int_{D(h)} |\nabla|u^0 - \bar{v}|^2 dx \right]^{1/2}. \tag{4.18}$$

Then from estimates (4.16), (4.18) and (4.14) it follows, that

$$\ell_{2,\Omega}(\bar{u}_C^0) \le \langle \{ \int_{D(h)} [|\nabla \bar{v}|^2 + \frac{1}{\varepsilon}|\mathrm{div}\,\bar{v}|^2]dx \}^{1/2} +$$

$$| \; M_1 \cdot h^{\frac{7+\varepsilon^2}{?}} \rangle \cdot \left\{ [\int_{D(h)} |\frac{\bar{v}}{\rho_{\partial\Omega}}|^2 dx]^{1/2} - 2\frac{1 + \varphi_h'}{1 - \varphi_h'} \cdot M_1 h^{\frac{7+\varepsilon^2}{2}} \right\}^{-1} \tag{4.19}$$

Further, as in the proof of Lemma 3.2, we use estimates (2.15) and (3.25)-(3.26) (for $\eta = \varepsilon^2$) and obtain

$$\ell_{2,\Omega}(\bar{u}_C^0) \le [(1 + \varphi_h') \cdot (\frac{1}{2} + \frac{\varepsilon^2}{2}) \cdot \varepsilon^{-1} + M_2] \cdot [(1 - \varphi_h')^1 \cdot \varepsilon^{-1} - M_2]^{-1}. \tag{4.20}$$

Therefore for sufficiently small $\varepsilon > 0$ and $h = h(\varepsilon) > 0$ estimate

$$\ell_{2,\Omega}(\bar{u}_C^0) \le \frac{1}{2} + M \cdot \varepsilon \tag{4.21}$$

holds with some $1 \le M < +\infty$, does not depending on ε. So, Theorem 3 is proved in the case $n = 2$. The general case $n \ge 2$ can be considered by the same scheme.

REFERENCES

[1] Babuska, I., Szabo, B. (1991), Finite element Analysis; New York, John Wiley, p. 368.

[2] Hardy, G. H., Littlewood, J. E., Polya, G. (1934), Inequalities; Cambridge Univ. Press., Cambridge, p. X11+ 314.

[3] Ladyzenskaja, O. A. (1970), Mathematical Questions of Viscous Non-compressible Fluid Dynamics; M. Nauka, p.228.

[4] Matskewich, T., Sobolevskii, P. E. (1994), The best possible constant in generalized Hardy's inequality for convex domain in R^n; Elliptic and Parabolic P. D. E.'s and Applications; Capri. September 19-23, summaries, pp. 13–18.

[5] Sobolevskii, P. E. (1996), Asymptotics of Stable Viscoelastic Fluid Motion (Oldroyd's Mathematical Model); Math. Nachr. 177, pp. 281–305.

[6] Teman, R. (1979), Navier-Stokes Equations: Theory and Numerical Analysis, Revized edition; Amsterdam, North-Holland, p. 519.

Semilinear Biharmonic Problems on $\mathbb{R}^N$

NIKOS M STAVRAKAKIS[1] Department of Mathematics, National Technical University, Zografos Campus, 157 80 Athens, Hellas. e-mail: nikolas@central.ntua.gr

1 INTRODUCTION

Our aim in this paper is to present resent results on the existence and the properties of solutions of the following semilinear biharmonic eigenvalue problem

$$(-\Delta)^2 u = \lambda g(x) f(u), \quad x \in \mathbb{R}^N, \qquad (1.1)_\lambda$$

$$0 < u(x), \quad \text{for all } x \in \mathbb{R}^N, \quad \lim_{|x| \to +\infty} u(x) = 0, \qquad (1.2)$$

where $\lambda \in \mathbb{R}^N$ and $N > 4$. Here we state the general hypothesis which will be assumed throughout the paper:

$(\mathcal{G})$ g is a smooth function, at least $C^{1,\alpha}(\mathbb{R}^N)$ for some $\alpha \in (0,1)$, such that $g \in L^{N/4}(\mathbb{R}^N) \bigcap L^\infty(\mathbb{R}^N)$ and $g(x) > 0$, on Ω^+, with measure of Ω^+, $|\Omega^+| > 0$.

$(\mathcal{F})$ $f : \mathbb{R} \longmapsto [0,\infty)$ is a smooth function such that $f(0) = 0$, $f'(0) > 0$, and $f(u) > 0$ for all $u \neq 0$. Also $f', f'' \in L^\infty$ and there is $k^* > 0$ such that $|f(s)| \leq k^*|s|$ for all $s \in \mathbb{R}$.
Also two different cases for g are going to be considered , that is, either

$(\mathcal{G}^+)$ $g(x) \geq 0$, almost everywhere in $\mathbb{R}^N$, or

$(\mathcal{G}^-)$ $g(x) < 0$, for $x \in \Omega^-$, with measure of Ω^-, $|\Omega^-| > 0$.

There is quite an extensive literature for the problem in the bounded domain case and the picture is fairly complete. We mention among others, for the equation the papers [5, Chen (1988)], [6, Clément et al. (1992)], [8, Dalmasso (1990)], [9,

[1]**Acknowledgments.** Part of this work was done by financial support from grant number ERBCHRXCT 930409 under the EC Human Capital and Mobility Scheme.

Dalmasso (1995)], [12, Edmunds et al. (1992)], [13, Fleckinger and Lapidus (1986)], [17, Gu (1991)]. We also mention the existence and nonexistence results in [22, Mitidieri (1993)], [26, Peletier et al. (1992)], [28, Pucci and Serrin (1986)], [29, Pucci and Serrin (1990)] on the subject.

In the case of unbounded domains the problem becomes more complicate as, in general, compact operators are not present. It is also unclear *a priori* in which function spaces solutions of $(1.1)_\lambda$ might lie. Nonlinear polyharmonic problems in unbounded domains, are studied recently by various authors, see for example in the radial case the results of [15, Furusho and Kusano (1994)], [19, Kusano et al. (1987a)], [23, Noussair et al. (1992a)], [24, Noussair et al. (1992b)], [35, Swanson and Yu (1993)], in the non-radial sub- (super-) linear the works in [2, Allegretto and Yu (1991)], [3, Bernis (1989)], [34, Swanson (1992)], and the results in [27, Peletier et al.(1995)] of the one dimensional problem. Also the fixed point theory is used in several cases as [11, Edelson and Vakilian (1989)], [18, Kusano et al. (1987)], [20, Kusano et al. (1988)]. Maximum principle results for the biharmonic equation in unbounded domain are obtained recently in [33, Stavrakakis and Sweers (1995)].

In all these papers the weight function is nonnegative. To our knowledge the only works, were the eigenvalue problem for the linear polyharmonic problem with indefinite weight-function is discussed, are those published by J. Fleckinger and her co-workers, see for example [10, Djellit (1992)], [13, Fleckinger and Lapidus (1986)] and the references there in. However, their weight-function is of a certain fractional type at infinity. For the case $m = 1$ we refer to [4, Brown and Stavrakakis (1996)] and the references there in. The same methods are used for a quasilinear eigenvalue problem in [14, Fleckinger et al. (1995)].

In Section 2 we shall discuss the existence of a positive principal eigenvalue (i.e., an eigenvalue corresponding to a positive eigenfunction) of $(2.1)_\lambda$. As in the case m=1 (see [4]) we construct a function space $\mathcal{V}_2$ with norm

$$|||u|||_2 = \{ \int_{\mathbb{R}^N} |\Delta u|^2 dx - \frac{\alpha}{2} \int_{\mathbb{R}^N} g|u|^2 dx \},$$

for an appropriate positive constant α to be chosen later. First, we show that the space $\mathcal{V}_2$ coincides with the standard space $\mathcal{D}^{2,2}$, i.e., the closure of the $C_0^\infty(\mathbb{R}^N)$ functions with respect to the norm $||u||_{\mathcal{D}^{2,2}}^2 = \int_{\mathbb{R}^N} |\Delta u|^2 dx$. By means of standard spectral methods the existence of the first eigenvalue λ_1 is achieved. Then the regularity, the uniform asymptotic behaviour at infinity as well as the positivity of its eigenfunctions are discussed.

In section 3, we study the existence of a global continuum of positive solutions branching out from the first eigenvalue of the semilinear biharmonic problem $(1.1)_\lambda$ (1.2). The proof is achieved by using standard local and global bifurcation theory in the "energy" space $\mathcal{D}^{2,2}$.

Notation: For simplicity we use the symbol $||.||_p$ for the norm $||.||_{L^p(\mathbb{R}^N)}$ and L^p for the space $L^p(\mathbb{R}^N)$, $1 \le p \le \infty$. The end of the proofs is marked by $\diamond$.

2 THE LINEARIZED PROBLEM : EXISTENCE OF NON ZERO PRINCIPAL EIGENVALUES

In this section we shall discuss the existence of nonzero simple principal eigenvalues for the linearized biharmonic problem

$$(-\Delta)^2 u = \lambda g(x) u, \quad x \in \mathbb{R}^N, \tag{2.1}_\lambda$$

$$0 < u(x), \quad \text{for all } x \in \mathbb{R}^N, \quad \lim_{|x| \to +\infty} u(x) = 0, \tag{2.2}$$

To simplify notation but without loss of generality we shall assume that $f'(0) = 1$ so that $(2.1)_\lambda$ becomes the linearisation of $(1.1)_\lambda$. The proofs of all results presented in this section are given in detail in [25, Peletier and Stavrakakis (1995)]. The natural space setting for the eigenfunctions of this problem, as we show next, will be the space $\mathcal{D}^{2,2}(\mathbb{R}^N)$, i.e., the closure of the $C_0^\infty(\mathbb{R}^N)$ functions with respect to the norm

$$||u||_{\mathcal{D}^{2,2}} = \left(\int_{\mathbb{R}^N} |\Delta u|^2 \, dx \right)^{1/2}.$$

It can be shown (see for example [34, Swanson (1992)]) that

$$\mathcal{D}^{2,2} = \left\{ u \in L^{\frac{2N}{N-4}}(\mathbb{R}^N) : |\nabla^2 u| \in L^2(\mathbb{R}^N) \right\}$$

and that there exists $K > 0$ such that for all $u \in \mathcal{D}^{2,2}$

$$||u||_{\frac{2N}{N-4}} \leq K \, ||u||_{\mathcal{D}^{2,2}}.$$

So that the space $\mathcal{D}^{2,2}(\mathbb{R}^N)$ is a *reflexive Banach space*. Our approach is based on the following inequality of generalized Poincare type

LEMMA 1 *Suppose* $g \in L^{N/4}(\mathbb{R}^N)$. *Then there exists* $a = \frac{1}{K||g||_{N/4}^{1/2}} > 0$ *such that*

$$\int_{\mathbb{R}^N} |\Delta u|^2 \, dx \geq a \int_{\mathbb{R}^N} |g||u|^2 \, dx,$$

for all $u \in C_0^\infty(\mathbb{R}^N)$.

Thus if $g \in L^{N/4}(\mathbb{R}^N)$ and $\alpha > 0$ is as in Lemma 1 then we can define an inner product on $C_0^\infty(\mathbb{R}^N)$ by

$$< u, v >_2 = \int_{\mathbb{R}^N} \Delta u \Delta v \, dx - \frac{a}{2} \int_{\mathbb{R}^N} g u v \, dx$$

Next we define $\mathcal{V}_2$ to be the completion of C_0^∞ with respect to the above product. The space $\mathcal{V}_2$ depends on the function g; it is natural to expect that $\mathcal{V}_2$ grows as $|g|$ becomes smaller. However, under condition $(\mathcal{G})$ we prove that $\mathcal{V}_2$ is independent of this function. In fact the space $\mathcal{V}_2$ is characterised by the following

LEMMA 2 *Suppose* $g \in L^{N/4}(\mathbb{R}^N)$. *Then* $\mathcal{V}_2 = \mathcal{D}^{2,2}$

Thus we may henceforth suppose that $|||.|||_2$, the norm in $\mathcal{V}_2$, coincides with the norm in $\mathcal{D}^{2,2}$ and that the inner product in $\mathcal{V}_2$ is given by

$$< u, v > = \int_{\mathbb{R}^N} \Delta u \Delta v \, dx.$$

Proceeding, as for example in [4], we define a bilinear form by

$$\beta(u, v) = \int_{\mathbb{R}^N} guv \, dx,$$

for all $u, v \in \mathcal{V}_2$. By the fact that $\mathcal{V}_2 \subseteq L^{\frac{2N}{N-4}}(\mathbb{R}^N)$ we obtain that β is *bounded* in $\mathcal{V}_2$. Hence by the Riesz Representation Theorem we can define a bounded linear operator M such that

$$\beta(u, v) = < Mu, v >, \text{ for all } u, v \in \mathcal{V}_2.$$

It is standard to check the following result

LEMMA 3 *Suppose* $g \in L^{N/4}(\mathbb{R}^N)$. *Then* M *is selfadjoint and compact.*

By means of standard spectral methods (see, for example [37, Weinberger (1974)]) we prove the existence of nonzero first eigenvalues and the main properties of them have. These are given next

THEOREM 4 *(i) Let g satisfies hypothesis $(\mathcal{G}^+)$. Then the equation $(2.1)_\lambda$ admits a positive first eigenvalue given by*

$$\lambda_1 = inf_{<Mu,u>=1}||u||^2_{\mathcal{D}^{2,2}}.$$

(ii) Let g satisfies $(\mathcal{G}^-)$. Then the equation $(2.1)_\lambda$ admits two first eigenvalues of opposite sign given by

$$\lambda_1^+ = inf_{<Mu,u>=1}||u||^2_{\mathcal{D}^{2,2}},$$

$$\lambda_1^- = inf_{<Mu,u>=-1}||u||^2_{\mathcal{D}^{2,2}}.$$

In both cases the associated eigenfunctions ϕ (resp. ϕ^+, ϕ^-) belong in $\mathcal{D}^{2,2}(\mathbb{R}^N)$.

Furthermore, there is an eigenfunction associated to λ_1 (resp. λ_1^+, λ_1^-), which does not change sign in $\mathbb{R}^N$.

When we have nonnegative weights the results are similar to the simple Laplacian problem with indefinite weight

COROLLARY 5 *Let $g(x) \geq 0$, everywhere in $\mathbb{R}^n$. Then all eigenfunctions associated to λ_1 are of constant sign, i.e. λ_1 is a principal eigenvalue of the equation $(2.1)_\lambda$.*

Using Agmon's theorem [1, Agmon (1959), Theorem 6.1] and Serrin's estimates from [31, Serrin (1964)] (see also [16, Gilbarg and Trudinger (1983), Theorem 9.19]) we can prove the L^σ character and the $C^{4,\alpha}_{loc}$ regularity as well as the asymptotic properties of the associated "first eigenfunctions " of $(2.1)_\lambda$. For these technics we refer also to [21, Luckhaus (1979)] and [36, Troisi (1987)]

THEOREM 6 *Suppose that* g *satisfies either* $(\mathcal{G}^+)$ *or* $(\mathcal{G}^-)$ *and* $u \in \mathcal{D}^{2,2}$ *is a solution of* $(2.1)_\lambda$*. Then*
(i) u *is a classical solution i.e* $u \in C^{4,\alpha}_{loc}(\mathbb{R}^N)$ *Moreover, for any* $x_0 \in \mathbb{R}^N$ *and* $p > 1$ *we have*

$$||u||_{W^{4,p}(B_1(x_0))} \leq C||u||_{L^{\frac{2N}{N-4}}(B_2(x_0))}, \tag{2.3}$$

where $C = C(x_0, \lambda, N, ||g||_\infty, |||u|||_2)$.

(ii) $D^\beta u(x)$ *decay uniformly to zero as* $|x| \to +\infty$*, for all* $|\beta| \leq 3$.

Having in mind the application of bifurcation theory for the study of problem $(2.1)_\lambda$, information concerning the dimension of the eigenspace associated to the principal eigenvalues of the linearized biharmonic problem $(2.1)_\lambda$ are of basic importance. The main result in this direction can be stated as follows

THEOREM 7 *Let* $g \geq 0$. *Then we have: (i) the eigenspace corresponding to the principal eigenvalue* λ_1 *is of dimension 1. (ii)* λ_1 *is the only eigenvalue of* $(2.1)_\lambda$ *to which corresponds a positive eigenfunction.*

The proof of this theorem is long and technical. We give only some of the necessary auxiliary lemmas, which also are of independent interest

LEMMA 8 *Suppose that* g *satisfies either* $(\mathcal{G}^+)$ *or* $(\mathcal{G}^-)$ *and* $u \in \mathcal{D}^{2,2}$. *Then there exists a sequence* R_n *with* $R_n \to \infty$ *as* $n \to +\infty$ *such that*

$$\lim_{n\to\infty} \int_{\partial B_{R_n}} \nabla u \frac{\partial \nabla u}{\partial n} dS = 0 = \lim_{n\to\infty} \int_{\partial B_{R_n}} u \frac{\partial \Delta u}{\partial n} dS = 0.$$

The next lemma gives the equivalent weak formulation of the polyharmonic problem $(2.1)_\lambda$;

LEMMA 9 *Suppose that* g *satisfies either* $(\mathcal{G}^+)$ *or* $(\mathcal{G}^-)$ *and* $w \in \mathcal{D}^{2,2}$ *be an eigenfunction of* $(2.1)_\lambda$ *corresponding to an eigenvalue* $\lambda > 0$. *Then*

$$\int_{\mathbb{R}^N} |\Delta w|^2 \, dx = \lambda \int_{\mathbb{R}^N} gw^2 \, dx > 0.$$

LEMMA 10 *Suppose that* g *satisfies either* $(\mathcal{G}^+)$ *or* $(\mathcal{G}^-)$ *and* $u \in \mathcal{D}^{2,2}$. *Then for some* $R_1 > 0$ *we have*

$$\int_{R_1}^{+\infty} \frac{dr}{M(r)} = +\infty,$$

where $M(r) = \int_{\partial B_r} |\nabla u(S)|^2 \, dS.$

Finally, denoting the following sum of multiple integrals by

$$Q(R) = \int_{B_R} \frac{|\phi|^2 |\Delta u|^2}{u^2} dx - 2 \int_{B_R} \frac{|\nabla \phi|^2 \Delta u}{u} dx - 2 \int_{B_R} \frac{|\phi|^2 |\nabla u|^2 \Delta u}{u^3} dx,$$

and the sum of surface integrals by

$$q(R) = 2 \int_{\partial B_R} \frac{\phi}{u} \nabla \phi \frac{\partial \nabla u}{\partial n} dS - \int_{\partial B_R} \frac{|\phi|^2}{u} \frac{\partial \Delta u}{\partial n} dS - \int_{\partial B_R} \frac{|\phi|^2}{u^2} \nabla u \frac{\partial \nabla u}{\partial n} dS.$$

we have a useful comparison result given below

LEMMA 11 *Let $g \geq 0$. Suppose that $\phi \in \mathcal{D}^{2,2}$ is an eigenfunction of $(2.1)_{\lambda_1}$ corresponding to the principal eigenvalue λ_1 (not neccesarily of constant sign) . Also let $u \in \mathcal{D}^{2,2}$ be a positive eigenfunction of $(2.1)_\lambda$ corresponding to an eigenvalue $\lambda > 0$. Then if $\lim_{R \to \infty} q(R) \neq 0$ there exist $d \in (0,1)$ and some R_1 large enought such that*

$$d\, Q(R) \leq -q(R), \quad \text{for all } R \geq R_1$$

3   THE SEMILINEAR BIHARMONIC PROBLEM : BIFURCATION RESULTS

In this section we shall obtain results on the existence of solutions for the nonlinear problem $(1.1)_\lambda$, (1.2), by considering bifurcation of solutions from the zero solution. The complete proofs of the results of this section can be found in the paper [32, Stavrakakis (1996)].

3.1   Local Bifurcation

To complete this aim we introduce the nonlinear operator $P : \mathbb{R} \times \mathcal{V}_2 \to \mathcal{V}_2$ through the relation

$$< P(\lambda, u), \phi >= \int_{\mathbb{R}^N} \Delta u \Delta \phi \, dx - \lambda \int_{\mathbb{R}^N} g f(u) \phi \, dx, \qquad (3.1)$$

for all $\phi \in \mathcal{V}_2$, where $<,>$ denotes the inner product in $\mathcal{V}_2$, i.c. in $\mathcal{D}^{2,2}$. Throughout this section we assume that f satisfies hypothesis $(\mathcal{F})$ and $g \geq 0$.

LEMMA 12 *The operator P is well defined by (3.1).*

Proof   For fixed $u \in \mathcal{D}^{2,2}$ we define the following functional

$$F(\phi) = \int_{\mathbb{R}^N} \Delta u \Delta \phi \, dx - \lambda \int_{\mathbb{R}^N} g f(u) \phi \, dx$$

for $\phi \in \mathcal{D}^{2,2}$. Since f satisfies $(\mathcal{F})$ then $f(u) \in L^{\frac{2N}{N-4}}(\mathbb{R}^N)$ and so for some constant K_1

$$
\begin{aligned}
|F(\phi)| &\leq ||\Delta u||_2 \, ||\Delta \phi||_2 + |\lambda| \, ||g||_{N/4} \, ||f(u)||_{\frac{2N}{N-4}} \, ||\phi||_{\frac{2N}{N-4}} \\
&\leq K_1 (||\Delta u||_2 + |\lambda| \, ||g||_{N/4} \, ||f(u)||_{\frac{2N}{N-4}}) \, ||\phi||_{\mathcal{V}_2}.
\end{aligned}
$$

So F is a bounded linear functional. Hence by the Riesz Representation Theorem we may define P as in (3.1). $\diamond$

LEMMA 13 *The operator* P *is continuous and for* $N = 5, 6, ..., 12$ *is Frechet differentiable with continuous Frechet derivatives given by*

$$
\begin{aligned}
< P_u(\lambda, u)\phi, \psi > &= \int_{\mathbb{R}^N} \Delta\phi\Delta\psi \, dx - \lambda \int_{\mathbb{R}^N} gf'(u)\phi\psi \, dx, \\
< P_\lambda(\lambda, u), \phi > &= - \int_{\mathbb{R}^N} gf(u)\phi \, dx, \\
< P_{\lambda u}(\lambda, u)\phi, \psi > &= - \int_{\mathbb{R}^N} gf'(u)\phi\psi \, dx
\end{aligned}
$$

for all $\phi, \psi \in \mathcal{D}^{2,2}$.

Proof For completeness we juct sketch the proof.
(i) To prove the continuity of P we have

$$
< P(\lambda, u) - P(\mu, w), v >
$$

$$
= < P(\lambda, u) - P(\lambda, w), v > + < P(\lambda, w) - P(\mu, w), v >
$$

$$
= |\int_{\mathbb{R}^N} (\Delta u - \Delta w) dx + \lambda \int_{\mathbb{R}^N} g(f(u) - f(w))v dx| + |\lambda - \mu| \, |\int_{\mathbb{R}^N} gf(w)v dx|
$$

$$
\leq \{1 + \lambda \sup_{\tau \in (0,1)} ||g||_{N/4} ||f'(u + \tau w)||_\infty\} ||u - w||_{\mathcal{D}^{2,2}} ||v||_{\mathcal{D}^{2,2}}
$$

$$
+ c|\lambda - \mu| ||g||_{N/4} ||w||_{\mathcal{D}^{2,2}} ||v||_{\mathcal{D}^{2,2}}.
$$

So P is continuous at any $(\lambda, u) \in \mathbb{R} \times \mathcal{D}^{2,2}$.
(ii) Let $u, h \in \mathcal{D}^{2,2}$ and

$$
P(\lambda, u, h) = P(\lambda, u + h) - P(\lambda, u) - P_u(\lambda, u)h.
$$

Then we have

$$
||P(\lambda, u + h) - P(\lambda, u) - P_u(\lambda, u)h||^2_{\mathcal{D}^{2,2}}
$$

$$
\leq |\lambda| \, |\int_{\mathbb{R}^N} g(f(u + h) - f(u) - f'(u)h) P(\lambda, u, h) dx|
$$

$$
\leq |\lambda| \sup_{t \in (0,1)} ||f''(u + th)||_\infty ||g||_q ||h||^2_p ||P(\lambda, u, h)||_p
$$

where $p = \frac{2N}{N-4}$ and $\frac{1}{q} + \frac{3}{p} = 1$ i.e. $q = \frac{2N}{12-N}$. Hence for $N = 5, 6, ..., 12$ we get that $q \geq \frac{N}{4}$ or that $g \in L^q(\mathbb{R}^N)$. Hence we have

$$||P(\lambda, u+h) - P(\lambda, u) - P_u(\lambda, u)h||_{\mathcal{D}^{2,2}}$$

$$\leq |\lambda| \sup_{t \in (0,1)} ||f''(u+th)||_\infty ||g||_q ||h||^2_{\mathcal{D}^{2,2}} = o(||h||_{\mathcal{D}^{2,2}}).$$

So P is Fréchet differentiable in $\mathbb{R} \times \mathcal{D}^{2,2}$.

(iii) Let $u, v, \phi \in \mathcal{D}^{2,2}$. Then we have

$$| < P_u(\lambda, u)h - P_u(\mu, v)h, \phi > |$$

$$\leq | < P_u(\lambda, u)h - P_u(\lambda, v)h, \phi > + < P_u(\lambda, v)h - P_u(\mu, v)h, \phi >$$

$$\leq |\lambda| |\int_{\mathbb{R}^N} g(f'(u) - f'(v))h\phi dx| + |\lambda - \mu| |\int_{\mathbb{R}^N} gf'(v)\phi dx|$$

$$\leq |\lambda| \sup_{t \in (0,1)} ||f''(tu + (1-t)v)||_\infty ||g||_q ||u - w||_{\mathcal{D}^{2,2}} ||h||_{\mathcal{D}^{2,2}} ||\phi||_{\mathcal{D}^{2,2}}$$

$$+ |\lambda - \mu| ||g||_{N/4} ||f'(v)||_\infty ||h||_p ||\phi||_p.$$

where $p = \frac{2N}{N-4}$ and $q = \frac{2N}{12-N}$ for $N = 5, 6, .., 12$. Hence we have that P_u is continuous in $\mathbb{R} \times \mathcal{D}^{2,2}$. $\diamond$

Consider the linear operator $P_u(\lambda_1, 0)$ where λ_1 is the principal eigenvalue of $(2.1)_\lambda$. We can easily prove that $P_u(\lambda_1, 0)$ is a bounded selfadjoint operator and that $P_u(\lambda_1, 0)\phi = 0$ if and only if $\phi \in \mathcal{V}_\in$ is a solution of $(2.1)_{\lambda_1}$. Therefore $N(P_u(\lambda_1, 0)) = [\phi]$ where ϕ is the principal eigenfunction of $(2.1)_{\lambda_1}$. Since $P_u(\lambda_1, 0)$ is selfadjoint, $R(P_u(\lambda_1, 0)) = [\phi]^\perp$, i.e., $\psi \in R(P_u(\lambda_1, 0))$ if and only if $< \psi, \phi > = 0$. Furthermore, we have that,

$$< P_{\lambda u}(\lambda_1, 0)\phi, \phi > = - \int_{\mathbb{R}^N} g \phi^2 dx < 0.$$

So we have proved that the operator P satisfies all the hypotheses of the local bifurcation theorem from a simple eigenvalue (see [7, Crandall and Rabinowitz (1971)]). So we have obtained the following result.

THEOREM 14 *There exists* $\epsilon_0 > 0$ *and continuous functions* $\eta : (-\epsilon_0, \epsilon_0) \to \mathbb{R}$ *and* $\psi : (-\epsilon_0, \epsilon_0) \to [\phi]^\perp$ *such that* $\eta(0) = \lambda_1$, $\psi(0) = 0$ *and every nontrivial solution of* $P(\lambda, u) = 0$ *in a small neighbourhood of* $(\lambda_1, 0)$ *is of the form* $(\lambda_\epsilon, u_\epsilon) = (\eta(\epsilon), \epsilon\phi + \epsilon\psi(\epsilon))$.

An easy consequence of the maximum principle applied to the equation $(1.1)_\lambda$, (1.2) is that for any solution u we have $u(x) > 0$ in $\mathbb{R}^N$. Conserning the asymptotic behavior of the solutions, inequality (2.3) implies that

$$||u||_{L^\infty(B_1(x))} \leq C ||u||_{L^{\frac{2N}{N-4}}(B_2(x))}$$

for all $x \in \mathbb{R}^N$ where $C = C(x, \lambda, N, ||g||_\infty, |||u|||_2)$.

3.2 Global Continuation

To discuss the global nature of the continuum of solutions bifurcating from $(\lambda_1, 0)$, we write the operator P in the form $P(\lambda, u) = u - \lambda R(u)$ where

$$< R(u), \phi > = \int_{\mathbb{R}^N} g(x) f(u(x)) \phi(x) \, dx, \quad \text{for all } \phi \in \mathcal{D}^{2,2}.$$

Also we assume that R satisfies the relation

$$R(u) = Mu + \mathcal{H}(u)$$

where M denotes the same linear operator as in section 2, i.e.,

$$< Mu, v > = \int_{\mathbb{R}^N} g u v \, dx, \quad \text{for all } u, v \in \mathcal{D}^{2,2},$$

and $\mathcal{H}(u) = O(|||u|||_2^\beta)$ as $|||u|||_2 \to 0$ for some $\beta > 1$.

It is shown in section 2 that λ_1 is an eigenvalue of L and by Theorem 7 the eigenspace associated with λ_1 has algebraic multiplicity 1. Also, as f is a Lipschitz function, it can be proved by modifying slightly the proof of Lemma 3 that R is a compact operator. Thus we can apply the classical global bifurcation theorem (see [30, Rabinowitz (1971)]) to obtain

THEOREM 15 *There exists a continuum $\mathcal{C}$ of nonzero solutions of $(1.1)_\lambda$, (1.2) bifurcating from $(\lambda_1, 0)$ which is, either (i) unbounded, or (ii) contains a point $(\lambda, 0)$, where $\lambda \neq \lambda_1$ is an eigenvalue of $(2.1)_\lambda$.*

Moreover, $\mathcal{C}$ has a connected subset $\mathcal{C}^+ \subset \mathcal{C} - \{(\eta(\epsilon), u_\epsilon) : -\epsilon_0 \leq \epsilon \leq 0\}$ for some $\epsilon_0 > 0$ such that $\mathcal{C}^+$ also satisfies one of the above alternatives. Close to the bifurcation point $(\lambda_1, 0)$, $\mathcal{C}^+$ consists of the curve $\epsilon \to (\eta(\epsilon), u_\epsilon)$, $0 < \epsilon \leq \epsilon_0$. The following lemmas describe the nature of solutions lying on $\mathcal{C}^+$.

LEMMA 16 *There exists $\lambda_* > 0$ such that $\lambda > \lambda_*$ whenever $(\lambda, u) \in \mathcal{C}^+$.*

In order to proceed further we must investigate the L^∞-closeness of solutions which are close in $\mathbb{R} \times \mathcal{D}^{2,2}$; since $\mathcal{D}^{2,2}$ does not embed in $L^\infty(\mathbb{R}^N)$, this is not immediately obvious. Actually, we can prove that there exist positive constants K_1 and K_2 such that

$$|u_\lambda(x) - u_\mu(x)| \leq K_1 |\lambda - \mu| + K_2 \|u_\lambda - u_\mu\|_{\mathcal{D}^{2,2}} \quad \text{for all } x \in \mathbb{R}^N$$

whenever μ is close to λ and $u_\lambda, u_\mu \in \mathcal{D}^{2,2}$ are solutions of $(1.1)_\lambda$, $(1.1)_\mu$ respectively.

LEMMA 17 *$\mathcal{C}^+$ contains no points of the form $(\lambda, 0)$, where $\lambda \neq \lambda_1$.*

Thus $\mathcal{C}^+$ must connect $(\lambda_1, 0)$ to ∞ in $\mathbb{R} \times \mathcal{D}^{2,2}$. The next theorem shows that $\mathcal{C}^+$ cannot become unbounded at a finite value of λ; in order to obtain this result we must impose some restrictions on f, g

THEOREM 18 *Suppose that for some* $\gamma \in (0,1)$ $f(s) \sim |s|^\gamma$ *at infinity and* $g \in L^p(\mathbb{R}^N)$, *where* $p < \frac{2N}{2N-(\gamma+1)(N-4)}$. *Then there exists a continuous function* $\mathcal{K}$: $\mathbb{R}^+ \to \mathbb{R}^+$ *such that* $|||u|||_2 \le \mathcal{K}(\lambda)$ *whenever* $(\lambda, u) \in \mathcal{C}^+$.

As an immediate consequence of the previous results we can give the following complete description of the continuum $\mathcal{C}^+$.

THEOREM 19 *Suppose that* $N = 5, 6, 7$, *for some* $\gamma \in (0,1)$, $f(s) \sim |s|^\gamma$ *at infinity and* $g \ge 0$, $g \in L^p(\mathbb{R}^N) \cap L^\infty(\mathbb{R}^N)$, *where* $1 < p < \frac{2N}{2N-(\gamma+1)(N-4)}$. *Then there exists a continuum* $\mathcal{C}^+ \subseteq \mathbb{R} \times \mathcal{D}^{2,2}$ *of solutions bifurcating from the zero solution at* $(\lambda_1, 0)$ *such that*
(i) if $(\lambda, u) \in \mathcal{C}^+$ *then* $\lambda > \lambda_* > 0$, $u \in L^\infty$ *and* $u(x) > 0$ *for all* $x \in \mathbb{R}^N$,
(ii) $\{\lambda : (\lambda, u) \in \mathcal{C}^+$ *for some* $u \in \mathcal{D}^{2,2}\} \supseteq (\lambda_1, \infty]$.
In particular $(1.1)_\lambda$, (1.2) *has a nontrivial solution* $u \in \mathcal{D}^{2,2}$ *such that* $u(x) > 0$ *for all* $x \in \mathbb{R}^N$ *whenever* $\lambda > \lambda_1$.

REFERENCES

[1] Agmon S. (1959). The L_p Approach to the Dirichlet Problem, Ann. Scuola Norm. Sup. di Pisa, 13, 405-448.

[2] Allegretto W. and Yu L. S. (1991). Decaying Solutions of $2m^{th}$ order Elliptic Problems, Can. J. Math. 43, 449-460.

[3] Bernis F. (1989). Elliptic and Parabolic Semilinear Problems without Conditions at Infinity, Arch. Rational Mech. Anal., 106, 217-241.

[4] Brown K. J. and Stavrakakis N. M. (1996). Global Bifurcation Results for a Semilinear Elliptic Equation on all of $\mathbb{R}^N$, Duke Math. Journ., to appear.

[5] Chen Zu-Chi (1988) Inequalities for Eigenvalues of a Class of Polyharmonic Operators, Applicable Anal. 27, 289-314.

[6] Clément Ph., de Figueiredo D. G. and Mitidieri E. (1992). Positive solutions of semilinear elliptic systems, Comm. in P.D.E., 17 (5-6), 923-940.

[7] Crandall M. and Rabinowitz P. H. (1971). Bifurcation from simple eigenvalues, J. Func. Anal., 8, 321-340.

[8] Dalmasso R. (1990). Problème de Dirichlet homogène pour une équation biharmonique semi-linéaire dans une boule, Bull. Sc.Math.,114, 123-137.

[9] Dalmasso R. (1995). Elliptic Equations of Order $2m$ in Annular Domains, Trans Am Math Soc, 347 (9) , 3575- 3585.

[10] Djellit A. (1992). Valeurs propres de problèmes elliptiques indèfinis sur des ouverts non-bornés, Ph.D. Thesis, Université Paul Sabatier, Toulouse.

[11] Edelson A. L. and Vakilian R. (1989). Bounded Entire Solutions of Higher Order Semilinear Equations, in Differential Equations (ed. C. M. Dafermos G. Ladas and G. Papanicolaou), Marcel Dekker, Inc. New York, pp. 207-214.

[12] Edmunds D. E., Fortunato D. and Jannelli E. (1992). Critical Exponents, Critical Dimensions and the Biharmonic Operator, Arch. Rat. Mech.An., 112, 269-289.

[13] Fleckinger J. and Lapidus M. L. (1986). Eigenvalues for Elliptic Boundary Value Problems with an Indefinite Weight Function, Trans. Amer. Math. Soc., 295, 305-324.

[14] Fleckinger J., Manasevich R., Stavrakakis N. M. and de Thelin F. (1995). Principal Eigenvalues for some Quasilinear elliptic Systems on $\mathbb{R}^N$, Advances in Diff Equ, to appear.

[15] Furusho Y. and Kusano T. (1994). Existence of Positive Entire Solutions for Higher Order Quasilinear Elliptic Equations, J.Math. Soc. Japan 46, 449-465.

[16] Gilbarg G. and Trudinger N. S. (1983) Elliptic Partial Differential Equations of Second Order (2nd Edition), Springer-Verlag, Berlin.

[17] Gu Yong-Geng (1991). The eigenvalue problems of elliptic equations of higher order, Acta Math. Scientia, 11, 361-368.

[18] Kusano T., Naito M. and Swanson C. A. (1987). Entire Solutions of a Class of Even Order Quasilinear Elliptic Equations, Math. Z. 195, 151-163.

[19] Kusano T., Naito M. and Swanson C. A. (1987) Radial Entire Solutions to Even Order Semilinear Elliptic Equations in the Plane, Proc. Roy. Soc. Edinb. 105A, 275-287.

[20] Kusano T., Naito M. and Swanson C. A. (1988). Radial Entire Solutions of Even Order Semilinear Elliptic Equations, Can. J. Math. 195, 1281-1300.

[21] Luckhaus S. (1979). Existence and Regularity of Weak Solutions to the Dirichlet Problem for Semilinear Elliptic Systems of Higher Order, J. Reine Angew. Math., 306, 192-207.

[22] Mitidieri E. (1993). A Rellich Type Identity and Applications, Comm. in PDE's, 18 (1-2), 125-151.

[23] Noussair E. S., Swanson C. A. and Yianfu Y. (1992). Transcritical Biharmonic Equations in $\mathbb{R}^N$, Funkcial. Ekvac. 35, No 3, 533-543.

[24] Noussair E. S., Swanson C. A. and Yianfu Y. (1992). Critical Semilinear Biharmonic Equations in $\mathbb{R}^N$, Proc. Roy. Soc. Edinb. 121A, 139-148.

[25] Peletier L. A. and Stavrakakis N.M. (1995). Existence and Simplicity of Principal Eigenvalues for a linear Polyharmonic problem on $\mathbb{R}^N$, preprint.

[26] Peletier L. A. and Van der Vorst R. C. A. M. (1992). Existence and Nonexistence of Positive Solutions of Nonlinear Elliptic Systems and the Biharmonic Equation, Diff. and Int. Equ.'s, 5, 747-767.

[27] Peletier L. A., Troy W.C. and Van der Vorst R. C. A. M. (1995). Stationary Solutions of a Fourth Order Nonlinear Diffusion Equation, preprint.

[28] Pucci P. and Serrin J. (1986). A General Variational Identity. Indiana Univ. Math. J., 35, 681-703.

[29] Pucci P. and Serrin J. (1990). Critical Exponents and Critical Dimensions for Polyharmonic Operators, J. Math. Pur. et Appl., 69, 55-83.

[30] Rabinowitz P. H. (1971). Some global results for nonlinear eigenvalue problems, J. Func. Anal., 7, 487-513.

[31] Serrin J. (1964). Local Behavior of Solutions of Quasilinear equations, Acta. Math., 111 , 247-302.

[32] Stavrakakis N. M. (1996). Global Bifurcation Results for a Semilinear Biharmonic Equation on all of $\mathbb{R}^N$, submitted.

[33] Stavrakakis N. M. and Sweers G. (1995). Positivity for a noncooperative system of elliptic equations on all of $\mathbb{R}^N$, submitted.

[34] Swanson C. A. (1992). The Best Sobolev Constant, Applic. Analysis, 47, 227-239.

[35] Swanson C. A. and Yu L. S. (1993). Radial Polyharmonic Problems in $\mathbb{R}^N$, J. Math. Anal. Appl. 174, 461-466.

[36] Troisi M. (1987). Un Theorema di Regolarizzazione per le Soluzioni dei Problemi Ellittici in Aperti non Limitati di $\mathbb{R}^n$, Rend Acc Naz Scie, 105 Vol XI, 127-137.

[37] Weinberger H. F.(1974). Variational Methods for eigenvalue approximation, CBMS Reg. Conf. Ser. in Appl. Math., 15.

Magnetic Field Wave Equations for TM-modes in Nonlinear Optical Waveguides

C. A. STUART, Département de Mathématiques, Ecole Polytechnique Fédérale de Lausanne, Lausanne, Switzerland

1   Introduction

In CGS units, Maxwell's equations for a dielectric medium are [1],

$$\nabla \wedge E = -\frac{1}{c}\partial_t B \quad (1) \qquad \nabla \wedge H = \frac{1}{c}\partial_t D \quad (2) \qquad \text{(ME)}$$

$$\nabla \cdot D = 0 \quad (3) \qquad \nabla \cdot B = 0 \quad (4)$$

where c is the speed of light in a vacuum and the medium is charge-free. The fields E, D, H and B are treated as functions of Cartesian coordinates $(x, y, z, t) \in \Re^4$ and e_i denotes the usual basis vector in $\Re^3$.

Assuming that the medium is non-magnetic we set $H \equiv B$. Then the remaining constitutive assumption about the medium should determine the electric displacement field, D, as a function of the electric field, E. In general, this is a complicated nonlinear relationship which for the moment we indicate roughly as $D = F(E)$.

The search for solutions of Maxwell's equations is usually based on the following electric field wave equation,

$$\nabla \wedge \nabla \wedge E = -\frac{1}{c^2}\partial_t^2 F(E) \qquad \text{(EFWE)}$$

which is obtained from (ME 1,2) and the constitutive relations $H = B$ and $D = F(E)$. Given a solution of this equation, the other fields can be defined so as to furnish a complete solution of Maxwell's equations. In cases where E has a simple form (some TE-modes, for example) this is the most natural and straight forward way to proceed. However in other cases (some TM-modes, for example), the magnetic fields have a simpler form than the electric fields and it would seem advantageous to use the magnetic field as the basic unknown. In the case of linear isotropic materials the constitutive relations are $H = B$ and $D = \varepsilon(x, y, z)E$ where ε is the dielectric constant. Thus $E = \frac{1}{\varepsilon}D$ and Maxwell's equations can be reduced to the study of the following magnetic field wave equation,

$$\nabla \wedge \frac{1}{\varepsilon} \nabla \wedge H = -\frac{1}{c^2} \partial_t^2 H. \qquad \text{(MFWE)}$$

Clearly anisotropic linear materials can be treated similarly using the dielectric tensor. For nonlinear materials (even isotropic ones), expressing E as a function of D is not such a simple matter since the constitutive relation expresses ε as a function of E and, perhaps for this reason, a purely magnetic field formulation of Maxwell's equations does not seem to have been used. Indeed, when comparing the study of TE-modes and TM-modes in planar optical waveguides, Stegeman describes the situation in the following way [[15], page 32]. "Extension of the analysis to TM nonlinear waves is complicated by the inherent structure of the fields; that is, they contain two electric field components $E_x(x)$ and $E_z(x)$ that are $\pi/2$ out of phase with one another. Although much of the early work involved various approximations, this case has recently been solved exactly. In fact, it is necessary to start with Maxwell's equations and not a wave equation. (Rigorously, no simple nonlinear wave equations exists for this case.)" The first aim of this paper is to show how such a wave equation can be derived in quite a general context which is appropriate for many problems in nonlinear optics. The particular situation being referred to by Stegeman is treated in Section 4 where the nonlinear wave equation is a single second order ordinary differential equation.

In nonlinear optics the discussion is often restricted to monochromatic fields and the constitutive relation $D = F(E)$ is taken in a special, but still nonlinear, form. (See the beginning of Section 2.) In this paper we only deal with this situation in the case of a homogeneous uni-axial medium, but the approach could be extended to cover heterogeneous, fully anisotropic materials and to allow for absorption. The treatment of non-monochromatic fields would seem to be much more difficult.

In Section 2 we present a general form, (CR), of the constitutive relation that is widely used in the analysis of nonlinear optical waveguides, [31] and [16]-[18]. The main subject of that section concerns the inversion of this relationship so as to express E as a function of D. Then in Section 3, we derive the nonlinear magnetic field wave equation that generalizes the linear one noted above. In particular, the special forms which are appropriate for the discussion of TM-modes propagating in planar and cylindrical wave-guides are presented. These configurations are of particular interest in nonlinear optics and previous work on them has been based on the system of two equations that arises in the electric field formulation [5]-[10] and [26]-[28], of the equations for TM-modes. An advantage of the present magnetic field formulation is that they are described by a single scalar second order (nonlinear) differential equation. For example, the equation (3.9) which we derive for cylindrical TM-modes has a relatively simple form compared to the equations (10 a,b) in [9] which arise in the EFWE formulation. A rigorous treatment of guidance for nonlinear TM-modes can be based on the equations derived in Section 3 and, in Section 4, we review what has been done so far.

Having given a summary of the contents of this paper, it is now appropriate to expand the earlier remarks about the nonlinear MFWE and about the study of TM-modes in nonlinear optics. The numerical approximation of a nonlinear MFWE has been undertaken recently; see for example [11] and [12]. However in that work the exact equation is never stated and the problem of expressing $\frac{1}{\varepsilon}$ as an explicit function of H is not addressed. Instead, a two-step iterative procedure is used so as to avoid inverting the constitutive relation. Since our aim is to obtain qualitative, as well as quantitative, information about true solutions we need the explicit, exact form of the nonlinear MFWE. The only case that we know of where this has been attempted previously is in the work of Leung [13] in the special case of a planar waveguide. He does obtain an exact (integrable) equation for TM-modes using a magnetic field formulation by inverting the function ε that appears in the isotropic version of the constitutive relation (CR). This does not lead to an inversion of the constitutive relation and his approach cannot be extended to more general situations. Even for planar waveguides, the equation which we formulate in Section 3 is simpler and more tractable. Apart from the papers that we have just mentioned, most of the work on TM-modes in planar or cylindrical waveguides has been based on a electric field formulation. A variety of approaches to the planar case have been proposed [5]-[10], [24], [26]-[28]. They are applicable to various levels of generality of the constitutive relation, but they rely on the hypothesis that the medium is composed of a small number of homogeneous layers.

2   Constitutive assumptions

All the e-m fields E, D, H and B which we consider can be expressed in the form

$$F(x, y, z, t) = F_T(x, y) \cos(kz - \omega t) + F_z(x, y) \sin(kz - \omega t) \qquad (2.1)$$

where $F_T, F_z : \Re^2 \longrightarrow \Re^3$ are such that $F_T.e_3 = 0$ and $F_z.e_i = 0$ for $i = 1, 2$.

Thus the fields constitute monochromatic waves propagating in the direction of e_3 with the transverse and axial components of the fields being out of phase by a quarter of a cycle.

For fields of this type the constitutive relation which we adopt will be discussed in two steps. First we deal with a medium having uni-axial anisotropy, [31], and then we show the simplifications which occur when the medium is actually isotropic. In both cases we start from the standard form of constitutive assumption ([4], [17], [32]) which is usually adopted in the literature concerning nonlinear optical wave-guides, but the main purpose here is to introduce an equivalent formulation which enables us to derive a magnetic field wave equation with the aim of simplifying the treatment of TM-modes.

2.1   Uni-axial medium

To introduce the constitutive assumption for a homogeneous uniaxial material, we choose the orthonormal basis $\{e_i : i = 1, 2, 3\}$ with e_3 in the direction of the axis of the medium.

(CR) $B = H$ and there exist two continuous functions $\varepsilon_i : [0.\infty)^2 \longrightarrow (0, \infty)$ such that

$$D_T(x, y) = \varepsilon_1(|\, E_T(x, y)\,|^2 /2, |\, E_z(x, y)\,|^2 /2) E_T(x, y)$$

$$D_z(x, y) = \varepsilon_2(|\, E_T(x, y)\,|^2 /2, |\, E_z(x, y)\,|^2 /2) E_z(x, y)$$

where $0 < A < \varepsilon_i(s_1, s_2)$ for $i = 1, 2$ and for $s_1, s_2 \geq 0$.

This means that with respect to the basis $\{e_i : i = 1, 2, 3\}$ the dielectric response tensor is represented by the diagonal matrix $\begin{pmatrix} \varepsilon_1 & 0 & 0 \\ 0 & \varepsilon_1 & 0 \\ 0 & 0 & \varepsilon_2 \end{pmatrix}$ where the elements ε_i are functions of the time-averages, $|\, E_T(x, y)\,|^2 /2$ and $|\, E_z(x, y)\,|^2 /2$, of the intensities of the transverse and axial components of the electric field E. Let us remark that the functions ε_i also depend upon the frequency ω of E, but since we shall always treat ω as a fixed parameter we do not exhibit this dependence explicitly. Furthermore in a heterogeneous medium the elements ε_i will also depend explicitly on x, y and z. Again we do not deal with this case but we emphasize that all of the subsequent development extends to cover heterogeneous materials with only notational changes and some obvious requirements about the smoothness of the dependance on (x, y, z).

According to (CR) the dielectric response of the medium is determined by a function $\varepsilon = (\varepsilon_1, \varepsilon_2) : [0, \infty)^2 \longrightarrow (0, \infty)^2$ about which we now make some further assumptions.

(V) There is a potential $\varphi \in C^1([0, \infty)^2)$ such that $\varepsilon = \nabla \varphi$.

This property of the dielectric response is used explicitly in [7] where it is assumed that $\partial_2 \varepsilon_1 = \partial_1 \varepsilon_2$ for all $s_1, s_2 \geq 0$. When (V) holds the potential, normalized so that $\varphi(0) = 0$, is given by

$$\varphi(s_1, s_2) = \int_0^1 \varepsilon_1(ts_1, ts_2)s_1 + \varepsilon_2(ts_1, ts_2)s_2 \quad dt \tag{2.2}$$

and we define an auxiliary function Φ by

$$\Phi(s_1, s_2) = \varphi(s_1^2/2, s_2^2/2). \tag{2.3}$$

(C) We suppose that $\Phi \in C^2(\Re^2)$ and that $D^2\Phi(s)$ is positive definite for all $s \in \Re^2$.

The positive definitiveness of $D^2\Phi(s)$ implies that $\partial_i \left[\varepsilon_i(s_1^2/2, s_2^2/2)s_i\right] > 0$ for $s \in \Re^2$ and $i = 1, 2$ and that Φ is strictly convex. Since $\Phi(0) = 0$ and $\nabla \Phi(0) = 0$, it follows that $\Phi(s) > 0$ for all $s \in \Re^2 \setminus \{0\}$. By (CR), $\|\nabla\Phi(s_1, s_2)\| \geq A\sqrt{(s_1^2 + s_2^2)}$ for all $(s_1, s_2) \in \Re^2$ and so from the conclusions of Theorem 26.4 to Lemma 26.7 of

[30] we find that $\nabla\Phi : \Re^2 \longrightarrow \Re^2$ is a diffeomorphism and $[\nabla\Phi]^{-1} = \nabla\Phi^*$ where Φ^* is the Legendre transform of Φ which can be defined by

$$\Phi^*(\tau) = \sup\left\{s.\tau - \Phi(s) : s \in \Re^2\right\} \text{ for } \tau = (\tau_1, \tau_2) \in \Re^2. \qquad (2.4)$$

From this formula we deduce that $\Phi^*(0,0) = 0$ and $\Phi^*(s_1,s_2) = \Phi^*(\mid s_1 \mid, \mid s_2 \mid)$. Since ε_1 and $\varepsilon_2 > 0$ we also have that $\nabla\Phi$ maps the four quadrants and half-axes onto themselves. Of course $\nabla\Phi(s_1,s_2) = \begin{pmatrix} \varepsilon_1(s_1^2/2, s_2^2/2)s_1 \\ \varepsilon_2(s_1^2/2, s_2^2/2)s_2 \end{pmatrix}$ and we now show that $\nabla\Phi^*$ has a similar structure. From (C) we deduce that $\Phi^* \in C^2(\Re^2)$ and that $D^2\Phi^*(t)$ is positive definite for all $t \in \Re^2$. See Corollary 4.2.10 of [29].

Define $\psi : [0,\infty)^2 \longrightarrow \Re$ by

$$\psi(t_1,t_2) = \Phi^*(\sqrt{2t_1}, \sqrt{2t_2}) \qquad (2.5)$$

so that

$$\Phi^*(t_1,t_2) = \psi(t_1^2/2, t_2^2/2). \qquad (2.6)$$

Clearly, $\psi \in C^2((0,\infty)^2)$ with $\partial_i\Phi^*(t_1,t_2) = \partial_i\psi(t_1^2/2, t_2^2/2)t_i$ and we define $(\gamma_1,\gamma_2) \in C^1((0,\infty)^2, \Re^2)$ by $\gamma_i = \partial_i\psi$. Then

$$\begin{pmatrix} t_1 \\ t_2 \end{pmatrix} = \begin{pmatrix} \varepsilon_1(s_1^2/2, s_2^2/2)s_1 \\ \varepsilon_2(s_1^2/2, s_2^2/2)s_2 \end{pmatrix}$$

$$\Leftrightarrow t = \nabla\Phi(s) \Leftrightarrow \nabla\Phi^*(t) = s \Leftrightarrow$$

$$\begin{pmatrix} s_1 \\ s_2 \end{pmatrix} = \begin{pmatrix} \gamma_1(t_1^2/2, t_2^2/2)t_1 \\ \gamma_2(t_1^2/2, t_2^2/2)t_2 \end{pmatrix}.$$

It follows that for $i = 1, 2$ and $t_1, t_2 > 0, \gamma_i(t_1,t_2) =$

$$1/\varepsilon_i(\gamma_1(t_1,t_2)^2 t_1, \gamma_2(t_1,t_2)^2 t_2) =$$

$$1/\varepsilon_i(\left[\partial_1\Phi^*(\sqrt{2t_1}, \sqrt{2t_2})\right]^2/2, \left[\partial_2\Phi^*(\sqrt{2t_1}, \sqrt{2t_2})\right]^2/2).$$

Using this formula, and recalling that $\nabla\Phi^*$ is continuous on $\Re^2$, we see that γ_i can be extended by continuity to $[0,\infty)^2$.

Let us note an additional property of γ_i. Since $D^2\Phi^*(t)$ is positive definite for all $t \in \Re^2, 0 < \partial_i^2\Phi^*(t_1,t_2) = \partial_i\left\{\gamma_i(t_1^2/2, t_2^2/2)t_i\right\}$ for $i = 1, 2$ and all $(t_1,t_2) \in \Re^2$.

Summarizing these remarks we can formulate the following result.

Theorem 2.1 *Suppose that the dielectric response in (CR) has the properties (V) and (C). Let Φ^* be defined by (2.4). Then $\Phi^* \in C^2(\Re^2)$ and $D^2\Phi^*(t)$ is positive definite for all $t \in \Re^2$. Furthermore $\nabla\Phi^*(t_1,t_2) = \begin{pmatrix} \gamma_1(t_1^2/2, t_2^2/2)t_1 \\ \gamma_2(t_1^2/2, t_2^2/2)t_2 \end{pmatrix}$ where (γ_1,γ_2) has the following properties*

(H) $(\gamma_1, \gamma_2) \in C^1((0,\infty)^2, \Re^2) \cap C([0,\infty)^2, \Re^2)$ *and for* $i = 1, 2$,

$$0 < \gamma_i(t_1, t_2) < 1/A,$$

$$\gamma_i(t_1^2/2, t_2^2/2) = 1/\varepsilon_i(s_1^2/2, s_2^2/2) \text{ where } t_i = \varepsilon_i(s_1^2/2, s_2^2/2)s_i$$

$$\text{and } \partial_i \left\{ \gamma_i(t_1^2/2, t_2^2/2)t_i \right\} > 0 \text{ for all } (t_1, t_2) \in \Re^2.$$

If the function ε in the constitutive relation (CR) has the properties (V) and (C), then (CR) can be expressed in the following way where the function γ has the properties (H).

(CR*) $B = H$ and

$$E_T(x, y) = \gamma_1(|D_T(x, y)|^2 /2, |D_z(x, y)|^2 /2)D_T(x, y)$$

$$E_z(x, y) = \gamma_2(|D_T(x, y)|^2 /2, |D_z(x, y)|^2 /2)D_z(x, y)$$

where γ is obtained from ε by the above construction.

Additional properties of the functions ε_i can easily be interpreted in terms of the functions γ_i.

1. Noting that $\gamma_1(t_1, 0) = 1/\varepsilon_1(\gamma_1(t_1, 0)^2 t_1, 0)$ and that $\gamma_1(t_1, 0)^2 t_1$ is strictly increasing, we see that $\gamma_1(t_1, 0)$ is a strictly decreasing function of t_1 when $\varepsilon(s_1, 0)$ is a strictly increasing function of s_1.

2. Suppose that there exist $\alpha \geq 0$ and $A > 0$ such that

    $$\varepsilon_2(s_1, s_2)/s_2^\alpha \to A \text{ as } s_2 \to \infty,$$
    uniformly for s_1 in bounded subsets of $[0, \infty)$.

    Then it follows that

    $$\gamma_2(t_1, t_2)t_2^\beta \to B \text{ as } t_2 \to \infty,$$
    uniformly for t_1 in bounded subsets of $[0, \infty)$

    where $\beta = \alpha/(1 + 2\alpha)$ and $B = (2^\alpha/A)^{1/(1+2\alpha)}$.

Finally we remark that in the case of a heterogeneous medium the fundamental relation between ε_i and γ_i becomes

$$\gamma_i(x, y, z, t_1^2/2, t_2^2/2) = 1/\varepsilon_i(x, y, z, s_1^2/2, s_2^2/2)$$
$$\text{where } t_i = \varepsilon_i(x, y, z, s_1^2/2, s_2^2/2)s_i.$$

2.2 Isotropic medium

The material described by (CR) is isotropic when the following additional symmetry, which we assume throughout this section, is present.

(I) There is a continuous function $\varepsilon : [0, \infty) \to (0, \infty)$ such that

$$\varepsilon_1(s_1, s_2) = \varepsilon_2(s_1, s_2) = \varepsilon(s_1 + s_2) \qquad \text{for all } s_1, s_2 \geq 0$$

where $0 < A \leq \varepsilon(s)$ for all $s \geq 0$.

In this case the dielectric response is completely determined by the scalar function ε and so the preceding discussion can be simplified. Furthermore the scalar character of the response is reflected in the equivalent form (CR*). To expose this we begin by observing that (I) implies that (V) is satisfied with

$$\varphi(s_1, s_2) = \int_0^{s_1+s_2} \varepsilon(t)dt \tag{2.7}$$

and so

$$\Phi(s_1, s_2) = \int_0^{\frac{1}{2}|s|^2} \varepsilon(t)dt \qquad \text{for } s = (s_1, s_2) \in \Re^2 \tag{2.8}$$

where $|s|^2 = s_1^2 + s_2^2$.

It is now easy to verify that the condition (C) is satisfied by an isotropic medium provided that the function ε has the following properties.

(A) $\varepsilon \in C^1\left((0, \infty)\right)$ with $\lim_{s \to 0} s\varepsilon'(s) = 0$ and $2s\varepsilon'(s) + \varepsilon(s)$ for all $s > 0$.

The last inequality (which can be expressed as $\left[\varepsilon(\frac{1}{2}s^2)s\right]'$ for all $s > 0$) ensures that, with $s = |(s_1, s_2)|$,

$$D^2\Phi(s_1, s_2) = \left[\begin{array}{cc} s_1^2\varepsilon'(\frac{1}{2}s^2) + \varepsilon(\frac{1}{2}s^2) & s_1 s_2 \varepsilon'(\frac{1}{2}s^2) \\ s_1 s_2 \varepsilon'(\frac{1}{2}s^2) & s_2^2 \varepsilon'(\frac{1}{2}s^2) + \varepsilon(\frac{1}{2}s^2) \end{array} \right]$$

is positive definite and we note that it is always satisfied if the material is self-focusing since in this case ε is non-decreasing and $\varepsilon(0) > 0$. Furthermore the same inequality implies that

$$f(s) \equiv \varepsilon(\frac{1}{2}s^2)s$$

is an increasing, odd diffeomorphism of $\Re$ onto itself. Setting $g = f^{-1}$, we find that

$$\Phi(s_1, s_2) = \int_0^{|s|} f(t)dt \qquad \text{for all } s = (s_1, s_2) \in \Re^2 \text{ and}$$

$$\Phi^*(\tau_1, \tau_2) = \int_0^{|\tau|} g(t)dt \qquad \text{for all } \tau = (\tau_1, \tau_2) \in \Re^2.$$

Thus

$$\psi(\tau_1, \tau_2) = \Gamma(\tau_1 + \tau_2) \text{ for } \tau_1, \tau_2 \geq 0$$

where $\Gamma(t) = \int_0^{\sqrt{2t}} g(u)du.$

Setting

$$\gamma(t) = \Gamma'(t) \text{ for all } t \geq 0$$

we have that

$$g(t) = \gamma(\frac{1}{2}t^2)t \text{ for all } t \geq 0$$

and that

$$\gamma_1(\tau_1, \tau_2) = \gamma_2(\tau_1, \tau_2) = \gamma(\tau_1 + \tau_2) \text{ for } \tau_1, \tau_2 \geq 0.$$

Thus

$$\gamma(\frac{1}{2}t^2)t = s \Leftrightarrow \varepsilon(\frac{1}{2}s^2)s = t$$

and

$$\gamma(\frac{1}{2}t^2) = \frac{1}{\varepsilon(\frac{1}{2}s^2)} \text{ where } t = \varepsilon(\frac{1}{2}s^2)s.$$

From (I) and (A) we see that the function $\gamma : [0, \infty) \to (0, \infty)$ has the following properties which correspond to the condition (H) for the more general uni-axial medium.

(K) $\gamma \in C^1((0, \infty)) \cap C([0, \infty))$ with $\gamma(0) = 1/\varepsilon(0), 2\tau\gamma'(\tau) + \gamma(\tau) > 0$ for all $\tau > 0$ and $\lim_{\tau \to 0} \tau\gamma'(\tau) = 0$.

Using the above formulae we find that ε and γ are related by the following identity which can be used to give a more direct definition of γ,

$$\varepsilon\left(u\gamma(u)^2\right)\gamma(u) = 1 \text{ for all } u > 0. \tag{2.9}$$

Setting $\xi(u, \gamma) = \varepsilon\left(u\gamma^2\right)\gamma - 1$, it follows from (I) and (A) that $\xi \in C^1((0, \infty) \times \Re)$ and the implicit function theorem defines a function $\gamma \in C^1((0, \infty))$ such that $\varepsilon\left(u\gamma(u)^2\right)\gamma(u) = 1$ for all $u > 0$. From this we can recover all of the properties of γ.

Furthermore additional properties of ε can easily be translated into information about γ. We record a few of these implications for future reference.

1. For $u > 0$ and $v = u\gamma(u)^2$, we have $u = v\varepsilon(v)^2$ and

$$\frac{d}{du}\gamma(u) = -\frac{\varepsilon'(v)}{\varepsilon(v)^2}\frac{dv}{du} = -\frac{\varepsilon'(v)}{\varepsilon(v)^3\left\{2v\varepsilon'(v) + \varepsilon(v)\right\}}.$$

Hence $\gamma' \leq 0$ on $(0, \infty) \Leftrightarrow \varepsilon' \geq 0$ on $(0, \infty)$.

2. If $\lim_{s \to \infty} \varepsilon(s)$ exists then $\lim_{\tau \to \infty} \gamma(\tau) = 1/\lim_{s \to \infty} \varepsilon(s)$.

3. If there exist $\sigma > 0$ and $L \neq 0$ such that

$$\lim_{s \to 0} \frac{\varepsilon(s) - \varepsilon(0)}{s^\sigma} = L \text{ then } \lim_{\tau \to 0} \frac{\gamma(\tau) - \gamma(0)}{\tau^\sigma} = -\frac{L}{\varepsilon(0)^{2(1+\sigma)}}$$

8

since

$$\frac{\gamma(u) - \gamma(0)}{u^\sigma} = \frac{\{\varepsilon(0) - \varepsilon(v)\}}{u^\sigma \varepsilon(0)\varepsilon(v)}$$

$$= -\frac{\{\varepsilon(v) - \varepsilon(0)\}}{v^\sigma} \frac{1}{\varepsilon(0)\varepsilon(v)} \left(\frac{v}{u}\right)^\sigma$$

and $\frac{u}{v} = \varepsilon(v)^2$.

Finally we remark that in a heterogeneous medium the function $\gamma(x, y, z, u)$ can be defined as the unique positive solution of

$$\varepsilon\left(x, y, z, u\gamma^2\right)\gamma = 1 \text{ for all } u > 0.$$

2.3 Examples

As a first example of the above procedure let us consider the Kerr law, [4], [17], [37], for a uniaxial material. We suppose that there are four positive constants α_T, α_z, P and Q such that

$\varepsilon_1(s_1, s_2) = \alpha_T + Ps_1 + Qs_2$

$\varepsilon_2(s_1, s_2) = \alpha_z + Qs_1 + Ps_2$ for all $s_1, s_2 \geq 0$.

Clearly $\varepsilon = \nabla\varphi$ where $\varphi(s_1, s_2) = \alpha_T s_1 + \alpha_z s_2 + Qs_1 s_2 + P(s_1^2 + s_2^2)/2$ and it is easy to check that $\Phi(s_1, s_2) = \varphi(s_1^2/2, s_2^2/2)$ satisfies the condition (C) provided that $Q/P \leq 3$. Referring to [4] or to page 246 of [8], we see that the ratio Q/P is $\frac{1}{3}$ when the nonlinearity of the dielectric response is due to electronic distortion, whereas it is 1 when this nonlinearity is caused by electrostriction. In a heterogeneous medium α_T, α_z, P and Q become functions of (x, y, z).

The medium described by this Kerr law is isotropic provided that $\alpha_T = \alpha_z$ and $P = Q$. In this case

$$\varepsilon(s) = \alpha + Ps \tag{2.10}$$

and so $\gamma(\tau)$ is the unique positive solution of the cubic equation

$$P\tau\gamma^3 + \alpha\gamma - 1 = 0.$$

However the Kerr law is at best an approximation which is valid at small to medium field strengths, [31], but which is unacceptable as the field strength becomes infinite since it implies unbounded growth of the dielectric response. In the past decade there has been increasing theoretical and experimental interest in the behaviour of materials under the assumption that the dielectric response remains bounded as the field strength becomes infinite. This property is usually referred to as saturation and for isotropic media two models have attracted considerable attention, [4], [19], [34], [38], [39], namely,

$$\varepsilon(s) = \alpha + \frac{RPs}{R + Ps} \tag{2.11}$$

and

$$\varepsilon(s) = \alpha + R \left[1 - e^{-Ps/R} \right]. \tag{2.12}$$

where α, P and R are positive constants. Both of these responses are approximated by the Kerr law as the field strength tends to zero. Some materials do not behave according to the Kerr law even at small field strengths, [19], [20], [35], and for them models of the following type have been used,

$$\varepsilon(s) = \alpha + \frac{RPs^\sigma}{R + Ps^\sigma} \tag{2.13}$$

where the extra parameter σ is also positive. For some semiconductors values of σ between 0 and 1 are appropriate. We observe that these three models for isotropic media all satisfy the conditions (I) and (A), and in fact represent self-focusing media since ε is an increasing function.

Reasonable models for the response of uni-axial materials with saturation can be generated from any isotropic response ε by setting

$$\varphi(s_1, s_2) = \alpha_T s_1 + \alpha_z s_2 + \int_0^{\sqrt{s_1^2 + 2Gs_1 s_2 + s_2^2}} \{\varepsilon(s) - \varepsilon(0)\} \, ds \tag{2.14}$$

where α_T, α_z and G are positive constants. Then the response is defined by

$$\varepsilon_i(s_1, s_2) = \partial_i \varphi(s_1, s_2) \text{ for } s_1, s_2 \geq 0. \tag{2.15}$$

We note that this construction generates the anisotropic Kerr law from the isotropic one by setting $\varepsilon(s) - \varepsilon(0) = Ps$ and $G = Q/P$. On the other hand, setting $G = 1$ in (2.14) we see that

$$\varphi(s_1, s_2) = [\alpha_T - \varepsilon(0)] s_1 + [\alpha_z - \varepsilon(0)] s_2 + \int_0^{s_1 + s_2} \varepsilon(s) \quad ds \tag{2.16}$$

and so the response will be isotropic if $G = 1$ and $\alpha_T = \alpha_z$.

Proposition 2.2 *Let the function ε satisfy the conditions (I) and (A) with $\varepsilon(s) \geq \varepsilon(0)$ for all $s \geq 0$. Then the response defined using the formulae (2.14) and (2.15) has the properties (V) and (C) provided that $G \leq 3$ and $\min \{\alpha_T, \alpha_z\} \geq \varepsilon(0)$.*

Proof. Clearly (V) holds and

$$\Phi(s_1, s_2) = \frac{1}{2} \{\alpha_T s_1^2 + \alpha_z s_2^2\} + \int_0^{B(s_1, s_2)} \{\varepsilon(s) - \varepsilon(0)\} \, ds$$

where

$$B(s_1, s_2) = \frac{1}{2} \{s_1^4 + 2Gs_1^2 s_2^2 + s_2^4\}^{\frac{1}{2}}.$$

For $(s_1, s_2) \neq (0,0)$,

$$
\begin{aligned}
&D^2 \Phi(s_1, s_2) \\
&= \begin{bmatrix} \alpha_T & 0 \\ 0 & \alpha_z \end{bmatrix} + \varepsilon'\left(B(s_1, s_2)\right) B(s_1, s_2) \frac{DB(s_1, s_2) \otimes DB(s_1, s_2)}{B(s_1, s_2)} \\
&\quad + \{\varepsilon\left(B(s_1, s_2)\right) - \varepsilon(0)\} D^2 B(s_1, s_2)
\end{aligned}
$$

where

$$
\begin{aligned}
&\frac{DB(s_1, s_2) \otimes DB(s_1, s_2)}{2B(s_1, s_2)} \\
&= \frac{1}{8B(s_1, s_2)^3} \begin{bmatrix} s_1^2(s_1^2 + Gs_2^2)^2 & s_1 s_2(s_1^2 + Gs_2^2)(Gs_1^2 + s_2^2) \\ s_1 s_2(s_1^2 + Gs_2^2)(Gs_1^2 + s_2^2) & s_2^2(Gs_1^2 + s_2^2)^2 \end{bmatrix}
\end{aligned}
$$

and

$$
\begin{aligned}
&D^2 B(s_1, s_2) \\
&= \frac{1}{8B(s_1, s_2)^3} \begin{bmatrix} s_1^6 + 3Gs_1^4 s_2^2 + 3s_1^2 s_2^4 + Gs_2^6 & 2s_1^3 s_2^3(G^2 - 1) \\ 2s_1^3 s_2^3(G^2 - 1) & Gs_1^6 + 3s_1^4 s_2^2 + 3Gs_1^2 s_2^4 + s_2^6 \end{bmatrix}
\end{aligned}
$$

Using (A), we see that $\Phi \in C^2(\Re^2)$ with $D^2 \Phi(0,0) = \begin{bmatrix} \alpha_T & 0 \\ 0 & \alpha_z \end{bmatrix}$.

Furthermore

$$
\begin{aligned}
&D^2 \Phi(s_1, s_2) \\
&= \begin{bmatrix} \alpha_T & 0 \\ 0 & \alpha_z \end{bmatrix} - \varepsilon(0) \frac{DB(s_1, s_2) \otimes DB(s_1, s_2)}{2B(s_1, s_2)} \\
&\quad + \{2\varepsilon'\left(B(s_1, s_2)\right) B(s_1, s_2) + \varepsilon\left(B(s_1, s_2)\right)\} \frac{DB(s_1, s_2) \otimes DB(s_1, s_2)}{2B(s_1, s_2)} \\
&\quad + \{\varepsilon\left(B(s_1, s_2)\right) - \varepsilon(0)\} \left\{ D^2 B(s_1, s_2) - \frac{DB(s_1, s_2) \otimes DB(s_1, s_2)}{2B(s_1, s_2)} \right\}
\end{aligned}
$$

where

$$
\begin{aligned}
&D^2 B(s_1, s_2) - \frac{DB(s_1, s_2) \otimes DB(s_1, s_2)}{2B(s_1, s_2)} \\
&= \frac{\{Gs_1^4 + (3 - G^2)s_1^2 s_2^2 + Gs_2^4\}}{8B(s_1, s_2)^3} \begin{bmatrix} s_2^2 & -s_1 s_2 \\ -s_1 s_2 & s_2^1 \end{bmatrix}.
\end{aligned}
$$

It follows that

$$
\frac{DB(s_1, s_2) \otimes DB(s_1, s_2)}{2B(s_1, s_2)} \geq 0
$$

and, since $\{Gs_1^4 + (3 - G^2)s_1^2 s_2^2 + Gs_2^4\} \geq 0$ for $0 < G \leq 3$,

$$
D^2 B(s_1, s_2) - \frac{DB(s_1, s_2) \otimes DB(s_1, s_2)}{2B(s_1, s_2)} \geq 0.
$$

Furthermore,

$$\begin{bmatrix} \alpha_T & 0 \\ 0 & \alpha_z \end{bmatrix} - \varepsilon(0)\frac{DB(s_1,s_2) \otimes DB(s_1,s_2)}{2B(s_1,s_2)}$$

is positive definite for $G \neq 1$ since the Euclidean matrix norm of $M(s_1,s_2) = \frac{DB(s_1,s_2) \otimes DB(s_1,s_2)}{2B(s_1,s_2)}$ is less than 1. In fact,

$$\begin{aligned} \det M(s_1,s_2) &= 0 \text{ and} \\ trace M(s_1,s_2) &= \frac{s_1^6 + 3Gs_1^4 s_2^2 + 3Gs_1^2 s_2^4 + s_2^6}{\{s_1^4 + 2Gs_1^2 s_2^2 + s_2^4\}^{\frac{3}{2}}}. \end{aligned}$$

Thus for all $(s_1,s_2) \neq (0,0)$,

$$\begin{aligned} \|M(s_1,s_2)\| &= trace M(s_1,s_2) \\ &\leq \sup\left\{\frac{1 + 3Gt + 3Gt^2 + t^3}{(1 + 2Gt + t^2)^{\frac{3}{2}}} : t > 0\right\} = \frac{(3G+1)^2}{2(G+1)^3} \\ &< 1 \text{ for } G \geq 0 \text{ and } G \neq 1. \end{aligned}$$

For $G = 1$, the result follows easily from (2.16).

3   Magnetic field wave equations

For monochromatic waves of the form (2.1) we show how Maxwell's equations with the constitutive assumption (CR) can be reduced to an equation for the magnetic field alone. This is done in four stages dealing in turn with the general situation, the general equation for TM-modes followed by the special case of TM-modes with planar and cylindrical symmetry. For isotropic media, an earlier version of this discussion was presented in [21]. Now we deal with uni-axial anisotropy, assuming throughout a constitutive relation of the form (CR) in which the dielectric the response has the properties (V) and (C).

Starting with an arbitrary (sufficiently smooth) field H in the form (2.1), we define the fields B, D and E as follows.

$$B = H$$

then

$$\begin{aligned} D(x,y,z,t) &= -\frac{c}{\omega^2}\nabla \wedge \partial_t H(x,y,z,t) \\ &= -\frac{c}{\omega}\nabla \wedge \{H_T \sin(kz - \omega t) - H_z \cos(kz - \omega t)\} \end{aligned}$$

and finally

$$E(x,y,z,t) = E_T(x,y)\cos(kz - \omega t) + E_z(x,y)\sin(kz - \omega t)$$

where

$$E_T(x,y) = \gamma_1(|D_T(x,y)|^2/2, |D_z(x,y)|^2/2)D_T(x,y)$$
$$E_z(x,y) = \gamma_2(|D_T(x,y)|^2/2, |D_z(x,y)|^2/2)D_z(x,y).$$

It follows from these definitions that the equations (ME$_2$) and (ME$_3$) are satisfied as is (CR*) and hence, by the properties of γ_i, so is (CR). Thus it is enough to choose H in such a way that the equations (ME$_1$) and (ME$_4$) hold. Before doing this it is convenient to introduce some notation which enables us to express D and E explicitly in terms of H. Let

$$H_T(x,y) = \frac{\omega}{c}\{h_1(x,y)e_1 + h_2(x,y)e_2\}$$

and

$$H_z(x,y) = \frac{\omega}{c}h_3(x,y)e_3.$$

where $h_i \in C^2(\Re^2)$ for $i = 1, 2, 3$. Then

$$D(x,y,z,t) = D_T(x,y)\cos(kz - \omega t) + D_z(x,y)\sin(kz - \omega t)$$

where

$$D_T(x,y) = [\partial_y h_3 + kh_2]e_1 - [\partial_x h_3 + kh_1]e_2$$
$$D_z(x,y) = -[\partial_x h_2 - \partial_y h_1]e_3$$

so that

$$|D_T(x,y)|^2 = [\partial_y h_3 + kh_2]^2 + [\partial_x h_3 + kh_1]^2$$
$$|D_z(x,y)|^2 = [\partial_x h_2 - \partial_y h_1]^2$$

and

$$E_T(x,y) = \tilde{\gamma}_1(x,y)\{[\partial_y h_3 + kh_2]e_1 - [\partial_x h_3 + kh_1]e_2\}$$
$$E_z(x,y) = \tilde{\gamma}_2(x,y)[-\partial_x h_2 + \partial_y h_1]e_3.$$

where

$$\tilde{\gamma}_i(x,y) = \gamma_i(\{[\partial_y h_3 + kh_2]^2 + [\partial_x h_3 + kh_1]^2\}/2, [\partial_x h_2 - \partial_y h_1]^2/2).$$

Thus the equation (ME$_1$) is satisfied provided that

$$\nabla \wedge \{E_T(x,y)\cos(kz - \omega t) + E_z(x,y)\sin(kz - \omega t)\}$$
$$= -\frac{\omega}{c}\{H_T\sin(kz - \omega t) - H_z\cos(kz - \omega t)\}$$

and, if H is a solution of this equation, then (ME$_4$) is automatically satisfied and the above construction furnishes a solution of (ME) with (CR). Using the expression for

E in terms of H the fundamental equation for the magnetic field becomes

$$\partial_y \left\{ \tilde{\gamma}_2 [\partial_x h_2 - \partial_y h_1] \right\} + k \tilde{\gamma}_1 [\partial_x h_3 + k h_1] = \left(\frac{\omega}{c}\right)^2 h_1$$

$$-\partial_x \left\{ \tilde{\gamma}_2 [\partial_x h_2 - \partial_y h_1] \right\} + k \tilde{\gamma}_1 [\partial_y h_3 + k h_1] = \left(\frac{\omega}{c}\right)^2 h_2 \qquad (3.1)$$

$$-\partial_x \left\{ \tilde{\gamma}_1 [\partial_x h_3 + k h_1] \right\} - \partial_y \left\{ \tilde{\gamma}_1 [\partial_y h_3 + k h_1] \right\} = \left(\frac{\omega}{c}\right)^2 h_3$$

or it can be expressed equivalently as

$$\partial_y \left\{ \tilde{\gamma}_2 [\partial_x h_2 - \partial_y h_1] \right\} + k \tilde{\gamma}_1 [\partial_x h_3 + k h_1] = \left(\frac{\omega}{c}\right)^2 h_1$$

$$-\partial_x \left\{ \tilde{\gamma}_2 [\partial_x h_2 - \partial_y h_1] \right\} + k \tilde{\gamma}_1 [\partial_y h_3 + k h_1] = \left(\frac{\omega}{c}\right)^2 h_2 \qquad (3.2)$$

$$\partial_x h_1 + \partial_y h_2 + k h_3 = 0.$$

In this way, (ME) with (CR) are reduced to a system of three equations for the three unknowns h_i which are functions of (x, y). Indeed h_3 can immediately be eliminated so as to leave a system of two equations for h_1 and h_2. However, on recalling the definition of the functions $\tilde{\gamma}_i$, we see that the system is highly nonlinear.

3.1   General TM-modes

The formulation above is particularly attractive when the solutions which are being sought are such that the magnetic field has a simpler form than the electric field. Here we discuss solutions having the form (2.1) with

$$H_z(x, y) \equiv 0. \qquad (3.3)$$

These solutions are called TM-modes since the condition (3.3) means that the magnetic field is transverse to the direction of propagation of the wave. For these modes, $h_3 \equiv 0$, and the system (2) becomes

$$\partial_y \left\{ \tilde{\gamma}_2 [\partial_x h_2 - \partial_y h_1] \right\} + k^2 \tilde{\gamma}_1 h_1 = \left(\frac{\omega}{c}\right)^2 h_1$$

$$-\partial_x \left\{ \tilde{\gamma}_2 [\partial_x h_2 - \partial_y h_1] \right\} + k^2 \tilde{\gamma}_1 h_1 = \left(\frac{\omega}{c}\right)^2 h_2 \qquad (3.4)$$

$$\partial_x h_1 + \partial_y h_2 = 0$$

where now

$$\tilde{\gamma}_i(x, y) = \gamma_i (k^2 [h_1^2 + h_2^2]/2, [\partial_x h_2 - \partial_y h_1]^2 /2).$$

In seeking TM-modes we must solve a system of three equations for the two unknown functions $h_1(x, y)$ and $h_2(x, y)$. The next two sections deal with situations where this is possible.

3.2   Planar TM-modes

The simplest case concerns TM-modes having the following special form,

$$h_1(x,y) \equiv 0 \text{ and } h_2(x,y) = u(kx)/k. \tag{3.5}$$

Indeed most of the literature on guided TM-modes is devoted to this situation. Using the above formulation, the problem is now reduced to finding $u \in C^2(\Re)$ such that

$$\{g_2(x)u'(x)\}' - g_1(x)u(x) + \lambda u(x) = 0 \tag{3.6}$$

where

$$\lambda = \left(\frac{\omega}{kc}\right)^2 \text{ and } g_i(x) = \gamma_i(u(x)^2/2, u'(x)^2/2).$$

In this case the e-m fields obtained from a solution of (3.6) are

$$
\begin{aligned}
H(x,y,z,t) &= B(x,y,z,t) = \left(\frac{\omega}{kc}\right)u(kx)\cos(kz - \omega t) \quad e_2 \\
D(x,y,z,t) &= u(kx)\cos(kz - \omega t)e_2 - u'(kx)\sin(kz - \omega t)e_3 \\
E(x,y,z,t) &= g_1(kx)u(kx)\cos(kz - \omega t)e_2 - g_2(kx)u'(kx)\sin(kz - \omega t)e_3.
\end{aligned}
$$

Note that the equation (3.6) can be written as

$$\{\partial_2 \Phi^*(u(r), u'(r))\}' - \partial_1 \Phi^*(u(r), u'(r)) + \lambda u(r) = 0. \tag{3.7}$$

3.3   Cylindrical TM-modes

Another family of TM-modes can be found by seeking solutions in the form

$$h_1(x,y) = -\left(\frac{u(kr)}{k}\right)\sin\theta \text{ and } h_2(x,y) = \left(\frac{u(kr)}{k}\right)\cos\theta \tag{3.8}$$

where (r,θ) are the usual polar coordinates in $\Re^2$. Modes of this type have a cylindrical symmetry appropriate for optical fibres and, in the electric field formulation, they have been studied numerically in [9], [24]. Using the formulation introduced above the problem is reduced to finding a single function $u \in C^2((0,\infty))$ such that

$$\left\{g_2(r)[u'(r) + \frac{u(r)}{r}]\right\}' - g_1(r)u(r) + \lambda u(r) = 0 \tag{3.9}$$

where

$$\lambda = \left(\frac{\omega}{kc}\right)^2 \text{ and } g_i(r) = \gamma_i(u(r)^2/2, [u'(r) + \frac{u(r)}{r}]^2/2).$$

In this case the e-m fields obtained from a solution of (3.9) are

$$H(x,y,z,t) = B(x,y,z,t) = \left(\frac{\omega}{kc}\right) u(kr)\cos(kz - \omega t)i_\theta$$

$$D(x,y,z,t) = u(kr)\cos(kz - \omega t)i_r - [u'(kr) + \frac{u(kr)}{kr}]\sin(kz - \omega t)i_z$$

$$E(x,y,z,t) = g_1(kr)u(kr)\cos(kz - \omega t)i_r$$
$$-g_2(kr)[u'(kr) + \frac{u(kr)}{kr}]\sin(kz - \omega t)i_z$$

where

$$i_r = \cos\theta e_1 + \sin\theta e_2, i_\theta = -\sin\theta e_1 + \cos\theta e_2, i_z = e_3.$$

Note that the equation (3.9) can also be written as

$$\left\{\partial_2\Phi^*\left(u(r), u'(r) + \frac{u(r)}{r}\right)\right\}' - \partial_1\Phi^*\left(u(r), u'(r) + \frac{u(r)}{r}\right) + \lambda u(r) = 0.$$

Of course these definitions hold for $r > 0$ but it is easy so check that the fields can be smoothly extended onto the axis $r = 0$ provided that u has the following properties.

$$\lim_{r\to 0} u(r) = \lim_{r\to 0}[u'(r) + \frac{u(r)}{r}]' = 0 \tag{3.10}$$

and both $\lim_{r\to 0} u'(r)$ and $\lim_{r\to 0}[u'(r) + \frac{u(r)}{r}]$ exist and are finite. $\tag{3.11}$

However for a solution of (3.9) these properties are not independent and, as the following result shows, it is sufficient to impose a seemingly weaker condition.

Proposition 3.1 *Let $u \in C^2\left((0,\infty)\right)$ be a solution of (3.9) such that $\lim_{r\to 0} u(r) = 0$ and u' is bounded on $(0,1)$. Then u has the properties (3.10) and (3.11).*

Proof.   Setting $u(0) = 0$, we have that $u \in C\left([0,\infty)\right)$. Let

$$v(r) = u'(r) + \frac{u(r)}{r}$$

and

$$\alpha = \sup\left\{|u'(r)| : 0 < r < 1\right\}.$$

Then, for $0 < R < r < 1$,

$$|u(r) - u(R)| = \left|\int_R^r u'(t)dt\right| \le \alpha(r - R)$$

and, letting $R \to 0$, we obtain

$$\left|\frac{u(r)}{r}\right| \le \alpha \quad \text{for} \quad 0 < r < 1.$$

This implies that $|v(r)| \leq 2\alpha$ for $0 < r < 1$. For $0 < r < 1$, the equation (3.9) can be written as

$$\{\partial_2 \Phi^* (u(r), v(r))\}' - \partial_1 \Phi^* (u(r), v(r)) + \lambda u(r) = 0$$

and hence

$$\partial_2^2 \Phi^* (u(r), v(r)) v'(r) = h(r)$$

where

$$
\begin{aligned}
h(r) &= \partial_1 \Phi^* (u(r), v(r)) - \lambda u(r) - u'(r)\partial_1\partial_2\Phi^* (u(r), v(r)) \\
&= u(r) \left\{ \gamma_1 \left(\frac{u(r)^2}{2}, \frac{v(r)^2}{2} \right) - \lambda - u'(r)v(r)\partial_1\gamma_2 \left(\frac{u(r)^2}{2}, \frac{v(r)^2}{2} \right) \right\}.
\end{aligned}
$$

Setting

$$w(r) = \left(\frac{u(r)^2}{2}, \frac{v(r)^2}{2} \right)$$

there exists $\beta > 0$ such that $|w(r)| \leq \beta$ for all $r \in (0, 1)$. Recalling the properties of Φ^*, we see that there exist constants L and K such that

$$\partial_2^2 \Phi^* (u(r), v(r)) \geq L > 0 \text{ and } |h(r)| \leq K \text{ for } 0 < r < 1.$$

Thus $|v'(r)| \leq K/L$ for all $r \in (0, 1)$ which ensures that $\lim_{r \to 0} v(r)$ exists and is finite. We set $v(0) = \lim_{r \to 0} v(r)$. Now by l'Hospital's rule,

$$\lim_{r \to 0} \frac{ru(r)}{r^2} = \lim_{r \to 0} \frac{(ru(r))'}{2r} = \frac{1}{2} \lim_{r \to 0} v(r) = \frac{1}{2} v(0)$$

and so $\lim_{r \to 0} \frac{u(r)}{r}$ exists and is equal to $\frac{1}{2} v(0)$.

Since $u'(r) = v(r) - \frac{u(r)}{r}$ for $r > 0$, we also have that $\lim_{r \to 0} u'(r)$ exists and is equal to $\frac{1}{2} v(0)$. Returning to the differential equation we now see that

$$\lim_{r \to 0} v'(r) = -\frac{v(0)}{2} \partial_1\partial_2\Phi^* (0, v(0)) / \partial_2^2 \Phi^* (0, v(0))$$

But $\partial_1 \Phi^* (0, t_2) = 0$ for all t_2 and so $\partial_1\partial_2\Phi^* (0, v(0)) = 0$, proving that $\lim_{r \to 0} v'(r) = 0$. $\square$

4 Existence of guided TM-modes

The guidance of light is usually achieved by exploiting the effect of variations in the refractive index due to inhomogeneity of the medium through which the beam is propagating, [40]. As can be understood from Snell's law, the favourable configuration consists of a region of high refractive index surrounded by layers of material having a lower refractive index. However it is also well-known that nonlinear effects can be used to enhance, or even to produce, guidance which in this case is often referred to as self-trapping since it is the light beam itself which induces the variation in refractive index keeping the light confined to a region near the axis of propagation, [15], [36]. For example, no guidance occurs in a homogeneous linear medium (where the refractive index is constant) whereas it will occur, at least for sufficiently intense beams, in a homogeneous, isotropic, self-focusing medium (where the refractive index is an increasing function of the intensity of the light passing through it). See [14], [22], [23], [25] for such results concerning TE-modes.

In this section we review some more recent work concerning the existence of guided TM-modes, [2], [3]. We assume throughout a constitutive relation of the form (CR) where the conditions (V) and (C) are satisfied. For planar and then cylindrical symmetry, the results establish additional properties of the dielectric tensor which are sufficient to ensure the existence of guided TM-modes within certain ranges of the wavelength.

4.1   Guided planar TM-modes

By a guided planar TM-mode we mean a solution $u \not\equiv 0$ of equation (3.6) which has the properties that

$$u \in H^1(\Re) \text{ and } \lim_{x \to \pm\infty} u(x) = \lim_{x \to \pm\infty} u'(x) = 0 \tag{4.1}$$

where $H^1(\Re)$ is the usual Sobolev space. In this section we discuss the existence of such solutions. The analysis is based on the observation that the function I defined by

$$I(p,q) = \Phi^*(p,q) - q\partial_2\Phi^*(p,q) - \lambda p^2/2$$

is a first integral for (3.6). Indeed, recalling that $\nabla\psi = \gamma = (\gamma_1, \gamma_2)$, we see that

$$I(p,q) = \psi(p^2/2, q^2/2) - \gamma_2(p^2/2, q^2/2)q^2 - \lambda p^2/2$$

and hence it is easy to verify that if u satisfies (3.6) then

$$\{I(u(x), u'(x))\}' = u'(x)\left[g_1 u(x) - \lambda u(x) - \{g_2 u'(x)\}'\right] = 0.$$

Since $\Phi^*(0,0) = 0$ and $0 < \gamma_2(p,q) < 1/A$, it follows that if u satisfies (3.6) and (4.1) then

$$I(u(x), u'(x)) = 0 \text{ for all } x \in \Re.$$

Hence the orbit of a guided TM-mode lies in the set $I^{-1}(0) \setminus \{(0,0)\}$ and, in view of the symmetries of I, it is sufficient to discuss the set

$$C = \{(p,q) : I(p,q) = 0 \text{ with } p \geq 0 \text{ and } q \geq 0\}.$$

For this we introduce some additional hypotheses about the dielectric response.

(S) (a) $\varepsilon_1(s_1, 0)$ is a strictly increasing function of s_1 on $[0, \infty)$, and

(b) there exist $B > 0$ and $\alpha \geq 0$ such that $\varepsilon_2(s_1, s_2)/s_2^\alpha \longrightarrow B$ as $s_2 \longrightarrow \infty$, uniformly for s_1 in bounded subsets of $[0, \infty)$.

We observe that, in the example of a Kerr material which was discussed in Section 2.3, the condition (S) is satisfied with $\alpha = 1$. On the other hand for realistic constitutive laws which model saturation, the value $\alpha = 0$ in (S)(b) is appropriate.

As we showed in Section 2.2, these properties of ε_i imply some analogous behaviour of the functions γ_i. In particular, $\lim_{s_1 \to \infty} \varepsilon_1(s_1, 0)$ exists and, setting

$$\varepsilon_1(\infty, 0) = \lim_{s_1 \to \infty} \varepsilon_1(s_1, 0),$$

we have that

$$0 \leq \gamma_1(\infty, 0) = 1/\varepsilon_1(\infty, 0)$$

where $\gamma_1(\infty, 0) = \lim_{t_1 \to \infty} \gamma_1(t_1, 0)$.

These properties of ε_i can be used to establish the following results [2].

Lemma 4.1 *Let the dielectric response have the properties (CR), (V), (C) and (S).*

If $\lambda \notin (\gamma_1(\infty, 0), \gamma_1(0, 0))$, then $C \cap \{(p, 0) : p > 0\} = \emptyset$.

If $\lambda \in (\gamma_1(\infty, 0), \gamma_1(0, 0))$, then there exist $p_\lambda > 0$ and $f \in C([0, p_\lambda]) \cap C^1((0, p_\lambda))$ such that

$f(0) = 0, f(p_\lambda) = 0, f(p) > 0$ *for all $p \in (0, p_\lambda)$*

and $C = \{(p, f(p)) : 0 \leq p \leq p_\lambda\}$.

Furthermore $\lim_{p \to 0} f'(p) = \sqrt{\frac{\gamma_1(0,0) - \lambda}{\gamma_2(0,0)}}$ and $\lim_{p \to p_\lambda} f'(p) = -\infty$.

From these properties of C we can deduce the following information about guided TM-modes.

Recall that $\gamma_1(0, 0) = 1/\varepsilon_1(0, 0)$ and that $\gamma_1(\infty, 0) = 1/\varepsilon_1(\infty, 0)$. By a guided planar TM-mode we mean a solution of (3.6) which satisfies (4.1).

Theorem 4.2 *Let the dielectric response have the properties (CR), (V), (C) and (S).*

If $\lambda \notin (\gamma_1(\infty, 0), \gamma_1(0, 0))$, then there is no guided planar TM-mode.

If $\lambda \in (\gamma_1(\infty, 0), \gamma_1(0, 0))$, then there exists a unique planar guided TM-mode u_λ such that $u_\lambda(0) = p_\lambda$ and $u'_\lambda(0) = 0$.

Furthermore, $u_\lambda(x) = u_\lambda(-x) > 0$ for all $x \in \Re$ and, for any $\mu < \sqrt{\frac{\gamma_1(0,0) - \lambda}{\gamma_2(0,0)}}$, $\lim_{x \to \infty} \exp(\mu x) u_\lambda(x) = 0$.

All guided TM-modes are of the form $\pm u_\lambda(x + \delta)$ for some $\delta \in \Re$ and some $\lambda \in (\gamma_1(\infty, 0), \gamma_1(0, 0))$.

4.2 Guided cylindrical TM-modes

By a guided cylindrical TM-mode we mean a solution $u \not\equiv 0$ of equation (3.9) which satisfies the regularity conditions (3.10) and (3.11) together with the following guidance conditions,

$$\int_0^\infty \{u(r)^2 + v(r)^2\} \quad r\,dr < \infty \qquad (4.2)$$

and

$$\lim_{r \to \infty} u(r) = \lim_{r \to \infty} v(r) = 0 \tag{4.3}$$

where $v(r) = u'(r) + \frac{u(r)}{r}$. For general modes of the form (2.1), the power, P, of the associated beam of light is defined to be the time-average of the energy flux,

$$\iint_{\Re^2} c(E \wedge H).e_3 \quad dxdy, \tag{4.4}$$

across a plane perpendicular to the direction of propagation. In the case of cylindrical TM-modes we obtain the expression

$$\begin{aligned} P &= \frac{\pi\omega}{k} \int_0^\infty g_1(kr)u(kr)^2 \quad rdr \\ &= \frac{\pi\omega}{k^3} \int_0^\infty \gamma_1(u(r)^2/2, v(r)^2/2)u(r)^2 \quad rdr. \end{aligned} \tag{4.5}$$

Henceforth we restrict our attention to the case of an isotropic self-focusing medium and we summarize the main results obtained in [3]. (The approach could be extended to deal with some uni-axial materials.) Thus we replace the conditions (V) and (C) by (I) and (A) and we assume in addition that

(T) $\varepsilon' \geq 0$ on $(0, \infty), \varepsilon(\infty) \equiv \lim_{s \to \infty} \varepsilon(s) < \infty$ and there exist positive constants σ and L such that

$$\lim_{s \to 0} \frac{\varepsilon(s) - \varepsilon(0)}{s^\sigma} = L.$$

These conditions mean that the material is self-focusing and that the dielectric response saturates as the field strength becomes infinite. As is shown in Section 2.2, they imply that, in addition to (K), the function γ has the following properties.

(G) $\gamma' \leq 0$ on $(0, \infty), \gamma(\infty) \equiv \lim_{t \to \infty} \gamma(t) > 0$ and

$$\lim_{t \to 0} \frac{\gamma(t) - \gamma(0)}{t^\sigma} = -K.$$

where $K = -L/\varepsilon(0)^{2(1+\sigma)}$.

The equation (3.9) can now be written as

$$\left\{ \gamma \left(\frac{1}{2} \left[u(r)^2 + v(r)^2 \right] \right) v(r) \right\}' - \gamma \left(\frac{1}{2} \left[u(r)^2 + v(r)^2 \right] \right) u(r) + \lambda u(r) = 0.$$

It turns out that the boundary conditions can be treated in a very convenient way by expressing the problem in terms of the new variable,

$$z(r) = \sqrt{r}u(r). \tag{4.6}$$

In fact, using z, guided cylindrical TM-modes are characterized by the following variational principle. On the Sobolev space

$$H_0^1(0,\infty) = \{z \in L^2(0,\infty) : z' \in L^2(0,\infty) \text{ and } z(0) = 0\}$$

we can define a C^1-functional by

$$J(z) = \int_0^\infty \Gamma\left(\frac{1}{2r}\left\{z(r)^2 + \left[z'(r) + \frac{z(r)}{2r}\right]^2\right\}\right) r\,dr \qquad (4.7)$$
$$- \frac{\gamma(0)}{2} \int_0^\infty z(r)^2 \, r\,dr.$$

where $\Gamma(t) = \int_0^\infty \gamma(\tau)\,d\tau$. For $d > 0$, we set

$$S(d) = \{z \in H_0^1(0,\infty) : \|z\| = d\} \qquad (4.8)$$

where $\|z\|^2 = \int_0^\infty z(r)^2 \, dr$. Every stationary point, z, of J restricted to $S(d)$ satisfies the Euler-Lagrange equation

$$J'(z)\varphi = \xi \int_0^\infty z(r)\varphi(r) \, dr \text{ for all } \varphi \in H_0^1(0,\infty) \qquad (4.9)$$

and as the next result asserts they correspond precisely to the solutions we are seeking.

Theorem 4.3 *Let the dielectric response have the properties (CR), (I), (A) and (T).*
   (i) $J \in C^1\left(H_0^1(0,\infty)\right)$.
   (ii) Let $(\xi, z) \in \Re \times H_0^1(0,\infty)$ satisfy (4.9) where $z \not\equiv 0$ and set $\lambda = \xi + \gamma(0)$, $u(r) = z(r)/\sqrt{r}$. Then $\lambda > \gamma(\infty)$ and (λ, u) is a guided cylindrical TM-mode. Conversely, if (λ, u) is a guided cylindrical TM-mode then $\lambda > \gamma(\infty)$, $z \in H_0^1(0,\infty)$ and (4.9) is satisfied with $\xi = \lambda - \gamma(0)$.
   (ii) If (λ, u) is a guided cylindrical TM-mode with $\lambda < \gamma(0)$ then u and v decay exponentially as $r \to \infty$.

The existence of guided cylindrical TM-modes is thus reduced to the search for stationary points of $J\mid_{S(d)}$ for $d > 0$. Since J is bounded below on $S(d)$, we can begin by trying to show that J attains its minimum on $S(d)$. However, as the next results show, although this is the case for all sufficiently large values of d, it is false for small values of d unless the exponent σ in (T) is less than 1.

Theorem 4.4 *Let the dielectric response have the properties (CR), (I), (A) and (T). For $d > 0$, set*

$$m(d) = \inf\{J(z) : z \in S(d)\}.$$

(i) Then $m(d) \geq -\frac{1}{2}[\gamma(0) - \gamma(\infty)]\,d^2$ and there exists $d_0 \geq 0$ such that, for all $d > d_0$, $m(d) < 0$ and there exists $z_d \in S(d)$ such that $J(z_d) = m(d)$. Furthermore, $z_d > 0$ on

$(0, \infty)$ *and there exists* λ_d *such that* (λ_d, u_d) *is a guided cylindrical TM-mode where* $u_d(r) = z_d(r)/\sqrt{r}$ *and* $\gamma(\infty) < \lambda_d < \gamma(0) + 2m(d)/d^2 < \gamma(0)$. *Also* $\lim_{d \to \infty} \lambda_d = \gamma(\infty)$ *and* $\lim_{d \to \infty} P_d = \infty$ *where* P_d *is the power of the beam associated with* (λ_d, u_d).

(ii) If $\sigma \in (0, 1)$ *we may choose* $d_0 = 0$ *and in this case* $\lim_{d \to 0} \lambda_d = \gamma(0)$ *and* $\lim_{d \to 0} P_d = 0$.

(iii) If $\sigma \geq 1$, *then* $\lambda \leq \gamma(0)$ *and there exists a constant* $d_1 > 0$ *such that* $\|u\| \geq d_1$ *for every guided cylindrical TM-mode* (λ, u). *The power of such modes is also bounded away from zero.*

Under slightly more restrictive assumptions on the dielectric response Ruppen, [41], as established the existence of non-positive guided cylindrical TM-modes by showing that there are stationary points of $J|_{S(d)}$ which are not minima.

Acknowledgement This work was partly supported by a grant from the Swiss Office Fédérale de l'Education et de la Science for the project No OFES 93.0190 Variational Methods in Nonlinear Analysis, in collaboration with the European Union programme Human Capital and Mobility.

5   References

[1] Born, M. and Wolf, E. : Principles of Optics, fifth edition, Pergamon Press, Oxford, 1975.

[2] Stuart, C. A. : Guided TM-modes in a self-focusing anisotropic dielectric, in Nonlinear Problems in Applied Mathematics, edited by T. S. Angell et al., SIAM Proceedings Series, Philadelpdia, 1995.

[3] Stuart, C. A. : Cylindrical TM-modes in a homogeneous self-focusing dielectric, to appear in Math. Models & Methods Appl. Sci. 1996.

[4] Mihalache, D., Bertolotti, M. and Sibilia, C. : Nonlinear wave propagation in planar structures, in Progress in Optics XXVII, edited by E. Wolf, Elsevier, 1987.

[5] Akmediev, N. N. : Nonlinear theory of surface polaritons, Sov. Phys., JETP. 57 (1983), 1111-1116.

[6] Mihalache, D., Stegeman, G. I., Seaton, C. T., Wright, E. M., Zanoni, R., Boardman, A. D. and Twardowski, T. : Exact dispersion relations for transverse magnetic polarized guided waves at a non-linear interface, Opt. Lett. 12 (1987), 187-189.

[7] Joseph, R. I. and Christodoulides, D. N. : Exact field decomposition for TM waves in nonlinear media, Opt. Lett. 10 (1987), 826-828.

[8] Ogusu, K. : TM waves guided by nonlinear planar waveguides, IEEE, Trans. Microwave Th. et Tech., 37 (1989), 941-946.

[9] Chen, Y. : TE and TM families of self-trapped beams, IEEE J. Quantum Elect., 27 (1991), 1236-1241.

[10] Yokota, M. : Guided transverse-magnetic waves supported by a weakly nonlinear slab waveguide, J. Opt. Soc. Am. B10 (1993), 1096-1101.

[11] Wang, X. H. and Cambrell, G. K. : Full vectorial simulation of bistability phenomena in nonlinear-optical channel wave-guides, J. Opt. Soc. Am. B10 (1993), 1090-1095.

[12] Wang, X. H. and Cambrell, G. K. : Simulation of strong nonlinear effects in optical waveguides, J. Opt. Soc.Am. B11 (1993), 2048-2055.

[13] Leung, K. M. : p-polarized nonlinear surface polaritons in materials with intensity-dependent dielectric function, Phy. Rev. B32 (1985), 5093-5101.

[14] Stuart, C. A. : Self-trapping of an electromagnetic field and bifurcation from the essential spectrum, Arch. Rational ;Mech. Anal., 113 (1991), 65-96.

[15] Stegeman, G. I. : Nonlinear guided wave optics, in Contemporary nonlinear optics, Edited by G. P. Agrawal and R. W. Boyd, Academic Press, Boston, 1992.

[16] Akhmanov, R.-V., Khorklov, R. V. & Sukhorukov, A. P. : Self-focusing, self-defocusing and self-modulation of laser beams, in Laser Handbook, edited by F.T. Arecchi & E.O. Schulz Dubois, North-Holland, Amsterdam, 1972.

[17] Svelto, O. : Self-focusing, self-trapping and self-phase modulation of laser beams, in Progress in Optics, Vol. 12, editor E. Wolf, North-Holland, Amsterdam, 1974.

[18] Reintjes, J. F. : Nonlinear Optical Processes, in Encyclopedia of Physical Science and Technology, Vol. 9, Academic Press, New York.

[19] Stegeman, G. I., Ariyant, J., Seaton, C. I., Shen, T.-P. & Moloney, J. V. : Nonlinear thin-film guided waves in non-Kerr media, Appl. Phys. Lett. 47 (1985), 1254-1256.

[20] Mathew, J. G. H., Lar, A. K., Heckenberg, N. R. & Galbraith, I. : Time resolved self-defocusing in InSb at room temperature, IEEE J. Quantum Elect. 21 (1985), 94-99.

[21] Stuart, C. A. : Magnetic field wave equation for nonlinear optical waveguides, Rapport Nr 11.93, Départ. Math., Ecole Polytechnique Fédérale de Lausanne, 1993..

[22] Ruppen, H.-J. : Multiple TE-modes for cylindrical, self-focusing waveguides, pre-print.

[23] Ruppen, H.-J. : Multiple TE-modes for planar, self-focusing waveguides, pre-print.

[24] Chen, Y. and Snyder, A. W. : TM-type self-guided beams with circular cross-section, Electr. Lett., 27 (1991), 565-566.

[25] Stuart, C .A. : Guidance properties of nonlinear planar waveguides, Arch. Rational Mech. Anal., 125 (1993), 145-200.

[26] Chen, Q. and Wang, Z. H. : Exact dispersion relations for TM waves guided by thin dielectric films bounded by nonlinear media, Opt. Letters, 18(1993), 260-262

[27] Kang, S.-W. : TM modes guided by nonlinear dielectric slabs, J. Lightwave Tech., 13(1995), 391-395

[28] Kushwaha, M. S. : Exact theory of nonlinear surface polaritons : TM case, Japanese J. Appl. Phys., 29(1995), 1826-1828

[29] Hiriart-Urruty, J. B. and Lemaréchal, C. : Convex Analysis and Minimisation Algorithms II, Springer-Verlag, Berlin, 1993

[30] Rockafellar, R. T. : Convex Analysis, Princeton Univ. Press, Princeton, 1970

[31] Landau, L. D., Lifshitz, E. M. and Pitaevskii, L.P. : Electrodynamics of Continuous

Media, 2nd edition, Landau and Lifshitz Course of Theoretical Physics, Vol. 8, Pergamon Press, Oxford, 1984.

[32] Marburger, J. M. : Self-focusing, Theory, Prog. Quant. Electr., 4 (1975), 35-110.

[33] Hayata, K., Koshiba, M. and Suzuki, M. : Finite-element solution of arbitrary nonlinear graded-index slab waveguides, Electr. Lttr., 23 (1987), 429-431.

[34] Chen, Y. : Effects of nonlinear saturation on vector spatial solitons, J. Opt. Soc. Am. B, 8 (1991), 2466-2469.

[35] Snyder, A. W. and Mitchell, D. J. : Spatial solitons of the power-law nonlinearity, Optics Letters, 18 (1993), 101-103.

[36] Wright, E. M., Heatley, D. R. and Stegeman, G. I. : Emission of spatial solitons from nonlinear waveguides, Physics Reports,194 (1990), 309-323.

[37] Boardman, A. D. and Twardowski, T. : Theory of nonlinear interaction between TE and TM waves, J. Opt. Soc. Am. B, 5 (1988), 523-528.

[38] Cuykendall, R. and Strobl, K. H. : Effect of soft saturation on nonlinear interface switching, Phy. Rev. A, 41 (1990), 352-358.

[39] Coutaz, J.-L. and Kull, M. : Saturation of the nonlinear index of refraction in semiconductor-doped glass, J. Opt. Soc. Am. B, 8 (1991), 95-98.

[40] Snyder, A. W. and Love, J. D. : Optical Wave Guide Theory, Chapman and Hall, London, 1983.

[41] Ruppen, H.-J. : Multiple cylindrical TM-modes for a homogeneous self-focusing dielectric, to appear in J. Diff. Equat..

ADDRESS : *Département de Mathématiques, EPFL, CH-1015 Lausanne, Switzerland*

On the Behavior of Positive Solutions for a Class of Semilinear Elliptic Systems

CECILIA S. YARUR [1]   Universidad de Santiago de Chile, Santiago, Chile

1 INTRODUCTION

The existence of a priori estimates for positive solutions of partial differential equations plays a fundamental role in many problems. Indeed, these estimates are very useful in proving existence and nonexistence results as well as to establish asymptotic properties of solutions, see Brézis and Lieb (1979), Brézis and Véron (1980), Clément et al. (1993), García-Huidobro et al., García-Huidobro and Yarur, Ni (1982), Ni and Serrin (1986) among others. Here we are interested on an priori estimate for nonnegative solutions (u, v) to the following system

$$\begin{array}{rcl} \Delta u & = & a(x)v^p \\ \Delta v & = & b(x)u^q \end{array} \quad in \quad \Omega \subset \mathbb{R}^N, \quad N \geq 3, \qquad (1.1)$$

where Ω is an exterior domain of $\mathbb{R}^N$, $p > 0, q > 0$ and $pq > 1$. The functions a and b are nonnegative functions in $L^\infty_{loc}(\Omega)$.

Our results are organized as follows. Section 2 is mainly devoted to state and prove an a priori estimate for the nonnegative solutions to (1.1).

This kind of results for the Thomas-Fermi equation

$$\Delta u = u^{3/2} \quad in \quad \Omega \subset \mathbb{R}^N, \quad N \geq 3,$$

was proved by Brézis and Lieb in Brézis and Lieb (1979), see also Véron (1981).

In section 3 we give some applications of the results in section 2. In particular, some nonexistence results in the whole space are given.

---
[1]Sponsored by FONDECYT grant 1961235 and DICYT

2   MAIN RESULTS

Throughout this section a and b are nonnegative functions in $L^\infty_{loc}(\Omega)$. Moreover, we will assume that there exist three constants α, β and c, with c positive, such that

$$
\begin{aligned}
a_p(|x|) &\geq c|x|^{-\alpha} \\
b_q(|x|) &\geq c|x|^{-\beta},
\end{aligned}
\qquad \text{for } |x| \text{ large}
\tag{2.1}
$$

where, as in Ni (1982), a_p and b_q are given by replacing f by a or b and ρ by p or q in

$$
f_\rho(r) = \left(\frac{1}{|S_{N-1}|} \int_{S_{N-1}} f(r\sigma)^{\frac{-1}{\rho-1}} \, d\sigma \right)^{1-\rho}
\qquad \text{for } \rho > 1
$$

and

$$
f_1(r) = \min_{\sigma \in S_{N-1}} f(r\sigma) \qquad \text{for } \rho = 1.
$$

We also need the spherical average of a function f, which is defined by

$$
\bar{f}(r) = \frac{1}{|S_{N-1}|} \int_{S_{N-1}} f(r\sigma) \, d\sigma,
$$

where $d\sigma$ denotes the invariant measure on the sphere

$$
S_{N-1} = \left\{ x \in {}^N \,/\, \sum_{i=1}^{N} x_i^2 = 1 \right\}.
$$

Here $|S_{N-1}|$ denotes the volume of the unit sphere.

Let us define γ_1 and γ_2 by

$$
\begin{aligned}
\gamma_1(\alpha, \beta) &= \frac{\alpha - 2 + (\beta - 2)p}{pq - 1}, \\
\gamma_2(\alpha, \beta) &= \frac{\beta - 2 + (\alpha - 2)q}{pq - 1},
\end{aligned}
\tag{2.2}
$$

and l_1, l_2 by

$$
\begin{aligned}
l_1 \gamma_1 (\gamma_1 + N - 2) &= l_2{}^p \\
l_2 \gamma_2 (\gamma_2 + N - 2) &= l_1{}^q.
\end{aligned}
$$

We write for our convenience $\gamma_1(\alpha, \beta)$ and $\gamma_2(\alpha, \beta)$, but they certainly depend also on p and q.

We observe that if $\gamma_i(\gamma_i + N - 2) > 0$ for $i = 1, 2$, then the pair (u, v), where $u(x) = l_1 |x|^{\gamma_1}$ and $v(x) = l_2 |x|^{\gamma_2}$, is a solution of

$$
\begin{aligned}
\Delta u &= |x|^{-\alpha} v^p \\
\Delta v &= |x|^{-\beta} u^q
\end{aligned}
\qquad in \quad R^N \setminus \{0\}, \quad N \geq 3.
$$

The above power functions $|x|^{\gamma_1}$ and $|x|^{\gamma_2}$ are crucial for a priori estimates results as we shall see later.

We next state our main result concerning to the superlinear case for the system (1.1).

THEOREM 2.1. Let $(u,v) \in (C\,(|x| \geq 1))^2$ be a positive solution of

$$\begin{aligned}\Delta u &\geq a(x)v^p \\ \Delta v &\geq b(x)u^q\end{aligned} \qquad \text{in} \qquad |x| \geq 1, \tag{2.3}$$

where $p \geq 1, q \geq 1$ and $pq > 1$. Assume that a and b are nonnegative functions defined in $|x| \geq 1$ and satisfying (2.1) for some α, β. Then

$$\begin{aligned}u(x) &\leq C|x|^{\gamma_1(\alpha,\beta)}, \\ v(x) &\leq C|x|^{\gamma_2(\alpha,\beta)},\end{aligned} \tag{2.4}$$

for some positive constant C and where γ_1 and γ_2 are given in (2.2). Moreover, if u and v are radially symmetric functions then conditions $p \geq 1$ and $q \geq 1$ may be replaced by $p > 0$ and $q > 0$.

In particular for the biharmonic we have the following

COROLLARY 2.1 Let $u \in C(|x| \geq 1)$ be a nonnegative solution of

$$(B) \qquad \begin{cases} \Delta^2 u = |x|^{-\beta}u^q, \\ \Delta u \geq 0 \end{cases}$$

for $q > 1$. Then,

$$u(x) \leq C|x|^{\frac{\beta-4}{q-1}},$$

for some positive constant C.

Theorem 2.1 can be improved if we are in the region of (α, β) given in Theorem 3.1 or in Theorem 3.2 of Yarur (1996). Also it can be improved in the following cases: when either $\gamma_1(\alpha,\beta) = 2 - N$ or $\gamma_2(\alpha,\beta) = 2 - N$ (see Yarur, preprint).

The proof of theorem 2.1 is based on the following result, which treats an a priori estimate for the one dimensional case.

THEOREM 2.2 Let (w_1, w_2) be a nonnegative solution of

$$\begin{aligned}\ddot{w}_1(s) &\geq c_1 s^{-\delta_1}w_2^p \\ \ddot{w}_2(s) &\geq c_2 s^{-\delta_2}w_1^q,\end{aligned} \tag{2.5}$$

for all s large. Assume that $p, q > 0$ and $pq > 1$. Then

$$\begin{aligned}w_1(s) &\leq cs^{\gamma_1(\delta_1,\delta_2)}, \\ w_2(s) &\leq cs^{\gamma_2(\delta_1,\delta_2)},\end{aligned} \tag{2.6}$$

for some positive constant c.

Proof. We divide the proof into three cases according to the sign of $\dot{w}_1$ and $\dot{w}_2$.

Case 1. Assume that one of w_1 and w_2 is increasing and the other is decreasing for all s large. Without lost of generality we can suppose that w_1 is an increasing function and w_2 is a decreasing one.

Integrating the first inequality of (2.5) from $s/2$ to s and using the assumptions on w_1 and w_2, that is, w_1 is increasing and w_2 is decreasing, we easily get

$$\dot{w}_1(s) \geq cs^{-\delta_1+1}w_2^p(s), \tag{2.7}$$

and integrating again from $s/2$ to s in (2.7) we have

$$w_1(s) \geq cs^{-\delta_1+2}w_2^p(s), \tag{2.8}$$

(along this paper c represents any positive constant).

Now, we integrate twice the second inequality of (2.5) from s to $2s$ to obtain

$$w_2(s) \geq cs^{-\delta_2+2}w_1^q(s). \tag{2.9}$$

The result follows now from (2.8) and (2.9).

Case 2. Assume now that w_1 and w_2 are both increasing functions of s, for all s large. Integrating twice from s to $2s$ in both inequalities of (2.5), we have

$$w_1(4s) \geq cs^{-\delta_1+2}w_2^p(s), \tag{2.10}$$

and

$$w_2(4s) \geq cs^{-\delta_2+2}w_1^q(s). \tag{2.11}$$

¿From the bounds (2.10) and (2.11), we obtain

$$w_1(16s) \geq cs^{-\gamma_1(pq-1)}w_1^{pq}(s) \tag{2.12}$$

and then rewrite (2.12) in the form

$$\frac{w_1(16s)}{(16s)^{\gamma_1}} \geq c\left(\frac{w_1(s)}{s^{\gamma_1}}\right)^{pq}. \tag{2.13}$$

Assume that there exists an s_0 such that

$$c^{\frac{1}{pq-1}}\frac{w_1(s_0)}{s_0^{\gamma_1}} > 1, \tag{2.14}$$

where c is the constant in (2.13). If condition (2.14) is satisfied we will prove the following

Claim: $\lim_{s\to\infty} w_1(s)/s^\beta = \infty$, for all $\beta > 0$.

Assuming that the claim is true we will get the following contradiction: Since from (2.11) we have that $\lim_{s\to\infty} w_2(s)/s^\beta = \infty$ for all $\beta > 0$, then w_1 and w_2 satisfy

$$\begin{aligned} \ddot{w}_1(s) &\geq w_2^{p'} \\ \ddot{w}_2(s) &\geq w_1^{q'}, \end{aligned} \tag{2.15}$$

for all s large, and for all p' and q' such that $p > p' > 0$, $q > q' > 0$. We can also assume that $p'q' > 1$ and thus, from Lemma 2.4 of Yarur (1996) we get, in particular, that w_1 is a bounded function, contradicting the hypothesis of case 2. Thus, for all s large enough

$$c^{\frac{1}{pq-1}}\frac{w_1(s)}{s^{\gamma_1}} \leq 1.$$

We prove now the claim. For $n \in $, let us choose, in (2.13) $s := 2^{4n}s_0$ and $y_n := c^{\frac{1}{pq-1}}\dfrac{w_1(2^{4n}s_0)}{(2^{4n}s_0)^{\gamma_1}}$. Therefore,

$$y_{n+1} \geq y_n{}^{pq}, \tag{2.16}$$

for all n. Iterating (2.16) leads to the estimate

$$y_{n+1} \geq y_0{}^{(pq)^{n+1}}$$

for all n. By the definition of y_0 and (2.14), we get that $y_0 > 1$. Therefore, for all $\beta > 0$

$$\lim_{n \to \infty} \frac{y_{n+1}}{\left(2^{4(n+1)} s_0\right)^{\beta}} = \infty. \tag{2.17}$$

Returning to the definition of y_n and since (2.17) is satisfied for all $\beta > 0$, we obtain

$$\lim_{n \to \infty} \frac{w_1\left(2^{4n} s_0\right)}{\left(2^{4n} s_0\right)^{\beta}} = \infty$$

for all $\beta > 0$. We prove next that $\lim\limits_{s \to \infty} \dfrac{w_1(s)}{s^{\beta}} = \infty$. Let s be sufficiently large and $n \in$ be such that $s \in \left[2^{4n} s_0, 2^{4(n+1)} s_0\right]$. Since $w_1(s)$ is nondecreasing, we have

$$\frac{w_1(s)}{s^{\beta}} \geq \frac{w_1\left(2^{4n} s_0\right)}{2^{4(n+1)\beta} s_0^{\beta}}, \tag{2.18}$$

which in turn implies that $\lim\limits_{s \to \infty} w_1(s)/s^{\beta} = \infty$ for all $\beta > 0$ and the claim follows.

Thus, we conclude that for all s

$$c^{\frac{1}{pq-1}} \frac{w_1(s)}{s^{\gamma_1}} \leq 1,$$

and from (2.10) we obtain the corresponding bound for w_2.

Case 3. w_1 and w_2 are both decreasing for all s sufficiently large. In this case arguing as above we get

$$w_1(s) \geq c s^{-\delta_1 + 2} w_2^p(4s), \tag{2.19}$$

and

$$w_2(s) \geq c s^{-\delta_2 + 2} w_1^q(4s). \tag{2.20}$$

Then, by replacing (2.20) into (2.19) we obtain

$$w_1(s) \geq c s^{-\gamma_1(pq-1)} w_1^{pq}(16s). \tag{2.21}$$

For $n \in$, let us choose $s = 2^{4n}$ in (2.21), and define $z_n := c^{\frac{1}{pq-1}} \dfrac{w_1(2^{4n})}{(2^{4n})^{\gamma_1}}$. Thus, there exists $n_0 \in$ such that

$$z_n \geq z_{n+1}{}^{pq} \tag{2.22}$$

for all $n \geq n_0$.

We now iterate (2.22). This leads to the estimate

$$z_{n+1} \leq (z_{n_0})^{a^{n+1-n_0}}$$

for all $n \geq n_0$, where $a = 1/pq$. This yields that $\{z_n\}_n$ is a bounded sequence and then, in a similar way as we did before to prove (2.18), we get $w_1(s)/s^{\gamma_1} \leq c z_n$ for $s \in \left[2^{4n} s_0, 2^{4(n+1)} s_0\right]$. Therefore, $w_1(s)/s^{\gamma_1}$ is a bounded function.

Proof of theorem 2.1. The proof of this result is based fundamentally on theorem 2.2 and on the subharmonic nature of the solutions to (2.3). As we did in Yarur (1996), see there the proof of their Theorem 3.1, we have

$$
\begin{aligned}
\bar{u}'' + \frac{N-1}{r}\bar{u}' &\geq cr^{-\alpha}\,\bar{v}^p \\
\bar{v}'' + \frac{N-1}{r}\bar{v}' &\geq cr^{-\beta}\,\bar{u}^q,
\end{aligned}
\tag{2.23}
$$

for all r large enough.

Let $s = r^{N-2}$ and let

$$
\begin{aligned}
w_1(s) &= s\bar{u}(r) \\
w_2(s) &= s\bar{v}(r).
\end{aligned}
$$

Then w_1 and w_2 satisfy

$$
\begin{aligned}
\ddot{w}_1(s) &\geq cs^{-\delta_1}w_2^p \\
\ddot{w}_2(s) &\geq cs^{-\delta_2}w_1^q
\end{aligned}
\qquad \left(\cdot = \frac{d}{ds}\right)
\tag{2.24}
$$

where

$$
\delta_1 = \frac{\alpha-2}{N-2} + p + 1,
$$

and

$$
\delta_2 = \frac{\beta-2}{N-2} + q + 1.
$$

It follows from theorem 2.2 that

$$
\begin{aligned}
w_1(s) &\leq cs^{\gamma_1(\delta_1,\delta_2)}, \\
w_2(s) &\leq cs^{\gamma_2(\delta_1,\delta_2)},
\end{aligned}
\tag{2.25}
$$

for some positive constant c. Thus, from the definition of w_1 and w_2, we get (2.4) for $\bar{u}$ and $\bar{v}$. To finish, we use the following mean value inequality for subharmonic functions (see Gilbarg and Trudinger (1977))

$$
u(x) \leq \frac{1}{|B_{\frac{|x|}{2}}(x)|} \int_{B_{\frac{|x|}{2}}(x)} u(y)\,dy,
$$

then

$$
u(x) \leq c|x|^{-N} \int_{\frac{|x|}{2}}^{\frac{3|x|}{2}} r^{N-1}\bar{u}(r)\,dr.
$$

Analogous for v. The conclusion follows easily from here.

3   APPLICATIONS

We begin this section by giving a nonexistence result of positive solutions (u,v) to (1.1). This result is contained in Yarur (1996) where a different proof is given.

THEOREM 3.1 Let $(u,v) \in \left(C\left(^N\right)\right)^2$ be a positive solution of

$$\begin{aligned} \Delta u &\geq a(x)v^p \\ \Delta v &\geq b(x)u^q, \end{aligned} \qquad \text{in} \quad {}^N \tag{3.1}$$

Let $p \geq 1, q \geq 1$ and $pq > 1$. Assume a and b are nonnegative functions defined in ^{N} and satisfying (2.1) with α, β such that

$$\min\{\gamma_1(\alpha, \beta), \gamma_2(\alpha, \beta)\} < 0 \tag{3.2}$$

Then

$$u \equiv 0 \text{ and } v \equiv 0.$$

Sketch of proof. We can assume that u and v are radially symmetric. Indeed, if this is not the case, $\bar{u}$ and $\bar{v}$ satisfy (3.1). Hence, using Lemma 2.1 in Benguria et al. (1994) and Theorem 2.1 the conclusion follows.

Next we apply the previous results to the biharmonic. Related results for the biharmonic can be found in Bidaut-Véron (preprint), Mitidieri (in press) and Soranzo (in press). Mitidieri proved, among others, nonexistence results for positive superharmonic solutions of superlinear biharmonic equations.

COROLLARY 3.1 Let $q > 1$ and $q(N - 4) < N - \beta$ and $u \in C^2\left({}^N\right)$ be a positive solution of

$$\Delta^2 u = b(x)u^q \qquad \text{in} \quad {}^N \tag{3.3}$$

Assume that b is a nonnegative function defined in ^{N} and satisfying

$$b_q(x) \geq c|x|^{-\beta}, \qquad \text{for all } |x| \text{ large.}$$

Then u is a super-harmonic function in ^{N}.

Proof. Let us define $v := \Delta u$. Then, the pair (u, v) is a solution for

$$\begin{aligned} \Delta u &= v \\ \Delta v &= b(x)u^q \end{aligned} \qquad \text{in} \quad {}^N . \tag{3.4}$$

Since v is a sub-harmonic function in ^{N} we get the following two possibilities for $\bar{v}$, either

(1)   There is a positive r_0 so that $\bar{v}(r) \geq 0$, for all r larger than r_0. Moreover, $\lim_{r \to \infty} r^{N-2}\bar{v}(r) = \infty$, or

(2) $\bar{v}(r) \leq 0$, for all $r > 0$.

Theorem 2.1 and the hypothesis on β imply that case 1 is impossible and then $\bar{v} \leq 0$. Repeating the above argument for the functions $v_y(x) := v(x + y)$ with $y \in {}^N$, we obtain that $\overline{v_y} \leq 0$ for all y. Then the conclusion follows.

Consider $B_1(0)$ to be the open unit ball centered at zero of ^{N}, $N \geq 3$. Let a and b are nonnegative functions in $L^\infty_{loc}(B_1(0)\backslash\{0\})$ such that

$$\begin{aligned} a_p(|x|) &\geq c|x|^{-\alpha} \\ b_q(|x|) &\geq c|x|^{-\beta} \end{aligned} \qquad \text{for all } x \text{ small,} \qquad (3.5)$$

for some positive constant c.

By using Kelvin transform and Theorem 2.1 we have the following estimate at zero.

THEOREM 3.2 Let $(u, v) \in (C(B_1(0)\backslash\{0\}))^2$ be a positive solution of

$$\begin{aligned} \Delta u &\geq a(x)v^p \\ \Delta v &\geq b(x)u^q \end{aligned} \qquad \text{in} \quad B_1(0)\backslash\{0\} \qquad (3.6)$$

where $p \geq 1, q \geq 1$, and $pq > 1$. Assume that a and b are nonnegative functions satisfying (3.). Then

$$\begin{aligned} u(x) &\leq C|x|^{\gamma_1(\alpha,\beta)}, \\ v(x) &\leq C|x|^{\gamma_2(\alpha,\beta)}, \end{aligned} \qquad (3.7)$$

for some positive constant C and where γ_1 and γ_2 are given in (2.2). Moreover, if u and v are radially symmetric functions then conditions $p \geq 1$ and $q \geq 1$ may be replaced by $p > 0$ and $q > 0$.

REFERENCES

M. F. Bidaut-Véron, Local behaviour of the solutions of a class of nonlinear elliptic systems, preprint.

H. Brézis, E. Lieb, Long range atomic potentials in Thomas-Fermi theory, *Comm. Math. Phys.* 65:231-246(1979).

R. Benguria, S. Lorca and C. Yarur, Nonexistence of positive solution of semilinear elliptic equations, *Duke Math. J.* 74:615-634(1994).

H. Brézis and L. Véron, Removable singularities of some nonlinear elliptic equations, *Arch. Rational Mech. Anal.* 75:1-6(1980).

Ph. Clément, R. Manásevich and E. Mitidieri, Positive solutions for a quasilinear system via blow-up, *Comm. in P.D.E.* 18:2071-2106(1993).

M. García-Huidobro , R. Manásevich R., E. Mitidieri and C. Yarur, Existence and Nonexistence of Positive Singular Solutions for a Class of Semilinear Elliptic Systems, to appear in *Arch. Rat. Mech & Anal.*

D. Gilbarg and N.S. Trudinger, Elliptic Partial Differential Equations of Second Order, Springer-Verlag, 1977.

M. García-Huidobro and C. Yarur, Classification of Positive Singular Solutions for a Class of Semilinear Elliptic Systems, to appear in *Advances in Differential Equations.*

E. Mitidieri, Nonexistence of positive solutions of semilinear elliptic systems in ^{N}, *Differential and Integral Equations,* in press.

W.M. Ni, On the elliptic equation $\Delta u + K(x)u^{(n+2)/(n-2)} = 0$, its generalizations, and applications in geometry, *Indiana Univ. Math. J.* 31:493-529(1982).

W.-M. Ni and J. Serrin, Non-existence theorems for singular solutions of quasilinear partial differential equations, *Comm. Pure Appl. Math.* 39:379-399(1986).

R. Soranzo, Isolated singularities of positive solutions of a superlinear biharmonic equation, *J. Potential Theory,* in press.

J.L. Vázquez and C. Yarur, Schroedinger equations with unique positive isolated singularities, *Manuscripta Math.* 67:143-163(1990).

L. Véron, Singular solutions of some nonlinear elliptic equations, *Nonlinear Analysis, Methods & Applications* 5:225-242(1981).

C. Yarur, Nonexistence of positive singular solutions for a class of semilinear elliptic systems, *E. Journal of Diff. Equat.* 1996:8:1-22(1996).

C. Yarur, A priori estimates for positive solutions for a class of semilinear and linear elliptic systems, preprint.